Elements of Finite–
Dimensional Systems
and Control Theory

Pitman Monographs and
Surveys in Pure and Applied Mathematics 37

Elements of Finite-Dimensional Systems and Control Theory

N. U. Ahmed

University of Ottawa

Longman Scientific & Technical

Copublished in the United States with
John Wiley & Sons, Inc., New York

Longman Scientific & Technical
Longman Group UK Limited
Longman House, Burnt Mill, Harlow
Essex CM20 2JE, England
and Associated Companies throughout the world.

Copublished in the United States with
John Wiley & Sons, Inc., 605 Third Avenue, New York, NY 10158

First published 1988

AMS Subject Classifications: (main) 93-XX, 49-XX, 60-XX
(subsidiary) 93-02, 49-02, 60H10

ISSN 0269–3666

British Library Cataloguing in Publication Data

Ahmed, N. U.
Elements of finite-dimensional systems
and control theory—(Pitman monographs
and surveys in pure and applied
mathematics, ISSN 0269-3666; 37).
1. Mathematical models
I. Title
003′.0724   QA401

ISBN 0-582-01373-9

Library of Congress Cataloging-in-Publication Data
Ahmed, N. U. (Nasir Uddin)
Elements of finite-dimensional systems and control theory.
(Pitman monographs and surveys in pure and applied
mathematics, ISSN 0269-3666; 37)
Bibliography: p.
Includes index.
1. System analysis 2. Control theory. I. Title.
II. Series.
QA402.A349   1988   003   87-21363
ISBN 0-470-20987-9 (USA only)

Typeset in 10/12 Times New Roman
Printed and Bound in Great Britain
at The Bath Press, Avon.

In memory of my uncle
Sardar Ahmad Ali

and

Dedicated to my mother
Amena Begum

Contents

Preface

Over the last two decades, enormous development has taken place in the subject we generally know as systems theory. It has expanded and matured in several directions such as systems governed by ordinary and functional differential equations, systems governed by partial differential equations, and stochastic systems.

Not only has an impressive body of theory grown in these areas but also substantial development has taken place in the direction of applications in various disciplines of engineering, economics, and social and management sciences. It is evident from the growing interest in the field that the next two decades will witness similar growth to that of the past. Most serious development, however, is likely to take place in the area of applications.

In order to keep pace with developments a researcher has to be in constant touch with current literature. An aspiring young researcher, however, is faced with a mountain of literature which, at times, may seem discouraging, and yet he must conquer it before he can begin to make fruitful contributions. Hence, in the very first year of graduate studies or during the senior undergraduate studies, he must be rapidly introduced to the field in order that he may appreciate the universality and strength of the discipline and may soon learn to enjoy its beauty. This book is a modest attempt to acquaint the young researcher rapidly with the basics of modern systems theory so that he may find it worthwhile and challenging to climb the mountain no matter how steep it may seem.

The book contains seven chapters. In chapter 1 we discuss the most common mathematical models used to represent physical systems, and include several examples illustrating the way in which mathematical models are constructed based on scientific or empirical reasonings. In chapters 2 and 3 we present the basic properties of linear and nonlinear systems addressing the questions of existence, uniqueness and continuous dependence of solutions. These results are used in the following chapters. In chapter 4 we study the question of stability based on the Lyapunov method. In chapter 5 we consider the questions of observability, controllability and stabilizability for linear and nonlinear systems. The subject of chapter 6 is optimal control, where we present LaSalle's bang-bang principle, Pontryagin's minimum principle and Bellman's principle of optimality including some existence theorems for optimal controls. The chapter is concluded with a section on system identification and numerical methods in optimal control. In chapter 7 we present the basic theory of stochastic systems and include some prominent results on stability, filtering, control and reliability.

With the background acquired from this book a young researcher in engineering, physical sciences, economics and management sciences will be able to make frutiful contributions in his discipline. A young researcher in mathematics will find the book a stepping stone towards further study and research in systems theory. Moreover, the author believes that the book will also be useful as a reference to practising engineers, scientists, economists and management specialists.

The background required for reading the book is advanced calculus. However, familiarity with real analysis is helpful but not essential. The book can be used as a text for senior undergraduate and first year graduate courses on systems theory. In final year undergraduate classes I have followed chapters 1 and 2 and the first few sections of each of chapters 4 and 5 (excluding controllability with constraints) and selected materials from chapter 6 (excluding proofs) emphasizing particularly the popular quadratic linear regulator theory. These I have covered in two semesters (one academic year) in that order. In a graduate course on system theory, I have covered chapter 3, the last parts of chapters 4 and 5 and chapter 6 in one semester. In the second semester chapter 7 was covered.

The author would like to thank his graduate students S. K. Biswas, T. E. Dabbous, S. S. Lim, T. Selami and P. Li for their help in the preparation of the manuscript. Special thanks go to Dr S. K. Biswas who took great pains in proof reading the entire manuscript and providing the necessary numerical results and preparing the figures. The author would like to thank his colleagues of the departments of electrical engineering and mathematics, and specially all the members of the systems science department at the University of Ottawa for their kind cooperation. Also the author would like to thank his wife and children for their patience and understanding. Finally thanks are due to the editors and staff of Longman for their expert cooperation in the production of the book.

N. U. Ahmed

Publisher's Acknowledgements

Thanks are due to Butterworth Scientific Limited for permission to reproduce, as figure 1.4 of this work, a figure from *Fuel* [134].

1

System models

1.0 Introduction

There are many natural phenomena that evolve with time beyond the reach of human influence. Though man is not the master of these natural events he can construct mathematical models that describe fairly accurately their future evolution with time. For example, the orbital motion of the planets around the sun, of the moons around their respective planets, and of the sun itself around the Galactic nucleus are all accurately predictable by modern astrophysics. Usually the mathematical models used to describe these natural phenomena are differential equations developed from the fundamental principles of mechanics.

Other than natural phenomena, there are events caused by man and his machines. Many of these events can be accurately described using appropriate mathematical models, and controls can be exercised to influence their behaviour – in particular, their future evolution. Examples are mechanical devices, electrical machines, chemical plants, nuclear (fission or fusion) reactors, hydrodynamical systems (aircraft, balloons, hovercraft, etc.) and thermodynamical systems like metallurgical and chemical plants involving thermal diffusion, fluid flow, and chemical kinetics etc. Again most of these physical objects are governed by fundamental laws of physics and hence can be accurately described by use of mathematical models like differential or integrodifferential equations, ordinary or partial, or combination thereof.

In recent years differential equations are also being used in the study of complex biological and biochemical phenomena. Inspired by the success of differential equations in the field of natural sciences, attempts have also been made, with considerable success, to develop dynamic models for management, economic, political and socio-economic systems.

It is clear from the above discussion that differential equations, of one kind or the other, play a key role in the study of evolution of the nature that surrounds us. One of the major objectives of systems theory is to unify all these apparently diverse concepts and physical phenomena into one single coherent body of knowledge. By the magic phrase 'system', we mean any object, natural or man made, that systematically obeys certain laws and regulations, deterministic or stochastic. More technically, a system is an object or entity whose evolution with time can be described without ambiguity by an appropriate mathematical formalism.

The state of the system is given by a function of time $x(t)$ which takes values from an abstract space X called the state space, as t takes values from an index set I contained in the real line R. If the state space X has finite dimension, then the system is called finite dimensional, and if its dimension is infinity then we have an infinite-dimensional system.

In this book we are concerned with finite-dimensional systems. Hence the state space can be considered to be embedded in a finite-dimensional Euclidean space. If the function $x(t)$, $t \in I$, is deterministic, then we have a deterministic system, and if it is a random process, then we have a stochastic system. For almost all man-made systems and for some natural systems, the state can be partially regulated by external agencies called inputs or controls which take values possibly from another abstract space U called the input space. In most cases the system state cannot be fully observed or measured due to technical limitations and hence we have a space of observation, say Z. In general one may wish the system to produce an output, which is not necessarily measurable, leading to the concept of an output space Y. Thus, in general, a system may be considered to be an entity characterized by several abstract spaces and mappings relating them, $\{X, U, Z, Y, I, T, g, h\}$, where

$$x(t) = T(t, t_0, x_0, u), \qquad t \geq t_0, \quad t, t_0 \in I$$

$$z(t) = g(t, x, u), \qquad t \in I$$

$$y(t) = h(t, x, u), \qquad t \in I$$

for some suitable functions (mappings) T, g and h with $u = \{u(t), t \in I\}$ being the input taking values $u(t) \in U$. In this book we consider $Z = Y$.

Before closing this section we wish to emphasize that system theory originates from applied sciences and engineering. It is unfortunate that the subject has been criticized of being impractical without any real foundation. In fact systems theory has been successfully applied in all branches of engineering, and social, economic and management sciences. It is a subject with great potential and its future is bright.

1.1 Linear, nonlinear and stochastic systems

Throughout this book we shall consider only finite dimensional systems which may be either deterministic or stochastic.

Linear systems (deterministic)

Deterministic controlled linear systems are governed by differential equations of the form

$$\dot{x} = A(t)x + B(t)u, \qquad t \in I, \tag{1.1.1}$$

with observation (output) given by

$$z = H(t)x + G(t)u, \qquad t \in I. \tag{1.1.2}$$

where $x(t)$ is the state of the system at time t and $u(t)$ is the input or control applied at the instant t. In general A, B, H and G are time dependent, taking values in the space of matrices of dimensions $(n \times n)$, $(n \times m)$, $(r \times n)$ and $(r \times m)$ respectively. Clearly A characterizes the dynamics of the uncontrolled system, and B the input structure of the system. The output structure of the system is described by H and G. Note that if $B \equiv 0$, no control can be exercised on the system (1.1.1), and similarly if H and G are identically zero, the system variables cannot be physically measured or observed. In many practical control problems, the control forces are limited and hence the system (1.1.1)–(1.1.2) must operate under certain input constraints. This is often specified by requiring that $u(t) \in U$ for all $t \in I$, where U is a subset of R^m specifying the available or admissible control forces. The ability of the control system in the task of regulation of the state trajectory is very much dependent on the system parameters A, B and the restraint set U. Given the present state $x(t) = \xi$, the admissible directions of motions at (t, ξ) is given by the set

$$R(t, \xi) \equiv \{\eta \in R^n : \eta = A(t)\xi + B(t)v,\ v \in U\}.$$

From this expression the significance of the role played by the plant matrix $A(t)$, the input matrix $B(t)$ and the restraint set U is evident. The ability of the system in the task of regulation of its output depends, of course, on all the members A, B, H, G and U.

The system (1.1.1) and (1.1.2) is said to be time invariant if all the variables A, B, H, and G are constant. In this case we write the system equations (1.1.1)–(1.1.2) as

$$\begin{aligned} \dot{x} &= Ax + Bu \\ z &= Hx + Gu. \end{aligned} \tag{1.1.3}$$

Nonlinear systems

Generally a nonlinear system is governed by a differential equation of the form

$$\dot{x} = f(t, x, u), \qquad u(t) \in U, \quad t \in I \tag{1.1.4}$$

or equivalently

$$\dot{x}_i = f_i(t, x, u), \qquad u(t) \in U, \quad t \in I, \quad i = 1, 2, \ldots, n,$$

where $f: I \times R^n \times U \to R^n$ is generally a nonlinear function. The output equation is given by

$$z = h(t, x, u) \tag{1.1.5}$$

where $h: I \times R^n \times U \to R^r$ is again, in general, a nonlinear function of its arguments. For time-invariant systems, both f and h are independent of time t.

Again the ability of the control system in its task of regulation is dependent on the set of admissible directions of motion,

$$R(t, \xi) \equiv \{\eta \in R^n : \eta = f(t, \xi, v), v \in U\}. \tag{1.1.6}$$

In chapters 2 and 3 we shall study the fundamental properties of linear and nonlinear systems.

Stochastic systems

Loosely speaking a stochastic system is governed by a differential equation of the form

$$\dot{x} = f(t, x) + \eta(t), \qquad t \in I, \tag{1.1.7}$$

where η is a random process, randomly perturbing the instantaneous directions of motion of the trajectory $x(t)$, $t \in I$. Usually η is considered to be a white-noise process which is the (generalized) time derivative of a more regular stochastic process called the Wiener process (or Brownian motion) $\{W(t), t \in I\}$. We shall study these processes rigorously in chapter 7. We shall see later that the Wiener process is an indexed family of Gaussian random vectors and that its time derivative does not exist in the classical sense. More rigorously, one writes (1.1.7) as

$$\mathrm{d}x = f(t, x)\,\mathrm{d}t + \mathrm{d}W(t), \tag{1.1.8}$$

which can be interpreted as follows: Given the current state $x(t) = \xi$ at time t, the state at a later time $(t + \Delta t)$ is given by

$$x(t + \Delta t) = \{\xi + f(t, \xi)\Delta t\} + \Delta W(t), \tag{1.1.9}$$

where $\Delta W(t) \equiv W(t + \Delta t) - W(t)$ is the increment of the Wiener process during the time interval $[t, t + \Delta t]$. The vector $\Delta W(t) \equiv W(t + \Delta t) - W(t)$ is Gaussian having zero mean and a well-defined variance–covariance matrix. It is clear from the expression (1.1.9) that, given the present state $x(t) = \xi$, in the absence of noise the state of the system at a later time $(t + \Delta t)$ is $\xi + f(t, \xi)\,\Delta t$. In other words, the second component in the expression (1.1.9) can be regarded as the product of local mean velocity times the elapsed time. The last component ΔW can be regarded as the random fluctuation about the mean position $(\xi + f(t, \xi)\Delta t)$. We shall see in chapter 7 that the evolution of the state process $x(t)$, $t \in I$, given by (1.1.9) and hence (1.1.8), is a well-defined Markov process.

The general model for a stochastic system is given by

$$\begin{aligned} dx &= f(t, x)\, dt + \sigma(t, x)\, dW, \qquad t \in I \\ x(0) &= x_0, \end{aligned} \tag{1.1.10}$$

where $f: I \times R^n \to R^n$, $\sigma: I \times R^n \to R(n \times d)$, the space of $n \times d$ matrices, $\{W(t), t \in I\}$ is a d-dimensional Wiener process and x_0 is a random n-vector. In view of the above discussion, the model of a controlled linear stochastic system is given by

$$dx = A(t)x\, dt + B(t)u\, dt + C(t)\, dW(t), \tag{1.1.11}$$

with observation

$$dz = H(t)x\, dt + G(t)u\, dt + K(t)\, dV(t), \tag{1.1.12}$$

where $\{W(t), t \in I\}$ and $\{V(t), t \in I\}$ are independent Wiener processes with values in R^{d_1} and R^{d_2} respectively. Note that W represents the dynamic noise and V the measurement noise. The matrices $\{A(t), B(t), C(t), t \in I\}$ are deterministic and have dimensions $(n \times n)$, $(n \times m)$ and $(n \times d_1)$ respectively. Similarly $\{H(t), G(t), K(t), t \in I\}$ are of dimensions $(r \times n)$, $(r \times m)$ and $(r \times d_2)$ respectively. Again, the controls being limited, one may demand that the process $\{u(t), t \in I\}$ takes values from a restraint set $U \subset R^m$. The processes x and z are the state and output histories (trajectories) of the system.

In general a nonlinear controlled stochastic system is given by the model,

$$\begin{aligned} dx &= f(t, x, u)\, dt + \sigma(t, x, u)\, dW \qquad &\text{(state)} \\ dz &= h(t, x, z, u)\, dt + \tilde{\sigma}(t, x, z, u)\, dV \qquad &\text{(output)}, \end{aligned} \tag{1.1.13}$$

where f, h, σ, $\tilde{\sigma}$ are suitable vector- and matrix-valued functions of their arguments. The functions f and h give the infinitesimal (local) mean called the drift, and the matrices σ and $\tilde{\sigma}$ are the measures of local dispersion (through $\sigma\sigma'$, $\tilde{\sigma}\tilde{\sigma}'$) called the diffusion.

1.2 Modelling of system dynamics

In this section we present some examples of dynamic systems illustrating the basic principles of modelling. We shall develop dynamic models for locomotives, satellites, electrical generation and transmission systems, ecological systems, basic economic systems, nuclear reactors and chemical kinetics, renewable resources and population etc. We shall also give an example illustrating how infinite-dimensional systems are approximated by finite-dimensional ones.

(a) *Dynamics of locomotives*

Consider a train of railway cars (see figure 1.1) consisting of n units including the engine. The masses of the cars starting from the engine are denoted by $\{M_n, M_{n-1}, \ldots, M_1\}$ and the stiffness and the damping coefficients of the coupling mechanisms are given by $\{K_n, B_n; K_{n-1}, B_{n-1}; \ldots; K_2, B_2\}$. The distances of the front ends of the cars including the engine are denoted by $\{S_n, S_{n-1}, \ldots, S_1\}$ measured with respect to a reference post. Let L denote the (common) distance between the front ends of the cars in the absence of any tension in the springs. Let μ be the coefficient of friction between the railway track and the wheels (with $\mu = 0$ when velocity $v = 0$), and g the acceleration due to gravity. Then, starting from the engine, one can write the dynamics of the train as follows: For the engine, we have

$$M_n\ddot{S}_n + K_n(S_n - S_{n-1} - L) + B_n(\dot{S}_n - \dot{S}_{n-1}) + \mu M_n g = f, \tag{1.2.1}$$

where f is the applied force. For the intermediate cars, we have

$$\begin{aligned} M_{\ell-1}\ddot{S}_{\ell-1} + K_{\ell-1}(S_{\ell-1} - S_{\ell-2} - L) + B_{\ell-1}(\dot{S}_{\ell-1} - \dot{S}_{\ell-2}) + \mu M_{\ell-1}g \\ = K_\ell(S_\ell - S_{\ell-1} - L) + B_\ell(\dot{S}_\ell - \dot{S}_{\ell-1}), \end{aligned} \tag{1.2.2}$$

for $\ell = n, n-1, \ldots, 3$, and for the last car, we have,

$$M_1\ddot{S}_1 + \mu M_1 g = K_2(S_2 - S_1 - L) + B_2(\dot{S}_2 - \dot{S}_1). \tag{1.2.3}$$

For the state vector, we define

$$\left.\begin{aligned} x_1 &= (S_2 - S_1 - L), & x_{n+1} &= \dot{S}_2 - \dot{S}_1, \\ x_2 &= (S_3 - S_2 - L), & x_{n+2} &= \dot{S}_3 - \dot{S}_2, \\ &\vdots & &\vdots \\ x_{n-1} &= (S_n - S_{n-1} - L), & x_{2n-1} &= \dot{S}_n - \dot{S}_{n-1}, \\ x_n &= S_n, & x_{2n} &= \dot{S}_n. \end{aligned}\right\} \tag{1.2.4}$$

Note that the quantities,

$$x_\ell \equiv (S_{\ell+1} - S_\ell - L), \qquad 1 \le \ell \le n-1, \tag{1.2.5}$$

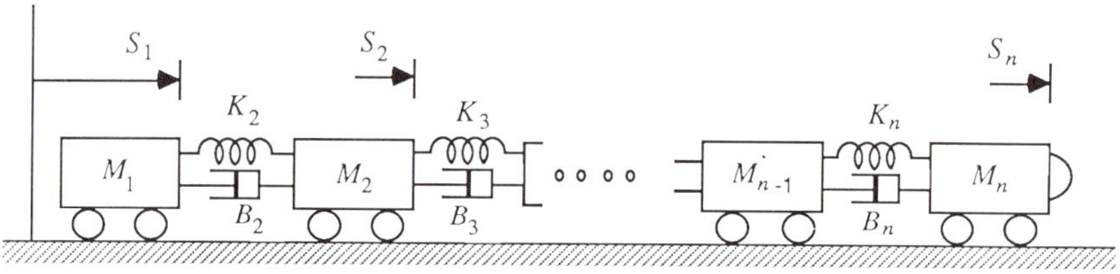

Figure 1.1 A train of railway cars

denote the deviations of the front-end distances from those in rest (unstretched) position. The time derivatives of these quantities give the velocities of approach or recession of any two adjacent cars. These relative velocities are indicators of longitudinal vibrations of the cars. Substituting (1.2.4) into equations (1.2.1)–(1.2.3) we obtain a system of $2n$ differential equations. For a train consisting of only four cars including the engine we have a system of eight differential equations given by

$$\left.\begin{aligned}
\dot{x}_1 &= x_5 \\
\dot{x}_2 &= x_6 \\
\dot{x}_3 &= x_7 \\
\dot{x}_4 &= x_8 \\
\dot{x}_5 &= -((M_1+M_2)/M_1M_2)K_2x_1 - ((M_1+M_2)/M_1M_2)B_2x_5 \\
&\quad + (K_3/M_2)x_2 + (B_3/M_2)x_6 \\
\dot{x}_6 &= -((M_2+M_3)/M_2M_3)K_3x_2 - ((M_2+M_3)/M_2M_3)B_3x_6 \\
&\quad + (K_4/M_3)x_3 + (B_4/M_3)x_7 + (K_2/M_2)x_1 + (B_2/M_2)x_5 \\
\dot{x}_7 &= -((M_3+M_4)/M_3M_4)K_4x_3 - ((M_3+M_4)/M_3M_4)B_4x_7 \\
&\quad + (K_3/M_3)x_2 + (B_3/M_3)x_6 + (1/M_4)f \\
\dot{x}_8 &= -(K_4/M_4)x_3 - (B_4/M_4)x_7 + (1/M_4)f - \mu g.
\end{aligned}\right\} \tag{1.2.6}$$

The friction coefficient μ is generally a function of the vehicle speed and is given by

$$\mu = \begin{cases} a+bv, & v \neq 0, \\ 0, & v = 0, \end{cases} \tag{1.2.7}$$

where v is the speed and a,b are constants depending on the properties of the track. For high-speed trains air friction which is proportional to v^2 must be added to the above expression. For low-speed trains one can assume μ to be constant.

For the n-car system, we have a $2n$th-order differential equation of the form

$$\dot{x} = Ax + bu + \xi, \tag{1.2.8}$$

where the matrix A can be partitioned into four $(n \times n)$ distinct block matrices given by

$$A = \begin{bmatrix} A_{11} & A_{12} \\ A_{21} & A_{22} \end{bmatrix}. \tag{1.2.9}$$

We note that $A_{11} \equiv 0$ is an $(n \times n)$ zero matrix, $A_{12} = I_n$ is an $(n \times n)$ identity matrix, and the matrices A_{21} and A_{22} arc also of dimension

$(n \times n)$ representing the stiffness and the damping operators respectively. The input matrix $b = (0, 0, 0, \ldots, 0, 1/M_n, 1/M_n)'$ is a $2n$-dimensional vector, $u = f$ is the applied force which may take either positive or negative values corresponding to acceleration or deceleration, and $\xi = (0, 0, \ldots, 0, -\mu g)'$ is a $2n$ vector with all the components except the last being zero. The vector $x = (x_1, x_2, \ldots, x_{2n})'$ is a $2n$-dimensional column vector representing the state of the train. Once the matrices A, b, ξ and the input $u = f$, including the initial state, are given, one can solve (1.2.8) for the state of the train. From x one can determine the state of vibration of the cars.

In modelling the train dynamics (1.2.1)–(1.2.8) we assumed flat terrain. If the terrain is not flat, then, given the profile of the track, one can modify the system (1.2.8) into the form

$$\dot{x} = Ax + bu + F(\theta), \tag{1.2.10}$$

where the vector F depends on the slope of the track and is given by

$$F(\theta) = g \begin{bmatrix} 0 \\ 0 \\ \vdots \\ 0 \\ (\sin\theta_1 - \sin\theta_2) + \mu(\cos\theta_1 - \cos\theta_2) \\ (\sin\theta_2 - \sin\theta_3) + \mu(\cos\theta_2 - \cos\theta_3) \\ \vdots \\ (\sin\theta_{n-1} - \sin\theta_n) + \mu(\cos\theta_{n-1} - \cos\theta_n) \\ (-\sin\theta_n - \mu\cos\theta_n) \end{bmatrix} \begin{matrix} \\ \\ \\ \\ \leftarrow (n+1)\text{th position} \\ \\ \\ \\ \\ \end{matrix} \tag{1.2.11}$$

with $\theta = (\theta_1, \ldots, \theta_n)$ being the slopes of the segments of the track occupied by the cars $\{1, 2, \ldots, n\}$. Since the slope depends on the position of the cars which, in turn, depends on x, using the known profile of the track, we can replace F by a suitable nonlinear function of x giving

$$\dot{x} = Ax + bu + G(x). \tag{1.2.12}$$

This model is useful for the operation of trains in mountainous regions and also for the study of roller-coaster dynamics. For smooth operation of the train one may wish to introduce a suitable measure of vibration and minimize it for passenger comfort by selecting an appropriate control. This is where optimal control theory is useful.

(b) *Satellite dynamics*

We consider here the attitude dynamics of a rigid body satellite in geosynchronous orbit. For communication and observatory satellites it is absolutely necessary to maintain a high degree of pointing accuracy. This requires an accurate model of the attitude dynamics of the satellite. We present here the dynamic model of a satellite equipped with reaction jets which are capable of exerting torques on the body of the satellite and thereby control its attitude.

According to Newtonian mechanics (conservation of angular momentum) the time rate of change of angular momentum must equal the external torques. Let H denote the angular momentum vector of the satellite and T the externally applied torque vector. Then

$$\frac{dH}{dt} = T, \tag{1.2.13}$$

where the differentiation with respect to time is taken with reference to some fixed inertial frame of reference, say, earth. Let w_b represent the body angular velocity with respect to a reference frame which is moving along the orbital path of the satellite, and let w_r denote the angular velocity of the reference frame itself with respect to the inertial frame. The angular momentum vector H for the satellite is given by the dot product

$$H = I \cdot (w_b + w_r), \tag{1.2.14}$$

where I is the inertia dyadic of the spacecraft body. Since a reference frame could always be selected to yield a diagonal inertia dyadic, we can assume without loss of generality that

$$I = I_x i_b i_b + I_y j_b j_b + I_z k_b k_b \tag{1.2.15}$$

where i_b, j_b and k_b denote the unit vectors in the x, y and z directions of the body coordinate system. Similarly we shall use $\{i_r, k_r, j_r\}$ to denote the unit vectors in the $\{x, y, z\}$ directions of the moving reference frame. Let

$$w_b = i_b p + j_b q + k_b r \tag{1.2.16}$$

and

$$w_r = -j_r w_0 \tag{1.2.17}$$

where w_0 is the angular velocity of the moving reference frame with respect to inertial space. By the usual rule ([63]) of time derivatives in the moving coordinate systems we have

$$\begin{aligned}\frac{d}{dt}H &= \frac{d}{dt}(I \cdot (w_b + w_r)) \\ &= I \cdot \left[\left(\frac{d}{dt}w_b\right)_b + \left(\frac{d}{dt}w_r\right)_b\right] + (w_b + w_r) \times [I \cdot (w_b + w_r)] \\ &= Q_1 + Q_2.\end{aligned} \tag{1.2.18}$$

Using the orthogonal transformation ([63]) from the reference frame to the body frame (assuming that the two coordinate systems have the same origin), given by

$$C_{\mathrm{b}}^{\mathrm{r}} = \begin{bmatrix} C\theta C\psi & C\theta S\psi & -S\theta \\ S\phi S\theta C\psi - C\phi S\psi & S\phi S\theta S\psi + C\phi C\psi & S\phi C\theta \\ C\phi S\theta C\psi + S\phi S\psi & C\phi S\theta S\psi - S\phi C\psi & C\phi C\theta \end{bmatrix} \tag{1.2.19}$$

where $C \equiv$ cosine, $S \equiv$ sine, and (ϕ, θ, ψ) are the Euler angles (attitude) of the satellite, any vector in the reference frame can be transferred to a corresponding vector in the body frame and vice versa. Hence

$$(w_{\mathrm{r}})_{\mathrm{b}} = C_{\mathrm{b}}^{\mathrm{r}} w_{\mathrm{r}} \tag{1.2.20}$$

is the angular velocity of the reference frame with respect to the body frame. Given that $w_{\mathrm{r}} = -j_{\mathrm{r}} w_0$, one can find, assuming $\{\phi, \theta, \psi\}$ are small, that

$$(w_{\mathrm{r}})_{\mathrm{b}} = -j_{\mathrm{b}} w_0. \tag{1.2.21}$$

Hence

$$\begin{aligned} Q_1 &\equiv I \cdot [i_{\mathrm{b}}\dot{p} + j_{\mathrm{b}}(\dot{q} - \dot{w}_0) + k_{\mathrm{b}}\dot{r}] \\ &= i_{\mathrm{b}} I_x \dot{p} + j_{\mathrm{b}} I_y (\dot{q} - \dot{w}_0) + k_{\mathrm{b}} I_z \dot{r}, \end{aligned} \tag{1.2.22}$$

and

$$\begin{aligned} Q_2 &\equiv [i_{\mathrm{b}}p + j_{\mathrm{b}}(q - w_0) + k_{\mathrm{b}}r] \times [i_{\mathrm{b}} I_x p + j_{\mathrm{b}} I_y (q - w_0) + k_{\mathrm{b}} I_z r] \\ &\equiv \det \begin{bmatrix} i_{\mathrm{b}} & j_{\mathrm{b}} & k_{\mathrm{b}} \\ p & q - w_0 & r \\ I_x p & I_y(q - w_0) & I_z r \end{bmatrix}. \end{aligned} \tag{1.2.23}$$

Hence

$$Q_2 = i_{\mathrm{b}}(I_z - I_y)(q - w_0)r + j_{\mathrm{b}}(I_x - I_z)pr + k_{\mathrm{b}}(I_y - I_x)p(q - w_0) \tag{1.2.24}$$

Using (1.2.18), (1.2.22) and (1.2.24) in (1.2.13), we obtain the momentum dynamics of the satellite given by

$$\begin{aligned} &I_x \dot{p} + (I_z - I_y)(q - w_0)r = T_x \\ &I_y(\dot{q} - \dot{w}_0) + (I_x - I_z)pr = T_y \\ &I_z \dot{r} + (I_y - I_x)p(q - w_0) = T_z \end{aligned} \tag{1.2.25}$$

where $T = i_{\mathrm{b}} T_x + j_{\mathrm{b}} T_y + k_{\mathrm{b}} T_z$ is the applied torque. For a geosynchronous satellite w_0 coincides with the spin velocity of the earth which is constant. Hence the equation (1.2.25) can be written as

$$\begin{aligned} I_x\dot{p} + (I_z - I_y)(q - w_0)r &= T_x \\ I_y\dot{q} + (I_x - I_z)pr &= T_y \\ I_z\dot{r} + (I_y - I_x)p(q - w_0) &= T_z. \end{aligned} \tag{1.2.26}$$

The Euler angles $\{\phi, \theta, \psi\}$ are related to the body angular velocities $\{p, q, r\}$ by the equation [63]

$$\begin{aligned} \dot{\phi} &= p + (q \sin \phi + r \cos \phi) \tan \theta \\ \dot{\theta} &= q \cos \phi - r \sin \phi \\ \dot{\psi} &= (q \sin \phi + r \cos \phi) \sec \theta. \end{aligned} \tag{1.2.27}$$

The set of six equations (1.2.26)–(1.2.27) describes the attitude dynamics of a geosynchronous satellite. For a satellite operating satisfactorily, the Euler angles can be assumed to be small and hence (1.2.27) can be approximated by

$$\begin{aligned} \dot{\phi} &= p \\ \dot{\theta} &= q \\ \dot{\psi} &= r. \end{aligned} \tag{1.2.28}$$

The equations (1.2.26) and (1.2.27) or (1.2.28) are widely used in the attitude control of modern satellites.

Since the reaction jets, providing the torque T, consume considerable amount of fuel, the need for fuel conservation is crucial. Hence, in present day satellites, momentum transfer devices like flywheels, gyrotorquers, etc. are also used in addition to the reaction jets. In fact, for fine controls (tuning), momentum transfer devices are used; and for course controls reaction jets are activated. This practice can lead to substantial fuel saving, thereby increasing the life of the satellites.

The dynamics of a satellite using both reaction jets and flywheel controls on all three axes is given by

$$\begin{aligned} I_x\dot{p} + (I_z - I_y)(q - w_0)r &= -C_x\dot{\Omega}_x + C_y\Omega_y r - C_z\Omega_z q + T_x \\ I_y\dot{q} + (I_x - I_z)pr &= -C_y\dot{\Omega}_y - C_x\Omega_x r + C_z\Omega_z p + T_y \\ I_z\dot{r} + (I_y - I_x)(q - w_0)p &= -C_z\dot{\Omega}_z - C_y\Omega_y p + C_x\Omega_x q + T_z, \end{aligned} \tag{1.2.29}$$

where C_x, C_y, C_z are the flywheel inertias and Ω_x, Ω_y, Ω_z are the flywheel angular velocities. Controls are exercised by acceleration or deceleration of the flywheels through $(\dot{\Omega}_x, \dot{\Omega}_y, \dot{\Omega}_z)$. Equation (1.2.29) is derived in the same way as before by taking for H the total angular momentum vector contributed by the satellite body and the flywheels given by

$$H = I \cdot (w_r + w_b) + J \cdot (w_r + w_b + w_G + w_g) \tag{1.2.30}$$

where w_G is the angular velocity (vector) of the gimbal with respect to the body and w_g is that of the gyro with respect to the gimbal frame.

Clearly the systems (1.2.26) and (1.2.27) can be written in the form $\dot{x} = f(x) + Bu$ where $x = (p, q, r, \phi, \theta, \psi)'$ and f, B and u are easily identified by simple inspection. The system is nonlinear in state but linear in control.

(c) *Electric power network*

Usually a modern power network consists of a large number of synchronous generators and loads interconnected through long transmission lines. The complete dynamics of such networks is determined by Kirchhoff's current and voltage laws which state that the algebraic sum of the currents at any node must be zero, and the algebraic sum of the branch voltages around any closed loop must vanish. For illustration, we shall consider the system shown in figure 1.2.

The dynamics of the synchronous generator is described by the flux linkage equations and voltage equations given below:

$$\begin{bmatrix} \psi_f \\ \psi_a \\ \psi_b \\ \psi_c \end{bmatrix} = \begin{bmatrix} L_{ff} & L_{fa} & L_{fb} & L_{fc} \\ L_{af} & L_{aa} & L_{ab} & L_{ac} \\ L_{bf} & L_{ba} & L_{bb} & L_{bc} \\ L_{cf} & L_{ca} & L_{cb} & L_{cc} \end{bmatrix} \begin{bmatrix} i_f \\ -i_{1a} \\ -i_{1b} \\ -i_{1c} \end{bmatrix} \tag{1.2.31}$$

$$\begin{bmatrix} V_f \\ V_a \\ V_b \\ V_c \end{bmatrix} = \frac{\mathrm{d}}{\mathrm{d}t} \begin{bmatrix} \psi_f \\ \psi_a \\ \psi_b \\ \psi_c \end{bmatrix} + \begin{bmatrix} R_f & 0 & 0 & 0 \\ 0 & -R_a & 0 & 0 \\ 0 & 0 & -R_a & 0 \\ 0 & 0 & 0 & -R_a \end{bmatrix} \begin{bmatrix} i_f \\ i_{1a} \\ i_{1b} \\ i_{1c} \end{bmatrix} \tag{1.2.32}$$

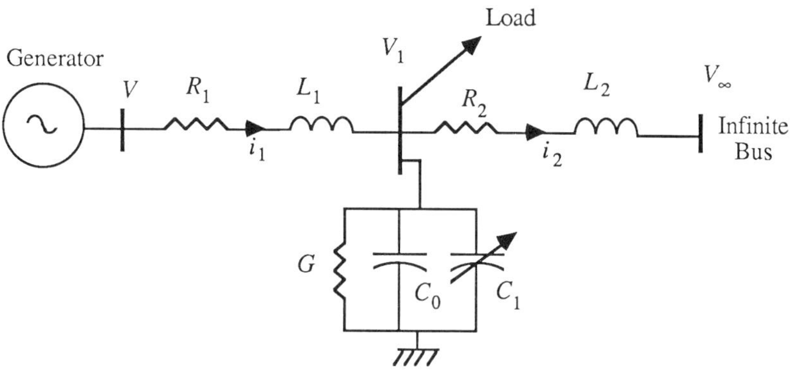

Figure 1.2 Schematic diagram of a small network

In the above equations the subscripts a, b and c refer to the variables in the three phases of the generator, and f corresponds to the field variables. The self and mutual inductances in (1.2.31) depend on the rotor or field position, and hence are functions of time. Denoting the rotor position by θ, these inductances are given by

$$\begin{aligned} L_{ff} &= L_f \\ L_{aa} &= L_s + L_m \cos 2\theta \\ L_{ba} &= L_{ab} = -M_s + L_m \cos 2(\theta + 120) \\ L_{fa} &= L_{af} = M_F \cos \theta. \end{aligned} \tag{1.2.33}$$

The rest of the inductances in (1.2.31) are obtained by replacing θ by $(\theta - 120)$ for phase b, and by $(\theta + 120)$ for phase c.

The transmission line equations are obtained easily using Kirchhoff's current and voltage laws. For a symmetrical network, the transmission line parameters are the same for all the three phases. In figure 1.2, only one of the phases is shown with the appropriate parameters. Denoting $\bar{L}_i = \mathrm{diag}(L_i, L_i, L_i)$, $\bar{R}_i = \mathrm{diag}(R_i, R_i, R_i)$, $i = 1, 2$, $\bar{C}_0 = \mathrm{diag}(C_0, C_0, C_0)$, $\bar{C}_1(t) = \mathrm{diag}(C_1(t), C_1(t), C_1(t))$, $\bar{G} = \mathrm{diag}(G, G, G)$, the dynamics of the transmission line could be described by

$$\begin{aligned} &\bar{L}_1 \frac{\mathrm{d}i_1}{\mathrm{d}t} + \bar{R}_1 i_1 + V_1 = V \\ &\bar{L}_2 \frac{\mathrm{d}i_2}{\mathrm{d}t} + \bar{R}_2 i_2 + V_\infty = V_1 \\ &i_1 = \bar{G}V_1 + \frac{\mathrm{d}Q}{\mathrm{d}t} + i_2 + I_L \\ &Q = (\bar{C}_0 + \bar{C}_1(t))V_1, \end{aligned} \tag{1.2.34}$$

where the line currents, node voltages and capacitor charges are three phase quantities, which, for brevity, are denoted without the subscripts a, b, c; that is $i_1 \equiv (i_{1a}, i_{1b}, i_{1c})$ and similarly for other variables. For node voltage regulation, generally the capacitor C_1 is considered as the control variable [102, 137].

The mechanical oscillations of the generator are described by the well-known swing equations. Denoting the torque angle by δ and per unit speed deviation by n, these equations are given by

$$\begin{aligned} &\frac{\mathrm{d}\delta}{\mathrm{d}t} = \omega_0 n \\ &\frac{\mathrm{d}n}{\mathrm{d}t} = \frac{1}{2H}[T_i - T_0 - Dn], \end{aligned} \tag{1.2.35}$$

where ω_0 is the synchronous speed, H is the moment of inertia, and T_i and T_0 are the input and output torques and D is the damping factor.

Equations (1.2.31)–(1.2.35) constitute a complete system representing the network shown in the figure. However, because of the presence of time-varying inductances and periodically varying currents and voltages, numerical and analytical treatment of these equations is extremely difficult. Fortunately, this problem could be greatly simplified by introducing a transformation defined by $x_{dq0} = Px_{abc}$, where $x_{abc} = (x_a, x_b, x_c)$ is a usual three phase variable, $x_{dq0} = (x_d, x_q, x_0)$ is the corresponding transformed variable and P is an orthogonal transformation matrix, commonly known as Park's transformation [60], given by

$$P = \sqrt{\frac{2}{3}} \begin{bmatrix} \cos\theta & \cos(\theta - 120°) & \cos(\theta + 120°) \\ -\sin\theta & -\sin(\theta - 120°) & -\sin(\theta + 120) \\ \dfrac{1}{\sqrt{2}} & \dfrac{1}{\sqrt{2}} & \dfrac{1}{\sqrt{2}} \end{bmatrix} . \tag{1.2.36}$$

Using this transformation for each of the three phase variables, the complete system dynamics can be represented by the following set of nonlinear differential equations:

$$\frac{\mathrm{d}}{\mathrm{d}t}\begin{bmatrix} i_f \\ i_{1d} \\ i_{1q} \\ n \\ \delta \\ Q_d \\ Q_q \\ i_{2d} \\ i_{2q} \end{bmatrix} = \begin{bmatrix} a_1 i_f + a_2 i_{1d} + a_3(1+n)i_{1q} + a_4 V_f + a_5 Q_d/C \\ a_6 i_f + a_7 i_{1d} + a_8(1+n)i_{1q} + a_9 V_f + a_{10} Q_d/C \\ a_{11}(1+n)i_f + a_{12}(1+n)i_{1d} + a_{13} i_{1q} + a_{14} Q_q/C \\ T_{in} - a_{15} i_f i_{1d} + a_{16} i_{1d} i_{1q} - a_{17} n \\ a_{18} n \\ i_{1d} - i_{2d} - (G/C)Q_d + (1+n)Q_q - I_{Ld} \\ i_{1q} - i_{2q} - (G/C)Q_q - (1+n)Q_d - I_{Lq} \\ [(1/C)Q_d - V_\infty \sin\delta - R_2 i_{2d} + (1+n)L_2 i_{2q}]/L_2 \\ [(1/C)Q_q - V_\infty \cos\delta - R_2 i_{2q} - (1+n)L_2 i_{2d}]/L_2 \end{bmatrix} \tag{1.2.37}$$

where $C = C_0 + C_1(t)$, and the coefficients a_1–a_{18} are constants depending on the direct and quadrature axis inductances of the generator and on line parameters. Note that for symmetrical networks under balanced operation the zero sequence components of currents and voltages are zero. The transformed system (1.2.37) is, in fact, a representation of the system as observed from a coordinate system which is in synchronism with the rotating electromagnetic field.

One common problem in power systems is quenching of generator oscillations following a disturbance, such as short circuit. For such problems the usual practice is to consider a detailed dynamics for the generator under study and to represent the rest of the power pool by an equivalent large generator (infinite bus). Then the dynamics (1.2.37) reduces to a fifth-order nonlinear system [101] of the form similar to the first five equations of (1.2.37). In this case the field voltage V_f is usually considered as the control variable. For multimachine networks, complete dynamics is developed following the same procedure.

(d) *Aquatic ecosystem*

Using the basic principle of cause and effect one can develop approximate dynamic models of the complex environment around us. Consider a body of water, in particular a lake, supporting many different marine lives interacting in a complex way subject to the surrounding environment. This is a system far more complex than those governed by well-understood Newtonian mechanics. However, a simple empirical model, developed on the basis of a cause-and-effect relation, can be useful for prediction and control of the state of the system.

The lake receives from its surroundings several inputs like

(1) phosphates and nitrates originating from the surrounding vegetation and washed into the lake by rain water;
(2) organic waste originating from city and cottage waste treatment plants and
(3) climatic conditions of the region such as temperature and illumination, etc.

Phosphates and nitrates are important nutrients (N) for plant growth and consequently their presence leads to the growth of algae (A) in the aqua system. Algae provide food for aquatic life like fish (F), and they use carbon dioxide (C) released by aquatic life, and other nutrients and sunlight to produce oxygen (Q), resulting in an increased concentration of dissolved oxygen in the water. Oxygen is vital for all marine life. Bacteria (B) uses oxygen and produces nutrients from the organic waste (W) and releases carbon dioxide. These in turn lead to further growth of algae, completing the cycle. Further, bacteria and other microorganisms (B) provide food for the fish which in turn are used for human consumption. At the lake bottom sludge (S) is produced from the organic waste by bacterial decomposition.

While a certain amount of algae and other plants are essential for a healthy lake, too many can lead to deterioration of water quality. Excessive algae die and pile up at the bottom consuming dissolved oxygen in the decay process. The extra oxygen demand leads to deoxygenation of the lower levels of water, resulting in the loss of fish and other aquatic lives. This leads to a vicious circle resulting in the deterioration of water quality and increased sludge build up. If unchecked, the lake will fill in from sludge build up and eventually turn into a swamp. Biologically this whole process is known as eutrophication.

In the above description we have already indicated by letters the variables that participate in the biochemical activities of the aquatic system. Denoting by e_n the natural runoffs and by W the rate of disposal from the waste treatment plants, one can write the system dynamics as

$$\left.\begin{aligned}
\frac{dN}{dt} &= -K_1A + K_2e_n + K_3u_1WB \\
\frac{dA}{dt} &= (K_4 - K_5A + K_6NC - K_7F)A - u_2A \\
\frac{dQ}{dt} &= K_8A - K_9B - K_{10}F - K_{11}S \\
\frac{dB}{dt} &= (K_{12} + K_{13}Q - K_{14}F - K_{15}B)B + u_3W \\
\frac{dF}{dt} &= (K_{16} - K_{17}F + K_{18}BQ + K_{19}AQ - K_{20}N)F - u_4F \\
\frac{dC}{dt} &= K_{21}F + K_{22}B - K_{23}A + K_{24}u_5WB \\
\frac{dS}{dt} &= (K_{25}A + K_{26}B + K_{27}F) + K_{28}e_n + K_{29}WB - u_6,
\end{aligned}\right\} \tag{1.2.38}$$

where all the interaction coefficients K_1–K_{29} are non-negative. The variables u_1–u_6 are control variables which can be used by regulating agencies to maintain certain desirable standards. We present a brief interpretation of equation (1.2.38). The first equation describes the growth of nutrient concentration in the aquatic system, depleting with increasing algae population and increasing with increasing runoffs (e_n) and waste disposal rates (W). The variable u_1 represents the control of the quality of treated waste as dumped into the system. The second equation describes the growth of algae population, where the first term represents intrinsic growth rate and the second term represents control (by removal). The growth rate in the absence of control, given by the factor within the parentheses, increases with the increase of concentration of nutrients and dissolved carbon dioxide and decreases with the increase of algae population (due to inhibition) and fish population (due to consumption). The fourth and the fifth equations have a similar interpretation. The third equation describes the state of concentration of dissolved oxygen increasing with the increase of algae population and decreasing with the increase of population of aquatic life (fish, bacteria, and other micro-organisms) and sludge build up. Similar explanation can be provided for the rest of the equations from the description given at the introduction. Note that u_4 represents fractional fishing rate and u_6 sludge removal rate.

We must emphasize that the model (1.2.38) is based entirely on empirical (intuitive) reasoning and must be modified as required to conform with the reality. Many of the coefficients, K_1–K_{29}, are quite fundamental and require basic scientific experiments to determine them. For example, K_1 represents the uptake rate of nutrients by algae, K_8 represents production rate of oxygen by algae, and K_{20} represents toxicity of the water caused by nutrients poisoning marine life, etc. In the section dealing with system identification we shall discuss procedures for approximate identification of the parameters K_1–K_{29}, avoiding basic scientific experiments. The model presented above was first developed in the paper [88] where optimal control was also considered.

(e) *Basic economic model*

The neoclassical growth model characterizes economic growth in an aggregative closed economy. The term 'aggregative' means that the economy produces a single homogeneous good called the output $Y(t)$ (goods and services) using two homogeneous factor inputs, capital K and labour L. The term 'closed' means that neither the input nor the output is imported or exported. All the output is used for domestic consumption and investment. In general

$$\begin{aligned} &\text{output} = \text{consumption} + \text{investment} \\ &\quad Y(t) = C(t) + I(t). \end{aligned} \tag{1.2.39}$$

Investment is used both to augment the capital and replace the depreciated capital. That is

$$I(t) = \frac{\mathrm{d}K(t)}{\mathrm{d}t} + \mu K(t) \tag{1.2.40}$$

where μ is the capital depreciation rate. The output is determined by an aggregated production function which depends on capital K and labour L.

$$Y = F(K, L). \tag{1.2.41}$$

In order that a given function F qualifies as a candidate for production function, it is necessary that it satisfies the following properties:

$$\text{(P1):}\quad \frac{\partial F}{\partial K}>0, \qquad \frac{\partial^2 F}{\partial K^2}<0 \tag{1.2.42}$$

$$\frac{\partial F}{\partial L}>0, \qquad \frac{\partial^2 F}{\partial L^2}<0 \tag{1.2.43}$$

$$\text{(P2):}\quad \lim_{k\to 0}\frac{\partial F}{\partial K}=\infty, \quad \lim_{k\to\infty}\frac{\partial F}{\partial K}=0 \tag{1.2.44}$$

$$\lim_{L\to 0}\frac{\partial F}{\partial L}=\infty, \quad \lim_{L\to\infty}\frac{\partial F}{\partial L}=0 \tag{1.2.45}$$

$$\text{(P3):}\quad F(\alpha K, \alpha L)=\alpha F(K, L) \qquad \text{for } \alpha>0. \tag{1.2.46}$$

Property (P1) signifies that production increases with increasing capital and labour but the marginal productivity diminishes. Similarly property (P2) means that the marginal productivity is very large at lower levels of labour and capital investment whereas at very high levels of investment further increase of capital or labour does not produce any significant gain in productivity. The property (P3) indicates that the production function must exhibit constant returns to scale.

The reader may easily verify that the function

$$F(K, L)\equiv \gamma K^{1/p}L^{1/q} \tag{1.2.47}$$

with

$$\gamma>0 \qquad \text{and} \qquad (1/p)+(1/q)=1, \qquad 1<p, \quad q<\infty, \tag{1.2.48}$$

satisfies all the required properties (P1)–(P3). Choosing $\alpha=(1/L)$ and defining

$$\begin{aligned} y&=(Y/L) \qquad \text{(per capita output)}\\ k&=(K/L) \qquad \text{(per capita investment)} \end{aligned} \tag{1.2.49}$$

it follows from (1.2.46) that

$$F(K/L, L/L)=\frac{1}{L}F(K, L)=\frac{Y}{L}. \tag{1.2.50}$$

Defining

$$f(k)\equiv F(k, 1), \tag{1.2.51}$$

observe that f relates the output per unit labour to the capital investment per unit labour. That is

$$y=f(k). \tag{1.2.52}$$

The function f satisfies the properties (P1)–(P3) with respect to the single variable k. Writing all the variables in per capita units, $y = Y/L$, $c = C/L$, $i = I/L$, and $k = K/L$, it follows from (1.2.39)–(1.2.41) that

$$y = c + i$$

$$i = \left(\frac{1}{L}\frac{\mathrm{d}}{\mathrm{d}t}K\right) + \mu k = \dot{k} + k(\dot{L}/L) + \mu k \tag{1.2.53}$$

$$y = f(k).$$

From these equations we obtain

$$\dot{k} = f(k) - c - \mu k - (\dot{L}/L)k. \tag{1.2.54}$$

If the labour force is assumed to grow exponentially, then

$$(\dot{L}/L) = \beta \geqslant 0, \tag{1.2.55}$$

and (1.2.54) takes the form

$$\dot{k} = f(k) - c - (\mu + \beta)k = f(k) - c - \lambda k, \tag{1.2.56}$$

where μ is the capital depreciation rate and β the capital dilation rate. For steady state, $\dot{k} = 0$, we have,

$$f(k) - \lambda k = c. \tag{1.2.57}$$

Whether certain level of consumption c is sustainable or not is determined from this equation (see figure 1.3).

For c sufficiently small there are two equilibrium points $p_1 \equiv (c, k_1)$, and $p_2 \equiv (c, k_2)$. For initial capital $k_0 < k_1$, $f(k_0) - \lambda k_0 - c < 0$ and $k \downarrow 0$.

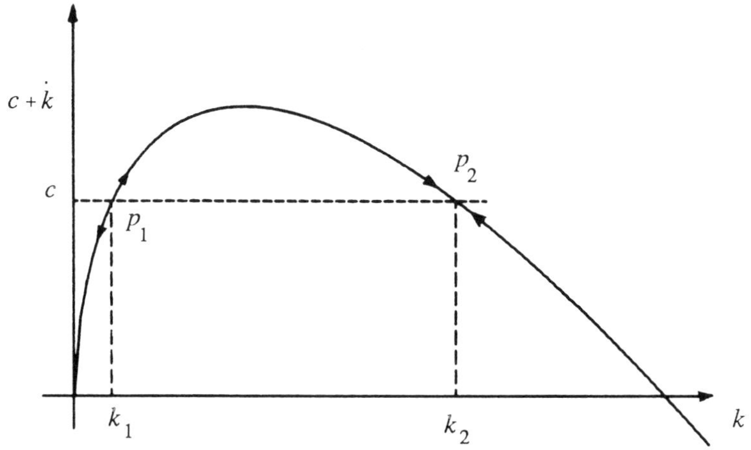

Figure 1.3 Economic growth model

For $k_1 < k_0 < k_2$, $f(k_0) - \lambda k_0 - c > 0$ and $k \uparrow k_2$. For $k_0 > k_2$, $f(k_0) - \lambda k_0 - c < 0$ and hence $k \downarrow k_2$. Therefore p_2 is a stable equilibrium point but p_1 is not. From this simple analysis we can observe that, for a given consumption level (determined by population and standard of living), an underdeveloped country receiving foreign aid (capital) below the critical level k_1 will sustain poverty. For sustained growth, the level of capital aid must exceed the critical level k_1; otherwise the foreign aid is deceptive.

Using Volterra's logistic model for population growth, given by

$$\dot{L} = (\beta - \gamma L)L, \tag{1.2.58}$$

one obtains the following system of equations,

$$\begin{aligned} \frac{dk}{dt} &= f(k) - c - \mu k - (\beta - \gamma L)k \\ \frac{dL}{dt} &= (\beta - \gamma L)L, \end{aligned} \tag{1.2.59}$$

where γ is the inhibition factor resulting from competition. We shall discuss this model further in stability studies. Optimal control problems for more elaborate economic models have been discussed by many authors, for example, see [91], [135].

(f) *Reactor dynamics and chemical kinetics*

We present two examples involving reaction kinetics.

(i) *Reactor dynamics.* Nuclear reactors are controlled by regulating the fission of fissile nuclei (^{235}U, ^{233}U, ^{239}Pu, etc.) by means of neutron chain reaction. A fissile nucleus, subject to neutron bombardment, breaks into fragments of different species of atoms and gives rise to several (2.5 on the average) new neutrons, some of which appear instantaneously, and a fraction β appear after varying lapses of time. Some of the neutrons are lost from the core through leakage and the rest contribute to further fission, leading to chain reaction. The fission fragments lose their kinetic energy to the surrounding atoms, thereby raising the temperature of the core. The resulting heat energy is then conveyed by coolants to heat exchangers, where it is finally converted into electrical power. The power generated in the core is directly proportional to neutron flux density, which is controlled by insertion or withdrawal of neutron absorbers (called control rods). Prompt neutrons and delayed neutrons are the two sources that contribute to the total neutron population in the reactor. There are about six species of atoms in the fission fragments that emit delayed neutrons. Therefore the dynamic state of the reactor is described by giving the neutron flux density and the population density of the six groups of delayed neutron emitters.

Let ℓ denote the (average) life time of prompt neutrons (from generation to absorption), N the population density of neutrons, R_i $(i = 1, 2, \ldots, 6)$ the population density of delayed neutron emitters, λ_i $(i = 1, 2, \ldots, 6)$ their decay rate.

Each fission gives rise to ν neutrons with $(1 - \beta)\nu$ being prompt neutrons and $\beta\nu$ delayed neutrons. The fraction $(1 - \beta)(\nu - k_{\text{eff}})$ leaks out leaving $(1 - \beta)k_{\text{eff}}$ for further fission. The population density of delayed neutron emitters being R_i with decay rate λ_i, the number of delayed neutrons emitted during the one life cycle period (ℓ) is

$$\sum_{i=1}^{6} R_i\lambda_i\ell/\text{cm}^3.$$

Hence the neutron population density increases from N to $(1 - \beta)k_{\text{eff}}N + \sum_i R_i\lambda_i\ell$ during the period ℓ. Therefore the time rate of change of population density is given by

$$\frac{dN}{dt} = \frac{1}{\ell}\left\{(1 - \beta)k_{\text{eff}}N + \sum_{i=1}^{6} R_i\lambda_i\ell - N\right\}. \tag{1.2.60}$$

Defining $k_{\text{ex}} \equiv (k_{\text{eff}} - 1)$, the excess reactivity, we have

$$\frac{dN}{dt} = \frac{N}{\ell}(k_{\text{ex}} - k_{\text{eff}}\beta) + \sum_{i=1}^{6} R_i\lambda_i. \tag{1.2.61}$$

Similarly from the delayed neutrons $\beta\nu$, the fraction $\beta(\nu - k_{\text{eff}})$ gets lost leaving βk_{eff} for further fission. Letting μ_i denote the fraction of the ith species, $(\beta k_{\text{eff}}\mu_i)N$ denotes the number of delayed neutron emitters of type i added to the population R_i during the period ℓ. The decay rate being λ_i, the number that disappears by decay, is $(R_i\lambda_i\ell)$ during the same period ℓ. Hence, in the balance, the population of species i increases from R_i to $R_i + (\beta k_{\text{eff}}\mu_i)N - R_i\lambda_i\ell$ during the period ℓ and consequently

$$\frac{dR_i}{dt} = \frac{1}{\ell}\{\beta k_{\text{eff}}\mu_i N - \lambda_i R_i\ell\}. \tag{1.2.62}$$

Assuming $k_{\text{eff}} \cong 1$, and defining the control variable $u = k_{\text{ex}}$, we have the system of equations

$$\begin{aligned}\frac{dN}{dt} &= \frac{N}{\ell}(u - \beta) + \sum_{i=1}^{6} \lambda_i R_i \\ \frac{dR_i}{dt} &= \frac{\beta\mu_i}{\ell}N - \lambda_i R_i, \qquad i = 1, 2, \ldots, 6.\end{aligned} \tag{1.2.63}$$

These are the equations used for control of nuclear reactors [61]. In addition, one must include the heat transfer equations to have a complete operational dynamics of the system. We must emphasize that for detailed analysis of the system one must use distributed parameter models (e.g. partial differential equations) for neutron transport, heat flow, and fluid flow, etc.

(ii) *Chemical kinetics.* In general (multiple) chemical reactions involve two or more reactants. Often binary reaction suffices to describe the reaction kinetics. Let $\{A_1, \ldots, A_n\}$ denote a set of n distinct chemicals participating in the reaction process with corresponding concentrations denoted by $\{x_1, \ldots, x_n\}$. Neglecting ternary and higher-order reactions, the general equation of evolution is given by

$$\frac{\mathrm{d}x_i}{\mathrm{d}t} = \sum_{j=1}^{N} a_{ij}x_j + \sum_{k,\ell} b_{ik\ell}x_k x_\ell, \qquad i = 1, 2, \ldots, n, \tag{1.2.64}$$

where, for $i \neq j$, a_{ij} is the transition rate $A_j \xrightarrow{a_{ij}} A_i$ and, for $i \neq \{k, \ell\}$, $b_{ik\ell}$ is the reaction rate $A_k + A_\ell \xrightarrow{b_{ikl}} A_i$. Assuming conservation of mass (that is, in the reaction process no mass is annihilated), we must have

$$\sum_{i=1}^{n} x_i(t) = \text{constant for all } t. \tag{1.2.65}$$

Hence

$$\begin{aligned} 0 &= \frac{\mathrm{d}}{\mathrm{d}t}\left(\sum_{i=1}^{n} x_i(t)\right) = \sum_{i=1}^{n} \frac{\mathrm{d}}{\mathrm{d}t} x_i(t) \\ &= \sum_{j=1}^{n}\left(\sum_{i=1}^{n} a_{ij}\right)x_j + \sum_{k,\ell}^{n}\left(\sum_{i}^{n} b_{ikl}\right)x_k x_\ell. \end{aligned} \tag{1.2.66}$$

Since $\{x_i(t), x_k(t), x_\ell(t), 1 \leqslant i,k,l \leqslant n\}$ are distinct and not identically zero, we must have

$$\sum_{i=1}^{n} a_{ij} = 0 \qquad \text{for all } 1 \leqslant j \leqslant n,$$

and (1.2.67)

$$\sum_{i=1}^{n} b_{ikl} = 0 \qquad \text{for all } 1 \leqslant k, \ \ell \leqslant n.$$

Clearly it follows from these relations that

$$\begin{aligned} a_{jj} &= -\sum_{i \neq j} a_{ij} \\ b_{kk\ell} + b_{\ell k\ell} &= -\sum_{i \neq \{k,\ell\}} b_{ik\ell}. \end{aligned} \tag{1.2.68}$$

These relations simply mean that gains of any one chemical must equal

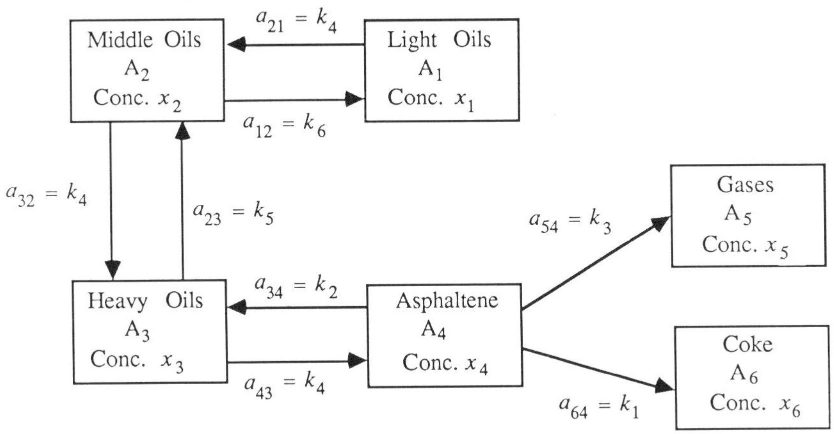

Figure 1.4 Model for bitumen cracking

the sum of the losses of the others (similar to Kirchhoff's laws). The reaction rates $\{a_{ij}, b_{ik\ell}\}$ generally depend on the temperature and pressure of the container.

For illustration we present a kinetic model for the thermal cracking of Athabasca Bitumen [134]. The bitumen begins to crack at relatively low temperatures (673 K and 620 kPa) producing a gasoline, high in aromatic content, together with gases and coke. The reaction cycle is shown in figure 1.4 (assuming only first-order reaction). The coke and the gases are the final stable products. Following the arrows one can easily write:

$$\frac{d}{dt}\begin{bmatrix} x_1 \\ x_2 \\ x_3 \\ x_4 \\ x_5 \\ x_6 \end{bmatrix} = \begin{bmatrix} a_{11} & a_{12} & 0 & 0 & 0 & 0 \\ a_{21} & a_{22} & a_{23} & 0 & 0 & 0 \\ 0 & a_{32} & a_{33} & a_{34} & 0 & 0 \\ 0 & 0 & a_{43} & a_{44} & 0 & 0 \\ 0 & 0 & 0 & a_{54} & a_{55} & 0 \\ 0 & 0 & 0 & a_{64} & 0 & a_{66} \end{bmatrix} \begin{bmatrix} x_1 \\ x_2 \\ x_3 \\ x_4 \\ x_5 \\ x_6 \end{bmatrix} \tag{1.2.69}$$

Using (1.2.68) we have $a_{11} = -a_{21}$, $a_{22} = -a_{12} - a_{32}$, $a_{33} = -a_{23} - a_{43}$, $a_{44} = -a_{34} - a_{54} - a_{64}$, $a_{55} = 0$ and $a_{66} = 0$. Hence

$$\begin{aligned} \dot{x}_1 &= -k_4x_1 + k_6x_2 \\ \dot{x}_2 &= k_4x_1 - k_6x_2 - k_4x_2 + k_5x_3 \\ \dot{x}_3 &= k_4x_2 - k_5x_3 - k_4x_3 + k_2x_4 \\ \dot{x}_4 &= k_4x_3 - k_2x_4 - k_3x_4 - k_1x_4 \\ \dot{x}_5 &= k_3x_4 \\ \dot{x}_6 &= k_1x_4, \end{aligned} \tag{1.2.70}$$

which can be directly verified following the arrows. This demonstrates that (1.2.64) is a general equation for chemical kinetics involving no more than binary reactions. The coefficients depend on temperature and pressure. Hence for fixed temperature and pressure, the system (1.2.69) is linear. By controlling these variables one can control the reaction dynamics and hence the production. Denoting these variables by $u = (T, P)$, equation (1.2.69) takes the form $\dot{x} = A(u)x$ which is obviously nonlinear and, in general, the equation (1.2.64) has the canonical form $\dot{x} = f(x, u)$.

(g) *Population dynamics with examples from fisheries and agriculture*

(i) *Fish population.* The population dynamics of a single species of fish in a given aquatic habitat can be simply written as [62, 106],

$$\dot{x} = f(x) - u \tag{1.2.71}$$

where x denotes the population density, f the growth function and u the fishing rate. The growth function f is given by the popular logistic function

$$f(x) = \begin{cases} \alpha x(1 - x/k), \\ \text{or} \\ \alpha x\left(1 - \dfrac{x}{k}\right)\left(\dfrac{x}{k_0} - 1\right), \end{cases} \tag{1.2.72}$$

where $\alpha > 0$ is the intrinsic growth rate and $k > 0$ is the environmental carrying capacity and $k_0 > 0$ is the viable population level. If the population exceeds k it will tend to diminish till it reduces to k, and if, to start with, it is below k, then the population will increase till it reaches the level k. The carrying capacity depends on several factors, the crucial ones being availability of food, nesting facilities and climatic conditions. Population level below k_0 leads to extinction of the species. Fish is a renewable resource and a staple food in many countries and certainly it requires proper management and regulation for continued supply. We shall consider the control problem later. The dynamics of multiple species can be expressed along the same lines, as shown in the next example.

(ii) *Agriculture.* Pests and weeds can cause colossal crop damage and hence, for control, it is important to understand their population dynamics. Consider an agricultural crop attacked by n different species of pests and weeds requiring m different chemicals (insecticide and herbicide) for control. Based on logistic arguments, the population dynamics can be modelled by the following system of differential equations:

$$\frac{dN_i}{dt} = a_i N_i + \sum_{j=1}^{n} b_{ij} N_i N_j + \sum_{k=1}^{m} c_{ik} u_k N_i, \qquad i = 1, 2, \ldots, n, \tag{1.2.73}$$

where N_i represents the population of the ith species and u_k the application rate of the kth chemical. The coefficients a_i, b_{ij}, c_{ik} are constants representing the natural growth rate, interspecies interaction, and chemical action on pests and weeds. The coefficients $b_{ij}=0$ if the species (i, j) are neutral, $b_{ij}>0$ if the species i preys upon j and $b_{ij}<0$ if the species j preys upon i. The coefficient $c_{ik}=0$ if the kth chemical is harmless to the species i and otherwise it is negative. By appropriate choice of insecticides and herbicides and their application rate one can control the population $N=(N_1 \cdots N_n)$. We shall consider the control problem in the sequel. For the population dynamics of multiple species of fish one may replace the third summation in (1.2.73) by $-u_i$. This model was originally developed by the author [83].

(h) *Finite-dimensional systems from infinite-dimensional ones*

In many situations systems governed by partial differential equations or more generally infinite-dimensional systems can be approximated by ordinary differential equations. Of course this is often done for numerical solution of partial differential equations. We consider here an example for illustration. Consider the normalized cantilever (near end fixed, far end free) beam equation [70],

$$\begin{aligned}
&\frac{\partial^2 y}{\partial t^2}+\frac{\partial^4 y}{\partial x^4}=f, \qquad x \in (0, 1), \quad t \geqslant 0, \quad \text{(i)}\\
&y(0, t)=0, \quad \text{(ii)}, \qquad \frac{\partial^2 y}{\partial x^2}(1, t)=0, \quad \text{(iv)}\\
&\frac{\partial y}{\partial x}(0, t)=0, \quad \text{(iii)}, \qquad \frac{\partial^3 y}{\partial x^3}(1, t)=0, \quad \text{(v)}
\end{aligned} \tag{1.2.74}$$

where $y(t, x)$ denotes the deflection of the beam at time t at the position $x \in [0, 1]$.

Defining $y_1 \equiv \partial y/\partial t$, $y_2 \equiv \partial^2 y/\partial x^2$, we obtain the equation 1.2.74(i) in the form

$$\begin{aligned}
\frac{\partial y_1}{\partial t} &= -\frac{\partial^2 y_2}{\partial x^2}+f\\
\frac{\partial y_2}{\partial t} &= \frac{\partial^2 y_1}{\partial x^2}.
\end{aligned} \tag{1.2.75}$$

Suppose the spatial domain $[0, 1]$ is partitioned into $(N-1)$ equal intervals of length (Δx) having N nodes and we wish to describe the state of the system by giving the values of $\{y_1, y_2\}$ at the nodes $\{0=x_1, x_2, \ldots, x_N=1\}$ labelled as $\{1, 2, \ldots, N\}$. That is, the state is described by the $2N$ variables $\{y_1(i), y_2(i), i = 1, 2, \ldots, N\}$ as functions of time. Using equation (1.2.75) and the central difference for approximating second partials, we obtain

$$\frac{d}{dt} y_1(i) = -\frac{1}{(\Delta x)^2} \{y_2(i+1) - 2y_2(i) + y_2(i-1)\} + f(i)$$

$$\frac{d}{dt} y_2(i) = \frac{1}{(\Delta x)^2} \{y_1(i+1) - 2y_1(i) + y_1(i-1)\},$$

$$i = 2, 3, \ldots, N-1. \quad (1.2.76)$$

For completing the set we must have four more equations for $y_1(1)$, $y_2(1)$, $y_1(N)$, $y_2(N)$ which can be determined by using the boundary conditions (1.2.74) (ii–v). From (ii) and (iv), we have

$$\frac{d}{dt} y_1(1) = 0, \qquad \frac{d}{dt} y_2(N) = 0. \quad (1.2.77)$$

For $y_2(1)$ and $y_1(N)$ we use interpolation and extrapolation and the equation (1.2.75). From (1.2.75), we have

$$\frac{dy_1(N)}{dt} = -\left.\frac{\partial^2 y_2}{\partial x^2}\right|_N + f(N). \quad (1.2.78)$$

Since

$$y_2(N-1) = y_2(N) - \left(\frac{\partial y_2}{\partial x}\right)(N)\Delta x + \frac{1}{2}\frac{\partial^2 y_2}{\partial x^2}(N)(\Delta x)^2 + \ldots,$$

it follows from the boundary conditions (iv) and (v) that the first two terms vanish leaving

$$\frac{\partial^2 y_2(N)}{\partial x^2} = \frac{2}{(\Delta x)^2} y_2(N-1) + o(\Delta x),$$

where $\lim_{\Delta x \to 0} o(\Delta x)/\Delta x = 0$. Using this in (1.2.78), we obtain

$$\frac{d}{dt} y_1(N) = -\frac{2}{(\Delta x)^2} y_2(N-1) + f(N). \quad (1.2.79)$$

Using the second equation of (1.2.75) and the approximation

$$y_1(2) = y_1(1) + \frac{\partial y_1}{\partial x}(1)\,\Delta x + \frac{1}{2}\frac{\partial^2 y_1}{\partial x^2}(1)\,(\Delta x)^2 + \cdots$$

along with the boundary conditions (ii) and (iii), we have

$$\frac{dy_2(1)}{dt} = \frac{2}{(\Delta x)^2} y_1(2). \quad (1.2.80)$$

Hence the complete system is given by (1.2.76), (1.2.77), (1.2.79) and (1.2.80). This is a $2n$th-order linear differential equation of the form $\mathrm{d}x/\mathrm{d}t = Ax + Bu$, and can be solved given the initial state. Accuracy of the solution can be improved by refinement of the grid size. In recent years renewed interest has been shown in the study of the beam dynamics due to its use in the design of flexible spacecraft.

1.3 Some remarks on stochastic models

The system models presented above have been assumed to be deterministic. However in all practical systems one could find some source of uncertainties. In the case of locomotive dynamics, one could add to f the aerodynamic forces which are considered random. This would change the model (1.2.8) into the form (1.1.11). In the case of satellite dynamics (1.2.26), the external torque T is given by $T = T_a + T_n$, where T_a is the control (applied) torque and T_n is the torque induced by the impact of micrometeorites or due to (on board) fuel sloshing etc. The equation (1.2.26) then takes the form of (1.1.11), or more precisely

$$\mathrm{d}x = f(x)\,\mathrm{d}t + Bu\,\mathrm{d}t + B\,\mathrm{d}W(t)$$

$$x = (p, q, r)'.$$

In the case of a power system network, governed by the equation (1.2.37), the input torque T_i and the field voltage V_f, are considered to be the control variables, while the electrical demand (or load) $I_L = (I_{Ld}, I_{Lq})'$ is a random process. Hence the system (1.2.37) takes the form of (1.1.10), or more precisely

$$\mathrm{d}x = f(t, x)\,\mathrm{d}t + Bu\,\mathrm{d}t + D\,\mathrm{d}W(t),$$

where x stands for the state vector of (1.2.37). The reader can easily identify f, B and D. For voltage regulation, the capacitor $C(t)$ $(\equiv C_0 + C_1(t))$ is also considered to be a control variable, and in this case the model (1.1.13) is more appropriate. More precisely $\mathrm{d}x = f(t, x, u)\,\mathrm{d}t + D\,\mathrm{d}W(t)$. For the aquatic system (1.2.38), randomness may arise from the environmental inputs (e_n, W) and the uncertainties due to variation of some of the coefficients, K_1–K_{29}, with the change of climatic conditions. Considering the economic model, the actual production may randomly differ from time to time even for the same level of investment of capital and labour. This uncertainty can be taken into account by adding a small noise term to (1.2.56) in the form

$$\mathrm{d}k = (f(k) - c - \lambda k)\,\mathrm{d}t + \varepsilon\,\mathrm{d}W$$

where ε is a small positive number. Similar modification can be introduced in all the subsequent models if necessary.

We conclude this chapter with the following comments. For modelling dynamic systems one must know the fundamental mechanism that governs the relationship between all the variables of concern. In almost all the physical systems these relations are determined by fundamental laws of physics and are better understood. On the other hand, for socio-economic systems and ecological problems, these interrelations are understood only at the qualitative level and must be determined by experimental observations. We shall see later that techniques for system identification and optimization are useful in determining the many unknown parameters that appear in the empirical models.

In this book we are concerned only with finite-dimensional systems. However many physical systems are governed by partial differential equations, for example, heat equation, wave equation, beam equation etc. [70, 55]. Another class of systems which has received special attention in recent years consists of systems governed by a coupled set of ordinary and partial differential equations. Two important examples in this class are flexible spacecraft [99, 100] and robots with flexible arms. Fortunately, very many of these systems can be approximated by ordinary differential equations, as illustrated by the example (1.2(h)).

Exercises

In order to have a better understanding of real physical systems, it is advisable to simulate their dynamics and study their response. The exercises below are intended for students who are familiar with computer programming. Various system models presented in this chapter could be simulated using standard techniques for numerical solution of ordinary differential equations. The reader may use any standard subroutine available in many computer libraries. Typical data for some of the models are given below, while for the rest they could be judiciously assumed or obtained from appropriate agencies. The instructor may use these models for student projects.

1.2.P1 Consider the train dynamics (1.2.6). Assuming that $M_i = 400$, $K_i = 300$, $B_i = 300$ for $i = 1, 2, \ldots, 4$ and

(a) $f(t) = \alpha S(t)$,

(b) $f(t) = \alpha(S(t) - 2S(t-50))$,

(c) $f(t) = \alpha(S(t) - S(t-10)) - \alpha(S(t-90) - S(t-100))$, where $\alpha > 0$ and

$$S(\xi) = \begin{cases} 0, & \xi \leq 0 \\ 1, & \xi > 0 \end{cases}.$$

Plot the variables $x_1(t)$–$x_8(t)$, $0 \leq t \leq 100$ and interpret the results.

1.2.P2 Carry out simulation experiments as in Problem 1.2.P1 for a sinusoidal (or any other) periodic track using the model (1.2.12).

1.2.P3 The moments of inertia of WHECON satellite are as follows: $I_x = 645$, $I_y = 100$, $I_z = 669$ slug ft^2. Using these data and $w_0 = 7.29 \times 10^{-5}$ rad/s, solve the satellite dynamics (1.2.26) for (a) $T_x = T_y = T_z = 0$, (b) $T_x = T_z = 0$ and $T_y(t) = \alpha(S(t) - S(t-1))$. Assume small initial perturbations.

1.2.P4 Using the model (1.2.47) for the production function write a program for solving the economic model equation (1.2.56) (a) for fixed but different values of $\{\lambda, c\}$, (b) for variable c, and (c) for fixed c, μ and variable β. Interpret all the results.

1.2.P5 Solve the economic growth model (1.2.59) for (a) $\gamma = 0$ (b) $\gamma = 0.01\beta$ and (c) $\gamma = 0.1\beta$ and give your interpretation of the results.

1.2.P6 Some typical data for the reactor kinetics equation (1.2.63) are as follows:

$$\beta = 0.76 \times 10^{-2}, \ \ell = 1.5 \times 10^{-3}$$

$$\lambda_i = 0.0125,\ 0.0315,\ 0.154,\ 0.456,\ 1.61,\ 13.85\ \mathrm{s}^{-1}$$

$$\mu_i = 0.032,\ 0.220,\ 0.282,\ 0.318,\ 0.115,\ 0.033$$

Solve the system (1.2.63) for $u = 0.01\beta S(t)$, $0.1\beta S(t)$, $\beta S(t)$.

1.2.P7 Using the models (1.2.72) for growth function, carry out a computer simulation for the population dynamics (1.2.71) for (a) $u \equiv 0$, (b) $u =$ a positive periodic function, (c) $u = 0.1x$ etc., and make critical analysis of the results. Assume that $k = 10$ and $k_0 = 1$ unit.

2

Linear systems

2.0 Introduction

In this chapter we consider multivariable linear time-varying systems. Since, generally, we shall always deal with multivariable systems in this book there is no need to repeat the phrase. Here we discuss the fundamental questions of existence and uniqueness of solutions and their regularity properties such as continuity and differentiability, etc. Then we consider the questions of continuous dependence of system response with respect to various parameters. These results are fundamental and they are used throughout the book indicating its immediate application to sensitivity studies. We also consider input–output stability (L_p-stability) and certain important algebraic properties of linear systems. In the final section we consider linear integral equations as they arise in feedback systems and certain tracking problems. The results presented here are obviously applicable to the special, but important, case of linear time-invariant systems.

2.1 Some elements from real analysis

In this section we present some basic definitions and terminology absolutely essential for reading the book. A reader familiar with these concepts can safely skip this section. For detailed study the reader may consult the book of Hewitt and Stromberg [65].

2.1.1 Linear space

A set X is called a linear space or a vector space over a field F if the following conditions are satisfied:

$$\left.\begin{array}{ll}\text{(L1)} & x + y \in X \text{ for all } x, y \in X \\ \text{(L2)} & \alpha x \in X \text{ for all } x \in X \text{ and } \alpha \in F.\end{array}\right\} \text{(closure)}$$

In other words the set X is closed under addition and scalar multiplication and further

(L3) $x + y = y + x$ (commutative) for all $x, y \in X$
(L4) $x + (y + z) = (x + y) + z$ (associative) for all $x, y, z \in X$
(L5) $x + y = x + z$ (equivalence) for all $x \in X$ implies $y = z$
(L6) $\alpha(x + y) = \alpha x + \alpha y$ for all $x, y \in X$ and $\alpha \in F$
(L7) $(\alpha + \beta)x = \alpha x + \beta x$ for all $x \in X$ and $\alpha, \beta \in F$
(L8) $\alpha\beta x = \alpha(\beta x)$ for all $x \in X$ and $\alpha, \beta \in F$
(L9) $1x = x$, 1 is the unit element of the field F
(L10) $0 \in X$ and $0 + x = x$ for all $x \in X$.

Example 2.1.1 Let F be any field and n a positive integer and define $X = F^n = F \times F \times F \times \cdots \times F$, the Cartesian product of n copies of F. For $x = (x_1, x_2, \ldots, x_n)$ and $y = (y_1, y_2, \ldots, y_n)$ in X and $\alpha \in F$ define $x + y = (x_1 + y_1, x_2 + y_2, \ldots, x_n + y_n)$ and $\alpha x = (\alpha x_1, \alpha x_2, \ldots, \alpha x_n)$. Then X is a vector space over F.

Example 2.1.2 Let F be any field and Λ any nonempty set, and let $X = F^\Lambda$. For $f \in X$ and $g \in X$, define $(f + g)(\lambda) = f(\lambda) + g(\lambda)$, $\lambda \in \Lambda$, and $(\alpha f)(\lambda) = \alpha f(\lambda)$, $\lambda \in \Lambda$. Then X is a vector space over the field F.

Remark 2.1.3 In Example 2.1.1 if $F = R(C)$, the field of real (complex) numbers, then $X = R^n(C^n)$. In Example 2.1.2 if Λ is any interval I of the real line and F is the field of real numbers then $X = F^\Lambda$ is the vector space of real valued functions defined on the interval I.

Example 2.1.4 Consider a large-scale fishing operation involving n different species of fish collected. To quantify the collection of each species a regulating agency may use any measure of weights (pounds, kilograms, etc.) and produce a vector $x = (x_1, x_2, \ldots, x_n)$ where each x_i represents the amount of collection of the ith species. If $x = (x_1, x_2, \ldots, x_n)$ and $y = (y_1, y_2, \ldots, y_n)$ represent the collections by vessels V_1 and V_2 respectively during any given period, then the total collection of the two vessels V_1 and V_2 is given by $x + y = (x_1 + y_1, x_2 + y_2, \ldots, x_n + y_n)$. If the vessel V_1 intensifies its operation by a factor α, then its collection over the same period is $\alpha x = (\alpha x_1, \alpha x_2, \ldots, \alpha x_n)$ and so on. The reader may satisfy himself by verifying ((L1)–(L10)) that here we have an example of a vector space.

Example 2.1.5 For linear electrical networks having n branches, the branch voltages or the branch currents constitute a linear vector space $X = R^n$.

Example 2.1.6 Let F be the field of scalars (real or complex) and $M = M(n \times m)$ the set of all $n \times m$ matrices with entries $\{a_{ij},\ i = 1, 2, \ldots, n,\ j = 1, 2, \ldots, m\} \in F$. Then M is a linear vector space.

Example 2.1.7 The set of all polynomials in t of degree equal or less than n is a vector space.

2.1.2 Metric space

A metric space is a nonempty set which is equipped with a measure of distance between its elements. Formally a pair $(X, \rho) = X_\rho$, where X is a nonempty set and ρ is a function mapping $X \times X$ to $\overline{R}_0 = [0, \infty]$, is called a metric space with metric ρ if it satisfies the following properties:

(M1) $\rho(x, y) \geqslant 0$ for all $x, y \in X$
(M2) $\rho(x, y) = 0$ if and only if $x = y$
(M3) $\rho(x, y) = \rho(y, x)$(symmetry) for all $x, y \in X$
(M4) $\rho(x, y) \leqslant \rho(x, z) + \rho(z, y)$(triangle inequality) for all $x, y, z \in X$.

Remark 2.1.8 A nonempty set X is called a *linear metric space* if it satisfies the following axioms:

(LM1) X is a linear (vector) space
(LM2) There exists a function $\rho: X \times X \to \overline{R}_0$ satisfying (M1)–(M4).

One can define different metrics on the same basic (nonempty) set X. Let ρ and σ be any two functions mapping $X \times X$ to $\overline{R}_0$ that satisfy the axioms of metric space (M1)–(M4). Then $X_\rho = (X, \rho)$ and $X_\sigma = (X, \sigma)$ are generally two different metric spaces.

Example 2.1.9 (A metric space that is not linear). The condition of weather is described by a meteorologist in terms of weather data such as (a) temperature T, (b) humidity H, (c) wind velocity V, (d) barometric pressure P, and (e) illumination I, etc. Let X be the set of all quintuples $\{T, H, V, P, I\}$. On this set we can define a metric

$$\rho_m(x, y) \equiv \max\{|T_1 - T_2|, |H_1 - H_2|, |V_1 - V_2|, |P_1 - P_2|, |I_1 - I_2|\}$$

where $x = (T_1, H_1, V_1, P_1, I_1)$, $y = (T_2, H_2, V_2, P_2, I_2)$. It is easy to verify that ρ_m satisfies all the properties (M1)–(M4) and hence (X, ρ_m) is a metric space and it is quite suitable for comparison of weather conditions of any two locations on earth. However, (X, ρ_m) is not a linear metric space since addition between elements of X does not make any physical sense. But it certainly can be considered as a subset of R^5 which is a linear space.

Example 2.1.10 (Linear metric space). $X \equiv R^n \equiv$ the set of all ordered n-tuples of real numbers. Define

$$\rho_1(x, y) \equiv \sum_{i=1}^{n} |x_i - y_i|, \; \rho_2(x, y) \equiv \left(\sum_{i=1}^{n} (x_i - y_i)^2\right)^{1/2},$$

and in general

$$\rho_q(x, y) \equiv \left(\sum_{i=1}^{n} |x_i - y_i|^q\right)^{1/q}, \; 1 < q < \infty,$$

$$\rho_\infty(x, y) \equiv \max\{|x_i - y_i|, i = 1, 2, \ldots, n\}.$$

$\{(X, \rho_q), 1 \leqslant q \leqslant \infty\}$ are all linear metric spaces over the field F of real numbers. Let $X \equiv E^n$ denote the set of all ordered n-tuples of complex numbers and $F = C$ the field of complex numbers. Define ρ_q, $1 \leqslant q \leqslant \infty$ as before with $|z|$ representing the modulus of the complex number. Again $(X, \rho_q) \equiv (E^n, \rho_q)$ is a linear metric space.

Example 2.1.11 $X \equiv$ the space of sequences of real or complex numbers

$$x = (x_1, x_2, x_3, \ldots), \qquad x_i \in F, \quad i = 1, 2, \ldots$$

with a function

$$\rho_q(x, y) \equiv \left(\sum_{i=1}^{\infty} |x_i - y_i|^q\right)^{1/q}, \qquad 1 \leqslant q \leqslant \infty,$$

provides a natural extension of the previous examples. The reader may like to verify that (X, ρ_q) is a metric space. That this is also a linear space is discussed later.

Example 2.1.12 Consider the set $X = \{(a_1, \ldots, a_n), a_i \in \{0, 1\}\}$ and take $F = \{0, 1\}$ and define

$$a + b = (a_1 \dotplus b_1, a_2 \dotplus b_2, \ldots, a_n \dotplus b_n)$$

where $a_i \dotplus b_i \equiv a_i + b_i (\text{mod } 2)$ and

$$\alpha a = (\alpha a_1, \alpha a_2, \ldots, \alpha a_n), \qquad \alpha \in F, \quad a \in X.$$

Define $\rho_H(a, b) \equiv$ number of positions where a differs with b. The reader may easily check that X is a vector space over F and that (X, ρ_H) is a metric space.

The metric ρ_H is called the Hamming metric and it is widely used in communication theory.

2.1.3 Normed vector space

Let X be any linear vector space over F and let N be a function defined on X to $\bar{R} = [-\infty, \infty]$ i.e., $N: X \to \bar{R}$ satisfying the following axioms:

(N1) $N(x) \geq 0$ for all $x \in X$
(N2) $N(x) = 0$ if and only if $x = 0$
(N3) $N(\alpha x) = |\alpha|\, N(x)$ for all $x \in X$, $\alpha \in F$
(N4) $N(x + y) \leq N(x) + N(y)$ for all $x, y \in X$.

The function $N(\cdot)$ is called the norm and it is usually denoted by $\|\cdot\|$. The pair $(X, N) \equiv (X, \|\cdot\|)$ is called a normed vector space.

Definition 2.1.13 (Cauchy sequence). Let $X \equiv (X, \rho)$ be a metric space furnished with the metric ρ. A sequence $\{x_n\} \in X$, $n = 1, 2, \ldots$, is called a Cauchy sequence if

$$\rho(x_{n+p}, x_n) \to 0 \qquad \text{as } n \to \infty$$

for any integer $p \geq I$.

Definition 2.1.14 (Complete metric space). A metric space $X = (X, \rho)$ is said to be complete if every Cauchy sequence has a limit point in X. That is, if $\{x_n\} \in X$ and $\lim_{n\to\infty} \rho(x_{n+p}, x_n) = 0$ for any $p \geq 1$, then there exists an $x^* \in X$ such that $\lim_{n\to\infty} \rho(x_n, x^*) = 0$.

2.1.4 Banach space

A normed vector space X, which is complete with respect to the metric induced by its norm, is called a Banach space. In other words, a normed vector space is a Banach space if every Cauchy sequence has a limit.

Suppose X is a Banach space and $\{x_n\} \in X$ is a Cauchy sequence, that is, $\lim_{n\to\infty} \|x_{n+p} - x_n\| = 0$ for any $p \geq 1$. Then there exists an $x^* \in X$ such that

$$\lim_{n\to\infty} \|x_n - x^*\| = 0.$$

Banach spaces are the most widely used normed vector spaces in modern systems and control theory.

2.1.5 Hilbert space

A linear vector space $X \equiv H$ is said to be a pre-Hilbert space (inner product space, scalar product space) if there exists a function $S: H \times H \to F \equiv (C/R)$, written as $S(x, y) = (x, y)$ satisfying the following properties:

(H1) $(x, y) = \overline{(y, x)}$, ($\bar{z} \equiv$ complex conjugate of z) for all $x, y \in H$,

(H2) $\left.\begin{array}{l}(x_1 + x_2, y) = (x_1, y) + (x_2, y)\\(x, y_1 + y_2) = (x, y_1) + (x, y_2)\end{array}\right\}$ for all $x_1, x_2, y_1, y_2, x, y \in H$,

(H3) $\left.\begin{array}{l}(\alpha x, y) = \alpha(x, y)\\(x, \alpha y) = \bar{\alpha}(x, y)\end{array}\right\}$ for all $x, y \in H$, $\alpha \in F$,

(H4) $(x, x) = \overline{(x, x)} = \|x\|^2 \geqslant 0$, with equality holding only if $x = 0$.

Clearly the scalar product induces a norm on H, and H is a normed space. The scalar product space H is called the Hilbert space if it is a Banach space with respect to the norm $\|\cdot\| = (\cdot, \cdot)^{1/2}$ induced by the inner product $(\cdot, \cdot)$.

One of the most important characteristics of a Hilbert space is that the parallelogram law holds:

$$\|x + y\|^2 + \|x - y\|^2 = 2\,\|x\|^2 + 2\,\|y\|^2$$

for all $x, y \in H$.

Later in this section we present some important examples of Banach and Hilbert spaces which will be useful in this book. For this purpose we shall need some important inequalities as described below.

Lemma 2.1.15 *Let $A, B > 0$ and $\alpha, \beta > 0$ with $\alpha + \beta = 1$. Then*

$$A^{\alpha}B^{\beta} \leqslant \alpha A + \beta B \tag{2.1.1}$$

Proof Since the logarithm function is concave one can easily verify by drawing a figure that

$$\log(\alpha A + \beta B) \geqslant \alpha \log A + \beta \log B = \log(A^{\alpha}B^{\beta}).$$

Taking the antilog the result follows. ■

Hölder's inequality

Let $p > 1$ and $(1/p) + (1/q) = 1$. Denote by ℓ_p^n the set of all n-tuples $x = (x_1, x_2, \ldots, x_n)$ such that

$$\|x\|_{\ell_p^n} = \left(\sum_{i=1}^{n} |x_i|^p\right)^{1/p} < \infty. \tag{2.1.2}$$

Similarly $y \in \ell_q^n$, if

$$\|y\|_{\ell_q^n} = \left(\sum_{i=1}^{n} |y_i|^q\right)^{1/q} < \infty. \tag{2.1.3}$$

Proposition 2.1.16 (Hölder's inequality). *For $x \in \ell_p^n$ and $y \in \ell_q^n$ the scalar product (or duality product),*

$$(x, y) = \sum_{i=1}^{n} x_i \bar{y}_i, \tag{2.1.4}$$

satisfies the following inequality

$$|(x, y)| \leq \|x\|_{\ell_p^n} \|y\|_{\ell_q^n}. \tag{2.1.5}$$

Proof Define

$$\alpha \equiv 1/p, \quad \beta \equiv 1/q; \qquad p > 1, \quad \frac{1}{p} + \frac{1}{q} = 1$$

and

$$A \equiv |x_i|^p / \|x\|_{\ell_p^n}^p$$

$$B \equiv |y_i|^q / \|y\|_{\ell_q^n}^q.$$

By Lemma 2.1.15 we have

$$\left(\frac{|x_i|}{\|x\|_{\ell_p^n}}\right)\left(\frac{|y_i|}{\|y\|_{\ell_q^n}}\right) \leq \frac{1}{p}\frac{|x_i|^p}{\|x\|_{\ell_p^n}^p} + \frac{1}{q}\frac{|y_i|^q}{\|y\|_{\ell_q^n}^q}. \tag{2.1.6}$$

Summing over the indices $i = 1, 2, \ldots, n$, it follows from (2.1.6) that

$$\sum_{i=1}^{n} \left(\frac{|x_i|\,|y_i|}{\|x\|_{\ell_p^n}\,\|y\|_{\ell_q^n}}\right) \leq \frac{1}{p}\frac{\sum_{i=1}^{n}|x_i|^p}{\|x\|_{\ell_p^n}^p} + \frac{1}{q}\frac{\sum_{i=1}^{n}|y_i|^q}{\|y\|_{\ell_q^n}^q} \leq \frac{1}{p} + \frac{1}{q} = 1. \tag{2.1.7}$$

Hence

$$\sum_{1}^{n} |x_i|\,|y_i| \leq \|x\|_{\ell_p^n}\,\|y\|_{\ell_q^n}. \tag{2.1.8}$$

Using (2.1.8) into (2.1.4) we have,

$$|(x, y)| \leq \sum_{i=1}^{n} |x_i|\,|y_i| \leq \|x\|_{\ell_p^n}\,\|y\|_{\ell_q^n} \tag{2.1.9}$$

which gives (2.1.5). ■

Remark 2.1.17 In case $p = 2$ (hence $q = 2$, also), (2.1.5) gives the Schwarz inequality.

Proposition 2.1.18 (Minkowski's inequality). *For $1 \leq p < \infty$, and $x, y \in \ell_p^n$,*

$$\|x + y\|_{\ell_p^n} \leq \|x\|_{\ell_p^n} + \|y\|_{\ell_p^n}. \tag{2.1.10}$$

Proof The case for $p = 1$ is obvious. We give a proof for $1 < p < \infty$. By definition,

$$\|x + y\|_{\ell_p^n} = \left(\sum_{i=1}^{n} |x_i + y_i|^p\right)^{1/p}. \tag{2.1.11}$$

Note that,

$$\begin{aligned}\sum_{i=1}^{n} (|x_i| + |y_i|)^p &= \sum_{i=1}^{n} (|x_i| + |y_i|)^{p-1}(|x_i| + |y_i|) \\ &= \sum_{i=1}^{n} (|x_i| + |y_i|)^{p-1} |x_i| + \sum_{i=1}^{n} (|x_i| + |y_i|)^{p-1} |y_i|\end{aligned}$$

and hence by Hölder's inequality (2.1.5), we have

$$\begin{aligned}\sum (|x_i| + |y_i|)^p &\leq \left(\sum (|x_i| + |y_i|)^{(p-1)q}\right)^{1/q} \\ &\quad \times \left\{\left(\sum |x_i|^p\right)^{1/p} + \left(\sum |y_i|^p\right)^{1/p}\right\}\end{aligned} \tag{2.1.12}$$

where $(1/p) + (1/q) = 1$.

Since $(p-1)q = p$, it follows from (2.1.12) that

$$\sum_{i=1}^{n} (|x_i| + |y_i|)^p \leq \left(\sum_{i=1}^{n} (|x_i| + |y_i|)^p\right)^{1/q} (\|x\|_{\ell_p^n} + \|y\|_{\ell_p^n}) \tag{2.1.13}$$

and hence

$$\left(\sum_{i=1}^{n} (|x_i| + |y_i|)^p\right)^{1/p} \leq (\|x\|_{\ell_p^n} + \|y\|_{\ell_p^n}). \tag{2.1.14}$$

Using (2.1.14) in (2.1.11), we have

$$\|x + y\|_{\ell_p^n} \leq \|x\|_{\ell_p^n} + \|y\|_{\ell_p^n} \tag{2.1.15}$$

which completes the proof. ■

We leave it to the reader to verify that

$$\|x\|_{\ell_\infty^n} \equiv \lim_{p \to \infty} \left(\sum_{i=1}^{n} |x_i|^p\right)^{1/p} = \max\{|x_i|, i = 1, \ldots, n\}. \tag{2.1.16}$$

Proposition 2.1.19 (Examples of Banach spaces). *ℓ_p^n, $1 \leq p \leq \infty$, are Banach spaces.*

Proof For $x \in \ell_p^n$, $y \in \ell_p^n$, it follows from Minkowski's inequality (Proposition 2.1.18) that $x + y \in \ell_p^n$. Similarly for $\alpha \in F$, $(\equiv R/C)$, $\alpha x \in \ell_p^n$ whenever $x \in \ell_p^n$. Thus, the closure properties, (L1)–(L2), are satisfied. The axioms (L3)–(L10) are obvious. Hence ℓ_p^n is a vector space. To prove that it is a normed space define

$$N(x) = \left(\sum_{i=1}^{n} |x_i|^p\right)^{1/p} \qquad \text{for } x \in \ell_p^n. \tag{2.1.17}$$

One can easily check that all the axioms, (N1)–(N4), of a normed space are satisfied for the given N. Hence ℓ_p^n, $1 \leq p \leq \infty$, are normed vector spaces. In fact, they are Banach spaces since the field $F(\equiv R/C)$ is a Banach space. ■

Remark 2.1.20 For $p = 2$, ℓ_2^n is a Hilbert space with the inner product $(x, y) = \sum_{i=1}^{n} x_i \bar{y}_i$.

Definition 2.1.21 (Linear independence). The elements of a vector space X are called vectors. The vectors $\{\xi_1, \xi_2, \ldots, \xi_m\} \in X$ are said to be linearly independent if

$$\sum_{i=1}^{m} \alpha_i \xi_i = 0 \text{ implies } \alpha_1 = \alpha_2 = \cdots = \alpha_m = 0. \tag{2.1.18}$$

However, if the equation is satisfied with α_i not all zero, then the vectors $\{\xi_1, \ldots, \xi_m\}$ are linearly dependent.

Definition 2.1.22 (Dimension of a vector space). If X contains a set of n linearly independent vectors and every set of $(n+1)$ vectors is linearly dependent, then X is said to be of dimension n. If the number of linearly independent vectors is not finite, then X is said to be of infinite dimension.

Any set of n linearly independent vectors in an n-dimensional vector space X constitutes a basis for X and each $x \in X$ has a unique representation $x = \sum_{i=1}^{n} \alpha_i \xi_i$ in terms of the basis $\{\xi_1, \xi_2, \ldots, \xi_n\}$.

A subset M of a linear (vector) space X over the field F is said to be a linear subspace of X if $x, y \in M$ implies $\alpha x + \beta y \in M$ for all $\alpha, \beta \in F$.

The vector space of all polynomials in t of degree $\leq n - 1$, and the vector spaces ℓ_p^n, $1 \leq p \leq \infty$, are all of dimension n. We shall now give some examples of infinite-dimensional spaces.

Let X denote the set of all infinite sequences $x = \{x_1, x_2, x_3 \ldots\}$ such that, for a given p, $1 \leq p \leq \infty$,

$$\left(\sum_{i=1}^{\infty} |x_i|^p\right)^{1/p} < \infty. \tag{2.1.19}$$

We denote this set by ℓ_p, $1 \leqslant p \leqslant \infty$ and prove that this is also a Banach space. In order to show that this is a vector space we must verify that $x, y \in \ell_p$ implies $x + y \in \ell_p$.

For this we need Minkowski's and hence the Hölder inequality. Hölder's inequality can be proved in the same way as before. For Minkowski's inequality, note that for *any* finite integer $k \geqslant 1$, we have

$$\left(\sum_{i=1}^{k} |x_i + y_i|\right)^{1/p} \leqslant \left(\sum_{i=1}^{k} |x_i|^p\right)^{1/p} + \left(\sum_{i=1}^{k} |y_i|^p\right)^{1/p}. \tag{2.1.20}$$

Since $x, y \in \ell_p$

$$\left(\sum_{i=1}^{k} |x_i|^p\right)^{1/p} \leqslant \left(\sum_{i=1}^{\infty} |x_i|^p\right)^{1/p} \equiv \|x\|_{\ell_p} < \infty,$$

and

$$\left(\sum_{i=1}^{k} |y_i|^p\right)^{1/p} \leqslant \left(\sum_{i=1}^{\infty} |y_i|^p\right)^{1/p} \equiv \|y\|_{\ell_p} < \infty,$$

for all positive integers k. Hence letting $k \to \infty$ in (2.1.20) we have

$$\|x + y\|_{\ell_p} \leqslant \|x\|_{\ell_p} + \|y\|_{\ell_p}. \tag{2.1.21}$$

Hence we can prove the following result.

Proposition 2.1.23 ℓ_p, $1 \leqslant p \leqslant \infty$, *are Banach spaces, and* ℓ_2, *furnished with the inner product*

$$(x, y) = \sum_{i=1}^{\infty} x_i \bar{y}_i, \tag{2.1.22}$$

is a Hilbert space.

Hölder's and Minkowski's inequalities also hold for function spaces. Let $I = (a, b)$ be any interval of the real line R and p any number satisfying $1 \leqslant p \leqslant \infty$. We shall denote by $L_p(I)$ the class of all real or complex-valued functions which are pth-power integrable in some general sense (Lebesgue integral, see [65], [64]) on the interval I. That is,

$$L_p(I) \equiv \left\{f : \int_I |f(\xi)|^p \, d\xi < \infty\right\}, \tag{2.1.23}$$

which is clearly a subset of F^I where F is the field of complex or real numbers.

Let q be a number conjugate to the number p in the sense that $(1/p) + (1/q) = 1$. Note that for $p = 1$, $q = \infty$, and for $p = \infty$, $q = 1$. Again for such function spaces, we can prove Hölder's and Minkowski's inequalities in a way similar to that used in Propositions 2.1.16 and 2.1.18.

Proposition 2.1.24 *Let* $1<p<\infty$ *and* $q=(p/p-1)$ *the corresponding conjugate number, and* $L_p(I)$ *and* $L_q(I)$ *denote the function spaces as described in* (2.1.23). *Then*
(a) *for* $f \in L_p(I)$ *and* $g \in L_q(I)$,

$$|\langle f, g\rangle_{L_p,L_q}| = \left|\int_I f(\xi)\bar{g}(\xi)\,\mathrm{d}\xi\right| \leq \left(\int_I |f(\xi)|^p\,\mathrm{d}\xi\right)^{1/p} \cdot \left(\int_I |g(\xi)|^q\,\mathrm{d}\xi\right)^{1/q} \qquad (2.1.24)$$

and

(b) *for* $f \in L_p(I)$, $g \in L_p(I)$

$$\left(\int_I |f(\xi)+g(\xi)|^p\,\mathrm{d}\xi\right)^{1/p} \leq \left(\int_I |f(\xi)^p\,\mathrm{d}\xi\right)^{1/p} + \left(\int_I |g(\xi)^p\,\mathrm{d}\xi\right)^{1/p}. \qquad (2.1.25)$$

Proof Clearly the inequality (2.1.24) holds trivially if any one of the integrals $\int_I |f(\xi)|^p\,\mathrm{d}\xi$ or $\int_I |g(\xi)|^q\,\mathrm{d}\xi$ vanishes. So we assume that they are strictly positive. For the proof of (a) we use the inequality (2.1.1) of Lemma 2.1.15. Define $\alpha=1/p$, $\beta=1/q$ and

$$A \equiv (|f(\xi)|^p/\|f\|^p_{L_p})$$
$$B \equiv (|g(\xi)|^q/\|g\|^q_{L_q})$$

where $\|f\|^p_{L_p} \equiv (\int_I |f(\xi)|^p\,\mathrm{d}\xi)$ and $\|g\|^q_{L_q} \equiv (\int |g(\xi)|^q\,\mathrm{d}\xi)$. Then by Lemma 2.1.15 we have

$$\frac{|f(\xi)|\,|\bar{g}(\xi)|}{\|f\|_{L_p}\,\|g\|_{L_q}} \leq \frac{1}{p}\left(\frac{|f(\xi)|^p}{\|f\|^p_{L_p}}\right) + \frac{1}{q}\left(\frac{|\bar{g}(\xi)|^q}{\|g\|^q_{L_q}}\right). \qquad (2.1.26)$$

Integrating over I, it follows from the above inequality that

$$\int_I \left(\frac{|f(\xi)|\,|\bar{g}(\xi)|}{\|f\|_{L_p}\,\|g\|_{L_q}}\right)\mathrm{d}\xi \leq \frac{1}{p}+\frac{1}{q}=1$$

and hence

$$\int_I |f(\xi)|\,|\bar{g}(\xi)\,\mathrm{d}\xi \leq \|f\|_{L_p}\,\|g\|_{L_q}. \qquad (2.1.27)$$

Again

$$\left|\int_I f(\xi)\bar{g}(\xi)\, \mathrm{d}\xi\right| \leqslant \int_I |f(\xi)|\, |\bar{g}(\xi)|\, \mathrm{d}\xi = \int_I |f(\xi)|\, |g(\xi)|\, \mathrm{d}\xi \tag{2.1.28}$$

and consequently

$$|\langle f, g\rangle_{L_p, L_q}| = \left|\int_I f(\xi)\bar{g}(\xi)\, \mathrm{d}\xi\right| \leqslant \|f\|_{L_p}\, \|g\|_{L_q}. \tag{2.1.29}$$

Minkowski's inequality (2.1.25) follows from (2.1.24) in the same way as in Proposition 2.1.18. ■

Remark 2.1.25 Note that for $p = 1$, the inequality (2.1.25) is obvious.

By $L_\infty(I)$ we mean the set of all measurable functions on I which are essentially bounded. A function $f \in L_\infty$ if there exists a positive number $\gamma < \infty$ such that the set $J \equiv \{t \in I\colon |f(t)| > \gamma\}$ has Lebesgue measure zero. The smallest number γ for which this is true is called the essential supremum of $|f(\xi)|$ over $\xi \in I$ and written as

$$\|f\|_{L_\infty} \equiv \text{ess sup}\{|f(\xi)|,\ \xi \in I\}.$$

Again the reader may verify that for finite interval I,

$$\|f\|_{L_\infty} = \lim_{p\to\infty} \left(\int_I |f(\xi)|^p\, \mathrm{d}\xi\right)^{1/p}.$$

From Minkowski's inequality (2.1.25) it follows that

(a) $f + g \in L_p$ whenever $f, g \in L_p$,

and

(b) $\alpha f \in L_p$ whenever $f \in L_p$ and $\alpha \in F$.

One can easily verify that the rest of the axioms (L3)–(L10) of a vector space are satisfied. Hence, the set L_p, $1 \leqslant p \leqslant \infty$, is a linear vector space.

Remark 2.1.26 We must mention that the results of Proposition 2.1.24 also hold for any measurable set $\Omega \subset R^n$ and not just for an interval $I \subset R$.

If any two elements f and g of the vector space L_p, $1 \leqslant p \leqslant \infty$, differ only on a set of measure zero (linear measure zero in case of an interval I, volume measure zero in case of $\Omega \subset R^n$), then, as far as the Lebesgue integral is concerned,

$$\int_J |f(\xi) - g(\xi)|^p\, \mathrm{d}\xi = 0, \tag{2.1.30}$$

for every subinterval $J \subset I$. Hence one may identify such elements and set $f \cong g$ (f is equivalent to g). Then, of course, the linear space L_p becomes a space of equivalence classes, in the sense that given any $f \in L_p$, we can add to it any element g from the set $M \equiv \{h \in L_p: \int |h|^p \, d\xi = 0\}$ without altering the integral $\int |f|^p \, d\xi$. Such a space is usually called a quotient space and is denoted by L_p/M. We avoid this symbol and continue to use the symbol L_p instead. Hence pointwise values $f(\xi)$, for $f \in L_p$, has no meaning. Considering L_p as the space of equivalence classes, it becomes a normed space if we define

$$N(f) \equiv \left(\int |f|^p \, d\xi \right)^{1/p}. \tag{2.1.31}$$

One can easily verify that N satisfies the axioms (N1)–(N4) of a normed space. Note, however, that, in case of (N2), $N(f) = 0$ does not imply that $f(\xi) = 0$ for all $\xi \in I$; it only means that $f = 0$ a.e. (almost everywhere) on I, that is $\{\xi \in I: f(\xi) \neq 0\}$ has Lebesgue measure zero or equivalently $\int |f(\xi)|^p \, d\xi = 0$. Using this norm we can define a metric on the L_p space as

$$\rho(f, g) \equiv N(f - g) \equiv \left(\int |f - g|^p \, d\xi \right)^{1/p}. \tag{2.1.32}$$

We state the following results without proof.

Proposition 2.1.27 *The L_p spaces, $1 \leq p \leq \infty$, are complete linear metric spaces with respect to the metric ρ given by* (2.1.32) *or equivalently Banach spaces with respect to the norm N given by* (2.1.31). *For $p = 2$, L_2 is a Hilbert space with respect to the inner product*

$$(f, g) \equiv \int f(\xi) \bar{g}(\xi) \, d\xi, \tag{2.1.33}$$

and the Hölder inequality reduces to the Schwarz inequality,

$$|(f, g)| \leq \|f\|_{L_2} \|g\|_{L_2}. \tag{2.1.34}$$

2.1.6 Continuous functions

One of the most important function spaces that we shall use frequently is the space of continuous functions.

Let (X, ρ) and (Y, σ) be any two metric spaces and f a function mapping X into Y denoted $f: X \to Y$.

Definition 2.1.28 (Continuity). The function $f: X \to Y$ is said to be continuous at $x_0 \in X$ if, for every $\varepsilon > 0$, there exists a δ (possibly dependent on ε and x_0), $\delta > 0$, such that

$$\sigma(f(x), f(x_0)) < \varepsilon \quad \text{whenever} \quad \rho(x, x_0) < \delta. \tag{2.1.35}$$

The function $f: X \to Y$ is said to be continuous if it is continuous everywhere in X in the above sense. The set of all such continuous functions is denoted by $C(X, Y)$.

In most of the applications related to systems governed by ordinary differential equations we shall be concerned with the space of continuous functions from $I \subset R$ to R^n denoted $C(I, R^n)$. It is easy to verify that

$$x + y \in C(I, R^n) \quad \text{whenever} \quad x, y \in C(I, R^n)$$

$$\alpha x \in C(I, R^n) \quad \text{whenever} \quad x \in C(I, R^n), \ \alpha \in R.$$

The rest of the axioms (L3)–(L10) of a vector space is obvious. Hence $C(I, R^n)$ is a linear space. Let $R(n \times n)$ denote the space of $n \times n$ matrices with real entries, then $C(I, R(n \times n))$ is also a linear space and for $x \in C(I, R^n)$ and $G \in C(I, R(n \times n))$, $Gx \in C(I, R^n)$ where $(Gx)(t) = G(t) \cdot x(t)$, $t \in I$.

Proposition 2.1.29 *The space $C(I, R^n)$ is a vector space and it is a Banach space with respect to the sup norm, and hence a complete metric space with respect to the metric*

$$\|x - y\| = \sup\{\rho(x(t), y(t)), t \in I\}, \tag{2.1.36}$$

where $\rho(\xi, \eta) \equiv |\xi - \eta| \equiv (\sum (\xi_i - \eta_i)^2)^{1/2}$ or any other equivalent metric on R^n, as given in Example 2.1.10.

In dealing with differential equations we invariably encounter the question of differentiation. Not all continuous functions are differentiable. The class of functions that is differentiable is known as the space of absolutely continuous functions.

Definition 2.1.30 (Absolute continuity). Let $I = [t_0, t_1] \subset R$ and $x \in C(I, R^n)$. The function x is said to be absolutely continuous on I if, for every partition, $t_0 \leq \tau_0 < \tau_1 < \tau_2 < \cdots < \tau_m \leq t_1$, of the interval I and for every $\varepsilon > 0$, there exists a $\delta > 0$ such that

$$\sum_{i=0}^{m-1} \rho(x(\tau_{i+1}), x(\tau_i)) < \varepsilon \quad \text{whenever} \quad \sum_{i=0}^{m-1} |\tau_{i+1} - \tau_i| < \delta,$$

where ρ is any suitable metric on R^n.

The class of absolutely continuous functions is denoted by $AC(I, R^n)$.

Clearly this is a subspace of $C(I, R^n)$. We conclude this section by stating the following result.

Proposition 2.1.31 [65]

(a) *If* $x \in AC(I, R^n)$, *then* $(\mathrm{d}/\mathrm{d}t)x \equiv \dot{x} \in L_1(I, R^n)$ *and*

$$x(t) = x(t_0) + \int_{t_0}^{t} \dot{x}(\theta)\, \mathrm{d}\theta \text{ for all } t \in I.$$

(b) *If* $y \in L_1(I, R^n)$ *and* x *is given by the indefinite integral*

$$x(t) = x(t_0) + \int_{t_0}^{t} y(\theta)\, \mathrm{d}\theta, \; t \in I,$$

then $x \in AC(I, R^n)$ *and* $\dot{x}(t) = y(t)$ *for almost all* $t \in I$.

Throughout this book we shall use $C^k(I, R^n)$ to denote the space of all R^n-valued functions which are continuously differentiable on I up to order k where k is any non-negative integer.

Another result, extensively used in the book, is the Lebesgue dominated convergence theorem which provides conditions under which interchange of the limiting operation and integration is permissible.

Proposition 2.1.32 (Dominated convergence theorem) [65, 64]. *Let* $\{g_n\}$ *be a sequence of Lebesgue integrable functions on* $I \subset R$ *and suppose there exists a Lebesgue integrable function* f *such that*
(a) $|g_n(t)| \leq |f(t)|$ *a.e. on* I
(b) $g_n(t) \to g_0(t)$ *a.e. on* I.
Then g_0 *is Lebesgue integrable on* I *and*

$$\lim_{n\to\infty} \int_I g_n(t)\, \mathrm{d}t = \int_I g_0(t)\, \mathrm{d}t. \tag{2.1.37}$$

Proposition 2.1.33 (Fatou's lemma) [65, 64]. *Let* $\{f_n\}$ *be a sequence of non-negative integrable functions such that* $\underline{\lim} f_n = f_0$ *a.e. Then if* $\underline{\lim} \int f_n\, \mathrm{d}t$ *is finite,* f_0 *is integrable and*

$$\int f_0(t)\, \mathrm{d}t \leq \underline{\lim} \int f_n\, \mathrm{d}t. \tag{2.1.38}$$

Frequently we shall encounter the notions of open, closed, and compact sets in both finite- and infinite-dimensional spaces. In infinite-dimensional Banach spaces there are in general three different notions of compactness: strong, weak, and weakstar. In any advanced study of system theory it is impossible to avoid these notions. Here we shall briefly present these concepts only in the context of Banach spaces

$C(I, R^n)$ and $L_p(I, R^n)$, $1 \leq p \leq \infty$ that we need. A serious reader may consult any standard text on analysis [65], [69].

First we need the following basic concepts.

Definition 2.1.34 (Open and closed sets). A set K in a metric space (M, ρ) is said to be open if for every $x \in K$, there exists an $\varepsilon > 0$ such that $N_\varepsilon(x) \equiv \{y \in M: \rho(x, y) < \varepsilon\} \subset K$. The set K is closed if it equals its closure $\bar{K}$.

Definition 2.1.35 (Compact sets). A set K in a metric space (M, ρ) is said to be relatively compact if every sequence $\{x_n\}$ from K has a convergent subsequence. The set K is said to be compact if it is relatively compact and closed (that is, the limit does not leave the set K).

A compact set in a metric space is always closed and bounded but the converse is not always true. However, in R^n a set K is compact if and only if it is closed and bounded.

The notion of connectedness is also very useful in the study of stability theory.

Definition 2.1.36 (Connected sets). An open (compact) set K in a metric space (M, ρ) is said to be connected if it cannot be expressed as the union of two disjoint open (compact) sets.

The following result characterizes compact sets in the Banach space $C(I, R^n)$.

Proposition 2.1.37 (Ascoli–Arzelà theorem). *For any closed bounded interval I, a set $K \subset C(I, R^n)$ is relatively compact (that is, $\bar{K}$ is compact) if and only if* (a) *K is bounded and* (b) *it is equicontinuous in the sense that, for any $\varepsilon > 0$, there exists a $\delta = \delta(\varepsilon) > 0$ such that $|x(t') - x(t)| < \varepsilon$ for all $x \in K$ and $t', t \in I$ satisfying $|t' - t| < \delta$.*

Definition 2.1.38 (Weak convergence and compactness). A set $K \subset L_p(I, R^n)$, $1 \leq p < \infty$, is said to be relatively weakly compact if, corresponding to every sequence $\{x_n\} \subset K$, there exists a subsequence $\{x_{n_l}\} \subset \{x_n\}$ and an $x \in L_p(I, R^n)$ such that

$$\lim_{l \to \infty} \int_I (x_{n_l}(t), g(t))\, dt = \int_I (x(t), g(t))\, dt$$

for every $g \in L_q(I, R^n)$ where $1 < q \leq \infty$ and $(1/p) + (1/q) = 1$. The set K is said to be weakly compact if the limit $x \in K$. The sequence x_{n_l} is said to be weakly convergent to x and is denoted by $x_{n_l} \xrightarrow{w} x$.

Proposition 2.1.39 *A set $K \subset L_p(I, R^n)$, $1<p<\infty$, is weakly compact if and only if it is* (a) *norm bounded and* (b) *weakly closed. Every bounded subset of $L_p(I, R^n)$, $1<p<\infty$, is relatively weakly compact.*

Proposition 2.1.40 (Dunford–Pettis) *A set $K \subset L_1(I, R^n)$ is relatively weakly sequentially compact if and only if it is* (a) *norm bounded and* (b) $\lim_{m(J)\to 0} \int_J |x(t)|\, dt = 0$ *uniformly with respect to $x \in K$. The set K is weakly sequentially compact if in addition to* (a) *and* (b) *it is also weakly closed.*

Definition 2.1.41 (Weakstar convergence). A sequence $\{x_n\} \subset L_\infty(I, R^n)$ is said to be weakstar(w^*)-convergent to $x \in L_\infty(I, R^n)$ if, for every $g \in L_1(I, R^n)$,

$$\lim_{n\to\infty} \int_I (x_n(t), g(t))\, dt = \int_I (x(t), g(t))\, dt.$$

Proposition 2.1.42 (w^*-compactness, Alaoglu). *A set $K \subset L_\infty(I, R^n)$ is w^*-compact if and only if it is* (a) *norm bounded and* (b) *w^*-closed. Any closed ball $B_r \equiv \{x \in L_\infty(I, R^n)\colon \|x\| \leq r\}$ is w^*-compact.*

The above result is a special case of Alaoglu's theorem and is extremely useful in control theory. In particular for any compact convex subset U of R^n, the set $\mathcal{U} \equiv \{u \in L_\infty(I, R^n)\colon u(t) \in U \text{ a.e.}\}$ is a w^*-compact subset of $L_\infty(I, R^n)$. If U is merely compact but not convex, then $\mathcal{U}$ is only relatively w^*-compact, that is, its w^*-closure is w^*-compact.

Proposition 2.1.43 *Any continuous function on a compact set attains both its maximum and minimum in the set.*

2.2 Existence and uniqueness of solutions of linear systems of differential equations

In this section we consider the questions of existence and uniqueness of solutions of linear differential systems of the form

$$\frac{dx}{dt} = A(t)x + f(t), \qquad t \in [0, T] = I$$

$$x(0) = x_0, \tag{2.2.1}$$

where f is a free term belonging to $L_1(I, R^n)$ and A is a $(n \times n)$ matrix valued function with entries $a_{ij} \in L_1(I, R)$ for every finite interval I, and

x_0 is a given element of R^n. We are looking for a vector-valued function $x = \{x(t), t \in I\}$ which is differentiable almost everywhere (a.e.) and satisfies the first equation a.e. on I, and at $t = 0$ it equals the given vector x_0. The question is, why should such a unique function exist for arbitrary A, f and x_0? Certainly the question is not trivial and must be answered before dealing with control problems.

2.2.1 Existence and uniqueness

For the solution of the existence (and uniqueness) problem we shall frequently use an inequality called Gronwall's lemma.

Lemma 2.2.1 (Gronwall). *Suppose*

$$\phi(t) \leqslant a + \int_{t_0}^{t} K(\theta)\phi(\theta)\,\mathrm{d}\theta, \qquad t \geqslant t_0, \tag{2.2.2}$$

where $a \geqslant 0$, $K \in L_1^{\mathrm{loc}}$ (*≡locally Lebesgue integrable functions*) *with* $K(t) \geqslant 0$ *a.e. and* $\phi(t) \geqslant 0$ *for all* $t \geqslant t_0$. *Then*

$$\phi(t) \leqslant a \exp \int_{t_0}^{t} K(\theta)\,\mathrm{d}\theta. \tag{2.2.3}$$

Proof Define $\psi(t) \equiv \int_{t_0}^{t} K(\theta)\phi(\theta)\,\mathrm{d}\theta$. Then $\dot{\psi}(t) = K(t)\phi(t)$ a.e. and

$$\phi(t) \leqslant a + \psi(t).$$

Since K is non-negative we have,

$$\dot{\psi}(t) = K(t)\phi(t) \leqslant K(t)(a + \psi(t)). \tag{2.2.4}$$

Hence, by virtue of the fact that ψ is monotone non-decreasing with $\psi(t_0) = 0$, we obtain, upon integration,

$$\int_{0}^{\psi(t)} \frac{\mathrm{d}\psi}{a + \psi} \leqslant \int_{t_0}^{t} K(\theta)\,\mathrm{d}\theta. \tag{2.2.5}$$

This leads to the inequality

$$\phi(t) \leqslant (\psi(t) + a) \leqslant a \exp \int_{t_0}^{t} K(\theta)\,\mathrm{d}\theta. \quad \blacksquare$$

In most of our applications the state space for the system (2.2.1) is taken as R^n. In finite-dimensional spaces, like R^n, we can choose any of the norms given for ℓ_p^n, $1 \leqslant p \leqslant \infty$. Most popular is the Euclidean norm corresponding to $p = 2$. In any case, for convenience of notation, we shall denote this norm by $|\cdot|$, that is, for $\xi \in R^n$, $|\xi|$ is the norm (length) of the vector ξ.

Let $\mathscr{L}(R^n)$ denote the set of all linear operators from R^n to R^n. One can easily verify that this is a linear vector space. It is convenient to introduce a norm (and hence a metric) on this space as follows. For $M \in \mathscr{L}(R^n)$ define

$$\|M\| = \max\{|M\xi|, |\xi| \leq 1\}. \tag{2.2.6}$$

Then

$$|M\xi| \leq \|M\| \, |\xi| \qquad \text{for all } \xi \in R^n. \tag{2.2.7}$$

It is clear that for all M, $L \in \mathscr{L}(R^n)$ and $\alpha \in R$,

(a) $\|M\| = 0$ iff $M = 0 \in \mathscr{L}(R^n)$,
(b) $|(M+L)(\xi)| = |M\xi + L\xi| \leq |M\xi| + |L\xi| \leq (\|M\| + \|L\|)\,|\xi|$
and hence

$$\|M + L\| \leq \|M\| + \|L\| \tag{2.2.8}$$

(c) $|\alpha M\xi| = |\alpha| \, |M\xi|$
and hence

$$\|\alpha M\| = |\alpha| \, \|M\|.$$

Therefore, we can conclude that the vector space $\mathscr{L}(R^n)$, with the norm defined by (2.2.6), is a normed vector space. In fact, R^n being a Banach space, $(\mathscr{L}(R^n), \|\cdot\|)$ is also a Banach space. Further, $\mathscr{L}(R^n)$ is also an algebra with respect to multiplication defined by

$$(K \cdot L)(\xi) = K(L\xi) \qquad \text{for } K, L \in \mathscr{L}(R^n)$$

and all $\xi \in R^n$. That is, for K, $L \in \mathscr{L}(R^n)$, $K \cdot L \in \mathscr{L}(R^n)$ and it satisfies the distributive and associative laws:

$$K \cdot (L + M) = K \cdot L + K \cdot M$$

$$K \cdot (L \cdot M) = (K \cdot L) \cdot M.$$

Clearly, for K, $L \in \mathscr{L}(R^n)$, we have

$$|(K \cdot L)(\xi)| = |K(L\xi)| \leq \|K\| \, |L\xi| \leq \|K\| \, \|L\| \, |\xi|. \tag{2.2.9}$$

Hence

$$\|K \cdot L\| \leq \|K\| \, \|L\|, \tag{2.2.10}$$

and, in particular, for any positive integer n,

$$\|K^n\| \equiv \|K \cdot K \cdot \cdots \cdot K\| \leq \|K\|^n. \tag{2.2.11}$$

In general, the results (2.2.6)–(2.2.11) also hold for general Banach spaces, like $L_p(\Omega)$, $\Omega \subset R^n$; ℓ_p, $1 \leq p \leq \infty$; $C(\Omega)$ etc. and not just R^n. In that case one uses the notation $(\mathscr{L}(X), \|\cdot\|)$ for a general Banach space X.

Returning to the case $(\mathscr{L}(R^n), \|\cdot\|)$, we note that every element $A \in \mathscr{L}(R^n)$ has a matrix representation

$$Ax = \begin{bmatrix} a_{11} & \cdots & a_{1n} \\ a_{n1} & \cdots & a_{nn} \end{bmatrix} \begin{bmatrix} x_1 \\ \vdots \\ x_n \end{bmatrix}. \tag{2.2.12}$$

This follows from the fact that if $\{e_i, i = 1, 2, \ldots, n\}$ is any (orthonormal) basis for R^n then, for any $x \in R^n$ and $y = Ax$, one can write

$$x = \sum_{j=1}^{n} (e_j, x)e_j \equiv \sum_{j=1}^{n} x_j e_j$$

and (2.2.13)

$$y = \sum_{j=1}^{n} (e_j, y)e_j \equiv \sum_{j=1}^{n} y_j e_j.$$

Hence, for $A \in \mathscr{L}(R^n)$,

$$Ax = \sum_{j=1}^{n} (e_j, x)Ae_j = \sum_{j=1}^{n} x_j Ae_j. \tag{2.2.14}$$

But

$$Ae_j = \sum_{k=1}^{n} (e_k, Ae_j)e_k. \tag{2.2.15}$$

Therefore,

$$\begin{aligned} y \equiv Ax &= \sum_{j=1}^{n} x_j \sum_{k=1}^{n} (e_k, Ae_j)e_k \\ &= \sum_{k=1}^{n} \left(\sum_{j=1}^{n} (e_k, Ae_j)x_j \right) e_k \end{aligned} \tag{2.2.16}$$

and hence it follows from (2.2.13) and (2.2.16) that

$$y_i = \sum_{j=1}^{n} (e_i, Ae_j)x_j, \qquad i = 1, 2, \ldots, n. \tag{2.2.17}$$

Thus, to each operator $A \in \mathscr{L}(R^n)$ there corresponds the matrix $\{a_{ij}, i, j = 1, 2, \ldots, n\}$ with $a_{ij} \equiv (e_i, Ae_j)$. Conversely, given any matrix $\{a_{ij}, i, j = 1, \ldots, n\}$, one can define a linear operator A in R^n through the expression (2.2.12). Thus, one can identify the space of linear operators in R^n with the space $R(n \times n)$ of $(n \times n)$ matrices having entries from the field of real numbers. This concept is expressed by saying that $\mathscr{L}(R^n)$ is isomorphic to $R(n \times n)$, written $\mathscr{L}(R^n) \cong R(n \times n)$.

Now we return to the questions of existence and uniqueness.

Theorem 2.2.2 *Consider the system* (2.2.1) *and suppose the elements of the matrix* $(R(n \times n))$*-valued function* $A \equiv \{A(t), t \geq 0\}$, *and those of the function* f *are locally* (*Lebesgue*) *integrable. Then for every* $x_0 \in R^n$, (a) *the system* (2.2.1) *has a unique solution* $x \in AC(I, R^n)$ *for every finite interval* $I \subset R_0 = [0, \infty]$, (b) *if both* A *and* f *are continuous on* I, *then* $x \in C^1(I, R^n)$ *and in general* (c) *if* $A \in C^k(I, R(n \times n))$ *and* $f \in C^k(I, R^n)$, *then* $x \in C^{k+1}(I, R^n)$.

Proof Let $I = [0, T]$, $T < \infty$. For each $t \in I$, define

$$\xi(t) \equiv x_0 + \int_0^t f(\theta)\, d\theta. \tag{2.2.18}$$

Since $f \in L_1(I, R^n)$ it follows from Proposition 2.1.31(b) that $\xi \in AC(I, R^n)$. We write the differential equation as an integral equation

$$x(t) = \xi(t) + \int_0^t A(\theta)x(\theta)\, d\theta, \qquad t \in I, \tag{2.2.19}$$

for the unknown x.

We use successive approximation, known as Picard approximation, to prove that the integral equation (2.2.19) has a unique solution and then we show that this is indeed the unique solution of (2.2.1). For the first approximation, define $x^0(t) = \xi(t)$, $t \in I$; then define, for $t \in I$,

$$\begin{aligned} x^1(t) &\equiv \xi(t) + \int_0^t A(\theta)x^0(\theta)\, d\theta \\ x^2(t) &\equiv \xi(t) + \int_0^t A(\theta)x^1(\theta)\, d\theta \end{aligned} \tag{2.2.20}$$

and in general

$$x^m(t) \equiv \xi(t) + \int_0^t A(\theta)x^{m-1}(\theta)\, d\theta, \qquad m = 1, 2, \ldots. \tag{2.2.21}$$

Thus, we have defined a sequence $\{x^m\} \in C(I, R^n)$. We show that this is a convergent sequence in $C(I, R^n)$. For this we must show that $\{x^m\}$ is a Cauchy sequence, that is,

$$\lim_{m \to \infty} \|x^{m+p} - x^m\| \equiv \lim_{m \to \infty} \left\{ \sup_{t \in I} |x^{m+p}(t) - x^m(t)| \right\} = 0,$$

for any finite integer $p \geq 1$. Since, by triangle inequality,

$$\|x^{m+p} - x^m\| \leq \sum_{k=m}^{m+p-1} \|x^{k+1} - x^k\|,$$

it suffices to prove that

$$\lim_{m\to\infty} \|x^{m+1} - x^m\| \to 0.$$

It follows from (2.2.20) and the definition of x^0 that

$$x^1(t) - x^0(t) = \int_0^t A(\theta)\xi(\theta)\,\mathrm{d}\theta,$$

hence

$$|x^1(t) - x^0(t)| \leq \int_0^t |A(\theta)\xi(\theta)|\,\mathrm{d}\theta \leq \int_0^t K(\theta)\,|\xi(\theta)|\,\mathrm{d}\theta, \tag{2.2.22}$$

where

$$K(t) \equiv \|A(t)\| \equiv \sup\{|A(t)\eta|,\ |\eta| \leq 1\}.$$

Since the elements of A are integrable on finite intervals, $K \in L_1^+(I)$. Since $\xi \in AC(I, R^n) \subset C(I, R^n)$, there exists a finite positive number $\beta < \infty$ such that

$$\|\xi\| = \sup\{|\xi(t)|,\ t \in I\} \leq \beta. \tag{2.2.23}$$

Hence

$$|x^1(t) - x^0(t)| \leq \beta \int_0^t K(\theta)\,\mathrm{d}\theta, \qquad t \in I. \tag{2.2.24}$$

Similarly

$$\begin{aligned} |x^2(t) - x^1(t)| &\leq \int_0^t \|A(\theta)\|\,|x^1(\theta) - x^0(\theta)|\,\mathrm{d}\theta \\ &\leq \beta \int_0^t K(\theta)h(\theta)\,\mathrm{d}\theta \end{aligned} \tag{2.2.25}$$

where

$$h(t) \equiv \int_0^t K(\theta)\,\mathrm{d}\theta, \qquad t \in I, \tag{2.2.26}$$

is a monotone nondecreasing function with $h(0) = 0$. By carrying out the integration (2.2.25) we have

$$\begin{aligned} |x^2(t) - x^1(t)| &\leq \beta \int_0^t h(\theta)\,\mathrm{d}h(\theta) = \beta \int_0^{h(t)} \xi\,\mathrm{d}\xi \\ &\leq \beta \frac{h^2(t)}{2}, \qquad t \in I. \end{aligned} \tag{2.2.27}$$

Similarly one can verify that

$$|x^3(t) - x^2(t)| \leq \int_0^t \|A(\theta)\| \, |x^2(\theta) - x^1(\theta)| \, d\theta$$

$$\leq \beta \int_0^t K(\theta) \frac{h^2(\theta)}{2} \, d\theta = \beta \int_0^t \frac{h^2(\theta)}{2} \, dh(\theta)$$

$$= \beta \frac{h^3(t)}{3!}, \qquad t \in I,$$

and in general

$$|x^{m+1}(t) - x^m(t)| \leq \beta \frac{(h(t))^{m+1}}{(m+1)!}.$$

Therefore

$$\|x^{m+1} - x^m\| = \sup_{t \in I} |x^{m+1}(t) - x^m(t)| \leq \frac{\beta(h(T))^{m+1}}{(m+1)!}. \tag{2.2.28}$$

Since $h(T) = \int_0^T K(\theta) \, d\theta < \infty$, it follows from the above inequality that $\lim_{m \to \infty} \|x^{m+1} - x^m\| = 0$. Hence $\{x^m\}$ is a Cauchy sequence in $C(I, R^n)$. Since $C(I, R^n)$ is a Banach space (see Proposition 2.1.29) there exists a unique $x^* \in C(I, R^n)$ such that $\|x^m - x^*\| \to 0$ as $m \to \infty$, that is, $x^m(t) \to x^*(t)$ uniformly in t on I. We show that x^* is a solution. By definition of the sequence $\{x^m\}$ we have

$$x^{m+1}(t) = \xi(t) + \int_0^t A(\theta) x^m(\theta) \, d\theta, \qquad t \in I. \tag{2.2.29}$$

Since $x^m(t) \to x^*(t)$ uniformly on I, we have, on taking the limit,

$$x^*(t) = \lim_m x^{m+1}(t) = \xi(t) + \lim_m \int_0^t A(\theta) x^m(\theta) \, d\theta. \tag{2.2.30}$$

At this point we use the Lebesgue dominated convergence theorem (Proposition 2.1.32) to justify that

$$\lim_m \int_0^t A(\theta) x^m(\theta) \, d\theta = \int_0^t A(\theta) x^*(\theta) \, d\theta, \qquad t \in I. \tag{2.2.31}$$

Define $g^m(t) \equiv A(t) x^m(t)$, $t \in I$, and note that, since $A \in L_1(I, R(n \times n))$, we can only claim that

(a) $g^m(t) \to g^*(t) \equiv A(t) x^*(t)$ a.e. on I.

Further, being a Cauchy sequence, $\{x^m\}$ is necessarily bounded in the sense that there exists a number $b < \infty$ such that

$$\sup_m \|x^m\| \equiv \sup_m \left\{ \sup_{t \in I} |x^m(t)| \right\} \leq b.$$

Hence,

(b) $|g^m(t)| = |A(t)x^m(t)| \leqslant \|A(t)\|\,|x^m(t)| \leqslant bK(t)$ for all $t \in I$.

Thus, by Proposition 2.1.32, (2.2.31) is justified and we have

$$x^*(t) = \xi(t) + \int_0^t A(\theta)x^*(\theta)\,\mathrm{d}\theta, \qquad t \in \delta I, \tag{2.2.32}$$

and hence x^* is a solution of the integral equation (2.2.19). We show that it is a solution of (2.2.1) (in some weak sense). Since ξ is absolutely continuous and $Ax^* \in L_1(I, R^n)$, by Proposition 2.1.31, it follows from (2.2.32) that x^* is absolutely continuous and hence it is differentiable almost everywhere on I. Therefore it follows from (2.2.18) and (2.2.32) that

$$\dot{x}^*(t) = f(t) + A(t)x^*(t) \qquad \text{a.e. on } I, \tag{2.2.33}$$

and not for all $t \in I$. (This is the weak sense.)

Further, letting $t \to 0$ in (2.2.32) and recalling (2.2.18) we have $x^*(0) = x_0$. Thus x^* satisfies (2.2.1) in the weak sense as discussed above. For uniqueness we show that x^* is the only solution. Suppose not, and let y^* be another solution, that is,

$$y^*(t) = \xi(t) + \int_0^t A(\theta)y^*(\theta)\,\mathrm{d}\theta, \qquad t \in I. \tag{2.2.34}$$

Subtracting (2.2.34) from (2.2.32) we have

$$x^*(t) - y^*(t) = \int_0^t A(\theta)(x^*(\theta) - y^*(\theta))\,\mathrm{d}\theta \tag{2.2.35}$$

and hence,

$$|x^*(t) - y^*(t)| \leqslant \int_0^t \|A(\theta)\|\,|x^*(\theta) - y^*(\theta)|\,\mathrm{d}\theta$$

or equivalently

$$\phi(t) \leqslant \int_0^t K(\theta)\phi(\theta)\,\mathrm{d}\theta, \tag{2.2.36}$$

where $K(t) \equiv \|A(t)\|$ and $\phi(t) = |x^*(t) - y^*(t)|$. Therefore, by Gronwall's lemma (Lemma 2.2.1), $\phi(t) \equiv 0$ and hence $x^*(t) \equiv y^*(t)$, $t \in I$, proving uniqueness. This proves part (a). Part (b) follows from (2.2.33), since $t \to A(t)x^*(t)$ is continuous whenever A is continuous. Hence $\dot{x}^* \in C(I, R^n)$ and consequently $x^* \in C^1(I, R^n)$. Part (c) is left as an exercise for the reader (see Problem 2.2.P10). ■

2.2.2 State-transition operator

Consider the homogeneous system with two different initial states $\eta, \beta \in R^n$ starting at time $s \geq 0$, written separately as

$$\dot{x} = A(t)x, \quad x(s) = \eta, \qquad t \geq s,$$

and (2.2.37)

$$\dot{y} = A(t)y, \quad y(s) = \beta, \qquad t \geq s.$$

Clearly, by the previous theorem, they have unique solutions. Let $x(t) = \phi(t, s; \eta)$ and $y(t) = \phi(t, s; \beta)$, $t \geq s$, denote their solutions. By adding the two equations term by term we also have

$$\dot{z} = A(t)z, \quad z(s) = \eta + \beta, \qquad t \geq s \tag{2.2.38}$$

with solution $z(t) = \phi(t, s; \eta + \beta)$, $t \geq s$. Clearly $z(t) = x(t) + y(t)$, $t \geq s$ and hence

$$\phi(t, s; \eta + \beta) = \phi(t, s; \eta) + \phi(t, s; \beta) \tag{2.2.39}$$

for $t \geq s \geq 0$. Thus, the mapping $\eta \to \phi(t, s; \eta)$ is a linear operation from $R^n \to R^n$ and hence for every pair s, t, satisfying $0 \leq s \leq t < \infty$, there exists an operator $\Phi(t, s) \in \mathscr{L}(R^n) \equiv R(n \times n)$ such that

$$\phi(t, s; \eta) = \Phi(t, s)\eta. \tag{2.2.40}$$

The operator-valued function $\Phi(t, s)$, $0 \leq s \leq t < \infty$, is called the state-transition operator since it transforms any given state ξ at time s into a unique state η at a later time t given by

$$\eta = \Phi(t, s)\xi. \tag{2.2.41}$$

In other words, Φ dictates the natural evolution of the system and is determined uniquely by the system operator $A = \{A(t), t \geq 0\}$ alone. Hence Φ plays a fundamental role in the study of the linear system (2.2.1).

The transition operator $\Phi(t, s)$, $0 \leq s \leq t \leq T < \infty$ satisfies the following properties:

(a) $t \to \Phi(t, s)$ is absolutely continuous from $[s, T]$ to $\mathscr{L}(R^n)$, $s \geq 0$ and $\Phi(s, s) = I$ (identity matrix) for all s,

(b) $\Phi(t, \theta)\Phi(\theta, s) = \Phi(t, s)$ for $0 \leq s \leq \theta \leq t$,

(c) $\dfrac{\partial}{\partial t}\Phi(t, s) = A(t)\Phi(t, s)$ for $0 \leq s < t \leq T < \infty$, (2.2.42)

(d) $\dfrac{\partial}{\partial s}\Phi(t, s) = -\Phi(t, s)A(s)$ for $0 \leq s < t < \infty$.

The first property follows from the fact that, given any $s \geq 0$, and

$\xi \in R^n$, $t \to x(t) = \Phi(t, s)\xi$ is absolutely continuous (see Theorem 2.2.2) and that $\xi = \lim_{t \downarrow s} x(t) = \lim_{t \downarrow s} \Phi(t, s)\xi$. Hence $\Phi(s, s) = I$.

Property (b) follows from uniqueness. Let $\xi \in R^n$ be the state given at time $s \geq 0$, then the states attained at later times θ and $t \geq \theta$ are uniquely determined by

$$x(\theta) = \Phi(\theta, s)\xi, \qquad \theta \geq s$$

$$x(t) = \Phi(t, s)\xi, \qquad t \geq s$$

$$x(t) = \Phi(t, \theta)x(\theta), \qquad t \geq \theta.$$

Hence

$$\Phi(t, s)\xi = \Phi(t, \theta)\Phi(\theta, s)\xi$$

for all $\xi \in R^n$ and, consequently, property (b) follows. For property (c), note that for any $s \geq 0$, and for every state $\xi \in R^n$ given at time s, it follows from the previous theorem that the solution $t \to x(t) \equiv \Phi(t, s)\xi$ is absolutely continuous. Hence, by Proposition 2.1.31, x is differentiable almost everywhere (a.e.) and, for $t > s$, we have

$$\frac{\partial}{\partial t}(\Phi(t, s)\xi) = \frac{\mathrm{d}}{\mathrm{d}t}x(t) = A(t)x(t) = A(t)\Phi(t, s)\xi, \qquad \text{a.e.}$$

and consequently, for $t > s \geq 0$,

$$\left(\frac{\partial}{\partial t}\Phi(t, s) - A(t)\Phi(t, s)\right)\xi = 0 \qquad \text{a.e.}$$

for all $\xi \in R^n$. Hence (c) follows. Property (d) follows from (b) and it is left as an exercise for the reader.

As a corollary of the preceding discussions we can deduce the following result characterizing the transition operator Φ.

Corollary 2.2.3

$$\Phi(t, s) = X(t)X^{-1}(s), \qquad t \geq s \geq 0 \tag{2.2.43}$$

where $X(t)$, $t \geq 0$, is the fundamental solution of the matrix differential equation

$$\frac{\mathrm{d}X}{\mathrm{d}t} = A(t)X, \qquad t \geq 0$$

$$X(0) = I. \tag{2.2.44}$$

Proof Let $e_i \in R^n$ denote the column vector with 1 at the ith entry and zero elsewhere, and let $\phi_i(t)$, $t \geqslant 0$, denote the unique solution of

$$\frac{\mathrm{d}x}{\mathrm{d}t} = A(t)x, \qquad t \geqslant 0, \qquad x(0) = e_i \tag{2.2.45}$$

for $i = 1, 2, \ldots, n$. Denote by $X(t)$ the matrix formed by using the vector $\phi_i(t)$ for its ith column, $i = 1, 2, \ldots, n$. Then $X(t)$ satisfies (2.2.44). Let $\xi \in R^n$ and let x denote the corresponding solution of

$$\dot{x}(t) = A(t)x, \qquad t \geqslant 0, \qquad x(0) = \xi. \tag{2.2.46}$$

Then, since $\xi = \sum_{i=1}^{n} (e_i, \xi)e_i = \sum \xi_i e_i$ it follows from linearity of the system that

$$x(t) = \sum_{i=1}^{n} \xi_i \phi_i(t) = X(t)\xi, \qquad t \geqslant 0. \tag{2.2.47}$$

That is, the solution of (2.2.46) is also given in terms of the fundamental solution of (2.2.44). Before we can prove the relation (2.2.43) we must show that $X(t)$ is non-singular for each $t \geqslant 0$. For this we consider the matrix differential equation

$$-\frac{\mathrm{d}Y}{\mathrm{d}t} = YA(t), \qquad Y(0) = I. \tag{2.2.48}$$

By our fundamental theorem (Theorem 2.2.1) this equation has a unique absolutely continuous (matrix-valued) solution $Y(t)$, $t \geqslant 0$. Differentiating the product $Y(t)X(t)$ we have,

$$\begin{aligned}\frac{\mathrm{d}}{\mathrm{d}t}[Y(t)X(t)] &= \left(\frac{\mathrm{d}}{\mathrm{d}t} Y(t)\right)X(t) + (Y(t))\left(\frac{\mathrm{d}}{\mathrm{d}t} X(t)\right)\\ &= -Y(t)A(t)X(t) + Y(t)A(t)X(t)\\ &= 0.\end{aligned}$$

Hence $Y(t)X(t)$ is a constant matrix and equals I since $Y(0) = X(0) = I$. Thus $X^{-1}(t) = Y(t)$ is defined for all $t \geqslant 0$. Returning to the proof of (2.2.43), let $x(s) = \eta$ be given for any $s > 0$. Since $X(s)$ is non-singular, for the given $\eta \in R^n$, there exists a unique initial state $x(0) = \xi \in R^n$ such that $\eta = X(s)\xi$ and $\xi = X^{-1}(s)\eta$. Hence, for all $t \geqslant s$, it follows from (2.2.47) that

$$x(t) = X(t)X^{-1}(s)\eta \tag{2.2.49}$$

given that $x(s) = \eta$. On the other hand, using the transition operator Φ, we have

$$x(t) = \Phi(t, s)\eta, \qquad t \geqslant s \geqslant 0. \tag{2.2.50}$$

Here $(\Phi(t, s) - X(t)X^{-1}(s))\eta = 0$ for all $t \geqslant s \geqslant 0$ and all $\eta \in R^n$ and consequently (2.2.43) follows. ■

Time-invariant system

For the time-invariant systems A is constant and in this case the transition operator is given by the exponential formula

$$\Phi(t, \tau) \equiv \mathrm{e}^{(t-\tau)A}, \qquad t \geqslant \tau. \tag{2.2.51}$$

One can directly prove this by use of Laplace transform or by the successive approximation technique as used in the proof of Theorem 2.2.2. Consider the system $\dot{x} = Ax$, $x(0) = x_0$, and write as an integral equation

$$x(t) = x_0 + \int_0^t Ax(\theta)\, \mathrm{d}\theta.$$

Set

$$x^0(t) \equiv x_0,$$

then

$$x^1(t) = x_0 + \int_0^t Ax^0(\theta)\, \mathrm{d}\theta = x_0 + tAx_0$$

$$x^2(t) = x_0 + \int_0^t Ax^1(\theta)\, \mathrm{d}\theta = x_0 + tAx_0 + \frac{t^2}{2}A^2x_0$$

and for n,

$$x^n(t) = \left(I + tA + \frac{t^2}{2}A^2 + \cdots + \frac{t^n}{n!}A^n\right)x_0. \tag{2.2.52}$$

Hence, letting $n \to \infty$, one has $x(t) = e^{tA}x_0$ and $\Phi(t, \tau) = e^{(t-\tau)A}$ as stated. The reader may use Laplace transform to verify this fact.

In general, using the transition operator $\Phi(t, \tau)$, we can immediately write the solution of the general time-varying system (2.2.1) as

$$\begin{aligned} x(t) &= \Phi(t, 0)x_0 + \int_0^t \Phi(t, \tau)f(\tau)\, \mathrm{d}\tau \\ &= X(t)x_0 + \int_0^t X(t)X^{-1}(\tau)f(\tau)\, \mathrm{d}\tau. \end{aligned} \tag{2.2.53}$$

This can be verified directly by differentiation. Since $t \to \Phi(t, \tau)$, $0 \leqslant \tau \leqslant t$, is absolutely continuous it follows from the properties (a) and (c) that

$$\dot{x}(t) = \frac{\partial}{\partial t}\Phi(t, 0)x_0 + \Phi(t, t)f(t) + \int_0^t \frac{\partial}{\partial t}\Phi(t, \tau)f(\tau)\,\mathrm{d}\tau \quad \text{a.e.}$$

$$= A(t)\Phi(t, 0)x_0 + f(t) + \int_0^t A(t)\Phi(t, \tau)f(\tau)\,\mathrm{d}\tau \quad \text{a.e.}$$

$$= A(t)x(t) + f(t), \qquad \text{a.e.}$$

Controlled systems

In the case of controlled systems we have

$$\begin{aligned} \dot{x} &= A(t)x + B(t)u(t) \qquad \text{(state equation)} \\ y(t) &= H(t)x + D(t)u(t) \qquad \text{(output equation)} \end{aligned} \tag{2.2.54}$$

where u is the control signal and is generally an element of some function space $\mathcal{U} \subset L_p(I, R^m)$, $1 \leqslant p \leqslant \infty$, and B is the control matrix which is an element of $L_q(I, R(n \times m))$ with $(1/p) + (1/q) = 1$ and I any finite interval. This choice of B and u guarantees the local integrability of Bu (see Hölder's inequality, Proposition 2.1.24) which is what is required by Theorem 2.2.2. In many cases, such as optimal control problems, the set of admissible controls is taken as

$$\mathcal{U} \equiv \{u \in L_\infty(I, R^m) \colon u(t) \in U \text{ a.e.}\}$$

where U is any closed bounded subset of R^m.

The output y is given by

$$y(t) = H(t)\Phi(t, 0)x_0 + \int_0^t H(t)\Phi(t, \theta)B(\theta)u(\theta)\,\mathrm{d}\theta + D(t)u(t), \qquad t \geqslant 0. \tag{2.2.55}$$

Later we shall have occasions to use these expressions in control problems.

2.3 Continuous dependence of solutions on system parameters and sensitivity

The question of continuous dependence of solutions is very important in the study of sensitivity, stability and optimal control.

2.3.1 Continuous dependence

Consider the system

$$\begin{aligned} \dot{x} &= A(t)x + f(t), \qquad t \geq s \\ x(s) &= \xi, \qquad\qquad\quad\;\; s \geq 0 \end{aligned} \tag{2.3.1}$$

and let us denote the solution by $x(t) = x(t, s, \xi)$.

Lemma 2.3.1 (Continuous dependence on s, ξ). *Under the assumptions of Theorem* 2.2.2 *the mapping* $(s, \xi) \to x(t, s, \xi)$, $t > s \geq 0$, *is continuous.*

Proof Let $s_1, s_2 \in [0, T]$, $T < \infty$, and $\xi_1, \xi_2 \in R^n$; and let $x_1(t) \equiv x_1(t, s_1, \xi_1)$, $t \geq s_1 \geq 0$, and $x_2(t) \equiv x_2(t, s_2, \xi_2)$, $t \geq s_2 \geq 0$ be the corresponding solutions. Without loss of generality we may assume that $s_1 < s_2$. Then

$$x_1(t) = \xi_1 + \int_{s_1}^{t} A(\theta)x_1(\theta)\,\mathrm{d}\theta + \int_{s_1}^{t} f(\theta)\,\mathrm{d}\theta, \qquad t \geq s_1, \tag{2.3.2}$$

and

$$x_2(t) = \xi_2 + \int_{s_2}^{t} A(\theta)x_2(\theta)\,\mathrm{d}\theta + \int_{s_2}^{t} f(\theta)\,\mathrm{d}\theta, \qquad t \geq s_2, \tag{2.3.3}$$

and hence,

$$\begin{aligned} x_1(t) - x_2(t) = &\left\{(\xi_1 - \xi_2) + \int_{s_1}^{s_2} A(\theta)x_1(\theta)\,\mathrm{d}\theta\right. \\ &\left. + \int_{s_1}^{s_2} f(\theta)\,\mathrm{d}\theta\right\} + \int_{s_2}^{t} A(\theta)(x_1(\theta) - x_2(\theta))\,\mathrm{d}\theta \end{aligned} \tag{2.3.4}$$

for $t \geq \max\{s_1, s_2\}$. Taking the norm on either side we have,

$$\begin{aligned} |x_1(t) - x_2(t)| \leq &\{|\xi_1 - \xi_2| + \int_{s_1}^{s_2} \|A(\theta)\|\,|x_1(\theta)|\,\mathrm{d}\theta \\ &+ \int_{s_1}^{s_2} |f(\theta)|\,\mathrm{d}\theta\} + \int_{s_2}^{t} \|A(\theta)\|\,|x_1(\theta) - x_2(\theta)|\,\mathrm{d}\theta \end{aligned} \tag{2.3.5}$$

and hence by use of Gronwall's lemma (Lemma 2.2.1) we obtain

$$\begin{aligned} |x_1(t) - x_2(t)| \leq &\left(\exp \int_{s_2}^{t} \|A(\theta)\|\,\mathrm{d}\theta\right)\left(|\xi_1 - \xi_2|\right. \\ &\left. + \int_{s_1}^{s_2} \|A(\theta)\|\,|x_1(\theta)|\,\mathrm{d}\theta + \int_{s_1}^{s_2} |f(\theta)|\,\mathrm{d}\theta\right) \end{aligned} \tag{2.3.6}$$

and further

$$\sup_{s_2 \leq t \leq T} |x_1(t) - x_2(t)| \leq \left(\exp \int_{s_2}^{T} \|A(\theta)\| \, d\theta\right)\left(|\xi_1 - \xi_2| + \int_{s_1}^{s_2} \|A(\theta)\| \, |x_1(\theta)| \, d\theta + \int_{s_1}^{s_2} |f(\theta)| \, d\theta\right). \quad (2.3.7)$$

From these one can deduce that, for every $\varepsilon > 0$, there exists a $\delta > 0$ such that, whenever $(s_2 - s_1) + |\xi_1 - \xi_2| < \delta$, $\sup_{s_2 \leq t \leq T} |x_1(t) - x_2(t)| < \varepsilon$. Indeed, since the entries of f and A are locally integrable ($f \in L_1(I, (R^n))$, $A \in L_1(I, \mathscr{L}(R^n))$, I = finite interval) and $\|x_1\| \equiv \sup_{s_1 \leq t \leq T} |x_1(t)| < \infty$, for every $\varepsilon > 0$, there exists a $\delta_1 > 0$ such that, whenever the length of the interval $J < \delta_1$ (in general Lebesgue measure of J denoted $\mu(J) < \delta_1$),

$$\int_J \|A(\theta)\| \, |x_1(\theta)| \, d\theta + \int_J |f(\theta)| \, d\theta < (\varepsilon/2) \exp - \int_0^T \|A(\theta)\| \, d\theta. \quad (2.3.8)$$

Define $\delta_2 \equiv (\varepsilon/2) \exp - \int_0^T \|A(\theta)\| \, d\theta$ and $\delta \equiv \min\{\delta_1, \delta_2\}$; then $(s_2 - s_1) + |\xi_1 - \xi_2| < \delta$ implies $\sup_{s_2 \leq t \leq T} |x_1(t) - x_2(t)| < \varepsilon$. Hence, for arbitrary $s_1, s_2 \in [0, T]$, we may conclude that $|s_1 - s_2| + |\xi_1 - \xi_2| < \delta$ implies that $\sup\{|x_1(t) - x_2(t)|, \max(s_1, s_2) \leq t \leq T\} < \varepsilon$. This implies the required continuity. ■

Next we consider continuous dependence on initial state and input.

Lemma 2.3.2 (Continuous dependence on ξ, f). *Consider the system*

$$\dot{x} = A(t)x + f(t), \qquad t \in [0, T], \quad T < \infty$$
$$x(0) = \xi \quad (2.3.9)$$

and suppose $A \in L_1(I, \mathscr{L}(R^n)) \equiv L_1(I, R(n \times n))$, *then* $(\xi, f) \to x(\cdot, \xi, f)$ *is continuous from* $R^n \times L_1(I, R^n)$ *to* $C(I, R^n)$.

A more general result that includes Lemma 2.3.2 as a special case is given below.

Lemma 2.3.3 (Continuous dependence on ξ, A, f). *Consider the system* (2.3.9) *for* $t \in I = [0, T]$ *and let* $x(t, \xi, A, f)$ *denote the solution corresponding to initial state* $\xi \in R^n$, *plant operator* $A \in L_1(I, \mathscr{L}(R^n))$ *and input* $f \in L_1(I, R^n)$. *Then* $(\xi, A, f) \to x(\cdot, \xi, A, f)$ *is continuous from* $R^n \times L_1(I, \mathscr{L}(R^n)) \times L_1(I, R^n)$ *to* $C(I, R^n)$.

Proof Define $x_1(t) \equiv x(t, \xi_1, A_1, f_1)$ and $x_2(t) \equiv x(t, \xi_2, A_2, f_2)$ for $t \in I$

and

$$(\xi_i, A_i, f_i) \in R^n \times L_1(I, \mathscr{L}(R^n)) \times L_1(I, R^n), \qquad i = 1,2.$$

Then one can easily verify that

$$|x_1(t) - x_2(t)| \leq |\xi_1 - \xi_2| + \int_0^t |f_1(\theta) - f_2(\theta)|\, d\theta$$
$$+ \int_0^t \|A_1(\theta) - A_2(\theta)\|\, |x_2(\theta)|\, d\theta + \int_0^t \|A_1(\theta)\|\, |x_1(\theta) - x_2(\theta)|\, d\theta. \tag{2.3.10}$$

Similarly,

$$|x_2(t)| \leq \left(|\xi_2| + \int_0^t |f_2(\theta)|\, d\theta\right) + \int_0^t \|A_2(\theta)\|\, |x_2(\theta)|\, d\theta. \tag{2.3.11}$$

Hence, by the Gronwall inequality

$$\|x_2\| \equiv \sup_{t \in I} |x_2(\theta)| \leq \left(\exp \int_0^T \|A_2(\theta)\|\, d\theta\right)\left(|\xi_2| + \int_0^T |f_2(\theta)|\, d\theta\right)$$
$$\leq b_2 < \infty. \tag{2.3.12}$$

Using this estimate in (2.3.10) one obtains

$$\|x_1 - x_2\| \leq \left(\exp \int_0^T \|A_1(\theta)\|\, d\theta\right)\left(|\xi_1 - \xi_2| + \int_0^T |f_1(\theta) - f_2(\theta)|\, d\theta\right.$$
$$\left. + b_2 \int_0^T \|A_1(\theta) - A_2(\theta)\|\, d\theta\right)$$
$$\leq \beta(|\xi_1 - \xi_2| + \|f_1 - f_2\|_{L_1(I, R^n)} + \|A_1 - A_2\|_{L_1(I, \mathscr{L}(R^n))}) \tag{2.3.13}$$

where

$$\beta = \max\{1, b_2\} \cdot \exp\left(\int_0^T \|A_1(\theta)\|\, d\theta\right).$$

This proves that $(\xi, A, f) \to x(\cdot, \xi, A, f)$ is a Lipschitz continuous mapping from $R^n \times L_1(I, \mathscr{L}(R^n)) \times L_1(I, R^n)$ to $C(I, R^n)$. ■

We shall write

$$\begin{aligned}
&x_k \xrightarrow{u} x \text{ if } \sup_{t \in I} |x_k(t) - x(t)| \to 0 && \text{as } k \to \infty, \\
&A_k \xrightarrow{s} A \text{ if } \int_I \|A_k(t) - A(t)\|_{\mathscr{L}(R^n)}\, dt \to 0 && \text{as } k \to \infty, \\
&f_k \xrightarrow{s} f \text{ if } \int_I |f_k(t) - f(t)|\, dt \to 0 && \text{as } k \to \infty.
\end{aligned} \tag{2.3.14}$$

Let $\mathscr{B}$ denote the Banach space $R^n \times L_1(I, \mathscr{L}(R^n)) \times L_1(I, R^n)$ furnished with the norm

$$\|\xi, A, f\|_{\mathscr{B}} \equiv |\xi|_{R^n} + \int_0^T \|A(t)\|_{\mathscr{L}(R^n)} \,\mathrm{d}t + \int_0^T |f(t)|_{R^n} \,\mathrm{d}t \tag{2.3.15}$$

for $(\xi, A, f) \in \mathscr{B}$. Then we can summarize our continuity results in the following theorem.

Theorem 2.3.4 *Consider the system* (2.3.9) *and suppose* $(\xi_k, A_k, f_k) \xrightarrow{s} (\xi_0, A_0, f_0)$ *in* $\mathscr{B}$, *then the corresponding sequence of solutions* $\{x_k\} = \{x(\cdot, \xi_k, A_k, f_k)\}$ *converge uniformly to the solution* x_0 *corresponding to the set* (ξ_0, A_0, f_0). *This is shown in the following diagram*:

$$\begin{array}{ccc} (\xi_k, A_k, f_k) & \xrightarrow{s} & (\xi_0, A_0, f_0) \\ \Downarrow & & \Downarrow \\ x_k & \xrightarrow{u} & x_0 \end{array}$$

where $\rightarrow$ *denotes convergence and* $\Rightarrow$ *denotes correspondence.*

The continuity result of Theorem 2.3.4 is rather demanding. There, it is required that the sequence of plant matrices $\{A_k\}$ and the input sequence $\{f_k\}$ converge in the mean. We shall see that we can obtain convergence of x_k to x_0 under certain weaker modes of convergence of the sequence $\{A_k, f_k\}$. For this we need a generalized version of Gronwall inequality.

Lemma 2.3.5 (Generalized Gronwall lemma). *Suppose*

$$\phi(t) \leq \alpha(t) + \int_0^t K(\theta)\phi(\theta) \,\mathrm{d}\theta \tag{2.3.16}$$

where $\alpha(t)$ *is continuous, bounded and* $\alpha(t) \geq 0$, $K \in L_1^{\mathrm{loc}}$ *with* $K(t) \geq 0$ *a.e., and* $\phi(t) \geq 0$. *Then for* $0 \leq t < \infty$,

$$\phi(t) \leq \alpha(t) + \int_0^t \left(\exp \int_\theta^t K(s) \,\mathrm{d}s\right) K(\theta)\alpha(\theta) \,\mathrm{d}\theta. \tag{2.3.17}$$

Proof Define $\psi(t) = \int_0^t K(\theta)\phi(\theta) \,\mathrm{d}\theta$; then ψ is monotone non-decreasing and absolutely continuous and hence differentiable almost everywhere giving $\dot{\psi}(t) = K(t)\phi(t)$. Therefore, it follows from the given inequality (2.3.16) that

$$\dot{\psi}(t) \equiv K(t)\phi(t) \leq K(t)(\alpha(t) + \psi(t)). \tag{2.3.18}$$

Defining

$$\psi(t)=\xi(t)\exp\int_0^t K(\theta)\,\mathrm{d}\theta, \tag{2.3.19}$$

where ξ is to be determined, we have,

$$\dot{\psi}(t)=\dot{\xi}(t)\exp\int_0^t K(\theta)\,\mathrm{d}\theta+K(t)\psi(t). \tag{2.3.20}$$

Hence, using (2.3.18), we obtain

$$\begin{aligned}\dot{\xi}(t)&=(\dot{\psi}(t)-K(t)\psi(t))\exp\left(-\int_0^t K(\theta)\,\mathrm{d}\theta\right)\\&\leqslant K(t)\alpha(t)\exp\left(-\int_0^t K(\theta)\,\mathrm{d}\theta\right).\end{aligned}$$

Therefore,

$$\xi(t)\leqslant\xi(0)+\int_0^t K(s)\alpha(s)\exp\left(-\int_0^s K(\theta)\,\mathrm{d}\theta\right)\mathrm{d}s. \tag{2.3.21}$$

Since $\psi(0)=0$, $\xi(0)$ must equal zero and hence it follows from (2.3.19) and (2.3.21) that

$$\psi(t)\leqslant\int_0^t\left(\exp\int_s^t K(\theta)\,\mathrm{d}\theta\right)K(s)\alpha(s)\,\mathrm{d}s \tag{2.3.22}$$

and consequently

$$\phi(t)\leqslant\alpha(t)+\psi(t)\leqslant\alpha(t)+\int_0^t\left(\exp\int_s^t K(\theta)\,\mathrm{d}\theta\right)K(s)\alpha(s)\,\mathrm{d}s,$$

proving the desired inequality (2.3.17). ■

Now we return to the question of continuity. We need the following definition.

Definition 2.3.6 (Weak convergence). A sequence $\{g_n\}\in L_1(I,R)$ is said to be weakly convergent to $g_0\in L_1$, denoted by $g_n\xrightarrow{w}g_0$, if, for every $h\in L_\infty(I,R)$,

$$(g_n,h)\equiv\int_I g_n(t)h(t)\,\mathrm{d}t\to\int_I g_0(t)h(t)\,\mathrm{d}t\equiv(g_0,h). \tag{2.3.23}$$

In particular, this means that for each $t\in I$,

$$\int_0^t g_n(\theta)\,\mathrm{d}\theta\to\int_0^t g_0(\theta)\,\mathrm{d}\theta. \tag{2.3.24}$$

For a sequence $\{A_k\} \in L_1(I, \mathscr{L}(R^n))$, $A_k \xrightarrow{w} A_0$ if each entry of $\{A_k\}$ converges weakly to the corresponding entry of A_0. Then clearly for each $t \in I = [0, T]$,

$$\int_0^t A_k(\theta)\, d\theta \to \int_0^t A_0(\theta)\, d\theta.$$

We shall prove that the sequence of solutions $\{x_k\} = \{x(\cdot, \xi_k, A_k, f_k)\}$ converges pointwise to the solution $x_0 = x(\cdot, \xi_0, A_0, f_0)$ whenever $\xi_k \to \xi_0$, $A_k \xrightarrow{w} A_0$, $f_k \xrightarrow{w} f_0$ and $\|A_k(\cdot)\|$ is dominated by an integrable function.

Theorem 2.3.7 *Suppose $(\xi_k, A_k, f_k) \xrightarrow{w} (\xi_0, A_0, f_0)$ in $\mathscr{B}$ and there exists a function $a \in L_1^+$ (non-negative integrable) such that $\|A_k(t)\| \leqslant a(t)$ a.e. on I. Then the sequence of solutions $x_k(\cdot) = x(\cdot, \xi_k, A_k, f_k)$ of (2.3.9) corresponding to $\{\xi_k, A_k, f_k\}$ converge, pointwise on I, to the solution x_0 of (2.3.9) corresponding to the data $\{\xi_0, A_0, f_0\}$.*

Proof It is easy to verify that

$$\phi_k(t) \leqslant \alpha_k(t) + \int_0^t \|A_k(\theta)\|\, \phi_k(\theta)\, d\theta, \qquad t \geqslant 0, \tag{2.3.25}$$

where

$$\phi_k(t) \equiv |x_k(t) - x_0(t)| \tag{2.3.26}$$

and

$$\alpha_k(t) \equiv |\xi_k - \xi_0| + \left|\int_0^t (f_k(\theta) - f_0(\theta))\, d\theta\right| + \left|\int_0^t (A_k(\theta) - A_0(\theta))x_0(\theta)\, d\theta\right|. \tag{2.3.27}$$

Using Gronwall's inequality (Lemma 2.3.5), it follows from (2.3.25) that, for all non-negative integers k,

$$0 \leqslant \phi_k(t) \leqslant \alpha_k(t) + \int_0^t \left(\exp \int_\theta^t \|A_k(s)\|\, ds\right) \|A_k(\theta)\|\, \alpha_k(\theta)\, d\theta, \qquad t \in I. \tag{2.3.28}$$

Since $\|A_k(t)\| \leqslant a(t)$, a.e. on I, for all k and $a \in L_1^+(I)$, we have, for all k,

$$0 \leqslant \phi_k(t) \leqslant \alpha_k(t) + C \int_0^t a(\theta)\alpha_k(\theta)\, d\theta, \qquad \text{for all } t \in I, \tag{2.3.29}$$

where

$$C \equiv \exp \int_I a(\theta)\,\mathrm{d}\theta.$$

Since $\xi_k \to \xi_0$, $A_k \xrightarrow{w} A_0$ and $f_k \xrightarrow{w} f_0$ and $x_0 \in C(I, R^n)$ it follows from (2.3.27) that there exists a constant C_1 such that $\sup_k \alpha_k(t) \leq C_1$ for all $t \in I$ and that $\alpha_k(t) \to 0$ for each $t \in I$. Hence, by the Lebesgue dominated convergence theorem (Proposition 2.1.32), it follows from (2.3.29) that, for each $t \in I$,

$$\lim_k \phi_k(t) \leq \lim_k \alpha_k(t) + C \lim_k \int_0^t a(\theta)\alpha_k(\theta)\,\mathrm{d}\theta = 0.$$

Thus $x_k(t) \to x_0(t)$ for each $t \in I$ and this is pointwise convergence. ■

2.3.2 Sensitivity

Consider the controlled system

$$\dot{x} = A(t, \alpha)x + B(t, \alpha)u, \qquad t \in I = [0, T]$$

$$x(0) = x_0.$$

In almost all practical problems, the matrices A and B may depend, in addition to time, on several physical parameters which may be constant or time varying. Let $\mathcal{P}$ denote the space of parameters with α's being elements of $\mathcal{P}$. A system analyst is interested to determine the impact on system performance caused by variation of the parameters from a nominal value $\alpha_0 \in \mathcal{P}$. The statement of 'performance' may mean transient response, stability, optimality, etc. In any case the question of sensitivity is interlinked with the questions of continuity and differentiability of solutions with respect to parameters. In the literature on sensitivity, the directional derivative of x with respect to α on $\mathcal{P}$ at α_0 is called the sensitivity operator.

It is convenient to introduce the following definitions, which will also be useful later.

Definition 2.3.8 (Gâteaux differentials) A function $f: X \to R$ is said to be Gâteaux differentiable at $x_0 \in X$ in the direction $h \in X$ if there exists a functional $\ell: X \times X \to R$ such that

$$\lim_{\varepsilon \to 0} \left\{ \frac{f(x_0 + \varepsilon h) - f(x_0)}{\varepsilon} \right\} = \ell(x_0, h). \qquad (2.3.30)$$

By definition, $h \to \ell(x_0, h)$ is homogeneous in the sense that $\ell(x_0, ah) = a\ell(x_0, h)$, $a \in R$; however, it may not be linear. The function

f is said to have linear Gâteaux derivative at x_0 in the direction h if $\ell(x_0, h) = (g(x_0), h)$ for some $g(x_0) \in X^*$, the dual of X. We call $g(x_0)$ the Gâteaux derivative of f at x_0 and denote it by $f'(x_0) \equiv g(x_0)$. Note if $X = R^n$ then $X^* = R^n$, and if $X = L_p$ then $X^* = L_q$, $p^{-1} + q^{-1} = 1$, $1 \leqslant p < \infty$.

The function f is said to be Gâteaux differentiable on X if it is so at every point of X.

Similarly, one can define Gâteaux differentials for operators.

Definition 2.3.9 (Gâteaux differential of an operator). An operator $F: X \to Y$ is said to be Gâteaux differentiable at $x_0 \in X$ in the direction $h \in X$ if there exists an $L = L(x_0, h) \in Y$ such that

$$\lim_{\varepsilon \to 0} \left\| \frac{F(x_0 + \varepsilon h) - F(x_0)}{\varepsilon} - L(x_0, h) \right\|_Y = 0. \tag{2.3.31}$$

Again $h \to L(x_0, h)$ is homogeneous but not necessarily linear. If it is linear in h then we write $L(x_0, h) = G(x_0)h$, where $G(x_0)$ is a linear operator from X to Y. The operator $G(x_0)$ is called the Gâteaux derivative of F at x_0 and written $F'(x_0) \equiv G(x_0)$.

Remark 2.3.10 In case the limit $L(x_0, h)$ exists only in the weak sense, that is,

$$\lim_{\varepsilon \to 0} \left(\frac{F(x_0 + \varepsilon h) - F(x_0)}{\varepsilon} - L(x_0, h), y^* \right) = 0 \tag{2.3.32}$$

for each $y^* \in Y^*$ (= dual of Y), then $L(x_0, h)$ is called the weak Gâteaux differential of F at x_0 in the direction h, and $G(x_0)$ is the corresponding Gâteaux derivative.

Definition 2.3.11 (Frechet differentials). A function $f: X \to R$ is said to be Frechet differentiable at $x_0 \in X$ if there exists an $L(x_0) \in X^*$ such that

$$\lim_{\|h\| \to 0} \left\{ \frac{f(x_0 + h) - f(x_0) - (L(x_0), h)}{\|h\|} \right\} = 0. \tag{2.3.33}$$

Again we shall use $f'(x_0)$ to denote $L(x_0)$. An operator $F: X \to Y$ is said to be Frechet differentiable at $x_0 \in X$ if there exists a bounded linear operator $G(x_0)$ from X to Y such that

$$\lim_{\|h\| \to 0} \left\| \frac{F(x_0 + h) - F(x_0) - G(x_0)h}{\|h\|} \right\|_Y = 0. \tag{2.3.34}$$

Let $\mathscr{P}$ denote the space of parameters and suppose it is a subset of a normed linear space. Consider the controlled linear system

$$L_\alpha : \begin{cases} \dfrac{dx}{dt} = A(t, \alpha)x + B(t, \alpha)u, & t \in I = [0, T] \\ x(0) = \xi \end{cases} \tag{2.3.35}$$

where $\alpha \in \mathscr{P}$ and u is a fixed control policy from an admissible set $\mathscr{U} \subset L_\infty(I, R^m)$. Let $x(\alpha) = \{x(t, \alpha), t \in I\}$ denote the solution trajectory corresponding to $\alpha \in \mathscr{P}$. Let $\{\alpha_k, \alpha_0\} \subset \mathscr{P}$ and suppose that whenever α_k converges to α_0, denoted $\alpha_k \to \alpha_0$ as $k \to \infty$,

$$A_k(\cdot) \equiv A(\cdot, \alpha_k) \xrightarrow{s} A(\cdot, \alpha_0) \equiv A_0(\cdot) \qquad \text{(in the mean) in } L_1(I, \mathscr{L}(R^n))$$

$$B_k(\cdot) \equiv B(\cdot, \alpha_k) \xrightarrow{s} B(\cdot, \alpha_0) \equiv B_0(\cdot) \qquad \text{in } L_1(I, \mathscr{L}(R^m, R^n)).$$

Then it follows from Theorem 2.3.4 that the corresponding sequence of solutions

$$x_k \equiv x(\cdot, \alpha_k) \to x(\cdot, \alpha_0) \equiv x_0 \qquad \text{in } C(I, R^n).$$

On the other hand if

$$A_k \xrightarrow{w} A_0 \quad \text{(weakly)} \qquad \text{in } L_1(I, \mathscr{L}(R^n))$$

$$B_k \xrightarrow{w} B_0 \quad \text{(weakly)} \qquad \text{in } L_1(I, \mathscr{L}(R^m, R^n))$$

then $x_k \to x_0$ pointwise.

In any case, under the given assumptions, the solution $x(\alpha)$ is continuously dependent on $\alpha \in \mathscr{P}$. For sensitivity studies we need more than just continuity, in fact, we need differentiability of the map $\alpha \to x(\alpha)$.

The following discussion is general and applies to both linear and non-linear problems.

If $x(\alpha_0) \equiv x_0$ denotes the solution corresponding to the nominal value $\alpha_0 \in \mathscr{P}$, then for any other $\alpha \in \mathscr{P}$, we can express $x(\alpha)$ in terms of $x(\alpha_0)$ and its Gâteaux differentials along the line segment joining α_0 and α. Suppose $\alpha_0 + \theta(\alpha - \alpha_0) \in \mathscr{P}$ for all $0 \leqslant \theta \leqslant 1$, then

$$\frac{d}{d\theta} x(\alpha_0 + \theta(\alpha - \alpha_0)) = x'(\alpha_0 + \theta(\alpha - \alpha_0), \alpha - \alpha_0) \tag{2.3.36}$$

where x' denotes the Gâteaux differential of x at $\alpha_0 + \theta(\alpha - \alpha_0)$ in the direction $(\alpha - \alpha_0)$. Integrating the above equation over $[0, 1]$ we have,

$$x(\alpha) = x(\alpha_0) + \int_0^1 x'(\alpha_0 + \theta(\alpha - \alpha_0), \alpha - \alpha_0)\, d\theta \tag{2.3.37}$$

where $x(\alpha), x(\alpha_0) \in C(I, R^n)$. If the Gâteaux differential is linear, then one can write

$$x(\alpha) = x(\alpha_0) + \int_0^1 S(\alpha_0 + \theta(\alpha - \alpha_0)) \cdot (\alpha - \alpha_0)\, d\theta \tag{2.3.38}$$

where, for the fixed arguments, $S(\alpha_0 + \theta(\alpha - \alpha_0))$ is a linear operator from $\mathscr{P}$ to $C(I, R^n)$. The operator $S(\alpha_0 + \theta(\alpha - \alpha_0))$ is called the sensitivity operator parametrized by points on the line segment joining α_0 and α. If α is sufficiently close to α_0 and the operator S is continuous in the neighbourhood of α_0, then, to a first approximation,

$$x(\alpha) \cong x(\alpha_0) + S(\alpha_0) \cdot (\alpha - \alpha_0). \tag{2.3.39}$$

In engineering literature, it is the operator $S(\alpha_0)$ which is known as the sensitivity operator. Note that (2.3.39) is only an approximation of the exact expression (2.3.38).

To each point $\alpha \in \mathscr{P}$ one can assign a measure of sensitivity of the system through the norm of the operator $S(\alpha)$, that is,

$$\nu(\alpha) \equiv \|S(\alpha)\| = \sup\{\|S(\alpha)\xi\|_{C(I,R^n)}, \|\xi\|_{\mathscr{P}} \leq 1\}. \tag{2.3.40}$$

If the solution $x(\alpha)$ is continuously Gâteaux differentiable on $\mathscr{P}$ then $\nu(\alpha)$ is also continuous on $\mathscr{P}$ and hence a system designer can select, if he has a choice, a parameter $\alpha^* \in \mathscr{P}$ at which the system is least sensitive to parameter variation, that is, $\nu(\alpha^*) \leq \nu(\alpha)$ for $\alpha \in \mathscr{P}$. At this point optimal control theory can be useful.

For the linear system (2.3.35) with a fixed control $u \in \mathscr{U}$ we can find the sensitivity operator as follows. Define

$$A_\varepsilon(t) \equiv A(t, \alpha_0 + \varepsilon(\alpha - \alpha_0)), \qquad A_0(t) \equiv A(t, \alpha_0)$$

$$B_\varepsilon(t) \equiv B(t, \alpha_0 + \varepsilon(\alpha - \alpha_0)), \qquad B_0(t) \equiv B(t, \alpha_0).$$

Theorem 2.3.12 (Sensitivity operator). *Suppose there exist*

$$A' \in L_1(I, \mathscr{L}(R^n)), \qquad B' \in L_1(I, \mathscr{L}(R^m, R^n))$$

such that as $\varepsilon \to 0$,

$$\frac{A_\varepsilon - A_0}{\varepsilon} \xrightarrow{w} A', \qquad A'(t) \equiv A'(t, \alpha_0, \alpha - \alpha_0)$$

and (2.3.41)

$$\frac{B_\varepsilon - B_0}{\varepsilon} \xrightarrow{w} B', \qquad B'(t) \equiv B'(t, \alpha_0, \alpha - \alpha_0).$$

Then the Gâteaux differential of $x(\alpha)$ *at* α_0 *in the direction* $(\alpha - \alpha_0)$ *exists and is given by the solution of the differential equation*

$$\begin{aligned} \dot{y} &= A_0(t)y + A'(t)x_0(t) + B'(t)u \\ y(0) &= 0. \end{aligned} \tag{2.3.42}$$

The sensitivity operator is given by the integral (Volterra) operator $y \equiv S(\alpha_0, \alpha - \alpha_0)$ *with*

$$y(t) \equiv \int_0^t K_0(t, \tau; \alpha(\tau) - \alpha_0(\tau))\, \mathrm{d}\tau, \qquad t \in I, \tag{2.3.43}$$

which is generally nonlinear. It is linear if the Gâteaux differentials A' *and* B' *are linear in* $(\alpha - \alpha_0)$ *and in that case*

$$y(t) \equiv \int_0^t K_0(t, \tau)(\alpha(\tau) - \alpha_0(\tau))\, \mathrm{d}\tau, \qquad t \in I. \tag{2.3.44}$$

where K_0 *is a suitable kernel given in the proof.*

Proof Let $x_0 = x(\alpha_0)$ and $x_\varepsilon \equiv x(\alpha_0 + \varepsilon(\alpha - \alpha_0)) \in C(I, R^n)$ denote the solutions of (2.3.35) corresponding to the parameters indicated. Then

$$\begin{aligned} \frac{x_\varepsilon(t) - x_0(t)}{\varepsilon} &= \int_0^t A_\varepsilon(\theta)\Big(\frac{x_\varepsilon(\theta) - x_0(\theta)}{\varepsilon}\Big)\, \mathrm{d}\theta \\ &\quad + \int_0^t \Big(\frac{A_\varepsilon(\theta) - A_0(\theta)}{\varepsilon}\Big)x_0(\theta)\, \mathrm{d}\theta \\ &\quad + \int_0^t \Big(\frac{B_\varepsilon(\theta) - B_0(\theta)}{\varepsilon}\Big)u(\theta)\, \mathrm{d}\theta, \qquad t \in I. \end{aligned} \tag{2.3.45}$$

Since $x_0 \in C(I, R^n)$ and $u \in \mathscr{U} \subset L_\infty(I, R^m)$ it follows from (2.3.41), (2.3.45) and the definition of Gâteaux differentials that all the terms in (2.3.45) have limits as $\varepsilon \downarrow 0$. Hence, the Gâteaux differential of x at α_0 in the direction $(\alpha - \alpha_0)$ is given by

$$\begin{aligned} x'(t, \alpha_0; \alpha - \alpha_0) &= \int_0^t A_0(\theta)x'(\theta, \alpha_0; \alpha - \alpha_0)\, \mathrm{d}\theta \\ &\quad + \int_0^t [A'(\theta)x_0(\theta) + B'(\theta)u(\theta)]\, \mathrm{d}\theta \end{aligned} \tag{2.3.46}$$

for all $t \in I$. Hence x' satisfies the differential equation (2.3.42) and it has the unique solution given by

$$\begin{aligned} y(t) &\equiv x'(t, \alpha_0; \alpha - \alpha_0) \\ &= \int_0^t \Phi_0(t, \tau)[A'(\tau, \alpha_0; \alpha - \alpha_0)x_0(\tau) + B'(\tau, \alpha_0; \alpha - \alpha_0)u(\tau)]\, \mathrm{d}\tau, \end{aligned}$$

(2.3.47)(a)

where Φ_0 is the transition operator corresponding to A_0. Comparing (2.3.43) with (2.3.47)(a) one can immediately identify the kernel $K_0(t, \tau; \eta)$. If A' and B' are linear Gâteaux differentials of A and B respectively, then x' is linear and hence

$$y(t) = S(\alpha_0)(\alpha - \alpha_0)(t)$$

$$\equiv \int_0^t \Phi_0(t, \tau) \sum_{i=1}^{p} (\alpha - \alpha_0)_i [A_i'(\tau, \alpha_0)x_0(\tau) + B_i'(\tau, \alpha_0)u(\tau)]\, d\tau$$

$$\equiv \int_0^t K_0(t, \tau) \cdot (\alpha - \alpha_0)(\tau)\, d\tau, \qquad t \in I, \tag{2.3.47)(b}$$

where $K_0(t, \tau)$ is easily identified from the above expression. ■

Remark 2.3.13 Note that the kernel $K_0(t, \tau)$, $0 \leqslant \tau \leqslant t \leqslant T < \infty$, depends on α_0, x_0, u, and A' and B' evaluated at α_0. Hence the dependence of K_0 reduces to that of the given u and α_0 under consideration.

In case $\mathcal{P} \subset L_\infty(I, R^p)$, the sensitivity operator is an integral (Volterra) operator; and in case $\mathcal{P} \subset R^p$, the operator $S(\alpha_0)$ reduces to a simple matrix which can be identified from

$$y(t) \equiv \left[\int_0^t K_0(t, \tau)\, d\tau\right](\alpha - \alpha_0). \tag{2.3.48}$$

The norm of the operator is given by

$$\nu(\alpha_0) = \|S(\alpha_0)\| \equiv \begin{cases} \sup_{t \in I} \int_0^t \|K_0(t, \tau)\|\, d\tau, & \alpha \text{ variable} \\ \sup_{t \in I} \left\|\int_0^t K_0(t, \tau)\, d\tau\right\|, & \text{for } \alpha \text{ constant.} \end{cases}$$

For complete insensitivity of the system to parameter variation one looks at the properties of the integral operator (2.3.47) (see Problems 2.3.P17, 2.3.P18).

2.4 Input–output stability (BIBO)

In this section we consider the question of input–output stability of linear systems. Loosely speaking a system is said to be input–output stable if bounded input produces bounded output (BIBO).

Consider the linear system

$$\begin{aligned} \dot{x} &= A(t)x + B(t)u, \qquad t \geqslant 0 \\ x(0) &= x_0 \end{aligned} \tag{2.4.1}$$

with the output

$$y(t) = H(t)x(t), \qquad t \geq 0, \tag{2.4.2}$$

where A, B and H are $(n \times n)$, $(n \times m)$ and $(r \times n)$ matrix-valued functions. Using the transition operator Φ, we can write

$$y(t) = h(t) + \int_0^t K(t, \theta)u(\theta)\,\mathrm{d}\theta, \qquad t \geq 0 \tag{2.4.3}$$

where

$$\begin{aligned} h(t) &\equiv H(t)\Phi(t, 0)x_0 \\ K(t, \theta) &\equiv H(t)\Phi(t, \theta)B(\theta). \end{aligned} \tag{2.4.4}$$

Clearly h takes values in R^r and K takes values from $\mathscr{L}(R^m, R^r)$. For distinction of dimensions we shall write $|\xi|_r$ and $\|M\|_{r,m}$ for the norms in R^r and $\mathscr{L}(R^m, R^r)$ respectively.

In the following theorem we shall present sufficient conditions that guarantee input–output stability in the L_p sense, $1 \leq p \leq \infty$.

Theorem 2.4.1 *Consider the evolution of the system* (2.4.1)–(2.4.2) *over the time horizon* $I = [0, T]$, $0 < T < \infty$, *and suppose there exist finite positive numbers* c_1 *and* c_2 *such that*

$$\begin{aligned} &\text{(a)} \quad \operatorname*{ess\,sup}_{t \in I} \int_0^t \|K(t, \theta)\|_{r,m}\,\mathrm{d}\theta = c_1 < \infty. \\ &\text{(b)} \quad \operatorname*{ess\,sup}_{\theta \in I} \int_\theta^T \|K(t, \theta)\|_{r,m}\,\mathrm{d}t = c_2 < \infty. \end{aligned} \tag{2.4.5}$$

Then, whenever $h \in L_p(I, R^r)$ *and* $u \in L_p(I, R^m)$, *the output* $y \in L_p(I, R^r)$ *for all* $1 \leq p \leq \infty$.

Proof We write (2.4.3) in the more compact form

$$y = h + Lu, \tag{2.4.6}$$

where L represents the integral operator. First we consider $p = \infty$, that is, $h \in L_\infty(I, R^r)$ and the input $u \in L_\infty(I, R^m)$. We must show that $y \in L_\infty(I, R^r)$. Since we are dealing with the linear space L_∞, it suffices to show that $Lu \in L_\infty(I, R^r)$. Clearly,

$$\begin{aligned} |(Lu)(t)|_r &\leq \int_0^t \|K(t, \theta)\|_{r,m}\,|u(\theta)|_m\,\mathrm{d}\theta \\ &\leq \left\{\operatorname*{ess\,sup}_{\theta \in I} |u(\theta)|_m\right\} \int_0^t \|K(t, \theta)\|_{r,m}\,\mathrm{d}\theta. \end{aligned}$$

Hence,

$$\left\{\operatorname*{ess\,sup}_{t\in I} |(Lu)(t)|_r\right\} \leq \left\{\operatorname*{ess\,sup}_{\theta\in I} |u(\theta)|_m\right\}\left\{\operatorname*{ess\,sup}_{t\in I} \int_0^t \|K(t,\theta)\|_{r,m}\,\mathrm{d}\theta\right\}$$

and consequently

$$\|Lu\|_{L_\infty(I,R^r)} \leq c_1 \|u\|_{L_\infty(I,R^m)}$$

and

$$\|y\|_{L_\infty(I,R^r)} \leq \|h\|_{L_\infty(I,R^r)} + c_1 \|u\|_{L_\infty(I,R^m)} < \infty.$$

Suppose now that $p=1$ and $h \in L_1(I, R^r)$ and write

$$z(t) = (Lu)(t) \equiv \int_0^t K(t,\theta)u(\theta)\,\mathrm{d}\theta.$$

Clearly

$$\begin{aligned}\int_0^T |z(t)|_r\,\mathrm{d}t &\leq \int_0^T \left(\int_0^t \|K(t,\theta)\|_{r,m} |u(\theta)|_m\,\mathrm{d}\theta\right)\mathrm{d}t \\ &\leq \int_0^T \left(\int_\theta^T \|K(t,\theta)\|_{r,m}\,\mathrm{d}t\right)|u(\theta)|_m\,\mathrm{d}\theta \\ &\leq \int_0^T \left\{\operatorname*{ess\,sup}_{\theta\in I}\int_\theta^T \|K(t,\theta)\|_{r,m}\,\mathrm{d}t\right\}|u(\theta)|_m\,\mathrm{d}\theta \\ &\leq c_2 \int_0^T |u(\theta)|_m\,\mathrm{d}\theta.\end{aligned}$$

That is, $\|z\|_{L_1(I,R^r)} \leq c_2 \|u\|_{L_1(I,R^m)}$, and hence $y \in L_1(I, R^r)$. Now we take the general case, $1<p<\infty$, and show that $z \in L_p(I, R^r)$ whenever $u \in L_p(I, R^m)$. We prove this by showing that for any $\xi \in L_q(I, R^r)$, $(1/p)+(1/q)=1$, $1<p, q<\infty$,

$$\left|\int_I (z(t), \xi(t))\,\mathrm{d}t\right| < \infty.$$

The fact that, under this condition, z must necessarily belong to $L_p(I, R^r)$ is justified later to conclude the proof. Define

$$\eta = \int_0^T (z(t), \xi(t))\,\mathrm{d}t = \int_0^T \left(\int_0^t K(t,\theta)u(\theta)\,\mathrm{d}\theta, \xi(t)\right)\mathrm{d}t, \tag{2.4.7}$$

and

$$c_t(\theta) = \begin{cases} 1, & \text{for } 0 \leq \theta \leq t \\ 0, & \text{for } \theta > t.\end{cases}$$

Then we can write

$$\eta = \int_0^T \int_0^T (c_t(\theta)K(t, \theta)u(\theta), \xi(t))\, d\theta\, dt$$

and hence

$$|\eta| \leqslant \int_0^T \int_0^T c_t(\theta)\, \|K(t, \theta)\|_{r,m}\, |u(\theta)|_m\, |\xi(t)|_r\, d\theta\, dt$$

$$\leqslant \int_0^T \int_0^T (c_t(\theta)\, \|K(t, \theta)\|_{r,m}^{1/p}\, |u(\theta)|_m)(c_t(\theta)\, \|K(t, \theta)\|_{r,m}^{1/q}\, |\xi(t)|_r)\, d\theta\, dt.$$

By Hölder's inequality,

$$|\eta| \leqslant \left(\int_0^T \int_0^T c_t(\theta)\, \|K(t, \theta)\|_{r,m}\, |u(\theta)|_m^p\, d\theta\, dt\right)^{1/p}$$

$$\times \left(\int_0^T \int_0^T c_t(\theta)\, \|K(t, \theta)\|_{r,m}\, |\xi(t)|_r^q\, d\theta\, dt\right)^{1/q}$$

$$\leqslant \left(\int_0^T \left[\int_\theta^T \|K(t, \theta)\|_{r,m}\, dt\right] \cdot |u(\theta)|_m^p\, d\theta\right)^{1/p}$$

$$\times \left(\int_0^T \left[\int_0^t \|K(t, \theta)\|_{r,m}\, d\theta\right] \cdot |\xi(t)|_r^q\, dt\right)^{1/q}.$$

$$\leqslant c_2^{1/p} c_1^{1/q}\, \|u\|_{L_p(I, R^m)} \cdot \|\xi\|_{L_q(I, R^r)}. \tag{2.4.8}$$

Since $u \in L_p(I, R^m)$ and $\xi \in L_q(I, R^r)$, $|\eta| < \infty$.

This being true for arbitrary $\xi \in L_q(I, R^r)$ it follows from (2.4.7) that z must belong to $L_p(I, R^r)$. Hence for $h \in L_p(I, R^r)$ and $u \in L_p(I, R^m)$,

$$y = h + Lu = h + z \in L_p(I, R^r).$$

This will conclude the proof once we justify the fact that $z \in L_p(I, R^r)$ whenever

$$\left|\int_I (z(t), \xi(t))\, dt\right| < \infty \tag{2.4.9}$$

for each $\xi \in L_q(I, R^r)$.

We prove the result for the scalar case and leave the vector case as a simple exercise for the reader. Define, for $z \not\equiv 0$,

$$\xi(t) \equiv \{(\text{sign}\, z(t))\, |z(t)|^{p-1}/\|z\|^{p-1}\}, \qquad t \in I, \tag{2.4.10}$$

and note that $(1/p) + (1/q) = 1$, $1 < p < \infty$, and hence $(p-1)q = p$ and consequently

$$\int_I |\xi(t)|^q\, dt = 1.$$

Thus ξ, as defined by (2.4.10), belongs to the unit sphere in L_q and for this ξ

$$\int_I z(t)\xi(t)\,dt = \left(\int_I |z(t)|^p\,dt/\|z\|_{L_p}^{p-1}\right) = \|z\|_{L_p}. \tag{2.4.11}$$

Hence, if $z \notin L_p$, then $\|z\|_{L_p} = \infty$ and for the given ξ, with $\|\xi\|_{L_q} = 1$, we will have $\int_I z(t)\xi(t)\,dt = \infty$ which contradicts our hypothesis that this integral is finite for each $\xi \in L_q$. Hence z must belong to L_p. ■

The above result tells us that L_p input results in L_p output for all $1 \leqslant p \leqslant \infty$, and the system is stable in this sense and the operator L is a bounded linear operator from $L_p(I, R^m)$ to $L_p(I, R^r)$.

Remark 2.4.2 If $t \to H(t)$ is continuous, then $t \to y(t) = H(t)x(t)$ is also continuous. Therefore, if $T = \infty$ and $h \in L_p(I, R^r)$ and $u \in L_p(I, R^m)$ for $1 \leqslant p < \infty$, then $y(t) \to 0$ as $t \to \infty$.

Remark 2.4.3 For a time-invariant system $K(t, \theta) = K(t - \theta)$, and hence the assumptions (a) and (b) of Theorem 2.4.1 reduce to the condition that

$$\int_0^T \|K(t)\|_{r,m}\,dt < \infty.$$

Algebra of linear systems

The major characteristics of the linear system (2.4.1)–(2.4.3) are all reflected in the integral operator

$$(Lu)(t) \equiv \int_0^t K(t, \theta)u(\theta)\,d\theta, \qquad t \in I, \tag{2.4.12}$$

where $L \in \mathscr{L}(L_p(I, R^m), L_p(I, R^r))$. For convenience of notation we shall not distinguish between the operator L and the corresponding kernel K and write

$$(Lu)(t) \equiv (Ku)(t) \equiv \int_0^t K(t, \tau)u(\tau)\,d\tau, \qquad t \in I. \tag{2.4.13}$$

For integers $s, r \geqslant 1$ define

$$V_{s,r} \equiv \{L \in \mathscr{L}(L_p(I, R^r), L_p(I, R^s)) : (Lu)(t) \equiv \int_0^t L(t, \theta)u(\theta)\,d\theta,$$
$$t \in I, \text{ with the kernel } L \text{ satisfying (2.4.5)}$$
$$\text{for some } c_1, c_2 \text{ finite}\}.$$

Clearly $V_{s,r}$ is a linear vector space – in fact, a Banach space of linear integral operators. For integers s, r, $m \geq 1$, consider the vector spaces $V_{s,r}$, $V_{r,m}$, $V_{s,m}$. We can define a generalized *convolution* operation between elements of $V_{s,r}$ and $V_{r,m}$ in that order. Let $K_1 \in V_{s,r}$ and $K_2 \in V_{r,m}$; then define the operator $K \equiv K_1 * K_2$ having the kernel

$$K(t, \theta) = \int_\theta^t K_1(t, \tau)K_2(\tau, \theta)\, d\tau, \qquad 0 \leq \theta \leq t \leq T. \tag{2.4.14}$$

It is easy to verify that the kernel K, and hence the corresponding operator, satisfies the required properties so that $K \in V_{s,m}$. Hence the following algebraic properties hold for all integers s, r, m, $\ell \geq 1$.

$$\begin{aligned} V_{s,r} + V_{s,r} &= V_{s,r} \\ V_{s,r} * V_{r,m} &= V_{s,m} \\ V_{s,r} * (V_{r,m} + V_{r,m}) &= V_{s,r} * V_{r,m} + V_{s,r} * V_{r,m} \\ V_{s,r} * (V_{r,m} * V_{m,\ell}) &= (V_{s,r} * V_{r,m}) * V_{m,\ell}. \end{aligned} \tag{2.4.15}$$

For a fixed integer $r \geq 1$, define $V_r = V_{r,r}$. Then $V_r \equiv \{V_r, +, \cdot, *\}$, with respect to addition between elements, multiplication by scalars and composition defined by (2.4.14), is a non-commutative algebra having a unit J (identity operator). All the following properties can be easily verified.

$$\begin{aligned} &\text{(a)}\quad \alpha K \in V_r \text{ for } \alpha \in R,\ K \in V_r \\ &\text{(b)}\quad J * K = K * J = K \text{ for all } K \in V_r \\ &\text{(c)}\quad K_1 + K_2 \in V_r \text{ for all } K_1, K_2 \in V_r \\ &\text{(d)}\quad K_1 * K_2,\ K_2 * K_1 \in V_r \text{ for all } K_1, K_2 \in V_r \\ &\text{(e)}\quad K_1 * (K_2 + K_3) = K_1 * K_2 + K_1 * K_3 \text{ for all } K_i \in V_r \\ &\text{(f)}\quad K_1 * (K_2 * K_3) = (K_1 * K_2) * K_3 \text{ for all } K_i \in V_r. \end{aligned} \tag{2.4.16}$$

Under certain conditions we can also define feedback operation in V_r. Consider the feedback system (see figure 2.1)

$$y = K_1 e$$

$$e = w + K_2 y,$$

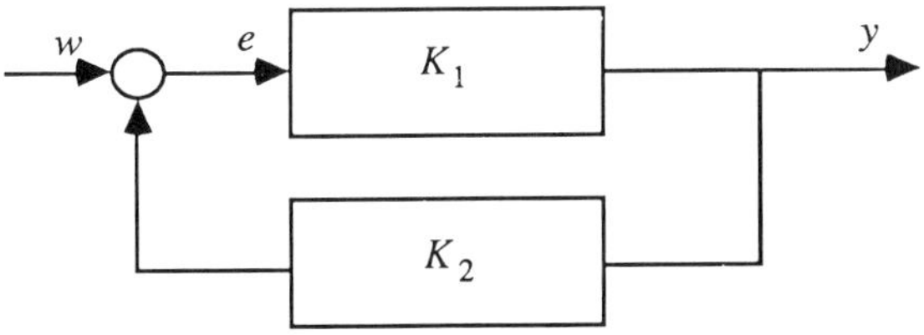

Figure 2.1

where K_1, $K_2 \in V_r$, y is the output, w is the input. From these equations we obtain an integral equation for e, given by,

$$e = w + K_2 * K_1 e = w + Re, \tag{2.4.17}$$

that is

$$e(t) = w(t) + \int_0^t R(t, \theta)e(\theta)\,\mathrm{d}\theta, \qquad t \in [0, T]$$

where

$$R(t, \theta) = \int_\theta^t K_2(t, \tau)K_1(\tau, \theta)\,\mathrm{d}\tau.$$

The vector $e = (J - R)^{-1}w$, provided the operator $(J - R)^{-1}$ exists. In the following section we shall see that, under certain conditions, $(J - R)^{-1} \in V_r$ also.

2.5 Linear integral equations arising from output feedback

In many engineering applications it is absolutely essential to provide feedback controls for automatic regulation of the system. In particular, the control is to be generated by output feedback, that is, available information. A typical model for such a system is given by

$$\begin{aligned} \dot{x} &= A(t)x + B(t)u \\ y &= H(t)x \\ u &= F(t)y + v \end{aligned} \tag{2.5.1}$$

where A, B, H, F are of appropriate dimension and v is the external input. Hence, the closed loop system is given by

$$\dot{x} = (A(t) + B(t)F(t)H(t))x(t) + B(t)v. \tag{2.5.2}$$

Thus, if the elements of A and BFH satisfy the requirements of Theorem 2.2.2, then there exists a closed-loop transition operator Φ_c which can be obtained by solving a system of integral equations as described below.

Let Φ denote the open-loop transition operator; then setting the external input $v \equiv 0$, we have

$$x(t) = \Phi(t, 0)x_0 + \int_0^t \Phi(t, \theta)B(\theta)F(\theta)H(\theta)x(\theta)\,\mathrm{d}\theta \tag{2.5.3}$$

and, in general, for all $\tau \geqslant 0$ and $\xi \in R^n$,

$$x(t) = \Phi(t, \tau)\xi + \int_\tau^t \Phi(t, \theta)B(\theta)F(\theta)H(\theta)x(\theta)\,\mathrm{d}\theta, \tag{2.5.4}$$

is the solution of (2.5.1) corresponding to $v \equiv 0$ and $x(\tau) = \xi$. Given that the elements of $(A + BFH)$ are locally integrable, it follows from Theorem 2.2.2 that there exists a transition operator Φ_c such that

$$x(t) = \Phi_c(t, \tau)\xi \qquad \text{for } t \geq \tau \geq 0. \tag{2.5.5}$$

Substituting this equation into (2.5.4) we have

$$\Phi_c(t, \tau)\xi = \Phi(t, \tau)\xi + \int_\tau^t \Phi(t, \theta)B(\theta)F(\theta)H(\theta)\Phi_c(\theta, \tau)\xi \, d\theta. \tag{2.5.6}$$

Since $\xi \in R^n$ is arbitrary, we have the integral equation

$$\Phi_c(t, \tau) = \Phi(t, \tau) + \int_\tau^t K(t, \theta)\Phi_c(\theta, \tau) \, d\theta \tag{2.5.7}$$

where

$$K(t, \theta) \equiv \Phi(t, \theta)B(\theta)F(\theta)H(\theta), \tag{2.5.8}$$

and $0 \leq \tau \leq t \leq T < \infty$ and we must solve for Φ_c. Using the compact notation

$$\Phi_c = \Phi + K\Phi_c \tag{2.5.9}$$

we see that $(J - K)\Phi_c = \Phi$ and hence

$$\Phi_c = (J - K)^{-1}\Phi \tag{2.5.10}$$

where J stands for the identity operator. The operator $(J - K)^{-1}$ is well defined through the Neumann series

$$\Phi_c = \Phi + \Gamma\Phi, \tag{2.5.11}$$

where Γ is called the resolvent operator with kernel

$$\Gamma(t, \theta) \equiv \sum_{n=1}^{\infty} K^{(n)}(t, \theta),$$

and

$$K^{(1)}(t, \theta) \equiv K(t, \theta)$$

$$K^{(2)}(t, \theta) \equiv \int_\theta^t K(t, \tau)K(\tau, \theta) \, d\tau$$

$$\vdots$$

$$K^{(n)}(t, \theta) \equiv \int_\theta^t K^{(n-1)}(t, \tau)K(\tau, \theta) \, d\tau$$

$$= \int_\theta^t K^{(n-s)}(t, \tau)K^{(s)}(\tau, \theta) \, d\tau$$

for $1 \leqslant s < n$. Note that the series (2.5.11) is easily obtained by successive approximation or equivalently by substitution of (2.5.9) into itself. A simple situation, under which the series converges on any finite interval $[0, T]$, $T < \infty$, is discussed below. Assuming that B, F and H are all bounded measurable matrix-valued functions it is clear that, for $T < \infty$, there exists a number M, possibly dependent on T, such that

$$\|\Phi(t, \theta)\| \leqslant M \quad \text{and} \quad \|K(t, \theta)\| = \|\Phi(t, \theta)B(\theta)F(\theta)H(\theta)\| \leqslant M < \infty, \tag{2.5.12}$$

for $0 \leqslant \theta < t \leqslant T$.

Under the assumption (2.5.12), the reader may easily verify that, for $0 \leqslant \theta \leqslant t \leqslant T$,

$$\left.\begin{aligned} &\|K^{(2)}(t, \theta)\| \leqslant M^2(t-\theta) \\ &\|K^{(3)}(t, \theta)\| \leqslant M^3(t-\theta)^2/2! \\ &\text{and} \\ &\|K^{(n)}(t, \theta)\| \leqslant M^n(t-\theta)^{n-1}/(n-1)!. \end{aligned}\right\} \tag{2.5.13}$$

Hence, for $0 \leqslant \theta \leqslant t \leqslant T$,

$$\|\Gamma(t, \theta)\| \leqslant M \exp M(t-\theta). \tag{2.5.14}$$

This proves the convergence of the series (2.5.11) and hence,

$$\begin{aligned} \|\Phi_c(t, \theta)\| &\leqslant \|\Phi(t, \theta)\| + \int_\theta^t \|\Gamma(t, s)\Phi(s, \theta)\| \, \mathrm{d}s \\ &\leqslant \|\Phi(t, \theta)\| + \int_\theta^t \|\Gamma(t, s)\| \, \|\Phi(s, \theta)\| \, \mathrm{d}s \\ &\leqslant M \, \mathrm{e}^{M(t-\theta)}, \qquad 0 \leqslant \theta \leqslant t \leqslant T. \end{aligned} \tag{2.5.15}$$

Thus, the solution Φ_c given by (2.5.11) is well defined on any finite interval, and for $v \equiv 0$,

$$x(t) = \Phi_c(t, 0)x_0 \tag{2.5.16}$$

and for arbitrary v we have,

$$x(t) = \Phi_c(t, 0)x_0 + \int_0^t \Phi_c(t, \theta)B(\theta)v(\theta) \, \mathrm{d}\theta, \tag{2.5.17}$$

which has similar form as the open-loop system. The stability of the closed-loop system depends on that of $\Phi_c(t, \theta)$. Note, however, that Φ_c is always norm bounded on bounded intervals and the system is always finite-time stable.

Integral equations also arise in certain tracking problems. Consider the open-loop system

$$\dot{x} = A(t)x + B(t)u, \qquad t \geqslant 0$$
$$x(0) = x_0, \tag{2.5.18}$$

let z be any absolutely continuous R^n-valued function, and suppose there is a $t_0 > 0$ such that $x(t_0) = z(t_0)$. The problem is to find a control u^*, from time $t \geqslant t_0$, such that $z(t) = x(t)$ for all $t \in [t_0, t_1]$. This is tracking, and it requires that

$$z(t) = x^*(t) = \Phi(t, t_0)z(t_0) + \int_{t_0}^{t} \Phi(t, \theta)B(\theta)u^*(\theta)\,\mathrm{d}\theta \tag{2.5.19}$$

for $t \in [t_0, t_1]$. Defining

$$w(t) \equiv z(t) - \Phi(t, t_0)z(t_0)$$

which is known, we have

$$w(t) = \int_{t_0}^{t} \Phi(t, \theta)B(\theta)u^*(\theta)\,\mathrm{d}\theta, \qquad t \in [t_0, t_1]. \tag{2.5.20}$$

Here u^* is unknown. The question of existence of u^* is crucial. Here we have a Volterra integral equation of the first kind which is always difficult to solve. Since w is absolutely continuous we can differentiate (2.5.20) to obtain the following integral equation for $t \in (t_0, t_1)$,

$$\dot{w}(t) = B(t)u^*(t) + \int_{t_0}^{t} A(t)\Phi(t, \theta)B(\theta)u^*(\theta)\,\mathrm{d}\theta. \tag{2.5.21}$$

Writing

$$\dot{w}(t) \equiv \xi(t)$$
$$B(t)u^*(t) = v^*(t)$$

we have

$$\xi(t) = v^*(t) + \int_{t_0}^{t} A(t)\Phi(t, \theta)v^*(\theta)\,\mathrm{d}\theta \tag{2.5.22}$$

which is an integral equation of the second kind and one can solve for v^* using the corresponding Neumann series. Once v^* is obtained, one looks for a solution of the algebraic equation

$$B(t)u = v^*(t). \tag{2.5.23}$$

Generally $B(t)$ is not invertible since, usually, it is a rectangular matrix and hence the problem may have one solution, infinitely many solutions or no solution at all. We shall discuss this question further in chapter 5.

Exercises

2.1.P1 Prove that in any metric space $(X, \rho) = X_\rho$

$$\rho(x, z) \geqslant |\rho(x, y) - \rho(y, z)| \qquad \text{for } x, y, z \in X_\rho.$$

2.1.P2 Verify that (X, ρ_H) of Example 2.1.12 is a linear metric space over $F = \{0, 1\}$.

2.1.P3 Show that a normed vector space is a linear metric space.

2.1.P4 Verify the parallelogram identity (see section 2.1.5) for L_2 space.

2.1.P5 Prove that for $n < \infty$

$$\left(\sum_{i=1}^{n} |x_i|^2\right)^{1/2} \leqslant \sum_{i=1}^{n} |x_i| \leqslant n^{1/2}\left(\sum_{i=1}^{n} |x_i|^2\right)^{1/2}$$

and hence the two normed spaces ℓ_2^n and ℓ_1^n are equivalent ($\ell_2^n \cong \ell_1^n$).

2.1.P6 Prove that $\ell_p^n \cong \ell_r^n$ for all $\infty \geqslant p, r \geqslant 1$.

2.1.P7 Prove that for any $\Omega \subset R^n$ with finite Lebesgue measure (think of volume measure) $L_p(\Omega) \subset L_r(\Omega)$ whenever $p > r \geqslant 1$. (*Hint*: Use Hölder's inequality.)

2.1.P8 Prove that a function f mapping (X, ρ) to (Y, σ) is continuous if the inverse image of every open set in (Y, σ) is an open set in (X, ρ).

2.1.P9 Verify that

$$\lim_n \int g_n(t)\, dt \neq \int \lim_n g_n(t)\, dt$$

for the following examples:

(a) $g_n(t) \equiv nC_{[0, 1/n]}(t)$, $t \in [0, 1]$

where

$$C_E(t) = \begin{cases} 1, & t \in E \\ 0, & \text{elsewhere.} \end{cases}$$

(b) $g_n(t) \equiv \exp(-(t-n)^2)$, $t \in (-\infty, +\infty)$.

Find why the Lebesgue dominated convergence theorem (Proposition 2.1.32) does not apply.

2.2.P10 Prove part (c) of Theorem 2.2.2.

2.2.P11 Show that

$$\frac{\partial}{\partial s}\Phi(t, s) = -\Phi(t, s)A(s) \qquad \text{for } 0 \leqslant s < t < \infty.$$

2.2.P12 Verify that the solution of equation (2.2.44) has the property,

$$\det X(t) = \det X(0) \exp \int_0^t \operatorname{trace} A(\theta)\, d\theta, \qquad t \geq 0,$$

and conclude that $X(t)$ is non-singular.

2.2.P13 The attitude dynamics of a spherical satellite could be described by the following second-order differential equation

$$\ddot{\theta} = u,$$

where θ denotes the attitude angle, and u the control torque. Describe the system in the state-space form and compute the state-transition operator $\Phi(t, \tau)$.

2.2.P14 Consider the time-varying system

$$\begin{bmatrix} \dot{x}_1 \\ \dot{x}_2 \end{bmatrix} = \begin{bmatrix} 0 & 0 \\ 2t & 0 \end{bmatrix} \begin{bmatrix} x_1 \\ x_2 \end{bmatrix} + \begin{bmatrix} 1 \\ 0 \end{bmatrix} u.$$

Show that the state transition operator of the system is given by

$$\phi(t, \tau) = \begin{bmatrix} 1 & 0 \\ t^2 - \tau^2 & 1 \end{bmatrix}.$$

(*Hint*: Use the property (2.2.42)(c) of the state-transition operator.)

2.2.P15 Prove that the series (2.2.52) converges uniformly on any finite interval $I \subset R$.

2.3.P16 Consider the system

$$\dot{x} = A(t)x + B(t)u, \qquad t \geq 0$$

$$x(0) = x_0$$

and suppose that the set of admissible controls is given by $\mathcal{U} \equiv L_2([0, T], R^m)$ and the entries of B are locally square integrable and those of A are locally integrable. Prove that there is a positive number $b < \infty$ such that for all $u_1, u_2 \in \mathcal{U}$

$$\|x(u_1) - x(u_2)\| \equiv \left\{ \sup_{t \in [0,T]} |x(t, x_0, u_1) - x(t, x_0, u_2)| \right\}$$

$$\leq b\, \|u_1 - u_2\|_{L_2([0,T];R^m)}.$$

2.3.P17 For nonzero kernel K_0 of the sensitivity operator $S(\alpha_0)$ (see equation (2.3.47)) is it possible for the system to be completely insensitive to parameter variation?

2.3.P18 Find conditions (except the trivial case) for complete insensitivity to parameter variation of the system (2.3.35). (*Hint*: use (2.3.47).)

2.4.P19 Justify Remarks 2.4.2, 2.4.3.

2.4.P20 For any positive integer k and real number $r > 0$, let

$$B_r(R^k) \equiv \{\xi \in R^k : |\xi| \leq r\}$$

denote the sphere of radius r. Consider the controlled system

$$\dot{x} = A(t)x + B(t)u, \quad t \geq 0$$

$$x(0) = x_0$$

$$u(t) \in B_\beta(R^m), \qquad 0 < \beta < \infty,$$

and suppose the elements of A and B are locally integrable. For $T < \infty$, give an estimate of the radius r of the sphere $B_r(R^n)$ such that $x(t) \in B_r(R^n)$ for all $t \in [0, T]$.

2.4.P21 Verify the relations (2.4.15), (2.4.16) and give an expression for the identity operator J.

2.5.P22 Using equation (2.5.9) and successive approximation prove the validity of (2.5.11).

2.5.P23 Following similar procedure as in 2.5P22 construct the solution for (2.5.22).

3

Nonlinear systems

3.0 Introduction

In the previous chapter we have considered linear systems. In many practical problems linearity is an exception rather than a rule. More often we have to deal with systems which are nonlinear, as seen in chapter 1.

In this chapter, again, we address ourselves to the fundamental questions of existence and uniqueness of solutions for nonlinear systems. After we have settled these problems, we consider the questions of continuous dependence of solutions on system parameters and inputs or controls. We also consider briefly the question of sensitivity. In the final section we consider some nonlinear integral equations that arise naturally from output feedback and study the questions of L_p stability as in chapter 2. The results presented here are of fundamental importance and used throughout the rest of the book.

3.1 Existence and uniqueness of solutions of nonlinear systems of differential equations

The system we consider is governed by the nonlinear differential equation in R^n,

$$\dot{x} = f(t, x), \tag{3.1.1}$$

where f is a nonlinear function mapping $I \times R^n$ to R^n, that is, $f: I \times R^n \to R^n$, and x is an n-vector-valued function defined on I, that is, $x: I \to R^n$, where I is any interval of the real line R. This model is sufficiently general to cover a wide range of physical systems.

There are several techniques for resolving the questions of existence (and uniqueness) of solutions of system (3.1.1). One of the most popular methods is the Picard approximation technique as discussed in section

2.2. A more general method is the so-called Banach fixed-point theorem which proves both existence and uniqueness results. Another more powerful method is provided by the Schauder fixed-point theorem which gives results on existence only (without uniqueness). There are other more powerful topological techniques which are beyond the scope of this book. However, the methods presented here cover a wide range of problems. In this section we shall discuss both the Banach and the Schauder fixed-point theorems to resolve the questions of existence of solutions of system (3.1.1). We shall find these methods also useful in resolving similar questions for stochastic systems in chapter 7.

3.1.1 Existence and uniqueness of solutions based on the Banach fixed-point theorem

Let X be a Banach space and T a continuous bounded (not necessarily linear) operator in X, that is, $T: X \to X$ and whenever $x_n \to x_0$ in X, $Tx_n \to Tx_0$ in X and for every bounded set $B \subset X$, $T(B) \equiv TB \equiv \{Tx, x \in B\}$ is a bounded subset of X. For convenience of notation we shall use the symbol $C_b(X)$ to denote the class of all continuous and bounded operators in X. The reader may think of X as being $C(I, R^n)$, $L_p(I, R^n)$, $L_p(\Omega)$, etc.

Definition 3.1.1 (Fixed point). An element $x \in X$ is said to be a fixed point of the operator T if $x = Tx$, that is, T leaves x invariant.

The concept of contraction mapping plays a fundamental role in the Banach fixed-point theorem.

Definition 3.1.2 (Contraction mapping). Let $T \in C_b(X)$ and X_0 a closed subset of X and suppose there is a constant $\rho \in [0, 1)$ such that

$$\|Tx - Ty\|_X \leq \rho \, \|x - y\|_X \qquad \text{for all } x, y \in X_0. \tag{3.1.2}$$

Then T is said to be a contraction operator on X_0.

The reader may note that the question of existence of a solution of the equation $y = Ty$ in X is equivalent to the question of existence of a fixed point of T in X.

With this background we can now present the Banach fixed-point theorem.

Theorem 3.1.3 (Banach fixed-point theorem). *Let $T \in C_b(X)$ and X_0*

a closed subset of X and suppose T satisfies the following properties:

(a) $TX_0 \subset X_0$ *(that is, X_0 is invariant under T)*
(b) *there exists a $\rho \in (0, 1)$ such that*
$\|Tx - Ty\| \leqslant \rho \|x - y\|$ *for all $x, y \in X_0$.*

Then T has a unique fixed point in X_0.

Proof Take any $x_0 \in X_0$ and define $x_1 = Tx_0$, $x_2 = Tx_1$, $x_3 = Tx_2$ and in general $x_n = Tx_{n-1}$, $n = 1, 2, \ldots, \infty$. Thus we have a sequence $\{x_n\}$ in X, and we show that it has a limit $x^* \in X_0$, and that x^* is a fixed point of T. Note that

$$x_n = Tx_{n-1} = T^n x_0, \tag{3.1.3}$$

where

$$T^n \equiv \underbrace{T \cdot T \cdot T \cdot \cdots \cdot T}_{n}$$

denotes the n-fold composition of the operator T. Then by assumption (b),

$$\begin{aligned}\|x_{n+1} - x_n\| = \|Tx_n - Tx_{n-1}\| &\leqslant \rho \|x_n - x_{n-1}\| \\ &\leqslant \rho^n \|x_1 - x_0\|,\end{aligned} \tag{3.1.4}$$

for any integer $n \geqslant 1$. Thus, for any integer $p \geqslant 1$,

$$\begin{aligned}\|x_{n+p} - x_n\| &= \left\| \sum_{k=n}^{n+p-1} (x_{k+1} - x_k) \right\| \\ &\leqslant \sum_{k=n}^{n+p-1} \|x_{k+1} - x_k\| \leqslant \left(\sum_{k=n}^{n+p-1} \rho^k \right) \|x_1 - x_0\| \\ &\leqslant \rho^n \left(\sum_{k=0}^{p-1} \rho^k \right) \|x_1 - x_0\| \\ &\leqslant (\rho^n/(1-\rho)) \|x_1 - x_0\|.\end{aligned} \tag{3.1.5}$$

Since T is a bounded operator and $x_0 \in X_0 \subset X$, it follows that $\|x_0\| < \infty$ and $\|x_1\| = \|Tx_0\| < \infty$, and consequently $\|x_1 - x_0\| \leqslant \|x_1\| + \|x_0\| < \infty$. Therefore, for ρ satisfying (b), it follows from (3.1.5) that

$$\lim_{n\to\infty} \|x_{n+p} - x_n\| = 0,$$

for any integer $p \geqslant 1$. Hence $\{x_n\}$ is a Cauchy sequence in X. Since X is a Banach space (see section 2.4.1) there exists an $x^* \in X$ such that $x_n \to x^*$ in X. In fact $x^* \in X_0$. Indeed, since X_0 is invariant under T, by assumption (a), and $x_0 \in X_0$, it is clear that the entire sequence $\{x_n\} \subset X_0$. Further, X_0 being closed, the limit point $x^* \in X_0$. We show that x^* is a fixed point of T. Clearly we can write

$$\begin{aligned} \|x^* - Tx^*\| &= \|x^* - x_n + x_n - Tx^*\| \\ &= \|x^* - x_n + Tx_{n-1} - Tx^*\| \\ &\leq \|x^* - x_n\| + \|Tx_{n-1} - Tx^*\| \\ &\leq \|x^* - x_n\| + \rho\, \|x_{n-1} - x^*\|, \end{aligned} \tag{3.1.6}$$

the last inequality following from the facts that $\{x_n\}$, $x^* \in X_0$ and that T is a contraction on X_0 by assumption (b). Since (3.1.6) is true for arbitrary $n \geq 1$, letting $n \to \infty$ we have $\|x^* - Tx^*\| = 0$. Hence $x^* = Tx^*$, proving that x^* is a fixed point of T. We prove uniqueness by contradiction. Suppose y^* is another fixed point of T in X_0, that is, $y^* = Ty^*$. Then

$$\|x^* - y^*\| = \|Tx^* - Ty^*\| \leq \rho\, \|x^* - y^*\|, \tag{3.1.7}$$

and hence $(1 - \rho)\, \|x^* - y^*\| \leq 0$. Since $0 \leq \rho < 1$, this is impossible if $x^* \neq y^*$. This proves uniqueness. The fixed point x^* is independent of the choice of x_0. Indeed let $x'_0 \in X_0$, $x'_0 \neq x_0$, and construct the sequence $x'_n = Tx'_{n-1}$, $n \geq 1$. Again $\{x'_n\}$ is a Cauchy sequence having a limit $x'_* \in X_0$ which is a fixed point of T. But by uniqueness of the fixed point $x'_* = x^*$. This completes the proof. ■

Remark 3.1.4 It is interesting to observe that according to the above result

$$\lim_{n\to\infty} T^n X_0 = \left\{\lim_{n\to\infty} T^n \zeta,\ \zeta \in X_0\right\} = \{x^*\} = x^*, \tag{3.1.8}$$

that is, no matter what the starting point is, one eventually ends up at the fixed point x^*.

As a corollary to Theorem 3.1.3 we can also prove an interesting result which is directly applicable to systems governed by differential equations.

Corollary 3.1.5 *If some power of the operator T is a contraction on $X_0 \subset X$, then T itself has a unique fixed point in X_0.*

Proof Suppose for some integer $n \geq 1$,

$$T^n = \underbrace{T \cdot T \cdot \cdots \cdot T}_{n}$$

is a contraction on X_0. Then by the previous theorem T^n has a unique fixed point, say $x^* \in X_0$. That is, $x^* = T^n x^*$. We show that x^* is a fixed point of T itself. Indeed

$$\begin{aligned} \|x^* - Tx^*\| &= \|T^n x^* - T(T^n x^*)\| \\ &= \|T^n x^* - T^n(Tx^*)\| \\ &\leq \rho_n \|x^* - Tx^*\|, \end{aligned} \tag{3.1.9}$$

where ρ_n is the contraction parameter of the operator T^n, that is,

$$\|T^n x - T^n y\| \leq \rho_n \|x - y\|, \tag{3.1.10}$$

with $0 \leq \rho_n < 1$. Hence $(1-\rho_n)\|x^* - Tx^*\| \leq 0$. Since $(1-\rho_n) > 0$, this implies that $\|x^* - Tx^*\| = 0$ and hence $x^* = Tx^*$, proving that x^* is also a fixed point of T. For uniqueness, let y^* be another fixed point of T. Then we have

$$\begin{aligned} x^* &= Tx^* = T^2 x^* = \cdots = T^n x^* \\ y^* &= Ty^* = T^2 y^* = \cdots = T^n y^*. \end{aligned} \tag{3.1.11}$$

Hence

$$\|x^* - y^*\| = \|T^n x^* - T^n y^*\| \leq \rho_n \|x^* - y^*\|$$

and consequently, by the same argument, $x^* = y^*$. ■

With the help of the above results we can prove, under certain assumptions, the existence and uniqueness of a solution of the initial-value problem (Cauchy problem),

$$\begin{aligned} \dot{x} &= f(t, x), \qquad t \geq 0 \\ x(0) &= x_0. \end{aligned} \tag{3.1.12}$$

As usual, we shall use $|\cdot|$ for norms in finite dimensional spaces, C^k, $k = 0, 1, 2, \ldots$, for the space of k-times continuously differentiable functions, AC for the space of absolutely continuous functions, and L_1^{loc} for the space of measurable functions which are locally integrable.

Theorem 3.1.6 (Existence and uniqueness). *Consider the system (3.1.12) and suppose that, for each fixed but arbitrary $\xi \in R^n$, the function $t \to f(t, \xi)$ is measurable on $R_0 = [0, \infty)$ and for almost all $t \in R_0$, $\xi \to f(t, \xi)$ is continuous on R^n and that there exists a non-negative measurable function $K \in L_1^{\mathrm{loc}}$ such that*

$$\begin{aligned} &\text{(a)} \quad |f(t, \xi)| \leq K(t)[1 + |\xi|] && \textit{for all } \xi \in R^n \\ &\text{(b)} \quad |f(t, \xi) - f(t, \eta)| \leq K(t)|\xi - \eta| && \textit{for all } \xi, \eta \in R^n. \end{aligned} \tag{3.1.13}$$

Then for each $x_0 \in R^n$, *the system* (3.1.12) *has a unique solution* $x^* \in AC$ *such that* $\dot{x}^*(t) = f(t, x^*(t))$ a.e. *and* $x^*(0) = x_0$. *Further, if f is continuous in all the variables, then* $x^* \in C^1$ *and* $\dot{x}^*(t) = f(t, x^*(t))$ *for all* $t \geqslant 0$.

Proof Let $\tau \geqslant 0$ be any finite number and consider the evolution of the system (3.1.12) over the finite interval $I = [0, \tau]$. Let X denote the Banach space $C(I, R^n)$ with the norm topology $\|x\| \equiv \sup\{|x(t)|, t \in I\}$. Define the operator T on X by

$$(Tx)(t) \equiv x_0 + \int_0^t f(\theta, x(\theta))\, d\theta, \qquad t \in I. \tag{3.1.14}$$

We wish to show that T has a unique fixed point in X. For this we must verify that $T \in C_b(X)$ and that some power of T is a contraction on X. If these conditions are met then, on the basis of Corollary 3.1.5, we shall be able to conclude that T has a fixed point in X. For $x \in X \equiv C(I, R^n)$, it follows from (3.1.13)(a) that

$$\begin{aligned} |(Tx)(t)| &\leqslant |x_0| + \int_0^t |f(\theta, x(\theta))|\, d\theta \\ &\leqslant |x_0| + \int_0^t K(\theta)[1 + |x(\theta)|]\, d\theta. \end{aligned}$$

Hence

$$\begin{aligned} \|Tx\| &= \sup\{|(Tx)(t)|, t \in I\} \\ &\leqslant |x_0| + \left(\int_0^\tau K(\theta)\, d\theta\right)(1 + \sup\{|x(\theta)|, \theta \in I\}) \\ &\leqslant \alpha_1 + \alpha_2(1 + \|x\|) \end{aligned} \tag{3.1.15}$$

with $\alpha_1 \equiv |x_0|$ and $\alpha_2 = \int_I K(\theta)\, d\theta$.

Since $x_0 \in R^n$ and $K \in L_1^{\text{loc}}$, α_1 and α_2 are finite and therefore it follows from (3.1.15) that T is a bounded operator on X. Similarly it follows from assumption (b) that, for $x, y \in X$,

$$\|Tx - Ty\| \leqslant \alpha_2 \|x - y\|, \tag{3.1.16}$$

which clearly implies continuity. Thus we have shown that $T \in C_b(X)$. It remains to prove that a certain power of T is a contraction on X. Note that

$$\begin{aligned} (T^2x)(t) &\equiv T((Tx))(t) \\ &= x_0 + \int_0^t f(\theta, (Tx)(\theta))\, d\theta, \qquad t \in I, \end{aligned}$$

and

$$(T^n x)(t) \equiv x_0 + \int_0^t f(\theta, (T^{n-1}x)(\theta))\,\mathrm{d}\theta, \qquad t \in I, \tag{3.1.17}$$

for all $n \geq 1$. By virtue of our assumption (3.1.13)(b), we have

$$|(Tx)(t) - (Ty)(t)| \leq \int_0^t K(\theta)\,|x(\theta) - y(\theta)|\,\mathrm{d}\theta, \qquad t \in I, \tag{3.1.18}$$

for all $x, y \in X$. Defining

$$d \equiv \sup\{|x(t) - y(t)|, t \in I\} \tag{3.1.19}$$

and $h(t) \equiv \int_0^t K(\theta)\,\mathrm{d}\theta$, it follows from (3.1.18) that

$$|(Tx)(t) - (Ty)(t)| \leq dh(t), \qquad t \in I. \tag{3.1.20}$$

Similarly

$$\begin{aligned}|(T^2x)(t) - (T^2y)(t)| &\leq \int_0^t |f(\theta, (Tx)(\theta)) - f(\theta, (Ty)(\theta))|\,\mathrm{d}\theta \\ &\leq \int_0^t K(\theta)\,|(Tx)(\theta) - (Ty)(\theta)|\,\mathrm{d}\theta \\ &\leq d \int_0^t K(\theta)h(\theta)\,\mathrm{d}\theta.\end{aligned}$$

Since $K \in L_1^{\mathrm{loc}}$, h is absolutely continuous and hence

$$\begin{aligned}|(T^2x)(t) - (T^2y)(t)| &\leq d \int_0^t h(\theta)\,\mathrm{d}h(\theta) \\ &\leq dh^2(t)/2, \qquad t \in I,\end{aligned}$$

and

$$|(T^3x)(t) - (T^3y)(t)| \leq dh^3(t)/3!, \qquad t \in I.$$

Repeating this procedure n times one can easily verify that

$$|(T^nx)(t) - (T^ny)(t)| \leq dh^n(t)/n!, \qquad t \in I. \tag{3.1.21}$$

Hence

$$\sup\{|(T^nx)(t) - (T^ny)(t)|, t \in I\} \leq dh^n(\tau)/n!,$$

that is

$$\|T^nx - T^ny\| \leq ((h^n(\tau))/n!)\,\|x - y\|. \tag{3.1.22}$$

Since $K \in L_1^{\mathrm{loc}}$, $h(\tau) < \infty$ and therefore, for sufficiently large n,

$$\rho_n \equiv ((h^n(\tau))/n!) < 1. \tag{3.1.23}$$

Thus T^n is a contraction on X for a sufficiently large positive integer n and hence, by Corollary 3.1.5, T itself has a unique fixed point in X. Note that here we have taken $X_0 = X$. Let x^* denote the fixed point of T, then $x^* = Tx^*$ and

$$x^*(t) = x_0 + \int_0^t f(\theta, x^*(\theta))\, d\theta, \qquad t \in I = [0, \tau]. \tag{3.1.24}$$

Since $x^* \in C(I, R^n)$, the function f^*, with values $f^*(t) \equiv f(t, x^*(t))$, belongs to $L_1(I, R^n)$ and hence, by virtue of Proposition 2.1.31, the integral in (3.1.24) is absolutely continuous and therefore differentiable almost everywhere on I. Consequently $x^* \in AC(I, R^n)$ and $\dot{x}^*(t) = f(t, x^*(t))$ a.e. on I and $x^*(0) = x_0$. Thus x^* is the unique solution of (3.1.12) in the almost everywhere sense.

If f is continuous in all the variables then $t \to f(t, x^*(t))$ is continuous and hence $\dot{x}^*(t) = f(t, x^*(t))$ is continuous for all $t \in I$ and $x^* \in C^1$. Thus we have proved that, on every finite interval, the system (3.1.12) has a unique solution. This completes the proof. ∎

The above result is also valid if the global Lipschitz condition is relaxed to a local one. That is, the hypothesis (3.1.13)(b) is replaced by

(b)′ For each $r > 0$, there exists a $K_r \in L_1^{\text{loc}}$ such that

$$|f(t, x) - f(t, y)| \leq K_r(t)\, |x - y| \qquad \text{for all } x, y \in S_r \equiv \{\xi \in R^n : |\xi| \leq r\}.$$

3.1.2 Existence of solutions based on the Schauder fixed-point theorem

In many practical problems the function f may not satisfy any of the global or local Lipschitz conditions. In that situation the Banach fixed point theorem does not hold. However, there is a more general result known as the Schauder fixed-point theorem. With the help of this theorem we can prove existence results without the Lipschitz condition. But unfortunately, in the absence of some form of Lipschitz condition, the Schauder fixed-point theorem does not guarantee uniqueness. In other words at the cost of uniqueness we have gained generality in the existence result.

In this section we shall prove the Schauder fixed-point theorem in an infinite-dimensional Banach space using the Brouwer fixed-point theorem, which holds for finite-dimensional spaces. Using the Schauder fixed-point theorem, we shall then prove the existence of a solution for the system (3.1.12).

To follow the programme outlined above, we shall need the following concepts.

Definition 3.1.7 (Finite covering). A set K in a metric space $X \equiv (X, \rho)$ is said to have a finite ε-covering if there exists a finite set of points $\{x_1, x_2, \ldots, x_m\}$ in K with $m \equiv m(\varepsilon)$ depending on ε such that

$$K \subset \bigcup_{i=1}^{m} N_\varepsilon(x_i)$$

where

$$N_\varepsilon(x_i) \equiv \{x \in X: \rho(x, x_i) < \varepsilon\}.$$

Using the above definition we can now introduce the concept of compactness which plays a fundamental role also in stability and control theory.

Definition 3.1.8 (Compactness). A set K in a metric space $X = (X, \rho)$ is said to be compact if, for every $\varepsilon > 0$, K has a finite ε-covering. A set K is relatively compact if its closure, $\bar{K}$, is compact.

This definition also implies that, for a compact set K, every sequence $\{x_n\} \subset K$ has a subsequence $\{x_{n_m}, m = 1, 2, \ldots\} \subset \{x_n\}$ that converges to a point x^* of K. Hence an equivalent definition of compactness is: A set $K \subset X$ is compact if every infinite sequence $\{x_n\}$ of K has a subsequence that has a limit in K. As examples, we note that a closed bounded set in an n-dimensional space is always compact while a closed bounded set in any of the Banach spaces $C(I, R^n)$, ℓ_p, $L_p(I, R^n)$, $1 \leqslant p \leqslant \infty$, is not necessarily compact.

Definition 3.1.9 (Compact operators). An operator T, mapping a Banach space X into a Banach space Y, is said to be compact or completely continuous if (a) it is continuous and (b) it maps every bounded subset of X into a relatively compact subset of Y.

For illustration we present an example of a compact operator. Let I be a closed bounded interval, Γ a continuous real-valued function on $I \times I$ and f a continuous real-valued function defined on the real line satisfying

$$|f(x)| \leqslant \alpha + \beta\,|x|, \qquad x \in R \tag{3.1.25}$$

for some $0 < \alpha$, $\beta < \infty$. Define the linear operator L by

$$(L\eta)(t) \equiv \int_I \Gamma(t, \theta)\eta(\theta)\, \mathrm{d}\theta, \qquad t \in I,$$

the nonlinear operator F by

$$(F\xi)(t) \equiv f(\xi(t))$$

and the operator $T \equiv LF$ by

$$(T\xi)(t) = \int_I \Gamma(t, \theta) f(\xi(\theta))\, d\theta, \qquad t \in I. \tag{3.1.26}$$

Then L is a continuous linear operator from $L_2(I)$ to $C(I)$ and F is a continuous bounded operator from $L_2(I)$ to $L_2(I)$. Thus, for $X \equiv L_2(I)$ and $Y \equiv C(I)$, we have a continuous bounded operator T from X into Y. We show that T is a compact operator. Since T is continuous it remains to show that for every bounded set $D \subset X$, TD is a relatively compact subset of Y. A subset E of $C(I)$ is relatively compact if and only if E is bounded (in norm) and equicontinuous, that is, for any $\varepsilon > 0$, there exists a $\delta > 0$ such that, for all t, $t + h \in I$, $|e(t+h) - e(t)| < \varepsilon$ uniformly with respect to $e \in E$ whenever $|h| < \delta$ (see Proposition 2.1.37).

Since $D \subset L_2(I)$ is bounded, there exists a finite $d > 0$ such that $\sup\{\|\xi\|_{L_2(I)}, \xi \in D\} \leqslant d < \infty$, and hence, by the Schwarz inequality,

$$\sup_{t \in I} |(T\xi)(t)| \leqslant \left\{\sup_{t \in I}\left(\int_I |\Gamma(t, \theta)|^2\, d\theta\right)^{1/2}\right\}(\alpha\mu^{1/2}(I) + \beta\, \|\xi\|_{L_2})$$
$$\leqslant \left\{\sup_{t \in I}\left(\int_I |\Gamma(t, \theta)|^2\, d\theta\right)^{1/2}\right\}(\alpha\mu^{1/2}(I) + \beta d)$$

for all $\xi \in D$, where $\mu(I)$ is the length (Lebesgue measure) of interval I. Hence TD is a bounded subset of Y. For equicontinuity, let t, $t + h \in I$ and $\xi \in D$, then

$$|(T\xi)(t+h) - (T\xi)(t)|$$
$$\leqslant \left(\int_I |\Gamma(t+h, \theta) - \Gamma(t, \theta)|^2\, d\theta\right)^{1/2} (\alpha\mu^{1/2}(I) + \beta d) \tag{3.1.27}$$

for all $\xi \in D$. Since Γ is continuous on I, it follows from the above inequality that for any $\varepsilon > 0$ we can find a $\delta = \delta(\varepsilon) > 0$ such that, whenever $|h| < \delta$,

$$\left(\int_I |\Gamma(t+h, \theta) - \Gamma(t, \theta)|^2\, d\theta\right)^{1/2} < (\varepsilon/(\alpha\mu^{1/2}(I) + \beta d)). \tag{3.1.28}$$

Hence for t, $t + h \in I$ and $|h| < \delta$, $|(T\xi)(t+h) - (T\xi)(t)| < \varepsilon$ uniformly with respect to $\xi \in D$. Thus TD is a bounded equicontinuous subset of $Y = C(I)$ and hence TD is relatively compact and therefore T is a compact operator. We state the following lemma without proof, assuming that X and Y are arbitrary Banach spaces.

Lemma 3.1.10. (Uniform limits of compact operators) [67]. *Let $\{T_n\}$ be a sequence of compact operators from X into Y and let $T: X \to Y$ be* *such that $T_n x \to Tx$ uniformly on bounded subsets of X. Then T is a compact operator.*

For an example, we consider again the operator T defined by (3.1.26) and assume that the kernel $\Gamma \in L_2(I \times I)$ instead of $C(I \times I)$. Since $\Gamma \in L_2(I \times I)$, there exists a sequence of kernels $\Gamma_n \in C(I \times I)$ such that

$$\int_{I \times I} |\Gamma(t, \theta) - \Gamma_n(t, \theta)|^2 \, d\theta \, dt \to 0 \qquad \text{as } n \to \infty. \tag{3.1.29}$$

Hence the operator T_n, defined by

$$(T_n \xi)(t) \equiv \int_I \Gamma_n(t, \theta) f(\xi(\theta)) \, d\theta, \tag{3.1.30}$$

is a compact operator from $X = L_2(I)$ to $Y = L_2(I)$. Then, by use of the Schwarz inequality, one can readily verify that

$$\|T_n \xi - T\xi\|_{L_2(I)} \leq \|\Gamma_n - \Gamma\|_{L_2(I \times I)} \cdot (\alpha \mu^{1/2}(I) + \beta \, \|\xi\|_{L_2(I)})$$

and hence for any bounded set $D \subset L_2(I)$ with $\sup\{\|\xi\|_{L_2(I)}, \xi \in D\} \leq d < \infty$, we have

$$\|T_n \xi - T\xi\|_{L_2(I)} \leq \|\Gamma_n - \Gamma\|_{L_2(I \times I)} (\alpha \mu^{1/2}(I) + \beta d) \tag{3.1.31}$$

for all $\xi \in D$. Using (3.1.29) we conclude that $T_n \xi \to T\xi$ uniformly on bounded subsets of $L_2(I)$. Thus T itself is a compact operator in X.

For the proof of the Schauder fixed-point theorem we shall need the following basic results.

Theorem 3.1.11 (Characterization of compact operators). *Let X, Y be Banach spaces with D a closed bounded subset of X and T a bounded continuous map from D to Y. Then T is a compact map if and only if it is the uniform limit of operators of finite rank (finite range).*

Proof Necessary condition: Suppose T is compact, then $\overline{TD}$ is a compact subset of Y. Hence, for every $\varepsilon > 0$, it has a finite ε-covering in the sense that there exists a finite set $\{y_i, i = 1, 2, \ldots, n(\varepsilon)\} \subset \overline{TD}$ so that

$$\overline{TD} \subset \bigcup_{i=1}^{n(\varepsilon)} N_\varepsilon(y_i) \tag{3.1.32}$$

where $N_\varepsilon(y_i) \equiv \{y \in Y: \|y - y_i\|_Y < \varepsilon\}$.

Let $\{\beta_i(y), i = 1, 2, \ldots, n(\varepsilon)\}$ be the partition of unity subordinate to the covering $\{N_\varepsilon(y_i), i=1, 2, \ldots, n(\varepsilon)\}$; that is,

(a) β_i is continuous on Y with support $\beta_i \subseteq N_\varepsilon(y_i)$, in other words, $\beta_i(y) \equiv 0$ for y outside $N_\varepsilon(y_i)$,

(b) $0 \leqslant \beta_i(y) \leqslant 1$, $y \in \overline{TD}$ for $i=1, 2, \ldots, n(\varepsilon)$,

and

(c) $\sum_{i=1}^{n(\varepsilon)} \beta_i(y) = 1$, $y \in \overline{TD}$.

Then define the sequence of operators

$$T_\varepsilon x \equiv \sum_{i=1}^{n(\varepsilon)} \beta_i(Tx) y_i, \qquad x \in D. \tag{3.1.33}$$

Clearly $x \to T_\varepsilon x$ is continuous from X to Y and has rank $n(\varepsilon) < \infty$. If $\beta_i(Tx) > 0$, then $Tx \in N_\varepsilon(y_i)$ and hence, by virtue of (a), (b) and (c),

$$\begin{aligned} \|T_\varepsilon x - Tx\|_Y &= \left\| \sum_{i=1}^{n(\varepsilon)} \beta_i(Tx) y_i - Tx \right\|_Y \\ &= \left\| \sum_{i=1}^{n(\varepsilon)} \beta_i(Tx) y_i - \sum_{i=1}^{n(\varepsilon)} \beta_i(Tx) Tx \right\|_Y \\ &= \left\| \sum_{i=1}^{n(\varepsilon)} \beta_i(Tx)(y_i - Tx) \right\|_Y \\ &\leqslant \sum_{i=1}^{n(\varepsilon)} \beta_i(Tx) \, \|y_i - Tx\|_Y \\ &\leqslant \varepsilon \text{ for all } x \in D. \end{aligned} \tag{3.1.34}$$

Since $\varepsilon > 0$ is arbitrary, it follows from (3.1.34) that

$$\lim_{\varepsilon \to 0} T_\varepsilon x = Tx \tag{3.1.35}$$

uniformly with respect to $x \in D$. Thus we have shown that if T is a compact map then it can be approximated by operators of finite rank. Sufficient condition: Since the operators T_ε have finite rank, they are compact maps. By (3.1.35), T is its uniform limit and hence, by Lemma 3.1.10, T is a compact operator. This proves the theorem. ■

For the proof of the celebrated Schauder fixed-point theorem we shall make use of the Brouwer fixed-point theorem, which holds only for finite-dimensional spaces. Here the concept of convexity plays an important role.

Definition 3.1.12 (Convex set). A set K is said to be convex if for each pair $\xi, \eta \in K$, $\alpha\xi + (1-\alpha)\eta \in K$ for all α satisfying $0 \leqslant \alpha \leqslant 1$.

Theorem 3.1.13 (Brouwer fixed-point theorem). *Let K be a closed bounded convex subset of R^n and suppose T is a continuous map from K into R^n satisfying $TK \subseteq K$. Then T has a fixed point in K.*

The theorem is intuitively clear though its proof is not trivial. Readers interested in the proof may consult [66, 67]. We shall take it for granted and prove the Schauder theorem.

Theorem 3.1.14 (Schauder fixed-point theorem). *Let K be a closed bounded convex subset of a Banach space X and T a compact map satisfying $TK \subseteq K$. Then T has a fixed point in K.*

Proof Since T is a compact map, it follows from Theorem 3.1.11 that T can be approximated by operators having finite-dimensional range. Let T_ε, $\varepsilon > 0$, be the ε-approximation of T with range contained in a finite-dimensional subspace X_ε of X. Then T_ε maps the compact convex set $K \cap X_\varepsilon$ into itself, that is,

$$T_\varepsilon(K \cap X_\varepsilon) \subseteq (K \cap X_\varepsilon).$$

Thus, by the Brouwer fixed-point theorem, for each $\varepsilon > 0$, there exists an $x_\varepsilon \in K \cap X_\varepsilon$ which is a fixed point of T_ε, that is, $T_\varepsilon x_\varepsilon = x_\varepsilon$. Since T and T_ε are compact maps, $\{T_\varepsilon x_\varepsilon\}$ is contained in a relatively compact set and hence there exists a subsequence $\{T_{\varepsilon_m} x_{\varepsilon_m}\} \subset \{T_\varepsilon x_\varepsilon\}$ and an x^* such that

$$\lim_{m\to\infty} x_{\varepsilon_m} = \lim_{m\to\infty} (T_{\varepsilon_m} x_{\varepsilon_m}) = x^*. \tag{3.1.36}$$

Since K is closed $x^* \in K$. We show that x^* is a fixed point of T. Clearly we can write

$$\begin{aligned} x^* - Tx^* &= x^* - x_{\varepsilon_m} + x_{\varepsilon_m} - Tx^* \\ &= (x^* - x_{\varepsilon_m}) + (T_{\varepsilon_m} x_{\varepsilon_m} - Tx^*) \\ &= (x^* - x_{\varepsilon_m}) + (T_{\varepsilon_m} x_{\varepsilon_m} - Tx_{\varepsilon_m}) + (Tx_{\varepsilon_m} - Tx^*), \end{aligned}$$

and hence

$$\|x^* - Tx^*\| \leqslant \|x^* - x_{\varepsilon_m}\| + \|Tx_{\varepsilon_m} - Tx^*\| + \|T_{\varepsilon_m} x_{\varepsilon_m} - Tx_{\varepsilon_m}\|. \tag{3.1.37}$$

Since $x_{\varepsilon_m} \to x^*$ and T is continuous, the first two terms on the right-hand side of (3.1.37) converge to zero as $m \to \infty$. For the third term we recall that, by Theorem 3.1.11, $T_{\varepsilon_m} \to T$ uniformly on bounded subsets of X, and that $\{x_{\varepsilon_m}\}$, being a convergent sequence, is contained in a bounded set. Hence

$$\lim_{m\to\infty} \|T_{\varepsilon_m} x_{\varepsilon_m} - Tx_{\varepsilon_m}\| = 0. \tag{3.1.38}$$

Since the left-hand expression in (3.1.37) is independent of m, upon taking the limit we have $\|x^* - Tx^*\| = 0$. Hence $x^* = Tx^*$, that is, x^* is a fixed point of T. This completes the proof. ■

As stated before, by use of the Schauder fixed-point theorem, we can prove the existence of solutions for the system

$$\begin{cases} \dot{x} = f(t, x) \\ x(0) = x_0 \end{cases} \tag{3.1.39}$$

without requiring the Lipschitz condition (3.1.13)(b) or its local version. This is given in the following theorem.

Theorem 3.1.15 (Existence without Lipschitz condition). *Consider the system* (3.1.39) *and suppose, for each* $\xi \in R^n$, *that* $t \to f(t, \xi)$ *is measurable and, for almost all* t, $\xi \to f(t, \xi)$ *is continuous and further there exists a non-negative* $K \in L_1^{\text{loc}}$ *such that*

$$|f(t, \xi)| \leq K(t)[1 + |\xi|], \qquad t \geq 0,$$

for all $\xi \in R^n$. *Then for each* $x_0 \in R^n$, *the system* (3.1.39) *has at least one solution which is absolutely continuous.*

Proof Let $a > 0$ and consider the interval $I_a = [0, a]$, and define the operator T on $C(I_a, R^n)$ by

$$(Tx)(t) \equiv x_0 + \int_0^t f(\theta, x(\theta))\, \mathrm{d}\theta, \qquad t \in I_a.$$

Choose $b \in (0, \infty)$ such that $|x_0| < b$ and define the set

$$B_b \equiv \{x \in C(I_a, R^n) : x(0) = x_0, \text{ and } \|x\|_{C(I_a, R^n)} \leq b\}.$$

Clearly B_b is a closed bounded convex subset of the Banach space $C(I_a, R^n)$. In order that we may apply Schauder theorem for the proof of the existence of a fixed point of T we must show that $TB_b \subseteq B_b$, for some suitable $a > 0$, and that T is a compact map. For this purpose let x be any element of B_b; then for $t \in I_a$,

$$\begin{aligned} |(Tx)(t)| &\leq |x_0| + \int_0^t K(\theta)[1 + |x(\theta)|]\, \mathrm{d}\theta \\ &\leq |x_0| + (1 + b) \int_0^t K(\theta)\, \mathrm{d}\theta. \end{aligned}$$

Hence,

$$\|Tx\|_a \equiv \sup_{t \in I_a} |(Tx)(t)| \leq |x_0| + (1 + b) \int_0^a K(\theta)\, \mathrm{d}\theta. \tag{3.1.40}$$

Since $|x_0| < b$ and $K \in L_1^{\mathrm{loc}}$, we can choose $a > 0$ so that

$$\int_0^a K(\theta)\, \mathrm{d}\theta < ((b - |x_0|)/(1+b)). \tag{3.1.41}$$

For such a choice of a, $\|Tx\|_a < b$ for every $x \in B_b$ and hence $TB_b \subseteq B_b$. For the proof of compactness of the operator T, we must show that it is continuous on B_b and TB_b is relatively compact. For continuity, let $\{x_n\} \in B_b$ and suppose $x_n \to x^*$ in B_b. Then for $t \in I_a$,

$$\begin{aligned} |(Tx_n)(t) - (Tx^*)(t)| &= \left| \int_0^t (f(\theta, x_n(\theta)) - f(\theta, x^*(\theta)))\, \mathrm{d}\theta \right| \\ &\leqslant \int_0^t |f(\theta, x_n(\theta)) - f(\theta, x^*(\theta))|\, \mathrm{d}\theta. \end{aligned} \tag{3.1.42}$$

Since $\xi \to f(t, \xi)$ is continuous for almost all t, it follows that $f(\theta, x_n(\theta)) \to f(\theta, x^*(\theta))$ a.e. on I_a. Further $|f(\theta, x_n(\theta))| \leqslant (1+b)K(\theta)$. Hence, by the Lebesgue dominated convergence theorem, it follows from (3.1.42) that

$$\|Tx_n - Tx^*\|_a \to 0 \qquad \text{as } n \to \infty.$$

This proves continuity. For relative compactness of TB_b, we note that for all $x \in B_b$, and t, $t+h \in I_a$ (without loss of generality we may assume $h > 0$),

$$\begin{aligned} |(Tx)(t+h) - (Tx)(t)| &\leqslant \int_t^{t+h} |f(\theta, x(\theta))|\, \mathrm{d}\theta \\ &\leqslant (1+b) \int_t^{t+h} K(\theta)\, \mathrm{d}\theta \end{aligned} \tag{3.1.43}$$

Since $K \in L_1^{\mathrm{loc}}$, for each $\varepsilon > 0$, there exists a $\delta = \delta(\varepsilon) > 0$, such that

$$\int_t^{t+h} K(\theta)\, \mathrm{d}\theta < (\varepsilon/(1+b)) \qquad \text{for } h < \delta,$$

and t, $t+h \in I_a$. Hence $|(Tx)(t+h) - (Tx)(t)| < \varepsilon$ uniformly with respect to $x \in B_b$ whenever $|h| < \delta$. Thus the set TB_b is bounded and equicontinuous and hence relatively compact. This shows that T has a fixed point $x \in B_b$ or equivalently x satisfies

$$x(t) = x_0 + \int_0^t f(\theta, x(\theta)))\, \mathrm{d}\theta \qquad \text{for } t \in I_a.$$

Since $x \in B_b$ and $f(t, x(t))$ is integrable on I_a, x is absolutely continuous on I_a. Clearly $x(a)$ is well defined and $|x(a)| \leqslant b$. Continuing the above procedure with an interval starting from a with initial state $x(a)$, the reader may verify that the solution can be continued for all $t \geq 0$. This completes the proof. ■

Remark 3.1.16 The above result proves existence under relaxed conditions at the loss of uniqueness. One may note that the set of all fixed points of T, denoted $X_b \equiv \{x \in B_b : x = Tx\}$, is a closed subset of $C(I_a, R^n)$.

If the function $f = f(t, x)$ is continuous on $I \times D$ where I is any open interval of the real line and D is any open connected subset of R^n, then a local existence result can be proved without the help of the Schauder fixed-point theorem. By a local solution here we mean that for any $(t_0, x_0) \in I \times D$, there exists an interval $I_\alpha \equiv (t_0 - \alpha, t_0 + \alpha) \subset I$ for which the equation $\dot{x} = f(t, x)$ has a C^1-solution x starting from $x(t_0) = x_0$ and lying in D for all $t \in I_\alpha$. For completeness we present this result below.

Theorem 3.1.17 (Local existence, continuous case). *Suppose f is continuous on $I \times D$ where $I \times D$ is any open connected subset of R^{n+1}. Then for each $(t_0, x_0) \in I \times D$ there exists a number $\alpha > 0$ possibly depending on $\{t_0, x_0\}$ such that the equation $\dot{x} = f(t, x)$ with $x(t_0) = x_0$ has at least one C^1 solution on $[t_0 - \alpha, t_0 + \alpha]$.*

Proof Let I_0 be any closed bounded subinterval of I with t_0 as its mid-point and Ω_0 any closed bounded connected subset of D containing x_0 in its interior. Let $d_0 \equiv d(x_0, \partial\Omega_0)$ denote the distance of x_0 from the boundary of Ω_0 and α_0 the distance of t_0 from the end points of I_0. Since f is continuous on $I \times D$ there exists a finite positive number M_0 such that

$$\sup\{|f(t, x)|, (t, x) \in I_0 \times \Omega_0\} \leq M_0$$

and f is uniformly continuous on $I_0 \times \Omega_0$. Define $\alpha = \min\{\alpha_0, d_0/M_0\}$ and consider the interval $[t_0 - \alpha, t_0 + \alpha]$. We show that the system has a C^1 solution on this interval. In fact it suffices to consider only the interval $I_\alpha = [t_0, t_0 + \alpha]$ as the same argument holds for $[t_0 - \alpha, t_0]$. Since f is uniformly continuous on $I_\alpha \times \Omega_0$, for every $\varepsilon > 0$, there exists a $\delta(\varepsilon) > 0$ such that, for $(t, y), (\tau, x) \in I_\alpha \times \Omega_0$,

$$|f(t, y) - f(\tau, x)| < \varepsilon$$

whenever $|t - \tau| < \delta$ and $|y - x| < \delta$. First we construct an ε-approximate solution φ_ε in the sense that

(a) $\varphi_\varepsilon(t_0) = x_0$,
(b) $\varphi_\varepsilon \in C(I_\alpha, R^n)$
(c) $\dot{\varphi}_\varepsilon(t)$ exists a.e. on I_α and
(d) $|\dot{\varphi}_\varepsilon(t) - f(t, \varphi_\varepsilon(t))| < \varepsilon$ for all $t \in I_\alpha$ except a finite set of points.

Partition the interval $I_\alpha = [t_0, t_0 + \alpha]$ into $n = n(\varepsilon)$ subintervals,

$$\Pi_n \equiv \{t_0 \equiv \tau_0 < \tau_1 < \tau_2 < \cdots < \tau_n \equiv t_0 + \alpha\},$$

such that the following conditions hold:

$$|\Pi_n| \equiv \max\{(\tau_{i+1} - \tau_i), 0 \leqslant i \leqslant n - 1\} < \delta$$

and

$$\max\{M_0(\tau_{i+1} - \tau_i), 0 \leqslant i \leqslant n - 1\} < \delta.$$

Then construct the family of vectors $\{x_i\}$ according to the following rule,

$$x_{i+1} = x_i + f(\tau_i, x_i)(\tau_{i+1} - \tau_i), \qquad 0 \leqslant i \leqslant n - 1$$

and note that $x_i \in \Omega_0$ and $|x_{i+1} - x_i| \leqslant M_0(\tau_{i+1} - \tau_i) < \delta$. Define the function φ_ε on I_α with $\varphi_\varepsilon(t_0) = x_0$ and

$$\varphi_\varepsilon(t) = x_i + f(\tau_i, x_i)(t - \tau_i), \qquad t \in (\tau_i, \tau_{i+1}], \quad 0 \leqslant i \leqslant n - 1.$$

We show that φ_ε is an ε-approximate solution of the initial-value problem $\dot{x} = f(t, x)$, $x(t_0) = x_0$, $t \in I_\alpha$. The function φ_ε as defined above is the Euler approximation and it is C^1 except at the points τ_i, $0 \leqslant i \leqslant n$. Clearly $\varphi_\varepsilon(t_0) = x_0$, $\varphi_\varepsilon \in C(I_\alpha, R^n)$, $\varphi_\varepsilon(t) \in \Omega_0$ for $t \in I_\alpha$ and $\dot{\varphi}_\varepsilon$ is uniquely defined at all points of I_α except at the points $\{\tau_i, 1 \leqslant i \leqslant n\}$. From the uniform continuity of f on $I_\alpha \times \Omega_0 \subset I_0 \times \Omega_0$ and the fact that $x_i \in \Omega_0$, $0 \leqslant i \leqslant n$, and $|x_{i+1} - x_i| < \delta$ it follows that

$$|f(t, \varphi_\varepsilon(t)) - f(\tau_i, x_i)| < \varepsilon \qquad \text{for } t \in (\tau_i, \tau_{i+1}].$$

Hence for any i from $0 \leqslant i \leqslant n - 1$ and $t \in (\tau_i, \tau_{i+1}]$ we have $|\dot{\varphi}_\varepsilon(t) - f(t, \varphi_\varepsilon(t))| < \varepsilon$. This shows that for each $\varepsilon > 0$, the problem, $\dot{x} = f(t, x)$, $x(t_0) = x_0$, has an ε-approximate solution φ_ε on I_α. Next we show that our problem has at least one solution which is obtained from φ_ε by a suitable limiting argument as $\varepsilon \downarrow 0$. For this purpose note that

$$\text{(a)} \quad \sup_{t \in I_\alpha} |\varphi_\varepsilon(t)| \leqslant |x_0| + \int_{t_0}^{t_0 + \alpha} |\dot{\varphi}_\varepsilon(\theta) - f(\theta, \varphi_\varepsilon(\theta))| \, d\theta + \int_{t_0}^{t_0 + \alpha} |f(\theta, \varphi_\varepsilon(\theta))| \, d\theta \leqslant |x_0| + (\varepsilon + M_0)\alpha < \infty,$$

and

$$\text{(b)} \quad \sup_{t \in I_\alpha} |\dot{\varphi}_\varepsilon(t)| \leqslant \sup_{t \in I_\alpha} |\dot{\varphi}_\varepsilon(t) - f(t, \varphi_\varepsilon(t))| + \sup_{t \in I_\alpha} |f(t, \varphi_\varepsilon(t))| \leqslant (\varepsilon + M_0) < \infty.$$

Now let $\{\varepsilon_k\}$ be a sequence of positive numbers such that $\lim_{k \to \infty} \varepsilon_k = 0$

and consider the family of functions $\{\varphi_k \equiv \varphi_{\varepsilon_k},\ k = 1, 2, \ldots\}$. By virtue of the properties (a) and (b), this is a bounded and equicontinuous subset of $C(I_\alpha, R^n)$ and hence by the Ascoli-Arzelà theorem (Proposition 2.1.37) there exists a subsequence $\{\varphi_{k_r}\} \subset \{\varphi_k\}$ and a $\varphi \in C(I_\alpha, R^n)$ such that $\varphi_{k_r}(t) \to \varphi(t)$ uniformly in t on I_α as $r \to \infty$. The function φ is a solution of our problem. Indeed $\varphi(t_0) = x_0$, and for $t \in I_\alpha$,

$$\varphi_{k_r}(t) = x_0 + \int_{t_0}^{t} \dot{\varphi}_{k_r}(\theta)\,\mathrm{d}\theta = x_0 + \int_{t_0}^{t} [\dot{\varphi}_{k_r}(\theta) - f(\theta, \varphi_{k_r}(\theta))]\,\mathrm{d}\theta + \int_{t_0}^{t} f(\theta, \varphi_{k_r}(\theta))\,\mathrm{d}\theta$$

and hence,

$$\sup_{t \in I_\alpha} \left| \varphi_{k_r}(t) - x_0 - \int_{t_0}^{t} f(\theta, \varphi_{k_r}(\theta))\,\mathrm{d}\theta \right| \leqslant (t - t_0)\varepsilon_{k_r} \leqslant \alpha\varepsilon_{k_r}.$$

Letting $r \to \infty$ we have

$$\lim_{r\to\infty} \varphi_{k_r}(t) = x_0 + \lim_{r\to\infty} \int_{t_0}^{t} f(\theta, \varphi_{k_r}(\theta))\,\mathrm{d}\theta \qquad \text{for each } t \in I_\alpha.$$

Since f is continuous and $\varphi_{k_r}(t) \to \varphi(t)$ uniformly on I_α it follows from this that φ satisfies the integral equation

$$\varphi(t) = x_0 + \int_{t_0}^{t} f(\theta, \varphi(\theta))\,\mathrm{d}\theta, \qquad t \in I_\alpha.$$

Further $\varphi(t) \in \Omega_0$ for $t \in I_\alpha$ and hence $|f(t, \varphi(t))| \leqslant M_0$, $t \in I_\alpha$, and therefore φ is differentiable. Differentiating the integral equation we obtain

$$\begin{cases} \dot{\varphi}(t) = f(t, \varphi(t)), & t \in I_\alpha \\ \varphi(t_0) = x_0. \end{cases}$$

Since $t \to f(t, \varphi(t))$ is continuous on I_α, φ is a C^1 solution of our problem. This completes the proof. ■

Remark 3.1.18 It follows from the above result that for any closed bounded subset Ω of D with $x_0 \in \operatorname{int} \Omega$, we can find a maximal interval I_Ω with t_0 as its mid-point such that the problem, $\dot{x} = f(t, x)$, $x(t_0) = x_0$, has a C^1 solution on I_Ω. Thus for every expanding sequence of compact sets $\{\Omega_n\}$ converging to the set D we have a sequence of maximal intervals $\{I_{\Omega_n} \equiv I_n\}$ of existence of C^1 solutions. If, as $\Omega_n \to D$, the interval I_n converges to $R = (-\infty, +\infty)$, then we have a global solution. On the other hand if $I_n \to I^*$ whose length is finite, then the solution x has finite escape time. For example the scalar equation $\dot{\xi} = \xi^r$ for $r > 1$ with $\xi_0 < 0$ has a finite escape time from the set $D \equiv R$.

3.2 Continuous dependence of solutions on system parameters, nonlinear semigroup and sensitivity

As indicated in chapter 2, the continuous dependence of solutions on system parameters plays a central role in the study of sensitivity, stability and control. Here in this section we consider this question for nonlinear systems.

3.2.1 Continuous dependence on initial data

Consider the system

$$\begin{cases} \dot{x} = f(t, x), & t \geqslant s \geqslant 0 \\ x(s) = \xi. \end{cases} \tag{3.2.1}$$

To indicate explicitly the dependence of the solution x on the initial data (s, ξ) we shall write $x(t) \equiv x(t, s, \xi)$. The domain of definition for this function is $\Delta \times R^n$ where $\Delta \equiv \{(t, s): 0 \leqslant s \leqslant t \leqslant \tau < \infty\}$.

Theorem 3.2.1 (Continuity with respect to initial data). *Under the assumptions of Theorem* 3.1.6, $(s, \xi) \to x(t, s, \xi)$ *is a continuous mapping from* $[0, t] \times R^n$ *to* R^n *for each fixed* $t > 0$, *and* $t \to x(t, s, \xi)$ *is continuous from* $[s, \infty)$ *to* R^n *for fixed* $(s, \xi) \in (0, \infty) \times R^n$.

Proof By Theorem 3.1.6, for a fixed data (s, ξ), $t \to x(t, s, \xi)$ is continuous for $t \geqslant s$. Thus it suffices to prove the continuity with respect to the initial data (s, ξ). Let $\{s_m, \xi_m\}$ be a sequence of initial data with $s_m \to s_0$ and $\xi_m \to \xi_0$ as $m \to \infty$. Let $\{x_m\}$ denote the solutions of the problems

$$\begin{cases} \dot{x} = f(t, x), & t \geqslant s_m \geqslant 0, \\ x(s_m) = \xi_m. \end{cases} \tag{3.2.2}$$

Without loss of generality we may assume that $\{s_m\}$ is an increasing sequence. Let x^0 denote the solution of the problem (3.2.1) corresponding to the data (s_0, ξ_0). Thus we have

$$\begin{aligned} x_m(t) &= \xi_m + \int_{s_m}^{t} f(\theta, x_m(\theta))\, d\theta, & t \geqslant s_m \\ x^0(t) &= \xi_0 + \int_{s_0}^{t} f(\theta, x^0(\theta))\, d\theta, & t \geqslant s_0. \end{aligned} \tag{3.2.3}$$

Applying Gronwall's lemma (Lemma 2.2.1) to the first equation, one can easily verify that

$$(1+|x_m(t)|) \leqslant (1+|\xi_m|) \exp \int_0^t K(\theta)\, d\theta. \tag{3.2.4}$$

Since $\{\xi_m\}$ is a convergent sequence in R^n, $\sup_m |\xi_m| \leqslant \beta$ for some $0<\beta<\infty$. Hence

$$\sup_m \{1+|x_m(t)|\} \leqslant (1+\beta) \exp \int_0^\tau K(\theta)\, d\theta, \qquad t \in [s_0, \tau]. \tag{3.2.5}$$

Using (3.2.3) one can also write

$$x_m(t) - x^0(t) = (\xi_m - \xi_0) + \int_{s_m}^{s_0} f(\theta, x_m(\theta))\, d\theta + \int_{s_0}^{t} [f(\theta, x_m(\theta)) - f(\theta, x^0(\theta))]\, d\theta$$

for $t \geqslant s_0$. Again by Gronwall's lemma we have,

$$|x_m(t) - x^0(t)| \leqslant \left(|\xi_m - \xi_0| + \int_{s_m}^{s_0} K(\theta)[1+|x_m(\theta)|]\, d\theta \right) \times \exp \int_{s_0}^{t} K(\theta)\, d\theta \qquad \text{for all } t \geqslant s_0. \tag{3.2.6}$$

Since $\xi_m \to \xi_0$, $s_m \to s_0$, and by (3.2.5), $\sup_{s_0 \leqslant t \leqslant \tau} \{\sup_m |x_m(t)|\} < \infty$, and $K \in L_1^{\text{loc}}$ it follows from (3.2.6) that $\lim_{m\to\infty} x_m(t) = x^0(t)$ uniformly on $[s_0, \tau]$. Thus we have shown that whenever $s_m \to s_0$ and $\xi_m \to \xi_0$, $x(t, s_m, \xi_m) \to x(t, s_0, \xi_0)$ uniformly on $[s_0, \tau]$. This proves continuity. ■

In the study of stability and control, it is often convenient to denote the solution $x(t) = x(t, s, \xi)$ by $T_{t,s}(\xi)$, where, for fixed (s, t), with $s \leqslant t$, one considers $T_{t,s}(\cdot)$ as the solution operator that takes the state ξ, given at time s, to a new state $T_{t,s}(\xi)$ at a later time $t \geqslant s$. This is similar to the transition operator $\Phi(t, s)$, $s \leqslant t$, given in section 2.2.2. The significant difference between the operators $\Phi(t, s)(\cdot)$ and $T_{t,s}(\cdot)$ is that the latter is nonlinear while the former is linear. However there are some important similarities which will be clear if one compares the properties of $T_{t,s}(\cdot)$ with those of $\Phi(t, s)(\cdot)$. By the previous theorem we have, for $0 \leqslant s \leqslant t \leqslant \tau < \infty$,

(a) $s \to T_{t,s}(\xi)$ is continuous from $[0, t]$ to R^n,
(b) $t \to T_{t,s}(\xi)$ is continuous from $[s, \tau]$ to R^n,
(c) $\xi \to T_{t,s}(\xi)$ is continuous from R^n to R^n.

Let $x(t, s, \xi)$, $s \geqslant 0$, $\xi \in R^n$, be the solution of (3.2.1) corresponding to the data (s, ξ). Let $s \leqslant \theta \leqslant t$, then, by the uniqueness of the solution, we have

$$x(t, \theta, x(\theta, s, \xi)) = x(t, s, \xi),$$

which is equivalent to

$$T_{t,\theta}(T_{\theta,s}(\xi)) = T_{t,s}(\xi). \tag{3.2.7}$$

Since this is true for all $\xi \in R^n$ we conclude that $T_{t,\theta}T_{\theta,s} = T_{t,s}$ for all $s \leqslant \theta \leqslant t$. Further by continuity of the solution we also have,

$$\lim_{t \downarrow s} T_{t,s}(\xi) = \lim_{t \downarrow s} x(t, s, \xi) = \xi. \tag{3.2.8}$$

Collecting these results we have,

$$\begin{aligned} &\text{(d)} \quad \lim_{t \downarrow s} T_{t,s} = I \text{ (identity operator)}, \\ &\text{(e)} \quad T_{t,\theta}T_{\theta,s} = T_{t,s} \qquad \text{for } s \leqslant \theta \leqslant t. \end{aligned} \tag{3.2.9}$$

Thus the transition operator $\{T_{t,s}, s \leqslant t\}$ satisfies the properties (a)–(e).

The reader may easily verify that for time-invariant systems the transition operator $T_{t,s}$ reduces to T_{t-s} or simply T_t, $t \geqslant 0$, and this is a one-parameter nonlinear semigroup of operators in R^n satisfying

$$\begin{aligned} &\text{(a)} \quad \lim_{t \downarrow 0} T_t = I \\ &\text{(b)} \quad T_{t+s} = T_t \cdot T_s. \end{aligned} \tag{3.2.10}$$

From the above discussion one can easily deduce the following result.

Corollary 3.2.2 *Let Γ be a closed bounded subset of R^n and $s \leqslant t$, then $T_{t,s}(\Gamma) \equiv \{T_{t,s}(\xi), \xi \in \Gamma\}$ is also a closed bounded subset of R^n and hence compact.*

As we shall see later this result is useful in optimal control theory.

We leave it as an exercise for the reader to verify that, under the hypothesis of Theorem 3.1.6, the operator $\xi \rightarrow T_{t,s}(\xi)$ is also continuous.

3.2.2 Continuous dependence on parameters

As we have seen in chapter 1, the dynamics of a physical system depends on various physical and empirical parameters. A question of critical importance is how does the system behave in response to variation of these parameters. Continuity, sensitivity and structural stability are the important issues here. We present some results on continuity and sensitivity.

Theorem 3.2.3 (Continuity with respect to parameters). *Let $\mathcal{P}$ be a closed bounded subset of R^p, $I \equiv [0, \tau]$, $\tau < \infty$, and $f: I \times R^n \times \mathcal{P} \to R^n$ such that for almost all $t \in I$, $(\xi, \alpha) \to f(t, \xi, \alpha)$ is continuous on $R^n \times \mathcal{P}$, and for fixed $(\xi, \alpha) \in \mathcal{P}$, $t \to f(t, \xi, \alpha)$ is measurable on I, and further it satisfies the hypothesis* (3.1.13) *of Theorem* 3.1.6 *uniformly with respect to $\alpha \in \mathcal{P}$. Then the solution $x(t) \equiv x(t, s, \xi, \alpha)$ of the problem*

$$\begin{cases} \dot{x} = f(t, x, \alpha), & t \geq s \\ x(s) = \xi, \end{cases} \tag{3.2.11}$$

is jointly continuous in all the variables $\{s, \xi, \alpha\} \in [0, t] \times R^n \times \mathcal{P}$.

Proof The proof is similar to that given for Theorem 3.2.1. Let $s_m \uparrow s_0$, $\xi_m \to \xi_0$ and $\alpha_m \to \alpha_0$ with $x_m(\cdot) \equiv x(\cdot, s_m, \xi_m, \alpha_m)$ being the solution of (3.2.11) corresponding to the data (s_m, ξ_m, α_m) and $x_0(\cdot) \equiv x(\cdot, s_0, \xi_0, \alpha_0)$ the solution corresponding to (s_0, ξ_0, α_0). Clearly $(s_0, \xi_0, \alpha_0) \in I \times R^n \times \mathcal{P}$. Taking the difference $(x_m(t) - x_0(t))$ and using Gronwall's lemma one can easily demonstrate that

$$\begin{aligned} \|x_m - x_0\| &= \sup\{|x_m(t) - x_0(t)|, \quad t \in [s_0, \tau]\} \\ &\leq C_1\{|\xi_m - \xi_0| + \int_{s_m}^{s_0} |f(\theta, x_m(\theta), \alpha_m)| \, d\theta \\ &\quad + \int_{s_0}^{\tau} |f(\theta, x_0(\theta), \alpha_m) - f(\theta, x_0(\theta), \alpha_0)| \, d\theta\} \end{aligned} \tag{3.2.12}$$

where $C_1 \equiv \exp \int_{s_0}^{\tau} K(\theta) \, d\theta$.

Since $\xi_m \to \xi_0$, the first term vanishes as $m \to \infty$. By assumption, f satisfies the growth condition (3.1.13)(a) uniformly with respect to $\alpha \in \mathcal{P}$, and consequently $\sup_m \|x_m\| < \infty$. Therefore the second term of (3.2.12) also vanishes as $s_m \to s_0$. For the last term, using the Lebesgue dominated convergence theorem (Proposition 2.1.32) and the continuity of f with respect to α on $\mathcal{P}$, one can verify that the third term also vanishes as $m \to \infty$. Thus $x_m \to x_0$ or equivalently $(s, \xi, \alpha) \to x(t, s, \xi, \alpha)$ is continuous. The continuity with respect to t is automatic. This completes the proof. ■

Sensitivity

Let R^p denote the space of parameters and $\mathcal{P}$ a closed, bounded, convex subset of R^p. Consider the system

$$\begin{cases} \dot{x} = f(t, x, \alpha_0(t)), & t \geq 0 \\ x(0) = \xi \end{cases} \tag{3.2.13}$$

where α_0 is any measurable function taking values from $\mathcal{P}$. The following theorem gives sufficient conditions for the existence of a sensitivity operator (see equation (2.3.39)) corresponding to variation of the parameters from the nominal trajectory $\alpha_0(t)$, $t \geq 0$. Here we are assuming that the parameters may be time varying.

Theorem 3.2.4 (Sensitivity). *Suppose*

(a) $f: I \times R^n \times \mathcal{P} \to R^n$ *is measurable in* t *on* $I = [0, \infty)$, *and continuous in the other variables,*
(b) $\nabla f \equiv f_x: I \times R^n \times \mathcal{P} \to \mathcal{L}(R^n)$ *is measurable in* $t \in I$, *and continuous in the rest of the variables,*
(c) *the Gâteaux differential of* f *with respect to the third variable in the direction* $(\alpha - \alpha_0)$ *denoted,* $f_\alpha(t, x, \alpha_0(t); \alpha(t) - \alpha_0(t))$, *is measurable in* t *and continuous in* x *for admissible* α *and* α_0.

Then the system (3.2.13) *has a sensitivity operator* $S(\alpha_0; \cdot)$ *given by*

$$S(\alpha_0; \alpha - \alpha_0)(t) \equiv \int_0^t \Phi_0(t, \theta) f_\alpha(\theta, x_0(\theta), \alpha_0(\theta); \alpha(\theta) - \alpha_0(\theta))\, d\theta, \quad t \in I. \tag{3.2.14}$$

In case the Gâteaux differential is linear,

$$\begin{aligned} S(\alpha_0; \alpha - \alpha_0)(t) &\equiv (S(\alpha_0) \cdot (\alpha - \alpha_0))(t) \\ &= \int_0^t \Phi_0(t, \theta) f_\alpha(\theta, x_0(\theta), \alpha_0(\theta))(\alpha - \alpha_0)(\theta)\, d\theta, \end{aligned} \tag{3.2.15}$$

where Φ_0 *is the transition operator corresponding to the matrix* $A_0(t) \equiv f_x(t, x_0(t), \alpha_0(t))$, $t \in I$.

Proof The proof is simple and left as an exercise for the reader. ■

Remark 3.2.5 Under the assumptions of Theorem 3.2.1 one can easily verify the existence of a constant C ($<\infty$) depending only on T, and the bounds of

$$\sup_{\tau_1 \leq t \leq T} |x(t, \tau_1, \xi_1)| \quad \sup_{\tau_2 \leq t \leq T} |x(t, \tau_2, \xi_2)| \quad \text{and} \quad \int_0^T K(\theta)\, d\theta$$

such that for $0 \leq \tau_1 \leq \tau_2 \leq t_1 \leq t_2 \leq T$,

$$|x(t_2, \tau_2, \xi_2) - x(t_1, \tau_1, \xi_1)| \leq C\left\{|\xi_2 - \xi_1| + \int_{\tau_1}^{\tau_2} K(\theta)\, d\theta + \int_{t_1}^{t_2} K(\theta)\, d\theta\right\}.$$

Hence the function $(t, s, \xi) \to x(t, s, \xi)$ is jointly continuous from $\Delta \times R^n$

to R^n. A similar conclusion holds in case of Theorem 3.2.3 asserting the joint continuity of $(t, s, \xi, \alpha) \to x(t, s, \xi, \alpha)$ on $\Delta \times R^n \times P$.

Remark 3.2.6 Since the sensitivity operator is a linear Volterra integral operator (equation (3.2.15)), complete insensitivity to parameter variation is impossible unless α_0 is such that

$$F_0(t) \equiv f_\alpha(t, x_0(t), \alpha_0(t)) \equiv 0 \quad \text{a.e. on } I.$$

Before closing this section we note that smoothness of solutions of differential equations depends on the smoothness of the data. More precisely one can easily prove the following result.

Corollary 3.2.7 (Smoothness of solutions). *Let $C^k(I \times R^n, R^n)$ denote the space of functions defined on $I \times R^n$ with values in R^n which are continuously differentiable up to order k. Consider the system*

$$\begin{cases} \dot{x} = f(t, x), & t \geq s, \quad t, s \in I, \\ x(s) = \xi \end{cases}$$

with $f \in C^k(I \times R^n, R^n)$. Then the solution $x(t, s, \xi)$ is C^{k+1} in t and C^k in the variables s and ξ.

3.3 Continuous dependence of solutions on controls (inputs)

In the study of optimal control problems, the question of continuous dependence of solutions on controls or inputs plays a significant role. In this section we present two such results.

First we shall prove the following general result. Consider the sequence of initial-value problems:

$$\begin{cases} \dot{x} = f_m(t, x), & t \geq s_m \geq 0, \\ x(s_m) = \xi_m, & I = [0, \tau], \quad \tau < \infty, \quad m = 0, 1, 2, 3, \ldots. \end{cases} \tag{3.3.1}$$

Theorem 3.3.1 *Let $\{f_m\}$ be a sequence of functions defined on $I \times R^n$ to R^n and suppose there exists a non-negative measurable function $K \in L_1^{\text{loc}}$ independent of m such that*

(a) $|f_m(t, x)| \leq K(t)[1 + |x|]$,
(b) $|f_m(t, x) - f_m(t, y)| \leq K(t)\, |x - y|$,
(c) $f_m(\cdot, x) \to f_0(\cdot, x)$ *weakly in $L_1(I, R^n)$ uniformly with respect to x in every bounded subset of R^n,*
(d) $s_m \uparrow s_0$ *and* $\xi_m \to \xi_0$ *as* $m \to \infty$.

Then the solution $x_m(t) = x(t, s_m, \xi_m, f_m)$ of the problem (3.3.1) *converges* *uniformly on I to the solution $x_0(t) = x(t, s_0, \xi_0, f_0)$ corresponding to the data (s_0, ξ_0, f_0).*

Proof By Theorem 3.1.6, the system (3.3.1) has a unique solution for each integer $m = 0, 1, 2, \ldots$. Hence

$$x_m(t) = \xi_m + \int_{s_m}^{t} f_m(\theta, x_m(\theta))\,\mathrm{d}\theta, \qquad m = 1, 2, \ldots \tag{3.3.2}$$

and

$$x_0(t) = \xi_0 + \int_{s_0}^{t} f_0(\theta, x_0(\theta))\,\mathrm{d}\theta. \tag{3.3.3}$$

Therefore for $t \geqslant s_0$,

$$|x_m(t) - x_0(t)| \leqslant \Psi_m(t) + \int_{s_0}^{t} K(\theta)\,|x_m(\theta) - x_0(\theta)|\,\mathrm{d}\theta, \tag{3.3.4}$$

where

$$\Psi_m(t) \equiv \eta_m + \ell_m(t) \tag{3.3.5}$$

$$\eta_m \equiv |\xi_m - \xi_0| + \int_{s_m}^{s_0} |f_m(\theta, x_m(\theta))|\,\mathrm{d}\theta, \tag{3.3.6}$$

$$\ell_m(t) \equiv \left| \int_{s_0}^{t} [f_m(\theta, x_0(\theta)) - f_0(\theta, x_0(\theta))]\,\mathrm{d}\theta \right|. \tag{3.3.7}$$

By virtue of assumption (a) it follows from Gronwall's lemma (Lemma 2.2.1) that

$$\sup_m \|x_m\| < \infty, \qquad \|x_0\| < \infty. \tag{3.3.8}$$

Hence $\sup_m \eta_m < \infty$ and $\lim_{m\to\infty} \eta_m = 0$.

Since $f_m(\cdot, x) \to f_0(\cdot, x)$ weakly in $L_1(I, R^n)$ uniformly with respect to x in every bounded subset of R^n, and $\|x_0\| < \infty$, it follows from (3.3.7) that $\ell_m(t) \to 0$ as $m \to \infty$ for each $t \in [s_0, \tau]$. Further note that

$$\sup_m \sup_{t\in[s_0,\tau]} \ell_m(t) < \infty. \tag{3.3.9}$$

Hence we conclude that

$$\sup_m \sup_{t\in[s_0,\tau]} \Psi_m(t) < \infty \tag{3.3.10}$$

and that

$$\lim_{m\to\infty} \Psi_m(t) = 0 \tag{3.3.11}$$

for all $t \in [s_0, \tau]$.

Applying the generalized Gronwall lemma (Lemma 2.3.5) to (3.3.4) we obtain, for $t \in [s_0, \tau]$,

$$|x_m(t) - x_0(t)| \leq \Psi_m(t) + \int_{s_0}^{t} \left(\exp \int_{\theta}^{t} K(s)\, \mathrm{d}s\right) \cdot K(\theta)\Psi_m(\theta)\, \mathrm{d}\theta. \tag{3.3.12}$$

Hence by virtue of (3.3.10), (3.3.11) and the Lebesgue dominated convergence theorem (Proposition 2.1.32) we have

$$\lim_{m\to\infty} |x_m(t) - x_0(t)| = 0$$

uniformly in $t \in [s_0, \tau]$. This completes the proof. ■

The above theorem will find application in optimal control theory, where

$$f_m(t, x) \equiv g(t, x, u_m(t)) \qquad \text{or} \qquad \int_U g(t, x, u)\mu_t^m(\mathrm{d}u),$$

with u_m representing a sequence of controls from an admissible class such as $L_\infty(I, U) \subset L_\infty(I, R^m)$ and μ^m representing a sequence of measure-valued controls with support U. We shall discuss this further when we study optimal control theory.

We present below another continuity theorem which is very useful in the study of necessary conditions of optimality.

Let I be a finite interval and M the class of Lebesgue measurable functions defined on I with values in a compact (closed bounded) set $U \subset R^m$. We define

$$\rho: M \times M \to [0, \infty)$$

by

$$\rho(u, v) = \mu\{t \in I: u(t) \neq v(t)\} \tag{3.3.13}$$

where μ denotes the Lebesgue measure on I. The function ρ defines a metric on M. Indeed it is clear that

(a) $\rho(u, v) \geq 0$ for all $u, v \in M$,
(b) $\rho(u, v) = 0$ if and only if $\mu\{t \in I: u(t) \neq v(t)\} = 0$, that is $u(t) = v(t)$ μ-a.e. on I,
(c) $\rho(u, v) = \rho(v, u)$.

It remains to verify the triangle inequality. Let $u, v, w \in M$, then it is clear that

$$\{t \in I: u(t) = v(t)\} \supset [\{t \in I: u(t) = w(t)\} \cap \{t \in I: w(t) = v(t)\}].$$

Hence

$$\{t \in I: u(t) \neq v(t)\} \subset [(t \in I: u(t) \neq w(t)\} \cup \{t \in I: w(t) \neq v(t)\}]$$

and consequently,

$$\mu\{t \in I: u(t) \neq v(t)\} \leq \mu\{t \in I: u(t) \neq w(t)\} + \mu\{t \in I; w(t) \neq v(t)\}. \quad (3.3.14)$$

Therefore, by definition of ρ, it follows from (3.3.14) that

$$\rho(u, v) \leq \rho(u, w) + \rho(w, v). \quad (3.3.15)$$

In fact $(M, \rho) \equiv M_\rho$ is a complete metric space.

With this preparation we can now prove the following result.

Theorem 3.3.2 *Consider the controlled system*

$$\begin{cases} \dot{x} = f(t, x, u(t)), & t \in I \equiv [0, \tau], \quad \tau < \infty \\ x(0) = \xi, \quad u \in M_\rho. \end{cases} \quad (3.3.16)$$

Suppose there exists a constant K_1 (possibly dependent on U) such that for all $w \in U$

(a) $|f(t, \eta, w)| \leq K_1[1 + |\eta|]$ *for all* $\eta \in R^n$,
(b) $|f(t, \eta_1, w) - f(t, \eta_2, w)| \leq K_1 |\eta_1 - \eta_2|, \ \eta_1, \eta_2 \in R^n$.

Then there exists a constant K_2, depending only on (K_1, τ, ξ), such that

$$\|x_u - x_v\| \equiv \sup_{t \in I} |x_u(t) - x_v(t)| \leq K_2 \rho(u, v), \quad (3.3.17)$$

for all $u, v \in M_\rho$ where x_u and x_v denote the responses of the system (3.3.16) corresponding to controls u and v.

Proof Under the given assumptions, the system (3.3.16) has a unique solution corresponding to any control from M_ρ. Let x_u and x_v denote the solutions corresponding to controls $u, v \in M_\rho$. Then

$$|x_u(t) - x_v(t)| \leq \int_0^t K_1 |x_u(\theta) - x_v(\theta)| \, d\theta + \int_0^t |f(\theta, x_v, u) - f(\theta, x_v, v)| \, d\theta. \quad (3.3.18)$$

Using Gronwall's inequality, it follows from (3.3.18) that

$$\|x_u - x_v\| \leq (\exp K_1\tau) \int_I |f(\theta, x_v(\theta), u(\theta)) - f(\theta, x_v(\theta), v(\theta))| \, d\theta.$$

Let $I^+ \equiv \{t \in I: u(t) \neq v(t)\}$; then

$$\|x_u - x_v\| \leq (\exp K_1\tau) \int_{I^+} |f(\theta, x_v(\theta), u(\theta)) - f(\theta, x_v(\theta), v(\theta))| \, d\theta. \tag{3.3.19}$$

Since for all $u \in M_\rho$,

$$\sup_{t\in I} \{(1 + |x_u(t)|\} \leq (1 + |\xi|) \exp K_1\tau, \tag{3.3.20}$$

it follows from (3.3.19) that

$$\|x_u - x_v\| \leq (1 + |\xi|)(2K_1 \exp 2K_1\tau)\mu(I^+). \tag{3.3.21}$$

By definition $\mu(I^+) = \rho(u, v)$ and hence for $K_2 \equiv (1 + |\xi|)(2K_1 \exp 2K_1\tau)$ it follows from (3.3.21) that

$$\|x_u - x_v\| \leq K_2\rho(u, v). \tag{3.3.22}$$

This completes the proof. ■

Remark 3.3.3 It follows from the above result that the input–output (state) map, $u \to x_u$, denoted by G (that is, $x = Gu$), is bounded and Lipschitz continuous from the metric space M_ρ to the Banach space $C(I, R^n)$, that is,

$$\|Gu - Gv\|_{C(I,R^n)} \leq K_2\rho(u, v). \tag{3.3.22$'$}$$

3.4 Nonlinear integral equations arising from output feedback and L_p stability

Many control systems consist of a physical plant and a set of measurement and control devices around the plant. Usually the control mechanism may consist of amplifiers and actuators containing nonlinearities, even though the plant itself may be linear. A typical model of such a system is given by

$$\begin{aligned} \dot{x} &= A(t)x + B(t)u && \text{(plant)} \\ y &= H(t)x && \text{(output)} \\ u(t) &= \lambda f(t, e(t)) && \text{(actuator)} \\ e(t) &= \gamma(t) - y(t) && \text{(error)} \end{aligned} \tag{3.4.1}$$

where x is the state vector with values in R^n, y the output vector with values in R^k, u the control vector with values in R^m, γ the command vector taking values in R^k and λ is a real number. Clearly the matrices A, B and H have dimensions $(n \times n)$, $(n \times m)$ and $(k \times n)$ respectively. Then by simple algebra one can easily verify that e satisfies the nonlinear Volterra integral equation, also called the Hammerstein equation,

$$e(t) = w(t) + \lambda \int_0^t K(t, \theta) f(\theta, e(\theta))\, d\theta, \tag{3.4.2}$$

where

$$w(t) \equiv \gamma(t) - H(t)\Phi(t, 0)x_0, \qquad t \in I$$
$$K(t, \theta) \equiv -H(t)\Phi(t, \theta)B(\theta), \qquad 0 \leq \theta \leq t < \infty$$

and Φ is the transition operator corresponding to the plant matrix A and x_0 is the initial state. We wish to prove L_p stability of the system (3.4.1) or equivalently (3.4.2) for strongly nonlinear function f. For convenience of analysis we shall write (3.4.2) in the abstract form

$$e = w + \lambda LFe = w + \lambda Ne \tag{3.4.3}$$

where

$$(Fe)(t) \equiv f(t, e(t)) \tag{3.4.4}$$

and

$$(Lz)(t) \equiv \int_0^t K(t, \theta) z(\theta)\, d\theta \tag{3.4.5}$$

$$N = LF. \tag{3.4.6}$$

the operator F, as defined above, is known as the Nemytskii operator.

For convenience we shall write $|\cdot|_k$ for norms in R^k and $\|\cdot\|_{k,m}$ for the operator norm of $(k \times m)$ matrices. The stability property of the system (3.4.2) obviously depends on the properties of the operators L, F and the actuator gain λ. We shall prove that, under certain conditions, the nonlinear operator N is compact in a suitable L_p space, and then using this fact we shall prove L_p stability of the system (3.4.2).

Lemma 3.4.1 (Continuity and boundedness of the operator F). *Let $I \subset [0, \infty]$ be an arbitrary interval and p_1, p_2 real numbers satisfying $1 < p_1, p_2 < \infty$. Suppose there exists a constant $\beta \geq 0$ and a function $g \in L_{p_2}^+(I)$ such that*

$$|f(t, \eta)|_m \leq g(t) + \beta\, |\eta|_k^{p_1/p_2} \qquad a.e.\ on\ I. \tag{3.4.7}$$

Then the operator F maps $L_{p_1}(I, R^k)$ to $L_{p_2}(I, R^m)$ and is continuous.

Proof By the use of Minkowski's inequality one can easily verify that

$$\|Fe\|_{L_{p_2}(I,R^m)} \leqslant \|g\|_{L_{p_2}(I)} + \beta(\|e\|_{L_{p_1}(I,R^k)})^{p_1/p_2}. \tag{3.4.8}$$

Hence F maps bounded subsets of $L_{p_1}(I, R^k)$ into bounded subsets of $L_{p_2}(I, R^m)$. Continuity also follows from this inequality. In fact the condition (3.4.7) is a necessary and sufficient condition for continuity and boundedness of the operator F [66]. ■

Note that the condition (3.4.7) allows strong nonlinearities in case p_1 is much larger than p_2. We shall now prove that the operator $N = LF$ is a compact operator in $L_{p_1}(I, R^k)$.

Lemma 3.4.2 (Compactness of N). *Suppose F satisfies the hypothesis of Lemma* 3.4.1, *and the kernel $K(t, \theta) \equiv -H(t)\Phi(t, \theta)B(\theta)$ of the linear operator L is Lebesgue measurable on the triangle*

$$\Delta \equiv \{(t, \theta): 0 \leqslant \theta \leqslant t;\ \theta, t \in I\}$$

and has the property

$$\hat{K}(\cdot) \in L^+_{p_1}(I)$$

where (3.4.9)

$$\hat{K}(t) \equiv \left(\int_0^t \|K(t, \theta)\|^{q_2}_{k,m}\, \mathrm{d}\theta\right)^{1/q_2}, \quad t \in I$$

and $(1/q_2) + (1/p_2) = 1$. *Then the operator $N = LF$ is a compact operator in* $L_{p_1}(I, R^k)$.

Proof To prove the compactness of the operator N, we must show that it is continuous and maps bounded sets into relatively compact sets all in the space $L_{p_1}(I, R^k)$. Continuity follows from the facts that, by Lemma 3.4.1, F is continuous from $L_{p_1}(I, R^k)$ to $L_{p_2}(I, R^m)$, and that L is a bounded linear operator from $L_{p_2}(I, R^m)$ to $L_{p_1}(I, R^k)$. In fact this follows from Hölder's inequality, giving

$$\left(\int_I |(LFe)(t)|^{p_1}\, \mathrm{d}t\right)^{1/p_1} \leqslant \left(\int_I |\hat{K}(t)|^{p_1}\, \mathrm{d}t\right)^{1/p_1} \|Fe\|_{L_{p_2}(I,R^m)}. \tag{3.4.10}$$

For compactness we must show that, for every bounded set $D_1 \subset L_{p_1}(I, R^k)$,

$$N(D_1) \equiv \{Ne,\ e \in D_1\} \equiv \{LFe,\ e \in D_1\}$$

is a relatively compact set in $L_{p_1}(I, R^k)$. We prove this by showing that every sequence $\{\eta_n\} \subset N(D_1)$ has a convergent subsequence. Since

$\eta_n \in N(D_1)$ there exists a sequence $\{e_n\} \subset D_1$ with $\eta_n = N(e_n) = Lz_n$ where $z_n = Fe_n$. Define $D_2 \equiv F(D_1) \equiv \{Fe, e \in D_1\}$. Since D_1 is a bounded set, by Lemma 3.4.1 D_2 is a bounded subset of $L_{p_2}(I, R^m)$, and hence, due to reflexivity of the Banach space L_{p_2} (see [65], p. 215), D_2 is relatively weakly sequentially compact. Hence there exists a subsequence $\{z_{n_m}\} \subset \{z_n\}$ and a $z_0 \in L_{p_2}(I, R^m)$ such that

$$z_{n_m} \to z_0 \quad \text{weakly} \quad \text{as } m \to \infty. \tag{3.4.11}$$

For convenience we let

$$\begin{aligned} \tilde{z}_m &\equiv z_{n_m} \\ \tilde{\eta}_m &\equiv \eta_{n_m} \equiv Lz_{n_m} \equiv LFe_{n_m}. \end{aligned} \tag{3.4.12}$$

Hence, by definition of weak convergence,

$$\tilde{\eta}_m(t) \equiv \int_0^t K(t, \theta)\tilde{z}_m(\theta)\,\mathrm{d}\theta \to \int_0^t K(t, \theta)z_0(\theta)\,\mathrm{d}\theta \equiv \eta_0(t), \tag{3.4.13}$$

for almost all $t \in I$. In other words,

$$\lim_{m\to\infty} [\tilde{\eta}_m(t) - \eta_0(t)] = 0 \quad \text{a.e. on } I,$$

and hence

$$\lim_{m\to\infty} |\tilde{\eta}_m(t) - \eta_0(t)|^{p_1} = 0 \qquad \text{a.e. on } I.$$

Further,

$$\begin{aligned} |\tilde{\eta}_m(t) - \eta_0(t)| &\leq \left(\int_0^t \|K(t, \theta)\|_{k,m}^{q_2}\,\mathrm{d}\theta\right)^{1/q_2} \|\tilde{z}_m - z_0\|_{L_{p_2}(I,R^m)} \\ &\leq \hat{K}(t)[\|\tilde{z}_m\| + \|z_0\|]. \end{aligned} \tag{3.4.14}$$

Since $\tilde{z}_m \in D_2$ and D_2 is a bounded set and $z_0 \in L_{p_2}(I, R^m)$ there exists a finite number $d_2 > 0$ such that $\|\tilde{z}_m\| \leq d_2$ and

$$|\tilde{\eta}_m(t) - \eta_0(t)| \leq (d_2 + \|z_0\|)\hat{K}(t) \equiv C\hat{K}(t) \qquad \text{a.e.}, \tag{3.4.15}$$

where $\hat{K} \in L_{p_1}^+(I)$. Hence, by the Lebesgue dominated convergence theorem (Proposition 2.1.32),

$$\int_I |\tilde{\eta}_m(t) - \eta_0(t)|^{p_1}\,\mathrm{d}t \to 0 \qquad \text{as } m \to \infty.$$

Since $\{\eta_n\}$ is an arbitrary sequence from $N(D_1)$ and D_1 is any bounded subset of $L_{p_1}(I, R^k)$, we have proved that $N(D_1)$ is relatively compact, and hence N is a compact operator in $L_{p_1}(I, R^k)$. This completes the proof. ■

Theorem 3.4.3 (L_p-stability). *Consider the feedback system* (3.4.1) *or equivalently the integral equation* (3.4.2) *and suppose the hypotheses of Lemma* 3.4.1 *and Lemma* 3.4.2 *hold. For any* $\alpha \in (0, \infty)$, *let*

$$a(\alpha) \equiv \sup\{\|Ne\|_{L_{p_1}(I,R^k)}: e \in S_\alpha\}$$

where

$$S_\alpha \equiv \{e \in L_{p_1}(I, R^k): \|e\|_{L_{p_1}(I,R^k)} \leq \alpha\},$$

and let λ *be sufficiently small so that*

$$b(\alpha) \equiv \alpha - |\lambda| a(\alpha) > 0.$$

Then, for every $w \in S_\beta$ *with* $0 < \beta < b(\alpha)$, *the integral equation* (3.4.2) *has at least one solution* $e^* \in S_\alpha$. *Further, if* $t \to H(t)$ *is continuous and* $\gamma \in C(I, R^k)$ *and* $I = [0, \infty)$, *then* $\lim_{t\to\infty} e^*(t) = 0$ *and hence the corresponding output* $y^*(t) \to \gamma(t)$ *as* $t \to \infty$.

Proof Since N is a compact operator in $L_{p_1}(I, R^k)$ (by Lemma 3.4.2), $a(\alpha) < \infty$ for each $0 \leq \alpha < \infty$. Thus for any such α one can choose λ sufficiently small so that $b(\alpha) > 0$. Then, for any positive number $\beta < b(\alpha)$, and $w \in S_\beta$, the operator $T_{w,\lambda}$, defined by

$$T_{w,\lambda}(e) \equiv w + \lambda N(e), \tag{3.4.16}$$

maps S_α into itself, that is, $T_{w,\lambda}(S_\alpha) \subseteq S_\alpha$. This follows from the fact that, for $e \in S_\alpha$,

$$\begin{aligned}\|T_{w,\lambda}(e)\|_{L_{p_1}(I,R^k)} &\leq \|w\| + |\lambda|\, \|N(e)\| \\ &\leq \beta + |\lambda|\, a(\alpha) < \alpha.\end{aligned}$$

Since N is a compact operator, it is clear from (3.4.16) that, for the given w and λ, $T_{w,\lambda}$ is also a compact operator. Thus by the Schauder fixed-point theorem (Theorem 3.1.14) there exists an $e^* \in S_\alpha$ such that $e^* = T_{w,\lambda}(e^*)$. That is, e^* is a solution of the integral equation (3.4.2) and $y^* = \gamma - e^*$ is the corresponding output. If $t \to H(t)$ is continuous, then $t \to y^*(t)$ is continuous; and if the command vector $t \to \gamma(t)$ is also continuous then $t \to e^*(t)$ is also continuous. Hence $e^* \in C(I, R^k) \cap S_\alpha$ and consequently, for $I = [0, \infty)$, $\lim_{t\to\infty} e^*(t) = 0$ and $y^*(t) \to \gamma(t)$ as $t \to \infty$. This completes the proof. ■

Remark 3.4.4 The above result shows that the system output y will asymptotically approach the command input γ. This is, of course, a desired characteristic of any feedback system.

The results of this section were extracted from the original paper of the author [74] where interested readers will find more general results.

Exercises

3.1.P1 Let (X, d) be a complete metric space and T a contraction operator on X. Prove that T has a unique fixed point in X.

3.1.P2 Prove the existence of a solution of the Cauchy problem (initial-value problem) (3.1.12) under the assumption that f is locally Lipschitz but satisfies the growth condition (a) of Theorem 3.1.6.

3.1.P3 Comment on Remark 3.1.4 and verify the fact (3.1.8) as stated there. Verify the last statement in Remark 3.1.16.

3.1.P4 If f satisfies only a local Lipschitz and local growth condition what can be said about the existence and, possibly, uniqueness of solutions of the Cauchy problem (3.1.1).

3.2.P5 Give sufficient conditions so that the solution $x(t, s, \xi)$ of the problem $\dot{x} = f(t, x)$, $x(s) = \xi$, is C^1 in all the variables (t, s, ξ).

3.2.P6 Verify the semigroup properties (3.2.10) for the nonlinear autonomous system $\dot{x} = f(x)$ using suitable assumptions of regularity for f.

3.2.P7 Prove the result of Corollary 3.2.2.

3.2.P8 How would you define continuity of the map $t \to T_{t,s}(\Gamma)$ for a closed bounded set $\Gamma \subset R^n$ [see also section 6.6].

3.2.P9 Justify the Remarks 3.2.5, 3.2.6.

3.2.P10 Prove the result of Corollary 3.2.7.

3.2.P11 Consider the system $\dot{x} = F(t, x) + B(t)u$, and suppose F satisfies the conditions for existence of solutions and the entries of B are locally integrable and $u \in \mathcal{U} \equiv$ {measurable functions with values in a compact set $U \subset R^m$}. Present two different modes of continuity $u \to x(u)$ with their comparison.

3.3.P12 Prove that the metric space M_ρ, with ρ given by (3.3.13), is complete.

3.3.P13 Consider the system

$$\dot{x} = f(x, u_n)$$

$$x(0) = \xi_n.$$

Find sufficient conditions under which $x(u_n) \equiv x_n \to x_0 \equiv x(u_0)$ whenever $u_n \to u_0$ in M_ρ and $\xi_n \to \xi_0$ in R^n.

3.3.P14 Generalize the result of Theorem 3.3.2 by replacing the Lipschitz condition by simple continuity.

3.4.P15 Modify the result of Lemma 3.4.2 to include the case for $p_2 = 1$ and hence obtain a corresponding stability result comparable to Theorem 3.4.3.

4

Stability

4.0 Introduction

One of the most important questions that must be taken into account in system design is stability. Obviously an unstable system cannot serve the purpose for which it has been designed. A system designed on the basis of a feasible scheme must be theoretically analysed to determine its stability before it is physically built. If it is found stable one may then look for other desirable features: for example, economics, reliability, optimality, etc. But if it fails to satisfy the basic stability requirement, then one must either modify the scheme or change the system parameters or both to obtain a stable system.

In actual practice system designers spend most of their time in theoretical analysis and computer simulation before they produce a final design that satisfies all the desirable qualities in addition to the stability requirement.

The question of stability also arises in socio-economic and ecological systems. For example, undue human interference in a natural environmental system may produce irreversible damage to an ecology. Erroneous government decisions may cause national and international economic disasters. It is not out of place to reflect that our very existence on the planet depends on the stability of its periodic motion in the solar system. These are all genuine questions of stability.

In this chapter we study this basic question and present methods of stability analysis as developed by Lyapunov in 1892 and extended and refined by many contributors in the field, notably LaSalle and Lefschetz in 1959 and many others.

4.1 Definitions of Lyapunov stability

In chapters 2 and 3, we have seen that the solutions of linear and nonlinear systems are continuous R^n-valued functions $\{x(t, t_0, x_0), t \geqslant t_0, x_0 \in R^n\}$ with respect to the variables $\{t, t_0, x_0\}$.

For linear (uncontrolled) systems we have

$$\dot{x} = Ax, \qquad \text{with solution} \qquad x(t, t_0, x_0) = e^{(t-t_0)A}x_0, \tag{4.1.1}$$

$$\dot{x} = A(t)x, \qquad \text{with solution} \qquad x(t, t_0, x_0) = X(t)X^{-1}(t_0)x_0, \tag{4.1.2}$$

and for nonlinear (uncontrolled) systems we have

$$\dot{x} = f(x) \qquad \text{with solution} \qquad x(t, t_0, x_0) \equiv T_{t-t_0}(x_0), \tag{4.1.3}$$

$$\dot{x} = f(t, x) \qquad \text{with solution} \qquad x(t, t_0, x_0) \equiv T_{t,t_0}(x_0), \tag{4.1.4}$$

where T_t is the nonlinear one-parameter semigroup (solution operator) as introduced in chapter 2 and T_{t,t_0} is a two-parameter semigroup. We also know (see chapter 2) that if f satisfies the growth condition

$$|f(t, x)| \leqslant K(t)[1 + |x|], \tag{4.1.5}$$

with K locally integrable, written, $K \in L_1^{\text{loc}}$, then

$$|x(t, t_0, x_0)| = |T_{t,t_0}(x_0)| < \infty, \qquad t \geqslant t_0,$$

for $t < \infty$. However, this does not guarantee that the solution $T_{t,t_0}(x_0)$ will not grow unbounded as $t \to \infty$.

This brings us to the question of definition of stability. At this stage, for the sake of simplicity of presentation, let us consider the nonlinear time-invariant system

$$S: \dot{x} = f(x).$$

Define the set

$$Z \equiv \{x \in R^n : f(x) = 0\} \subset R^n,$$

and note that it is the set of rest states for the system S. In other words, if the system enters the set Z at some time, say, $\tau \geqslant 0$, that is, $T_\tau(x_0) \in Z$, then the system may come to rest at $T_\tau(x_0) \equiv x^*$ and remain there until perturbed by some external forces. In many applications such as engineering or economics, x^* may be viewed as an ideal operating state of a machine or economic equilibrium. The question is, how does the system behave if it is perturbed from the ideal rest state x^*? Would the system state trajectories stay near x^*, or move away from it and attain another equilibrium, or continue to diverge from x^* forever or eventually return to x^*? Let x^* be any element of Z and define $\tilde{x} \equiv x - x^*$, where x is any other solution of the system S.

Clearly, for constant x^*

$$\dot{\tilde{x}} = \dot{x} - \dot{x}^* = f(x) - f(x^*) = f(\tilde{x} + x^*) \equiv f^*(\tilde{x}).$$

where $f^*(0) = 0$ since $f(x^*) = 0$. Thus, the question of stability of the rest state x^* of S is entirely equivalent to the question of stability of the system

$$S^*: \dot{\tilde{x}} = f^*(\tilde{x}),$$

with respect to the zero state (0).

Similarly, suppose the system S has a periodic solution (limit cycle) denoted by x_p, that is,

$$\dot{x}_p(t) = f(x_p(t)), \qquad t \geq 0.$$

Define $\tilde{x} \equiv x - x_p$, where x is any other solution of the system S, and note that

$$\dot{\tilde{x}}(t) = f(x(t)) - f(x_p(t)) = f(x_p(t) + \tilde{x}) - f(x_p(t)).$$

Defining

$$f^*(t, \xi) \equiv f(x_p(t) + \xi) - f(x_p(t)),$$

we can rewrite the above system as

$$S^*: \dot{\tilde{x}}(t) = f^*(t, \tilde{x}(t)),$$

where, by virtue of continuity of f, we have $f^*(t, 0) = 0$. Thus again, the question of stability of the periodic motion x_p is equivalent to the question of stability of the origin in R^n. Thus we can always arrange matters so that, without loss of generality, we may consider from the outset that $f(0) = 0$, thereby reducing the questions of stability of periodic motions or of a given equilibrium to the question of stability of the 0 state. This observation allows us to formulate all our definitions of stability with reference to the zero state.

Let Ω be an open connected set in R^n containing the origin. (An open set is said to be connected if it cannot be described as the union of two disjoint open sets.) Let $B_a \equiv \{x \in R^n: |x| < a, a > 0\}$, denote the open ball in R^n of radius a.

Definition 4.1.1 (Lyapunov stability). The system

$$\dot{x} = f(t, x), \qquad f(t, 0) = 0,$$

is said to be stable with respect to the zero state if for every $\beta > 0$, so that $B_\beta \subset \Omega$, there exists an $\alpha = \alpha(\beta, t_0) > 0$ such that $T_{t,t_0}(B_\alpha) \subset B_\beta$ for all $t \geq t_0$, that is, every trajectory originating from the ball B_α stays in the ball B_β for all time.

If Ω is a bounded set we have local stability and if $\Omega \equiv R^n$ we have global stability.

Example 4.1.2 Consider the system

$$\dot{x}_1 = x_2, \qquad \dot{x}_2 = -x_1.$$

Define $E(t) = \frac{1}{2}(x_1^2(t) + x_2^2(t)) = \frac{1}{2}|x(t)|^2$. Clearly

$$\dot{E}(t) = x_1\dot{x}_1 + x_2\dot{x}_2 = x_1x_2 - x_2x_1 = 0.$$

Hence the system is stable, since $x(t) \in B_\beta$, for any $\beta > 0$ whenever $x(0) \in B_\alpha \equiv B_\beta$.

Definition 4.1.3 (Instability). The origin (0-state) is said to be unstable if it is not stable in the sense of Definition 4.1.1, or equivalently, the origin is unstable if for any given $\beta > 0$ with $B_\beta \subset \Omega$, there is no $\alpha > 0$ such that $T_{t,t_0}(B_\alpha) \subset B_\beta$ for all $t \geqslant t_0$.

Example 4.1.4 Consider the system

$$\dot{x}_1 = x_1, \qquad \dot{x}_2 = x_2.$$

Clearly for $E(t)$ as given in Example 4.1.2, $E(t) = E(0)e^{2t}$. Let x_{10} and x_{20} denote the initial states. Clearly $E(t)$ grows without bound with t and hence the system is unstable.

Mere stability in the sense of Definition 4.1.1 is not satisfactory for many engineering or physical systems: for example, an aircraft or any other machine must eventually return to its normal operating state (rest state) after the perturbing forces have disappeared. An appropriate definition bearing this significance is given below.

Definition 4.1.5 (Asymptotic stability). The system is said to be asymptotically stable with respect to the origin (0-state) in the region Ω if it is stable in the sense of Definition 4.1.1 and further

$$\lim_{t \to \infty} T_{t,t_0}(x_0) = 0,$$

for every $x_0 \in \Omega$.

If, in Definition 4.1.5, Ω is only a bounded set, we have local asymptotic stability and if Ω can be taken equal to R^n we have global asymptotic stability.

Example 4.1.6 Consider the system $\dot{x}_1 = -x_1$, $\dot{x}_2 = -x_2$. For the same E as in Example 4.1.2, we have $\dot{E} = -2E$ and hence $E(t) = E(0)e^{-2t}$, implying global asymptotic stability of the origin.

Definition 4.1.7 (Exponential stability). The system is said to be exponentially stable with respect to the zero state, or equivalently, the origin is exponentially stable if it is asymptotically stable and $T_{t,t_0}(x_0) \to 0$ exponentially as $t \to \infty$.

The system of Example 4.1.6 is also exponentially stable.

Definition 4.1.8 (Lagrange stability). The system is said to be Lagrange stable if the solution remains bounded for all $t \geq 0$, that is, $\sup_{t \geq t_0} |T_{t,t_0}(x_0)| < \infty$. This is equivalent to BIBO as discussed in chapters 2 and 3.

4.2 Asymptotic stability of linear systems

We have discussed the importance of asymptotic stability of physical systems. In this section we study this problem for both linear time-invariant and linear time-varying systems. We present them in two subsections.

4.2.1 Time-invariant systems

We consider the question of asymptotic stability of the linear time-invariant system

$$S: \dot{x} = Ax. \tag{4.2.1}$$

It is known that stability of this system is determined by the properties of the roots of the characteristic equation

$$\Phi(s) \triangleq \det(sI - A) = 0, \qquad s \in C \equiv \text{complex plane}.$$

The function Φ is a polynomial in s of degree n and has at most n distinct roots. It is well known that if all the characteristic roots of A have negative real parts, then the system is asymptotically stable. Except in trivial cases, it usually requires a digital computer to determine the roots of the equation $\Phi(s) = 0$. However, fortunately, for detecting stability it is not essential to find the roots. One can use the classical Routh–Hurwitz criterion that detects stability from the sign of the entries of the first column of the Routh–Hurwitz matrix. For stability, these elements must be different from zero and of identical sign [21], [30], [10].

However, there is a more general and useful way to determine the stability of the system S. This is given in the following theorem.

Theorem 4.2.1 *The system $S: \dot{x} = Ax$, is asymptotically stable (with respect to the zero state) if, and only if, the algebraic equation (called the*

Lyapunov equation)

$$A'Y + YA = -\Gamma, \tag{4.2.2}$$

has a real, symmetric and positive definite solution Y for some real, symmetric and positive definite matrix Γ.

Proof (If): Suppose the Lyapunov equation

$$A'Y + YA = -\Gamma$$

has a real, symmetric, positive definite solution Y for some real, symmetric and positive definite matrix Γ. We must show that the system S is asymptotically stable. For any solution $x = \{x(t), t \geq 0\}$ of the system S define

$$V(t) = (Yx(t), x(t)). \tag{4.2.3}$$

Clearly, for all $t \geq 0$,

$$\dot{V}(t) = \frac{\mathrm{d}}{\mathrm{d}t} V(t) = (Y\dot{x}, x) + (Yx, \dot{x})$$
$$= ((A'Y + YA)x(t), x(t)) = -(\Gamma x(t), x(t)).$$

Integrating we have

$$\int_0^T \dot{V}\,\mathrm{d}t = -\int_0^T (\Gamma x(t), x(t))\,\mathrm{d}t, \qquad T \geq 0,$$

or equivalently for all $T \geq 0$

$$V(T) - V(0) = -\int_0^T (\Gamma x(t), x(t))\,\mathrm{d}t.$$

Since $Y > 0$, $V(T) \geq 0$ for all $T \geq 0$ and consequently

$$\int_0^T (\Gamma x(t), x(t))\,\mathrm{d}t \leq V(0) \qquad \text{for all } T \geq 0.$$

Since $\Gamma > 0$, there exists a $\gamma > 0$ such that $(\Gamma\xi, \xi) \geq \gamma\,|\xi|^2$, where γ may be taken as the smallest eigenvalue of Γ. Therefore,

$$\gamma \int_0^T |x(t)|^2\,\mathrm{d}t \leq V(0) \qquad \text{for all } T \geq 0,$$

and consequently

$$\int_0^\infty |x(t)|^2\,\mathrm{d}t \leq (V(0)/\gamma) \leq (\|Y\|\,|x_0|^2/\gamma) < \infty.$$

This implies that

$$\lim_{T\to\infty}\int_T^\infty |x(t)|^2\,\mathrm{d}t = 0.$$

Combining this with the fact that $t \to x(t)$ is continuous, being the solution of a differential equation, one may conclude that

$$\lim_{t\to\infty} |x(t)| = 0,$$

and hence $x(t)$ converges to the zero state asymptotically.

(Only if): Given that the system $S: \dot{x} = Ax$ is asymptotically stable we must show that, for any real, symmetric matrix $\Gamma > 0$, the Lyapunov equation $A'Y + YA = -\Gamma$ has a real, symmetric, positive definite solution Y. Consider the matrix differential equation

$$\frac{\mathrm{d}Z}{\mathrm{d}t} = A'Z + ZA,$$
$$Z(0) = \Gamma. \tag{4.2.4}$$

One can easily verify that the solution of this equation is given by

$$Z(t) = \mathrm{e}^{tA'}\Gamma\mathrm{e}^{tA}. \tag{4.2.5}$$

We show that

$$Z_\infty \triangleq \int_0^\infty Z(t)\,\mathrm{d}t \equiv \int_0^\infty \mathrm{e}^{tA'}\Gamma\mathrm{e}^{tA}\,\mathrm{d}t,$$

is a solution of the Lyapunov equation

$$A'Y + YA + \Gamma = 0,$$

and that it is real, symmetric and positive definite. Since the system S is asymptotically stable all the eigenvalues of A have negative real parts (i.e. A is a stability matrix). Hence the integral Z_∞ is well defined. Integrating the matrix differential equation, we have, for each $T \geqslant 0$,

$$\int_0^T \left(\frac{\mathrm{d}Z}{\mathrm{d}t}\right)\mathrm{d}t = \int_0^T A'Z(t)\,\mathrm{d}t + \int_0^T Z(t)A\,\mathrm{d}t,$$

or equivalently

$$Z(T) - \Gamma = A'\left(\int_0^T Z(t)\,\mathrm{d}t\right) + \left(\int_0^T Z(t)\,\mathrm{d}t\right)A.$$

Since A is a stability matrix $\lim_{T\to\infty} Z(T) = 0$ and hence

$$-\Gamma = A'Z_\infty + Z_\infty A.$$

Thus

$$Y \equiv Z_\infty = \int_0^\infty Z(t)\,\mathrm{d}t = \int_0^\infty \mathrm{e}^{tA'}\Gamma\mathrm{e}^{tA}\,\mathrm{d}t,$$

is a solution of the Lyapunov equation. Since Z is real so also is Y, and since Γ is symmetric Y is also symmetric. For the proof of positivity of Y we compute

$$(Y\xi, \xi) = \int_0^\infty (Z(t)\xi, \xi)\,\mathrm{d}t = \int_0^\infty (\mathrm{e}^{tA'}\Gamma\mathrm{e}^{tA}\xi, \xi)\,\mathrm{d}t$$

$$= \int_0^\infty (\Gamma e^{tA}\xi, \mathrm{e}^{tA}\xi)\,\mathrm{d}t \geqslant \gamma \int_0^\infty |\mathrm{e}^{tA}\xi|^2\,\mathrm{d}t.$$

Since e^{tA} is non-singular, $\mathrm{e}^{tA}\xi \equiv 0$ if and only if $\xi = 0$. Hence $(Y\xi, \xi) > 0$ for all $\xi \neq 0$ implying positivity of Y. This completes the proof. ■

According to the above theorem, the question of asymptotic stability of the system $\dot{x} = Ax$ is equivalent to the question of existence of a real symmetric positive definite solution of the algebraic equation (4.2.2). Given a matrix Γ and the system matrix A, the equation (4.2.2) represents a system of linear algebraic equations of the form

$$FY = -\Gamma, \tag{4.2.6}$$

where F is a linear transformation from the space of $(n \times n)$ square matrices into itself. The linear operator F is completely determined by the matrix A. If $\{\lambda_1, \lambda_2, \ldots, \lambda_n\}$ are the eigenvalues of the matrix A, each repeated as many times as its multiplicity requires, then the eigenvalues of the operator F can be represented in the form

$$\mu_{ij} = \lambda_i + \lambda_j, \qquad 1 \leqslant i, j \leqslant n. \tag{4.2.7}$$

Therefore, if the matrix A has no eigenvalue that is zero and there are no two eigenvalues that are opposites ($\lambda_i \neq -\lambda_j$, for $1 \leqslant i, j \leqslant n$), then the operator F is non-singular. In this case the matrix Γ determines the matrix Y uniquely and if Γ is symmetric so also is Y. Thus, if A is a stability matrix (i.e. A has all its eigenvalues with negative real parts) the condition

$$\mu_{ij} = \lambda_i + \lambda_j \neq 0 \qquad \text{for } 1 \leqslant i, j \leqslant n, \tag{4.2.8}$$

is always satisfied. In this situation for any real, symmetric, positive definite matrix Γ the equation (4.2.2) or equivalently (4.2.6) has a unique, real, symmetric, positive definite solution Y. This fact is practically very useful since in Theorem 4.2.1 one can then take any (and not just some) real, symmetric, positive definite matrix Γ, for example,

$\Gamma = I$ and check if

$$A'Y + YA = -I$$

has a solution with the stated properties.

We will have occasion to apply this result in establishing stability of weakly nonlinear systems in section 4.3. We shall also meet this once again in section 5.3 in the study of stabilizing controls.

4.2.2 Time-varying systems

For a class of linear time-varying systems we can prove similar results as in section 4.2.1. We consider the system

$$\dot{x} = A(t)x, \tag{4.2.9}$$

where the elements of the system matrix $A(t)$ are locally integrable. We know from chapter 2 that the solution for this system is given by

$$x(t) = \Phi(t, 0)x_0, \qquad x_0 \equiv \text{initial state},$$

where

$$\Phi(t, \tau) = X(t)X^{-1}(\tau), \qquad \tau \leqslant t, \tag{4.2.10}$$

and X is the fundamental solution of

$$\frac{\mathrm{d}X}{\mathrm{d}t} = A(t)X$$
$$X(0) = I. \tag{4.2.11}$$

Clearly, if $M \equiv \sup_{t \geqslant 0} \|X(t)\| < \infty$, then $|x(t)| \leqslant M\,|x_0| < \infty$ for all $t \geqslant 0$ and the system is obviously Lagrange stable. For asymptotic stability of the zero state we must have

$$\lim_{t \to \infty} \|X(t)\| = 0. \tag{4.2.12}$$

We show, in this section, that linear systems with periodic coefficients can be reduced to time-invariant systems by a suitable transformation. Once this is accomplished the stability properties of the time-varying system can be deduced from those of the transformed time-invariant system by using the results of section 4.2.1.

Recall that the transition operator $\Phi(t, \tau)$, $\tau \leqslant t$, satisfies the differential equation

$$\frac{\partial}{\partial t}\Phi(t, s) = A(t)\Phi(t, s) \qquad \text{for } s \leqslant t,$$
$$\Phi(s, s) = I. \tag{4.2.13}$$

Suppose the matrix $t \to A(t)$ is a periodic function of period $T > 0$, that is, $A(t) = A(t+T)$ for all t. Replacing t by $t+T$ in equation (4.2.13) and using the periodicity of A one obtains

$$\begin{aligned} &\frac{\partial}{\partial t}\Phi(t+T, s) = A(t)\Phi(t+T, s), \qquad \text{for } s \leq t, \\ &\Phi(s, s) = I. \end{aligned} \tag{4.2.14}$$

Hence we conclude that if $\Phi(t, s)$ is the transition matrix so also is $\Phi(t+T, s)$. From the semigroup property

$$\Phi(t, s) = \Phi(t, \theta)\Phi(\theta, s), \qquad s \leq \theta \leq t$$

it follows that

$$\Phi(t+T, 0) = \Phi(t+T, T)\Phi(T, 0). \tag{4.2.15}$$

From the periodicity of the matrix A we also know that the solution

$$x(t) = \Phi(t, t_0)x_0, \qquad t_0 \leq t,$$

starting from the state x_0 at time t_0 and the solution

$$y(t) = \Phi(t+T, t_0+T)x_0, \qquad t_0 \leq t,$$

starting from the same state x_0 at time $t_0 + T$ are equal for all $t \geq t_0$. Therefore

$$\Phi(t, t_0)x_0 = x(t) = y(t) \equiv \Phi(t+T, t_0+T)x_0,$$

for all $t \geq t_0$ and all $x_0 \in R^n$ and consequently

$$\Phi(t, t_0) = \Phi(t+T, t_0+T) \qquad \text{for all } t \geq t_0. \tag{4.2.16}$$

In particular, for $t_0 = 0$, we have $\Phi(t, 0) = \Phi(t+T, T)$ and hence it follows from equation (4.2.15) that

$$\Phi(t+T, 0) = \Phi(t, 0)\Phi(T, 0), \qquad t \geq 0. \tag{4.2.17}$$

Since the transition operator Φ is nonsingular $\ln \Phi(T, 0)$ is well defined and hence the matrix

$$B \equiv \frac{1}{T}\ln \Phi(T, 0) \tag{4.2.18}$$

is also well defined.

Define the matrix

$$P(t) \equiv \Phi(t, 0)e^{-tB}. \tag{4.2.19}$$

Clearly this matrix is nonsingular. We show that it is periodic of period T.

Indeed, by virtue of equations (4.2.17) and (4.2.18) we have

$$\begin{aligned} P(t+T) &= \Phi(t+T, 0)\mathrm{e}^{-(t+T)B} \\ &= \Phi(t, 0)\Phi(T, 0)\mathrm{e}^{-(t+T)B} \\ &= \Phi(t, 0)\mathrm{e}^{TB}\mathrm{e}^{-(t+T)B} \\ &= \Phi(t, 0)\mathrm{e}^{-tB} = P(t), \end{aligned} \tag{4.2.20}$$

for all t. Thus we can express the transition matrix $\Phi(t, s)$ as

$$\Phi(t, s) = P(t)\mathrm{e}^{(t-s)B}P^{-1}(s). \tag{4.2.21}$$

With the help of the matrix $P(t)$, we can convert the periodic system $\dot{x} = A(t)x$ into a system with constant coefficients. Formally we have the following result.

Lemma 4.2.2 *Consider the periodic system* (4.2.9) *with* $t \to A(t)$ *periodic of period* T. *Then* $P(t)$, *as defined by* (4.2.19), *transforms the time-varying system into a time-invariant system*

$$\dot{y} = By, \tag{4.2.22}$$

where B *is given by* (4.2.18).

Proof Define $x(t) = P(t)y(t)$. Substituting this into equation $\dot{x} = A(t)x$ we have

$$P\dot{y} = (AP - \dot{P})y.$$

Since P is non-singular

$$\dot{y} = P^{-1}(AP - \dot{P})y. \tag{4.2.23}$$

Differentiating (4.2.19) once and noting that B commutes with e^{-tB}, we have

$$\begin{aligned} \dot{P}(t) &= \left(\frac{\partial}{\partial t}\Phi(t, 0)\right)\mathrm{e}^{-tB} - \Phi(t, 0)\mathrm{e}^{-tB}B \\ &= A(t)\Phi(t, 0)\mathrm{e}^{-tB} - \Phi(t, 0)\mathrm{e}^{-tB}B \\ &= A(t)P(t) - P(t)B. \end{aligned} \tag{4.2.24}$$

Hence, it follows from (4.2.23) and (4.2.24) that

$$\dot{y} = P^{-1}(AP - AP + PB)y = By.$$

This completes the proof. ■

With the help of the above lemma we can prove that the time-varying periodic system

$$S_p: \dot{x}(t) = A(t)x(t) \tag{4.2.25}$$

is stable if and only if the time-invariant system

$$S_c: \dot{y}(t) = By(t) \tag{4.2.26}$$

is stable. We present this in the following theorem.

Theorem 4.2.3 *The periodic system S_p is stable (asymptotically stable) if and only if the time-invariant system S_c is stable (asymptotically stable).*

Proof By the previous lemma $x(t) = P(t)y(t)$. Since $t \to P(t)$ is continuous and periodic there exists a finite positive number M_p such that

$$\sup_{t \geq 0} \|P(t)\| = M_p < \infty.$$

Hence

$$|x(t)| \leq \|P(t)\| \; |y(t)| \leq M_p \; |y(t)|. \tag{4.2.27}$$

Therefore, if the time-invariant system S_c is stable (asymptotically stable) so also is the time-varying system S_p. It remains to show the converse.

Since $t \to P(t)$ is continuous, periodic and non-singular, $t \to P^{-1}(t)$ is also continuous and periodic. Hence

$$y(t) = P^{-1}(t)x(t), \qquad t \geq 0$$

is well defined and there exists a positive number M_q such that

$$\sup_{t \geq 0} \|P^{-1}(t)\| = M_q < \infty.$$

Consequently

$$|y(t)| \leq \|P^{-1}(t)\| \; |x(t)| \leq M_q \; |x(t)|. \tag{4.2.28}$$

This inequality implies that if the periodic system S_p is stable (asymptotically stable) so also is the time-invariant system S_c. ■

Remark 4.2.4 The reader may wish to verify (see Problem 4.2.P4) that the matrix-valued function $P^{-1}(t)$ satisfies the differential equation

$$\begin{cases} \dot{Q} = -QA(t) + BQ \\ Q(0) = I. \end{cases} \tag{4.2.29}$$

Note that the transformed system S_c: $\dot{y} = By$ is asymptotically stable if and only if the eigenvalues of the constant matrix B have negative real parts. The eigenvalues of B have negative real parts if and only if the eigenvalues of $\Phi(T, 0)$ have real parts less than 1. Indeed recalling that

$$B = \frac{1}{T} \ln \Phi(T, 0),$$

we have

$$\Phi(T, 0) = e^{TB}.$$

If λ_i is an eigenvalue of B corresponding to the eigenvector, say, ξ_i, then $\mu_i \equiv e^{T\lambda_i}$ is an eigenvalue of $\Phi(T, 0)$ with corresponding eigenvector ξ_i. This follows from the fact that

$$\Phi(T, 0)\xi_i \equiv e^{TB}\xi_i = \sum_{n=0}^{\infty} \frac{T^n}{n!} B^n \xi_i$$

$$= \sum_{n=0}^{\infty} \frac{T^n}{n!} (\lambda_i)^n \xi_i = e^{T\lambda_i}\xi_i = \mu_i \xi_i.$$

Hence

$$\lambda_i = \frac{1}{T} \ln \mu_i. \tag{4.2.30}$$

Thus if $\operatorname{Re} \mu_i < 1$, then $\operatorname{Re} \lambda_i < 0$ and conversely. Therefore, in principle, one can determine the stability of the periodic system $\dot{x} = A(t)x$ if one can determine whether or not the eigenvalues of the matrix $\Phi(T, 0) = X(T)$ have real parts less than unity.

We shall illustrate the above results by an example. In this example we shall discover an interesting fact that even though the eigenvalues of the matrix $A(t)$ of a time-varying system are all constant and have negative real parts, yet the system is not asymptotically stable.

Example 4.2.5 Consider the time-varying system $\dot{x} = A(t)x$, where

$$A(t) = \begin{bmatrix} -2\cos^2 t & -(1 + 2\sin t \cos t) & 0 & 0 \\ 1 - 2\sin t \cos t & -2\sin^2 t & 0 & 0 \\ -k & 0 & k - 2\cos^2 t & -(1 + 2\sin t \cos t) \\ 0 & -k & 1 - 2\sin t \cos t & k - 2\sin^2 t \end{bmatrix}$$

Clearly $A(t)$ is periodic with period $T=2\pi$. Note that

$$\det(\eta I - A(t)) = (\eta+1)^2(\eta-k+1)^2$$

and hence the eigenvalues are $\eta_1=-1$, $\eta_2=-1$, $\eta_3=k-1$ and $\eta_4=k-1$ (all constant). For $k<1$, all the eigenvalues are negative for all t, yet we shall see that the system is not asymptotically stable or even stable. By the previous theory (Theorem 4.2.3) we know that the stability of the periodic system is determined by the eigenvalues of B where

$$B=\frac{1}{T}\ln\Phi(T,0)=\frac{1}{T}\ln X(T),$$

and $X(t)$ is the fundamental solution of $\dot{X}(t)=A(t)X(t)$ with $X(0)=I$. One can easily check that X is given by

$$X(t)=\begin{bmatrix} e^{-2t}\cos t & -\sin t & 0 & 0 \\ e^{-2t}\sin t & \cos t & 0 & 0 \\ (e^{-2t}\cos t)(1-e^{kt}) & (e^{kt}-1)\sin t & e^{(k-2)t}\cos t & -e^{kt}\sin t \\ (e^{-2t}\sin t)(1-e^{kt}) & (1-e^{kt})\cos t & e^{(k-2)t}\sin t & e^{kt}\cos t \end{bmatrix}$$

and for $t=T=2\pi$,

$$\det(\mu I - X(2\pi)) = (\mu - e^{-4\pi})(\mu-1)(\mu - e^{2(k-2)\pi})(\mu - e^{2k\pi}),$$

with eigenvalues for $X(2\pi)$ given by $\mu_1=e^{-4\pi}$, $\mu_2=1$, $\mu_3=e^{2(k-2)\pi}$ and $\mu_4=e^{2k\pi}$. Hence the eigenvalues of B, given by $\lambda_i=(1/2\pi)\ln\mu_i$, are $\lambda_1=-2$, $\lambda_2=0$, $\lambda_3=k-2$ and $\lambda_4=k$. Therefore, the system is unstable if $k>0$ even though the eigenvalues of $A(t)$ are negative for $k<1$. The system is stable in the Lyapunov sense if $k<0$ but certainly not asymptotically stable, even though it may appear to be so from the eigenvalues η_i, $i=1,2,3,4$. Hence one can conclude that the eigenvalues of the system matrix $A(t)$ of a linear time-varying system do not determine its stability in any sense.

4.3 Lyapunov's first method: stability of mildly nonlinear systems

In certain situations, the stability or instability of a mildly nonlinear system with respect to an equilibrium state (critical or rest state) can be determined from that of its linear approximation. This is, in essence, the first method of Lyapunov.

4.3.1 Time-invariant system (nonlinear)

Consider the system $S: \dot{x}=f(x)$, with $x^* \in R^n$ a rest state ($f(x^*)=0$) and suppose f is sufficiently smooth, for example, $f \in C^2$, so that we can represent f as

$$f(x) = (\operatorname{grad} f(x^*)) \cdot (x - x^*) + g(x - x^*),$$

where g is another continuous n-vector function of its argument near the origin. Defining $e \equiv (x - x^*)$ as the deviation from x^*, and $A = \operatorname{grad} f(x^*)$, the system S can be written as

$$E: \frac{\mathrm{d}e}{\mathrm{d}t} = Ae + g(e). \tag{4.3.1}$$

Let $o(\varepsilon)$ denote a real-valued nonnegative function of the variable $\varepsilon \geqslant 0$ satisfying the property

$$\lim_{\varepsilon \downarrow 0} \frac{o(\varepsilon)}{\varepsilon} = 0. \tag{4.3.2}$$

Definition 4.3.1 (Mildly nonlinear systems). The system S or equivalently the system E is said to be mildly nonlinear if the vector function f or equivalently the function g satisfies the inequality

$$|f(\xi + x^*)| \leqslant a\,|\xi| + o(|\xi|),$$

or

$$|g(\xi)| = o(|\xi|), \tag{4.3.3}$$

for all ξ in an open neighbourhood of the origin, where the constant a is possibly dependent on x^*.

In the following theorem we show that if A is a stability matrix (that is, $\dot{y} = Ay$, is asymptotically stable) and g is mildly nonlinear, then the original system is locally asymptotically stable near x^*.

Theorem 4.3.2 (Stability of mildly nonlinear systems). *The system $S: \dot{x} = f(x)$ is asymptotically stable near its rest state x^* or equivalently the system $E: \dot{e} = Ae + g(e)$ is asymptotically stable in the neighbourhood of the origin if* (a) *A is a stability matrix and* (b) *g is mildly nonlinear.*

Proof Since A is a stability matrix, it follows from Theorem 4.2.1 that for each real symmetric positive definite matrix Γ there exists a unique real symmetric positive definite solution Y of the algebraic equation

$$A'Y + YA = -\Gamma. \tag{4.3.4}$$

Using any such Y, along any trajectory e of the system (E): $\dot{e} = Ae + g(e)$, define

$$V(t) = (Ye(t), e(t)). \tag{4.3.5}$$

Then one can easily verify that

$$\dot{V}(t) = -(\Gamma e(t), e(t)) + 2(Ye(t), g(e(t))). \tag{4.3.6}$$

Since $\Gamma > 0$ there exists a $\gamma > 0$ such that $(\Gamma\xi, \xi) \geqslant \gamma\,|\xi|^2$ for all $\xi \in R^n$. Hence, integrating the above equation over the interval $[0, T]$, $T < \infty$, one obtains

$$V(T) + \gamma \int_0^T |e(t)|^2\, dt \leqslant V(0) + 2\,\|Y\| \int_0^T |e(t)|\,|g(e(t))|\, dt. \tag{4.3.7}$$

Rearranging the above inequality, we have

$$V(T) + \int_0^T \left(\gamma - 2\,\|Y\|\,\frac{|g(e)|}{|e|}\right) |e(t)|^2\, dt \leqslant V(0) < \infty, \tag{4.3.8}$$

for all $T \geqslant 0$. Define the set

$$N_\gamma \equiv \left\{\xi \in R^n : \frac{2\,\|Y\|\,|g(\xi)|}{|\xi|} < \gamma\right\}, \tag{4.3.9}$$

and note that it is a non-empty neighbourhood of the origin. Then for every $e_0 \in N_\gamma$, it follows from the inequality (4.3.8), by virtue of continuity of $e(t) = e(t, e_0)$, positive definiteness of Y and the property $|g(\xi)| = o(|\xi|)$, that $e(t) \in N_\gamma$ for all $t \geqslant 0$. Since $V(T) \geqslant 0$ for all $T \geqslant 0$ and $e(t) \in N_\gamma$ for all $t \geqslant 0$ and $V(0) < \infty$ it follows from the inequality

$$\int_0^T \left(\gamma - 2\,\|Y\|\,\frac{|g(e(t))|}{|e(t)|}\right) |e(t)|^2\, dt \leqslant V(0) < \infty, \qquad T \geqslant 0, \tag{4.3.10}$$

that, unless $\lim_{t\to\infty} e(t) = 0$, the inequality is violated after finite time T. Therefore, the system is asymptotically stable near the origin. ■

For illustration of the above result we present an example.

Example 4.3.3 (Economic equilibria). The dynamics of a simplified economic model are given by

$$\begin{aligned} \frac{dk}{dt} &= f(k) - \mu k - c - k(\beta - \gamma L), \\ \frac{dL}{dt} &= \beta L - \gamma L^2 \end{aligned} \tag{4.3.11}$$

where k represents capital investment (*per capita*) and L is the labour supply. The function $f(k)$ is the normalized production function (*per capita*), μ is the capital depreciation rate, c is (*per capita*) consumption rate, β is the growth rate of the labour force and γ is the competition factor. Setting

$$\begin{aligned} f(k) - \mu k - c - k(\beta - \gamma L) &= 0 \\ \beta L - \gamma L^2 &= 0, \end{aligned} \tag{4.3.12}$$

one obtains the equilibria. Disregarding the trivial case ($L = 0$) we have

$$\begin{aligned} f(k_0) - \mu k_0 - c &= 0 \\ L_0 &= \beta/\gamma \end{aligned} \tag{4.3.13}$$

giving all the non-trivial equilibrium states. The production function $f(k)$ is a strictly concave monotone increasing function satisfying the following properties:

$$f(0) = 0, \qquad f'(k) > 0, \qquad f''(k) < 0,$$

and (4.3.14)

$$\lim_{k \downarrow 0} f'(k) = \infty, \qquad \lim_{k \uparrow \infty} f'(k) = 0.$$

Thus, if c is sufficiently small, equation (4.3.13) has two solutions. In other words there are two equilibria. Defining $e_1 = k - k_0$ and $e_2 = L - L_0$ one can rewrite the system equation (4.3.11) as

$$\begin{aligned} \frac{\mathrm{d}}{\mathrm{d}t} e_1 &= -\mu e_1 + \gamma k_0 e_2 + f(k_0 + e_1) - f(k_0) + \gamma e_1 e_2 \\ \frac{\mathrm{d}}{\mathrm{d}t} e_2 &= -\beta e_2 - \gamma e_2^2. \end{aligned} \tag{4.3.15}$$

Approximating

$$f(k_0 + e_1) - f(k_0) = f'(k_0) e_1 + \tfrac{1}{2} f''(k_0) e_1^2 + o(e_1^2),$$

we can rewrite (4.3.15) in the canonical form

$$\frac{\mathrm{d}e}{\mathrm{d}t} = Ae + g(e), \tag{4.3.16}$$

where

$$A = \begin{bmatrix} f'(k_0) - \mu & \gamma k_0 \\ 0 & -\beta \end{bmatrix}, \qquad g(e) = \begin{bmatrix} g_1(e_1, e_2) \\ g_2(e_1, e_2) \end{bmatrix},$$

with g_1 and g_2 containing only nonlinear terms of second and higher orders. The eigenvalues of A are given by $\lambda_1 = f'(k_0) - \mu$, $\lambda_2 = -\beta$. Since k_0 has two possible values, it is clear that the one, for which $f'(k_0) > \mu$, is unstable. This is the lower level of investment at which the economy cannot sustain itself. On the other hand, at the upper level of investment, for which $f'(k_0) < \mu$, the system is asymptotically stable and a small perturbation is easily absorbed by the economy. If the consumption level (rate) is increased to its maximum possible value c^* for which the equation, $f(k) - \mu k - c^* = 0$, has only one solution k^*, then there is only one equilibrium (k^*, L_0) and this is unstable.

Example 4.3.4 (Epidemic model). The dynamic model for the spread of an epidemic disease is given by

$$\begin{cases} \dot{x}_1 = -ax_1 - bx_1x_2 \\ \dot{x}_2 = -cx_2 + bx_1x_2 \\ \dot{x}_3 = ax_1 + cx_2 \end{cases} \tag{4.3.17}$$

where x_1 represents that part of the population which is susceptible to the disease, x_2 represents the infectives and x_3 is the part that is isolated from the other two groups by immunization or death. Since the population x_3 does not affect the first two groups one may ignore the third equation. With the third equation removed one may easily verify that the system is asymptotically stable near the origin. Note that in this case also $|g(x)| \leqslant |b|\,|x|^2 = o(|x|)$. We leave the details as an exercise.

4.3.2 Time-varying system (nonlinear)

We show that under suitable assumptions the nonlinear time-varying system

$$\dot{x} = A(t)x + f(t, x), \qquad t \geqslant 0, \tag{4.3.18}$$

is asymptotically stable with respect to the zero state.

Theorem 4.3.5 (Stability of mildly nonlinear periodic systems). *Suppose $t \to A(t)$ is periodic of period T, and $x \to f(t, x)$ is mildly nonlinear uniformly in $t \geqslant 0$. Then the system* (4.3.18) *is asymptotically stable in the neighbourhood of the zero state if all the eigenvalues $\{\mu_i\}$ of $\Phi(T, 0) = X(T)$ have modulus less than one, and unstable if at least one μ_i has modulus greater than unity.*

Proof Using the periodic matrix $P(t)$ as introduced in section 4.2.2 (see equation (4.2.19)) and the transformation of coordinates $x = P(t)y$, it follows from the system equation (4.3.18) that

$$\dot{y} = P^{-1}(AP - \dot{P})y + P^{-1}f(t, Py). \tag{4.3.19}$$

But we know from section 4.2.2 (see equation (4.2.24)) that $P^{-1}(AP - \dot{P})$

$= B$ is a constant matrix. Hence

$$\dot{y} = By + P^{-1}(t)f(t, P(t)y), \qquad t \geqslant 0. \tag{4.3.20}$$

Also recall that $e^{TB} = \Phi(T, 0) = X(T)$ and hence if the eigenvalues $\{\mu_i\}$ of $X(T)$ have modulus less than one, then the eigenvalues $\{\lambda_i\}$ of B have negative real parts. Thus B is a stability matrix and as a result it follows from Theorem 4.2.1 that the Lyapunov equation $B'Y + YB = -\Gamma$ has a positive definite solution Y for each $\Gamma > 0$. Defining $V(t) = (Yy(t), y(t))$, as before, one can verify as in the previous Theorem that, for all $T \geqslant 0$,

$$\int_0^T \left(\gamma - \frac{2\,\|Y\|\,\|P^{-1}(t)\|\,|f(t, P(t)y)|}{|y(t)|}\right)|y(t)|^2\,\mathrm{d}t \leqslant V(0). \tag{4.3.21}$$

Since $\xi \to f(t, \xi)$ is mildly nonlinear uniformly in $t \geqslant 0$, that is $|f(t, \xi)| = o(|\xi|)$ independently of t, and $t \to P(t)$, $t \to P^{-1}(t)$ are periodic and continuous with $\sup_{t \geqslant 0} \|P(t)\| = M_p$, $\sup_{t \geqslant 0} \|P^{-1}(t)\| = M_q$, it follows from the above inequality that, for all $T \geqslant 0$,

$$\int_0^T \left(\gamma - \frac{2\,\|Y\|\,M_q o(M_p\,|y(t)|)}{|y(t)|}\right)|y(t)|^2\,\mathrm{d}t \leqslant V(0) < \infty. \tag{4.3.22}$$

Hence it follows from similar arguments as in the proof of Theorem 4.3.2 that there exists a neighbourhood of the origin in which the system (4.3.20) is asymptotically stable and consequently the system (4.3.18) is also asymptotically stable near the zero state. If the modulus of an eigenvalue μ_i of $\Phi(T, 0) \equiv X(T)$ is greater than one, then the corresponding eigenvalue of B has positive real part and hence the system is unstable. ■

For illustration we present an example.

Example 4.3.6

$$\begin{cases} \dot{\xi}_1 = -\xi_2 + \xi_1(1 - \xi_1^2 - \xi_2^2) \\ \dot{\xi}_2 = \xi_1 + \xi_2(1 - \xi_1^2 - \xi_2^2) \end{cases} \tag{4.3.23}$$

The system has a periodic solution $\xi_1 = \cos t$, $\xi_2 = \sin t$. Expanding around the periodic solution (limit cycle) we obtain the linear variational equation

$$\frac{\mathrm{d}e}{\mathrm{d}t} = A(t)e, \tag{4.3.24}$$

where

$$A(t) = \begin{bmatrix} -2\cos^2 t & -1-2\sin t\cos t \\ 1-2\sin t\cos t & -2\sin^2 t \end{bmatrix}.$$

The matrix $A(t)$ is periodic with period $T = 2\pi$. One can easily check that the fundamental (matrix) solution of the above equation is given by

$$X(t) = \begin{bmatrix} e^{-2t}\cos t & -\sin t \\ e^{-2t}\sin t & \cos t \end{bmatrix}. \tag{4.3.25}$$

Hence by equations (4.2.18) and (4.2.30)

$$B = \frac{1}{T}\ln X(T) = \frac{1}{2\pi}\ln X(2\pi)$$

and

$$\text{ev}(B) = \frac{1}{2\pi}\ln \text{ev}(X(2\pi)) \tag{4.3.26}$$

or

$$\lambda_i = \frac{1}{2\pi}\ln \mu_i,$$

where ev(B) represents the eigenvalues of B.

For stability

$$\text{Re}\,\lambda_i = \text{Re ev}(B) \leqslant 0,$$

or equivalently

$$\text{Re}\,\mu_i \leqslant 1.$$

Since $\det(\mu I - X(2\pi)) = (\mu - 1)(\mu - e^{-4\pi})$, we have $\mu_1 = 1$, $\mu_2 = e^{-4\pi}$. Hence the variational equation (4.3.24), $\dot{e} = A(t)e$, is merely stable but not asymptotically stable. We will see, in Example 4.5.21, that the periodic solution is actually asymptotically stable.

4.4 Lyapunov's second method (fundamental stability theorems)

One of the most powerful techniques for detecting stability or instability of a system is given by the Lyapunov theory, in that one solves the stability problem without actually solving the associated differential equations. This is one of the reasons why the second method is also known as the 'direct method'. In fact, according to this method, the whole stability question is reduced to the questions of existence of certain special functions, called Lyapunov functions, and their methods of construction.

In this section, we present the basic stability and instability theorems of Lyapunov. For illustration we also present some examples. First, we shall consider time-invariant systems and then state results for time-varying systems whose proofs are very much similar to those of their time-invariant counterparts.

We consider the time-invariant system $\dot{x} = f(x)$, $x(t) \in R^n$, $t \geq 0$, and recall that the question of stability of an arbitrary rest (equilibrium) state is equivalent to the question of stability of the zero state after appropriate translation if required. Hence without loss of generality we may consider $f(0) = 0$ and formulate the stability problems with respect to the zero state.

Let Ω be an open subset of the state space R^n and $0 \in \Omega$ (interior of Ω).

Definition 4.4.1 (Positive definite functions). A function V mapping Ω to R is said to be positive definite if it has the following properties:

(a) $V \in C^1(\Omega)$, that is, V is a real-valued once continuously differentiable function on Ω.
(b) $V(0) = 0$.
(c) $V(x) > 0$ for $x \in \Omega \backslash \{0\}$, that is, V has an isolated minimum at the origin.

Some simple examples of positive definite functions are given below.

(e1) $V(x) = (\Gamma x, x), \qquad x \in R^n,$ (4.4.1)

where Γ is a positive definite $(n \times n)$ matrix, is a positive definite function on R^n.

(e2) $V(x) = \begin{cases} x^5 \text{ for } x \geq 0 \\ x^8 \text{ for } x \leq 0 \end{cases} \qquad x \in R$ (4.4.2)

is positive definite on R.

(e3) Suppose Q is any real-valued smooth function on $\Omega \subset R^n$ containing the origin in its interior. Let Q' and Q'' denote the first and second partials of Q, that is,

$$Q' = \text{grad } Q = \nabla Q$$
$$Q'' = \nabla^2 Q.$$

If Q admits a representation of the form

$$Q(x) = Q(0) + (Q'(0), x) + \tfrac{1}{2}(Q''(0)x, x) + o(\|x\|^2) \qquad (4.4.3)$$

with $Q(0)=0$, $Q'(0)=0$ and $Q''(0)$ positive definite, then Q is a positive function near the origin. In general, suppose Q has representation of the form

$$Q(x) = Q_0 + Q_1(x) + Q_2(x) + \cdots + Q_m(x) + Q_{m+1}(x), \tag{4.4.4}$$

where each Q_k is a homogeneous polynomial in $x=(x_1, x_2, \ldots, x_n)$ of degree exactly k. If $Q_k \equiv 0$ for $0 \leqslant k \leqslant m-1$ and Q is positive definite near the origin, then m must be even. However, the converse need not be true. For example $Q(x) = x_1^4 - x_2^4$ is not positive definite near the origin in R^2.

Definition 4.4.2 (Lyapunov function). A function V defined on Ω to R is said to be a Lyapunov function for the system $\dot{x} = f(x)$, $f(0)=0$, if

(a) V is positive definite on Ω,

and

(b) $\dot{V}(x) \equiv (f(x), V_x) = \Sigma f_i(x) V_{x_i} \leqslant 0$, on Ω.

Lyapunov stability theorems

Lyapunov stability theorems are, in fact, generalizations of the concept that if the energy of the system near an equilibrium point decreases with increasing time, then the system is stable. Thus the Lyapunov function, as defined above, is a sort of generalized 'energy function', though it may not represent the actual energy of the system.

With this background we are now prepared to study the basic stability theorems.

Theorem 4.4.3 (Local stability). *Consider the autonomous system*

$$S: \dot{x} = f(x), \qquad \text{with } f(0) = 0,$$

and suppose there exists an open neighbourhood Ω of the origin in R^n and a Lyapunov function V on Ω.

Then the system S is stable in the Lyapunov sense (see Definition 4.1.1) *with respect to the zero state.*

Proof Let β be a positive number such that the ball $B_\beta \equiv \{x \in R^n: \|x\| < \beta\} \subset \Omega$. Define, for $k>0$, the set $E_k \equiv \{x \in \Omega: V(x) < k\}$. Since V is continuous and $V(x) > 0$ for $x \in \Omega \backslash \{0\}$ with $V(0)=0$, E_k is a nonempty, open, and connected set containing the origin. Consequently there exists a $k^* > 0$, so that E_{k^*}, with its boundary $\partial E_{k^*} \equiv \{x \in \Omega: V(x) = k^*\}$, is contained in B_β; that is, $\bar{E}_{k^*} \equiv E_{k^*} \cup \partial E_{k^*} \subset B_\beta$. Since

$0 \in E_{k^*}$ we can choose an α sufficiently small such that the ball $B_\alpha \subset E_{k^*}$. Now take any point $x_0 \in B_\alpha$ and let $T_t(x_0) \equiv x(t, x_0)$, $t \geqslant 0$, be the trajectory starting from the state x_0. Then

$$V(x(t, x_0)) = V(x_0) + \int_0^t \dot{V}(x(\theta, x_0))\, d\theta \qquad \text{for } t \geqslant 0. \tag{4.4.5}$$

Since $V \in C^1(\Omega)$ and $t \rightarrow x(t, x_0)$ is continuous, it is clear that $\dot{V}(x(t, x_0)) \leqslant 0$ as long as $T_t(x_0) \in \Omega$, and consequently $V(x(t, x_0)) \leqslant V(x_0)$. By our choice $x_0 \in B_\alpha \subset E_{k^*}$ and hence $V(x(t, x_0)) \leqslant V(x_0) < k^*$. Therefore $T_t(x_0) \equiv x(t, x_0) \in E_{k^*} \subset B_\beta$ for all $t > 0$, which implies stability in the Lyapunov sense (Definition 4.1.1). ■

Note that α depends on k^* which in turn depends on β and hence $\alpha = \alpha(\beta)$. The proof given above can be easily visualized from the diagram (figure 4.1). The next theorem gives sufficient conditions for asymptotic stability of the zero state ($0 \in R^n$).

Theorem 4.4.4 (Local asymptotic stability). *If the system S is stable with respect to the zero state in the sense of Theorem* 4.4.3 *and further* $\dot{V} < 0$ *on* $\Omega\backslash\{0\}$, *then the system is also asymptotically stable, i.e.* $\lim_{t\rightarrow\infty} T_t(x_0) = 0$ *for each* $x_0 \in \Omega$.

Proof Since the system is stable in the sense of Theorem 4.4.3, for every $\beta > 0$ there exists an $\alpha = \alpha(\beta) > 0$ such that $T_t(x_0) \in B_\beta$ for all $x_0 \in B_\alpha$. We must show that $T_t(x_0) \rightarrow 0$ as $t \rightarrow \infty$. Since $V(x) > 0$ for $x \in \Omega\backslash\{0\}$ and $V(0) = 0$ it suffices to show that $V(T_t(x_0)) \rightarrow 0$ as $t \rightarrow \infty$.

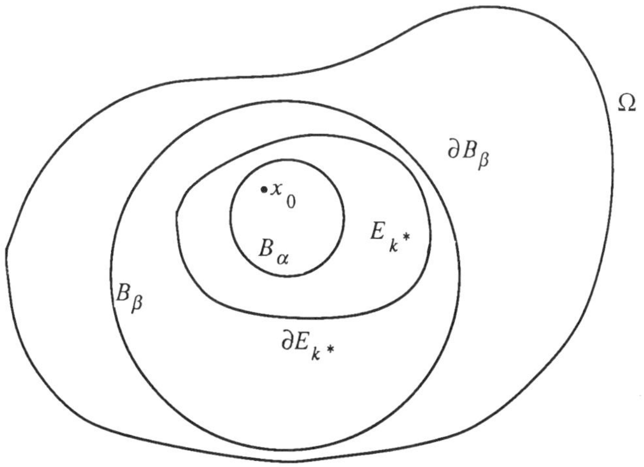

Figure 4.1

Suppose not. Then there exists an $r \in (0, \beta]$ such that $V(T_t(x_0)) \geqslant r$ for all $t \geqslant 0$ and consequently $T_t(x_0) \in E_r'$ for all $t \geqslant 0$, where $E_r = \{x \in \Omega: V(x) < r\}$. Since $\dot{V}(x) = (f(x), V_x)$ and $f(0) = 0$, $\dot{V}(0) = 0$ and by hypothesis $\dot{V}(x) < 0$ for $x \in \Omega \backslash \{0\}$. Further, $\dot{V}(x)$ being continuous on $\Omega(V \in C^1(\Omega))$ it is clear that $\max\{\dot{V}(x), x \in E_r' \cap \bar{B}_\beta\}$ exists and since $0 \notin E_r' \cap \bar{B}_\beta$, there exists an $\eta > 0$ such that $\max\{\dot{V}(x), x \in E_r' \cap B_\beta\} = -\eta < 0$. Therefore, since $T_t(x_0) \in E_r' \cap \bar{B}_\beta$, for all $t \geqslant 0$, $\dot{V}(T_t(x_0)) \leqslant -\eta$ for all $t \geqslant 0$. Hence we conclude from the equality

$$V(T_t(x_0)) = V(x_0) + \int_0^t \dot{V}(T_\theta(x_0))\, \mathrm{d}\theta, \tag{4.4.6}$$

that

$$0 < r \leqslant V(T_t(x_0)) \leqslant V(x_0) - \eta t \qquad \text{for all } t \geqslant 0.$$

But this is impossible and hence our hypothesis that $\lim_{t \to \infty} V((T_t(x_0)) \neq 0$ is false. This completes the proof. ■

In certain situations it may be easier to detect instability rather than stability. Therefore, at least for the purpose of detection (stability or instability), instability theorems may prove to be useful. In the process of system design if a particular configuration proves to be unstable one may try to modify the design. In the following we present a couple of results of this nature. The proofs are quite simple.

Theorem 4.4.5 (Instability). *If in some neighbourhood $\Omega_0 \subset \Omega$ of the origin there exists a function V which is positive definite on Ω_0 and further $\dot{V} > 0$ on $\Omega_0 \backslash \{0\}$ then the zero state is unstable.*

Proof Let $\beta > \alpha > 0$ be arbitrary so that $B_\alpha \subset B_\beta \subset \Omega_0$. Take $x_0(\neq 0) \in B_\alpha$ and let $T_t(x_0)$ denote the corresponding solution trajectory. Since $\dot{V}(x) > 0$ for $x \in \Omega_0 \backslash \{0\}$ and $t \to T_t(x_0)$ is continuous, it follows from the equation

$$V(T_t(x_0)) = V(x_0) + \int_0^t \dot{V}(T_\theta(x_0))\, \mathrm{d}\theta,$$

that $V(T_t(x_0)) > V(x_0)$ for all $t > 0$. Then $T_t(x_0) \neq 0$ for all $t > 0$, which in turn implies that $\dot{V}(T_t(x_0)) > 0$ for all $t > 0$, and hence it follows from the above equation that $V(T_t(x_0))$ is a monotonically increasing function of $t > 0$. Consequently $T_t(x_0)$ cannot have a limit (as $t \to \infty$) in B_β; it must eventually hit the boundary ∂B_β. This is instability; that is, the zero state is unstable. ■

An interesting instability result given by Četaev is illustrated in figure 4.2.

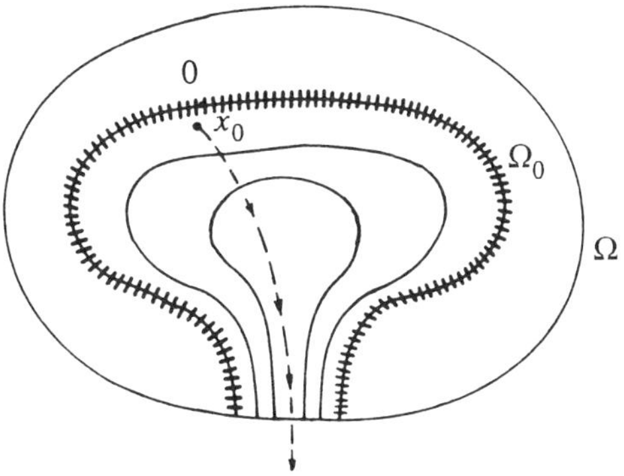

Figure 4.2

Suppose Ω is a neighbourhood of the zero state (0) and Ω_0 an open subset of Ω and V a function defined on Ω_0 such that

(a) $V \in C^1(\Omega_0)$,
(b) $V(x), \dot{V}(x) > 0$ for $x \in \Omega_0$,
(c) $V(x) = 0$ for $x \in \partial\Omega_0 \cap \Omega$ (shown cross-hatched in figure 4.2),
(d) $0 \in \partial\Omega_0 \cap \Omega$.

Then, if $x_0 \in \Omega_0$, near the zero state, $V(T_t(x_0))$ must increase with time (due to (b)) and the solution trajectory $T_t(x_0)$ must leave Ω_0. But $T_t(x_0)$ can not cross $\partial\Omega_0 \cap \Omega$ (due to (c)), since in that case $V(T_\tau(x_0)) = 0$ at some time $\tau > 0$. Since V is continuous this will contradict the fact that $t \rightarrow V(T_t(x_0))$ is an increasing function. Hence $T_t(x_0)$ must leave Ω_0 and Ω through the bottle neck $\partial\Omega_0 \cap \partial\Omega$ as shown. Thus we have shown that if V satisfies conditions (a)–(d) then the origin (0 state) is unstable. ■

Stability for time-varying systems

Consider the time-varying system

$$\dot{x} = f(t, x), \qquad t \geqslant 0, \tag{4.4.7}$$

where $f: R_0 \times \Omega \rightarrow R^n$ is measurable in t on $R_0 = \{t: t \geqslant 0\}$ and continuous in x on Ω. For each $x_0 \in \Omega$, let $x(t) \equiv x(t, t_0, x_0) \equiv T_{t,t_0}(x_0)$ denote the solution trajectory (path) corresponding to the initial state x_0.

Let $W(x)$, $x \in \Omega$, be a positive definite function as given in Definition 4.4.1. For time-varying systems we define $V(t, x)$, $t \geqslant 0$, $x \in \Omega$, to be a positive definite function if it satisfies the following properties:

(a) $V \in C^1$, that is, once continuously differentiable on $R_0 \times \Omega$,
(b) $V(t, 0) = 0$ for all $t \geqslant 0$,
(c) $V(t, x) \geqslant W(x)$ for all $t \geqslant 0$ and $x \in \Omega$.

Clearly the total time derivative of a positive definite function V along any solution path is given by

$$\dot{V}(t, x(t)) = \frac{\partial V}{\partial t} + \sum_{i=1}^{n} V_{x_i}(t, x(t)) f_i(t, x(t)). \tag{4.4.8}$$

We can now define a Lyapunov function as follows:

Definition 4.4.6 For time-varying systems a function V satisfying properties (a)–(c) and

(d) $\dot{V} \leqslant 0$,

is called a Lyapunov function.

With this background we can now present stability results for time-varying systems which are similar to those of time-invariant systems.

Theorem 4.4.7 (Local stability, time-varying systems). *Consider the system*

$$T: \dot{x} = f(t, x), \qquad f(t, 0) = 0, \quad t \geqslant 0,$$

and suppose there exists an open neighbourhood Ω of the origin in R^n and a Lyapunov function V on $R_0 \times \Omega$. Then the system T is stable with respect to the origin.

The proof is similar to that of Theorem 4.4.3. For asymptotic stability we have a result similar to Theorem 4.4.4.

Theorem 4.4.8 (Local asymptotic stability). *If the system T is stable with respect to the zero state in the sense of Theorem* 4.4.7 *and further if $V < 0$ on $R_0 \times \Omega \backslash \{0\}$, then the system is also asymptotically stable, i.e. $T_{t,t_0}(x_0) \to 0$ as $t \to \infty$ for each $t_0 \in R_0$ and $x_0 \in \Omega$.*

Again the proof is analogous to that of Theorem 4.4.4.

To obtain conditions for global stability we need some minor variations in the statements of the above theorems.

Theorem 4.4.9 (Global stability/asymptotic stability). *If in Theorem* 4.4.3 (4.4.4) *and Theorem* 4.4.7 (4.4.8) *one can take $\Omega = R^n$ and, in addition, if $V(x) \to \infty$ as $|x| \to \infty$, then the system is globally stable (globally asymptotically stable).*

The proof is left as an exercise for the reader.

For illustration of the preceding results we present here a few examples.

Example 4.4.10 Consider the second order-system

$$m\ddot{\xi} + b\dot{\xi} + k\xi = 0, \tag{4.4.9}$$

with mass m, viscous damping b and stiffness coefficient k, all assumed positive. Setting $x_1 = \xi$, $x_2 = \dot{\xi}$, we have the state equation

$$\begin{aligned} \dot{x}_1 &= f_1(x_1, x_2) \equiv x_2 \\ \dot{x}_2 &= f_2(x_1, x_2) \equiv -(k/m)x_1 - (b/m)x_2, \end{aligned} \tag{4.4.10}$$

where $(x_1, x_2) = (0, 0)$ is the only rest state. Define $V(x_1, x_2) = \frac{1}{2}kx_1^2 + \frac{1}{2}mx_2^2$ which represents the sum of potential and kinetic energies of the system. Computing $\dot{V}$, we have

$$\dot{V} = (V_x f) = V_{x_1} f_1 + V_{x_2} f_2 = -bx_2^2 \leqslant 0.$$

Clearly, V satisfies all the conditions (a) and (b) of Definition 4.4.2, and therefore is a Lyapunov function for the system; hence by Theorem 4.4.3 the system is stable. In fact, the system is asymptotically stable with respect to the zero state. To justify this we must show that (see Theorem 4.4.4) $\dot{V} < 0$ for all $x \in R^2 \backslash \{0\}$. Indeed if $x_2 = 0$, but $x_1 \neq 0$, then

$$\left(\frac{\mathrm{d}x_2}{\mathrm{d}x_1}\right)\Bigg|_{\substack{x_1 \neq 0 \\ x_2 = 0}} = \frac{(-k/m)x_1 - (b/m)x_2}{x_2}\Bigg|_{\substack{x_1 \neq 0 \\ x_2 = 0}} = \begin{cases} -\infty, & x_1 > 0 \\ +\infty, & x_1 < 0 \end{cases}$$

which means that no part of the trajectory, $x(t) = (x_1(t), x_2(t))'$, can spend a positive length of time on the x_1-axis away from the origin. In other words all trajectories must leave the x_1-axis instantly. Hence $\dot{V} < 0$ for all $(x_1, x_2) \in R^2 \backslash \{0\}$, and the system is asymptotically stable.

Note that the damper is a dissipator of energy with the rate of dissipation given by $\dot{V} = -bx_2^2$. One can feel this by touching the damper of a car after driving on a rough road.

Example 4.4.11 Consider Van der Pol's equation given by

$$\ddot{\xi} + \beta(\xi^2 - 1)\dot{\xi} + k\xi = 0, \tag{4.4.11}$$

with $k > 0$ and $\beta < 0$. The state equation is

$$\begin{cases} \dot{x}_1 = f_1(x_1, x_2) = x_2 \\ \dot{x}_2 = f_2(x_1, x_2) = -kx_1 - \beta(x_1^2 - 1)x_2, \end{cases} \tag{4.4.12}$$

having the unique rest state $x^* = (0, 0)'$. Define $\Omega \equiv \{(x_1, x_2) \in R^2: |x_1| < 1\}$ and $V(x) \equiv V(x_1, x_2) = (\frac{1}{2})kx_1^2 + (\frac{1}{2})x_2^2$. Computing $\dot{V}$, we have

$$\dot{V}(x) = \beta(1 - x_1^2)x_2^2.$$

Clearly V is a Lyapunov function for the system on Ω and $\dot{V}<0$ on $\Omega\backslash\{0\}$. Hence by Theorem 4.4.4, the system is asymptotically stable with respect to the origin in the neighbourhood Ω.

Example 4.4.12 Consider the system

$$\ddot{\xi}+b\dot{\xi}+g(\xi)=0, \tag{4.4.13}$$

with $b>0$, $g(\xi)\xi>0$ for $\xi\neq 0$ and $g(0)=0$. The state equation is

$$\begin{cases}\dot{x}_1=x_2\\ \dot{x}_2=-g(x_1)-bx_2,\end{cases} \tag{4.4.14}$$

with $0\in R^2$ being the only rest state. For the Lyapunov function, let us try

$$V(x_1,x_2)=(\tfrac{1}{2})x_2^2+\int_0^{x_1} g(\xi)\,\mathrm{d}\xi.$$

Computing $\dot{V}$ we have

$$\dot{V}=g(x_1)x_2+x_2(-g(x_1)-bx_2)=-bx_2^2\leqslant 0.$$

Clearly $V\in C^1(R^2)$, $V(0)=0$, $V(x)=V(x_1,x_2)>0$ for all $x\in R^2\backslash\{0\}$ and $\dot{V}\leqslant 0$. Hence V is a Lyapunov function for the system and the zero state is stable. In fact, the zero state is asymptotically stable. Indeed, by using similar arguments as given in Example (4.4.10), justified by the relation

$$\left(\frac{\mathrm{d}x_2}{\mathrm{d}x_1}\right)\Bigg|_{\substack{x_1\neq 0\\ x_2=0}}=\frac{-g(x_1)-bx_2}{x_2}\Bigg|_{\substack{x_1\neq 0\\ x_2=0}}=\pm\infty,$$

one arrives at the conclusion that $\dot{V}<0$ off the origin, and the asymptotic stability follows. The system is actually globally asymptotically stable.

Example 4.4.13 Consider the system of Example 4.4.11 with $\beta>0$. Then V given there is positive definite and $\dot{V}>0$ for $(x_1,x_2)\in\Omega\backslash\{0\}$. Hence by the instability Theorem 4.4.5 the origin is unstable.

Example 4.4.14 (Stabilization of satellites). The dynamics of a satellite in geosynchronous orbit is given by

$$\begin{cases}I_x\dot{p}+(I_z-I_y)qr=T_1,\\ I_y\dot{q}+(I_x-I_z)pr=T_2,\\ I_z\dot{r}+(I_y-I_x)pq=T_3,\end{cases} \tag{4.4.15}$$

along with Euler equations for the attitudes, where I_x, I_y, I_z are the moments of inertia along the x, y, z directions respectively and (p,q,r) are the angular velocities and T_1, T_2, T_3 are the external torques. In the absence of external torques, the system is conservative and keeps wobbling around, which is undesirable for certain applications. For example, communication satellites must maintain pointing accuracy and any undesirable motion must be stalled. For a Lyapunov function one can take the total kinetic energy

$$V(p, q, r) = \tfrac{1}{2}(I_x p^2 + I_y q^2 + I_z r^2),$$

giving

$$\dot{V} = T_1 p + T_2 q + T_3 r. \tag{4.4.16}$$

Choosing $T_1 = -u_1(p)$, $T_2 = -u_2(q)$, $T_3 = -u_3(r)$, where $u_i(\xi)\xi > 0$ for $\xi \neq 0$, we have

$$\dot{V} = -u_1(p)p - u_2(q)q - u_3(r)r < 0 \qquad \text{for } (p, q, r) \neq (0, 0, 0). \tag{4.4.17}$$

Clearly with this control the system is globally asymptotically stable with respect to the 0 state. For proportional control one takes $u_i(\xi) = k_i \xi$, and for bang-bang control one has $u_i(\xi) = k_i \operatorname{sign} \xi$ with $k_i > 0$.

4.5 Domain of stability, limit sets, invariant sets and stability of sets

For practical systems asymptotic stability is more important than mere stability. The reason is very simple. Machines, operating in a given state of performance, must (eventually!) return to that state after the disturbance has been removed. For example, a synchronous generator must operate at a specified frequency and terminal voltage, and it is absolutely necessary that the machine returns to that state following the removal of the perturbation which may have been caused by temporary short circuit or otherwise. A highway bridge, subjected to vibration caused by a passing vehicle, must return to its stationary state after the vehicle has passed. Similarly often it is desirable that an economy returns to its equilibrium following the removal of the cause of disturbance.

In the previous section we have studied stability of nonlinear systems in the immediate neighbourhood of their equilibrium states (rest states or critical states). Those are local properties except for linear systems. In the case of linear systems, local and global stability concepts are equivalent. But nonlinear systems can be locally stable and yet globally unstable. Hence the stability results of the previous sections may not be completely satisfactory for practical applications. In the case of large perturbations the system may very well become unstable even though theoretically it is locally stable. Thus, for practical systems, it is absolutely essential to know the size of perturbations that may be tolerated without causing serious instability. In other words, it is necessary to determine the size of the domain of stability around the equilibrium states. In this section we present stability theorems which are useful for estimating the domain of asymptotic stability. In this connection, LaSalle's invariance principle plays a central role. The results we present here are broadly extensions of Lyapunov theory discovered by LaSalle, Lefschetz and others [14], [15].

For the results presented here we make use of the basic properties of the transition semigroup $\{T_t, t \in R\}$, also called the solution operator, as established in section 3.2.

Let Ω be an open connected set in R^n (could be the entire space R^n) and suppose all the solutions of the system governed by the differential equation

$$S: \dot{x} = f(x), \qquad t \in R,$$

lie in Ω. That is, for each $x_0 \in \Omega$, $T_t(x_0) \in \Omega$ for all $t \in R$. This of course eliminates the possibility of finite escape (explosion) time. However, this can be taken care of by defining the maximal interval of existence of solutions for each initial state $x_0 \in \Omega$. Indeed, for each $x_0 \in \Omega$, define $I(x_0) \equiv (a(x_0), b(x_0)) \subset R$ to be the maximal interval of existence of solution of the system S corresponding to the initial state x_0 satisfying the properties

(a) $T_t(x_0) \in \Omega$ for all $t \in I(x_0)$,

(b) $\lim_{t \to b(x_0)} T_t(x_0)$, $\lim_{t \to a(x_0)} T_t(x_0) \in \partial\Omega$.

However, we will avoid this cumbersome notation by assuming that $\Omega = R^n$ and that the $\lim_{t \to \pm\infty} |T_t(x_0)|$ may assume finite or infinite values. The concept of limit points is extremely useful here. It is the set of all the points in Ω to which any of the solutions of the system S converge as $t \to \pm\infty$. Formally, we define this as follows.

Definition 4.5.1 (Positive limit point). A point $y \in \Omega$ is called a positive limit point corresponding to an initial state $\xi \in \Omega$ if there exists a sequence $\{t_n\} \in R_+ = [0, \infty)$ such that $t_n \to +\infty$ as $n \to \infty$ and $T_{t_n}(\xi) \to y$ as $n \to \infty$.

From this we obtain the definition for positive limit set.

Definition 4.5.2 (Positive limit set). A positive limit set, corresponding to an initial state ξ, is the set of all positive limit points of ξ denoted

$\Gamma^+(\xi)$. That is

$$\Gamma^+(\xi) = \{y \in \Omega\text{: there exists a sequence } \{t_n\} \in R_+ = [0, \infty) \text{ with } t_n \to \infty \text{ as } n \to \infty \text{ and } T_{t_n}(\xi) \to y \text{ as } n \to \infty\}.$$

Similarly, by reversing the flow of time one can define negative limit points and negative limit sets denoted $\Gamma^-(\xi)$, $\xi \in \Omega$. If, for all sequences $\{t_n\} \subset R_+ = [0, \infty)$, with $t_n \to \infty$, $T_{t_n}(\xi) \to y$, then $\Gamma^+(\xi) = \{y\}$ is a one point set. Roughly speaking, if the solution trajectory, $x(t, \xi) = T_t(\xi)$, $t \geq 0$, spirals indefinitely around a set $D \subset \Omega$, then D is its positive limit set. In case of second-order systems D is a closed curve in R^2 representing a periodic solution and is known as the limit cycle. For a set $K \subset \Omega$, $\Gamma^+(K) = \bigcup\{\Gamma^+(\xi), \xi \in K\}$ is the set of positive limit points originating from the set K. One can easily verify that

$$\Gamma^+(\xi) = \bigcap_{t>0} \overline{\{T_\theta(\xi), \theta \geq t\}} \tag{4.5.1}$$

where the bar stands for the closure of the underlying set.

Definition 4.5.3 (Invariant set). With reference to the system $\dot{x} = f(x)$, or equivalently its transition operator T_t, $t \in R$, a set $K \subset \Omega$ is said to be positively invariant if for each $\xi \in K$, $T_t(\xi) \in K$ for all $t \in [0, b(\xi))$, negatively invariant if $T_t(\xi) \in K$ for all $t \in (a(\xi), 0]$, weakly invariant if $T_t(\xi) \in K$ for all $t \in (a(\xi), b(\xi))$, and invariant if it is weakly invariant and $(a(\xi), b(\xi)) = (-\infty, +\infty)$.

Recall that the maximal interval of existence of solution $(a(\xi), b(\xi))$ for the system $\dot{x} = f(x)$, with $x(0) = \xi$, may be finite or infinite. In case the interval is finite, the solution $T_t(\xi)$ reaches the boundary of Ω or escapes to ∞ as t approaches one of the end points of $a(\xi)$ or $b(\xi)$.

Some useful properties of the limit sets are given in the following results.

Theorem 4.5.4

(a) *For each $\xi \in \Omega$, the positive limit set $\Gamma^+(\xi)$ is a closed, invariant set.*
(b) *If the solution trajectory $\{T_t(\xi), t \geq 0\}$ is contained in a bounded subset of Ω, then $\Gamma^+(\xi)$ is also compact and connected.*

Proof (a) For closure we must show that if the sequence $\{y_n\} \in \Gamma^+(\xi)$ and $y_n \to y_0$, then $y_0 \in \Gamma^+(\xi)$. Since $\{y_n\} \in \Gamma^+(\xi)$, for each positive integer n, there exists a sequence $\{t_k^{(n)}, k = 1, 2, \ldots\} \in R_+ = [0, \infty)$ such that $\lim_{k\to\infty} t_k^{(n)} = \infty$ and $\lim_{k\to\infty} T_{t_k^{(n)}}(\xi) = y_n$. Hence for each n there exists an integer k_n such that

$$|T_{t_{k_n}^{(n)}}(\xi) - y_n| < \frac{1}{n}.$$

Define $\tau_n \equiv t_{k_n}^{(n)}$. Using the triangle inequality we have

$$|T_{\tau_n}(\xi) - y_0| \leq |T_{\tau_n}(\xi) - y_n| + |y_n - y_0| \leq \frac{1}{n} + |y_n - y_0|.$$

Since $y_n \to y_0$, it follows from the above inequality that $T_{\tau_n}(\xi) \to y_0$ as $n \to \infty$. Consequently, by definition of the limit set, $y_0 \in \Gamma^+(\xi)$ and hence $\Gamma^+(\xi)$ is closed. For invariance, we must show that, for every $y \in \Gamma^+(\xi)$, $T_t(y) \in \Gamma^+(\xi)$. Let $y \in \Gamma^+(\xi)$, then by definition of limit sets, there exists a sequence $\{t_n\} \in R_+$ such that $t_n \to \infty$ and $T_{t_n}(\xi) \to y$ as $n \to \infty$. Since, for each t fixed, $z \to T_t(z)$ is a continuous mapping from $R^n \to R^n$ (see section 3.2), we have

$$\lim_{n\to\infty} T_t(T_{t_n}(\xi)) = T_t\Big(\lim_{n\to\infty} T_{t_n}(\xi)\Big) = T_t(y).$$

On the other hand, by the semigroup property,

$$T_t T_s = T_{t+s}, \qquad T_t(T_{t_n}(\xi)) = T_{t+t_n}(\xi).$$

Since $\Gamma^+(\xi)$ is closed and $T_{t+t_n}(\xi)$ is a convergent sequence,

$$T_t(y) = \lim_{n\to\infty} T_{t+t_n}(\xi) = z \in \Gamma^+(\xi).$$

This proves that for each $y \in \Gamma^+(\xi)$, $T_t(y) \in \Gamma^+(\xi)$, i.e., $T_t(\Gamma^+(\xi)) \subset \Gamma^+(\xi)$ for all $t \geq 0$.

(b) For each $\xi \in \Omega$, for which the solution $\{T_t(\xi), t \geq 0\}$ is bounded, the corresponding (positive) limit set $\Gamma^+(\xi)$ must necessarily be bounded since it is the limit of such bounded sequences. Combining this with the result (a) we conclude that $\Gamma^+(\xi)$ is a closed bounded subset of Ω and hence compact. We prove connectedness by establishing a contradiction. Suppose $\Gamma^+(\xi)$ is not connected. Then there exist two disjoint compact sets P, Q such that $\Gamma^+(\xi) = P \cup Q$. Since P, Q are disjoint there exist ε, $\eta > 0$ such that $\overline{N_\varepsilon(P)} \cap \overline{N_\eta(Q)} = \varnothing$ (empty) where $\overline{N_\varepsilon(P)}$ denotes the closure of the ε-neighbourhood of the set P, and $\overline{N_\eta(Q)}$ has similar meaning. Let $p \in P$ and $q \in Q$; then by definition there exists a pair of time sequences $\{t_n, \tau_n\} \subset R_+$ such that $T_{t_n}(\xi) \to p$ and $T_{\tau_n}(\xi) \to q$. We can choose p, q and $\{t_n, \tau_n\}$ such that $t_n < \tau_n$ for all n and both t_n, $\tau_n \to \infty$ as $n \to \infty$. Then we choose another sequence $\{s_n\} \in R_+$ such that $t_n < s_n < \tau_n$ and $T_{s_n}(\xi) \in \partial\overline{N_\eta(Q)}$ (boundary of $\overline{N_\eta(Q)}$) for all n. Since Q is compact, both $\overline{N_\eta(Q)}$ and $\partial\overline{N_\eta(Q)}$ are compact; and hence there exists a subsequence $\{n_k\} \subset \{n\}$ and a point $\gamma \in \partial\overline{N_\eta(Q)}$ such that $T_{s_{n_k}}(\xi) \to$

$\gamma \in \partial \overline{N_\eta(Q)}$. On the other hand, by definition of the limit set $\Gamma^+(\xi)$, $\gamma \in \Gamma^+(\xi)$. This is a contradiction since $\gamma \in \Gamma^+(\xi)$ but it is not in $P \cup Q$. ■

Remark 4.5.5 Similar results hold for negative limit sets $\Gamma^-(\xi)$. Note that in the definitions for positive (or negative) limit sets it is not required that $\lim_{t\to\infty} T_t(\xi)$ exists. The limit sets are actually the limits of all convergent subsequences if, indeed, any exist. In fact, there may be no subsequences having limits and hence $\Gamma^+(\xi)(\Gamma^-(\xi))$ may very well be empty depending on the system and hence on the semigroup $\{T_t, t \in R\}$ and $\xi \in \Omega$. However, if for a given $\xi \in \Omega$, the solution trajectory $\{T_t(\xi), t \geq 0\}$ remains bounded, then $\Gamma^+(\xi)$ is non-empty and also, as $t \to \infty$, $T_t(\xi) \to \Gamma^+(\xi)$, in the sense that

$$\lim_{t\to\infty} \rho(T_t(\xi), \Gamma^+(\xi)) = 0,$$

where, for $x \in R^n$ and $G \subset R^n$,

$$\rho(x, G) = \inf\{|x - y|, y \in G\}. \tag{4.5.2}$$

With this background we can now study the invariance principle of LaSalle [14] and present stability results that can be used for estimating the domain of stability.

Theorem 4.5.6 (LaSalle invariance). *Let $V: \Omega_0 \to R$, with $V \in C^1(\Omega_0)$, where Ω_0 is any subset of Ω containing the origin, and define the set $E_\ell \equiv \{x \in \Omega_0: V(x) < \ell\}$ with $\ell \geq 0$. Suppose there exists an $\ell > 0$ for which E_ℓ is a bounded set and*

(a) *$V(x) > 0$ on $E_\ell \backslash \{0\}$*
(b) *$\dot{V}(x) \leq 0$ on E_ℓ.*

Let $F \equiv \{x \in E_\ell: \dot{V}(x) = 0\}$ and M be the largest invariant subset of F. Then every solution $T_t(\xi)$, starting from $\xi \in E_\ell$, tends to M, i.e.,

$$T_t(E_\ell) \to M \qquad \text{as } t \to \infty.$$

Proof Let $x_0 \in E_\ell$ be arbitrary and $T_t(x_0)$, $t \geq 0$, the corresponding solution of $\dot{x} = f(x)$. Then, since $\dot{V} \leq 0$ on E_ℓ,

$$V(T_t(x_0)) = V(x_0) + \int_0^t \dot{V}(T_\theta(x_0))\, d\theta \leq V(x_0) < \ell.$$

Thus $t \to V(T_t(x_0))$ is non-increasing, $T_t(x_0) \in E_\ell$ for all $t \geq 0$, and, due to assumption (a), $V(T_t(x_0)) \geq 0$ for all $t \geq 0$. Therefore, $V(T_t(x_0))$ has a limit, say ℓ_0, as $t \to \infty$. That is,

(i) $\lim_{t\to\infty} V(T_t(x_0)) = \ell_0 < \ell$.

Since V is continuous on E_ℓ, we also have

(ii) $\lim_{t\to\infty} V(T_t(x_0)) = V\left(\lim_{t\to\infty} T_t(x_0)\right)$.

Since $T_t(x_0) \in E_\ell$, $t \geqslant 0$, and E_ℓ is a bounded set, it follows from our previous discussion that $\Gamma^+(x_0)$ is non-empty and

(iii) $\lim_{t\to\infty} T_t(x_0) \in \Gamma^+(x_0)$.

Hence from (i)–(iii) it follows that

(iv) $V(\xi) = \ell_0$ for all $\xi \in \Gamma^+(x_0)$.

Consequently V is constant on $\Gamma^+(x_0)$ and hence $\dot{V}(\xi) = 0$ for $\xi \in \Gamma^+(x_0)$. This implies that $\Gamma^+(x_0) \subset F$. Since $\Gamma^+(x_0)$ is an invariant set (see Theorem 4.5.4) and, by hypothesis, M is the largest invariant subset of F, it follows that

(v) $\Gamma^+(x_0) \subset M$.

Hence

$$T_t(x_0) \to M \qquad \text{as } t \to \infty.$$

Since this is true for every $x_0 \in E_\ell$ we have also proved that

$$T_t(E_\ell) \to M \qquad \text{as } t \to \infty. \quad \blacksquare$$

Remark 4.5.7 Note that it follows from (iv), as given in the proof of the above theorem, that

$$\Gamma^+(x_0) \subset V^{-1}(\ell_0) \equiv \{x \in E_\ell : V(x) = \ell_0\}$$

and hence combining with (v) we have

$$\Gamma^+(x_0) \subset M \cap V^{-1}(\ell_0). \tag{4.5.3}$$

Consequently $T_t(x_0) \to M \cap V^{-1}(\ell_0)$ as $t \to \infty$. From this result and the fact that $\Gamma^+(x_0)$ is a connected set (by Theorem 4.5.4) we can conclude that, if $M \cap V^{-1}(\ell_0)$ is a set of isolated points, then $\Gamma^+(x_0)$ consists of only one of these points. This shows that a Lyapunov function can be used to establish the existence of equilibrium points for the system $\dot{x} = f(x)$.

Example 4.5.8 The above remark is well illustrated by the following example.

$$\begin{cases} \dot{x}_1 = -x_1 g_1(x_2), & g_1(\xi) > 0, \quad \xi \neq 0 \quad \text{and} \quad g_1(0) = 0, \\ \dot{x}_2 = -x_2 g_2(x_1), & g_2(\xi) < 0, \quad \xi \neq 0 \quad \text{and} \quad g_2(0) = 0. \end{cases} \tag{4.5.4}$$

Taking $V = \frac{1}{2}x_1^2 + \frac{1}{2}x_2^2$, one has

$$\dot{V} = -x_1^2 g_1(x_2) - x_2^2 g_2(x_1),$$

$$F = \{(x_1, x_2): \dot{V}(x_1, x_2) = 0\} = \{(x_1, x_2): x_2 = 0\} \cup \{(x_1, x_2): x_1 = 0\},$$

and $M = F$. Also $V^{-1}(\ell_0) = \{p_1, p_2, p_3, p_4\}$ where

$$p_1 = (0, (2\ell_0)^{1/2}), \quad p_2 = (0, -(2\ell_0)^{1/2}),$$
$$p_3 = ((2\ell_0)^{1/2}, 0), \quad p_4 = (-(2\ell_0)^{1/2}, 0).$$

Since $\Gamma^+(x_0) \subset M \cap V^{-1}(\ell_0) = \{p_1, p_2, p_3, p_4\}$ it is one of these points to which $T_t(x_0)$ converges to as $t \to \infty$.

For asymptotic stability a result similar to Theorem 4.5.6 hold.

Theorem 4.5.9 *If condition* (*b*) *of Theorem* 4.5.6 *is replaced by* (b′): $\dot{V}(x) < 0$ *on* $E_\ell \backslash \{0\}$, *then the zero state is asymptotically stable and the domain of asymptotic stability contains the set* E_ℓ.

Proof The proof follows immediately from the previous theorem. Indeed, in this case, the set F consists of only one point $\{0\}$ and hence, being a subset of F, $M = \{0\}$. Since, for each $x_0 \in E_\ell$, $\Gamma^+(x_0)$ is not empty and $\Gamma^+(x_0) \subset M$, we have $\Gamma^+(x_0) = \{0\}$. That is, $T_t(x_0) \to 0$ as $t \to \infty$ for each $x_0 \in E_\ell$. Thus, E_ℓ is contained in the region of asymptotic stability. ■

Remark 4.5.10 For application of the above results in estimating the size of the domain of stability one can choose any arbitrary function V from $C^1(R^n)$ which is positive in a neighbourhood of the origin of R^n with $V(0) = 0$. Then one checks if $\dot{V}$ is negative semidefinite (definite) on that set. If so, one has an estimate of the size of the domain of stability (asymptotic stability). If not, one tries for a smaller domain until one discovers that there is no non-empty set satisfying the above properties. In other words if, for a given V, E_ℓ is empty for each $\ell > 0$, one must try with a new V function. Clearly the size of the estimated domain depends on the selection of the V function.

We illustrate this by a few examples.

Example 4.5.11 Consider the general second-order equation

$$\ddot{\xi} + f(\xi)\dot{\xi} + g(\xi) = 0 \tag{4.5.5}$$

where f and g are even and odd functions respectively. Define $F(x) = \int_0^x f(\xi)\,d\xi$ and $G(x) = \int_0^x g(\xi)\,d\xi$. Then an equivalent system is

$$\begin{cases} \dot{x} = y - F(x) \\ \dot{y} = -g(x). \end{cases} \tag{4.5.6}$$

The rest state is given by the solution of

$$\begin{cases} y - F(x) = 0 \\ \quad g(x) = 0. \end{cases} \tag{4.5.7}$$

If $g = 0$ only at $x = 0$, then the rest state is $p_0 = (0, 0)$. We wish to estimate the domain of stability of the rest state p_0. Try the total energy function

$$V(x, y) = \tfrac{1}{2}y^2 + G(x).$$

Then $\dot{V} = -yg(x) + g(x)(y - F(x)) = -g(x)F(x)$. Now for the estimation of the size of the domain we find ℓ such that on

$$E_\ell = \{(x, y): \tfrac{1}{2}y^2 + G(x) < \ell\},$$

$V \geqslant 0$ and $\dot{V} \leqslant 0$. That is, we find E_ℓ so that

$$\begin{cases} 0 \leqslant \tfrac{1}{2}y^2 + G(x) < \ell, \qquad (x, y) \in E_\ell \\ \quad g(x)F(x) \geqslant 0. \end{cases} \tag{4.5.8}$$

From the second inequality, we find a positive number α such that

$$g(x)F(x) \geqslant 0 \quad \text{for } |x| < \alpha.$$

Then from the first inequality we choose a positive number ℓ such that

$$\tfrac{1}{2}y^2 + G(x) < \ell \qquad \text{for } |x| < \alpha. \tag{4.5.9}$$

Clearly this means that ℓ must be chosen so that

$$G(x) < \ell \qquad \text{for } |x| < \alpha. \tag{4.5.10}$$

Since $G(0) = 0$ and G is continuous, one can always find an $\ell^* = \ell(\alpha)$ satisfying the given inequalities. For example, for $g(x) = 4x^3$, $\ell^* = \ell(\alpha) = \alpha^4$ will satisfy the requirement. In any case,

$$F = \{(x, y) \in E_{\ell^*}: \dot{V} = -g(x)F(x) = 0\} = \{x = 0\} \equiv y\text{-axis},$$

and one can verify that F does not contain any arc of a solution. The only invariant set we have in F is the point $p_0 = (0, 0)$, giving $M = \{p_0\}$. Hence all solutions starting from E_{ℓ^*} must converge to p_0.

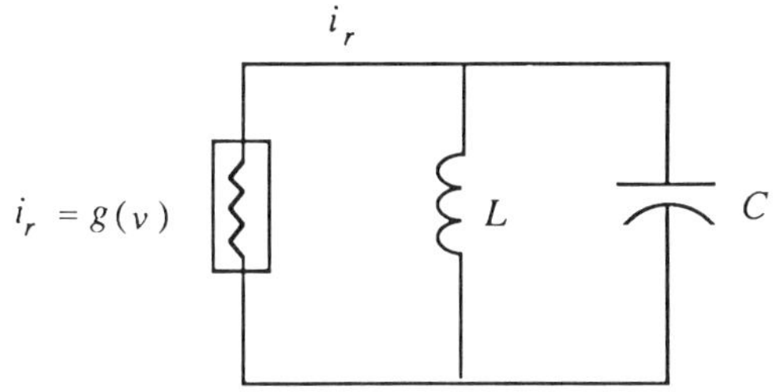

Figure 4.3

Example 4.5.12 Consider the nonlinear electrical circuit shown in figure 4.3 with a nonlinear resistor or device whose voltage–current characteristic is given by

$$i_r = g(v). \tag{4.5.11}$$

The differential equations characterizing the circuit are given by

$$\begin{cases} \dot{x}_1 = (1/L)x_2, \\ \dot{x}_2 = -(1/C)(x_1 + g(x_2)), \end{cases} \tag{4.5.12}$$

where x_1 is the current through the inductor L and x_2 is the voltage across the capacitor C.

For V, the total energy in the circuit, we have

$$V(x_1, x_2) = (L/2)x_1^2 + (C/2)x_2^2, \qquad L, C > 0,$$
$$\dot{V} = -x_2 g(x_2).$$

Suppose $g(\xi) = \sin \xi$, then

$$\dot{V} \leq 0 \qquad \text{for } |x_2| < \pi.$$

Then, we choose ℓ so that $V < \ell$ implies $|x_2| < \pi$. Indeed, for $\ell = (C/2)\pi^2$, this condition is met. In this case

$$M = F = \{(x_1, x_2): \dot{V} = 0\} = \{p_0\}, \qquad p_0 = (0, 0),$$

and hence for $\ell = (C/2)\pi^2$, the ellipse

$$E_\ell = \left\{(x_1, x_2): \frac{L}{2}x_1^2 + \frac{C}{2}x_2^2 < \frac{C}{2}\pi^2\right\} \tag{4.5.13}$$

is contained in the domain of asymptotic stability of the zero state.

Example 4.5.13 Consider the system

$$\ddot{\xi} + \alpha\dot{\xi} + (4\beta\xi^3 + 5\xi^4) = 0, \qquad \alpha, \beta > 0, \tag{4.5.14}$$

and define $x_1 = \xi$, $x_2 = \dot{\xi}$ as the state variables. The state equation is

$$\begin{cases} \dot{x}_1 = x_2 \\ \dot{x}_2 = -(\alpha x_2 + 4\beta x_1^3 + 5x_1^4), \end{cases} \tag{4.5.15}$$

for which the rest states are $p_0 = (0, 0)$ and $p_1 = (-4\beta/5, 0)$. Using the linearized equation around the states p_0 and p_1 one can easily verify that the state p_0 is asymptotically stable and p_1 unstable (saddle point). We wish to find an estimate of the size of the domain of asymptotic stability of the zero state p_0. Again we can choose the total energy for the V-function

$$V(x_1, x_2) = \tfrac{1}{2}x_2^2 + (\beta x_1^4 + x_1^5),$$

giving $\dot{V} = -\alpha x_2^2 \leqslant 0$.

Since we must exclude the point p_1 we take $\ell = V(p_1) = (\tfrac{1}{4})(4\beta/5)^5$ and define

$$\begin{aligned} E_\ell &= \{(x_1, x_2): V(x_1, x_2) < \ell\} \\ &= \{(x_1, x_2): \tfrac{1}{2}x_2^2 + (\beta x_1^4 + x_1^5) < \tfrac{1}{4}(4\beta/5)^5\}. \end{aligned} \tag{4.5.16}$$

Here,

$$F = \{(x_1, x_2) \in E_\ell : \dot{V} = 0\} = \{x_1\text{-axis}\} \cap E_\ell \text{ and } M = \{p_0\}$$

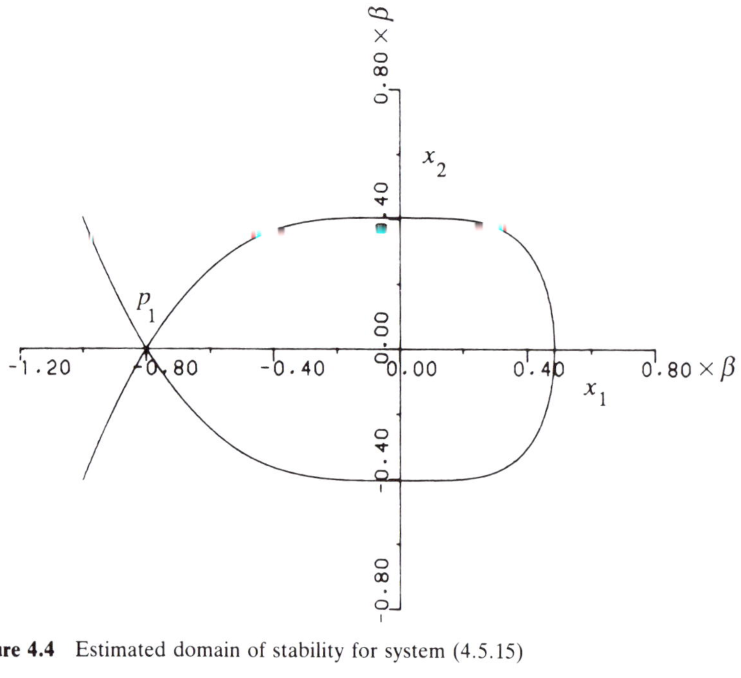

Figure 4.4 Estimated domain of stability for system (4.5.15)

is the largest invariant subset of F and hence any trajectory starting from within E_ℓ tends to the origin p_0 as $t \to \infty$. Thus, the set E_ℓ given above is in the domain of asymptotic stability of the zero state as shown in figure 4.4. Any trajectory starting on the left of p_1 will diverge to ∞ in finite or infinite time.

Remark 4.5.14 For a given system the domain of asymptotic stability is fixed. However, their estimates depend very much on the selection of the V-function. A poor selection tends to be conservative, giving a poor estimate, i.e. the estimated domain may be much smaller than its actual size.

Stability of sets

So far we have considered stability of equilibrium points. Since a system may possess periodic solutions to which it may eventually converge, or from which it may diverge following a perturbation, it is of interest to consider stability of sets. For example, the question of stability of limit cycles for second-order systems is a subject of significant interest to system analysts. In general, a periodic motion in R^n consists of a closed bounded and hence compact set. Therefore, we will limit our attention to the question of stability of compact sets only.

Let Ω be an open connected subset of R^n and K any subset of Ω. We define the distance of $x \in \Omega$ from the set K as

$$\rho(x, K) = \inf\{|x - \xi|, \xi \in K\}. \tag{4.5.17}$$

An ε-neighbourhood of K is defined as

$$N_\varepsilon(K) = \{x \in R^n: \rho(x, K) < \varepsilon\}, \qquad \varepsilon > 0.$$

If K is compact and $\rho(x, K) = 0$, then $x \in K$, and if K is not compact, then $\rho(x, K) = 0$ implies that either $x \in K$ or $x \in \partial K$, boundary of K.

For further development, we need the following definitions.

Definition 4.5.15 (Stability of compact sets). A compact set $K \subset \Omega$ is said to be stable if for each $\varepsilon > 0$ there exists a $\delta = \delta(\varepsilon) > 0$ such that whenever $\xi \in N_\delta(K)$, $T_t(\xi) \in N_\varepsilon(K)$ for all $t \geq 0$.

The concept of 'attractor' also plays a significant role in the study of stability. It resembles the concept of 'black hole' in astrophysics.

Definition 4.5.16 (Attractor, asymptotic stability). A compact set $K \subset \Omega$ is said to be a local attractor if there exists an $\varepsilon > 0$ such that for each $\xi \in N_\varepsilon(K)$, $T_t(\xi) \to K$ as $t \to \infty$. The set K is said to be a global attractor if $T_t(\xi) \to K$ for all $\xi \in \Omega$. A compact set K is said to be asymptotically stable if it is stable and an attractor. For a compact set K, the set

$$E(K) \equiv \{\xi \in \Omega : T_t(\xi) \to K \text{ as } t \to \infty\},$$

is called the domain of attraction of K.

It is not difficult to verify the following properties:

(a) if a compact set $K \subset \Omega$ is stable, then it is positively invariant, i.e., $T_t(\xi) \in K$ for all $t \geq 0$ whenever $\xi \in K$.
(b) the domain of attraction $E(K)$ is an open set containing K.
(c) the set $E(K)$, its boundary $\partial E(K)$ and its complement $\Omega \backslash E(K)$ are all positively invariant.

With this introduction we can now present stability results for compact sets.

Theorem 4.5.17 (Stability of compact sets). *Let D be a compact subset of Ω. Suppose there exists an open set $G \subset \Omega$ containing D and a function $V \in C^1(G)$ such that*

(a) *$V(x) > 0$ on $G \backslash D$ and $V(x) \geq 0$ on D,*
(b) *$\sup\{V(\xi), \xi \in D\} \equiv d \leq V(x)$ for $x \in G \backslash D$, and*
(c) *$\dot{V}(x) \leq 0$ on G.*

Then the set D is stable.

Proof Let $\varepsilon > 0$ so that $N_\varepsilon(D) \subset G$. Since V is C^1 on G it follows from (a) and (b) that there exists a positive number $r > d$ such that the set $E_r = \{x \in G : V(x) < r\}$ contains D and is contained in $N_\varepsilon(D)$. Since E_r is an open set containing D, there exists a $\delta > 0$ (dependent on r and hence ε) such that $N_\delta(D) \subset E_r$. Then for any $\xi \in N_\delta(D)$, it follows from the equation

$$V(T_t(\xi)) = V(\xi) + \int_0^t \dot{V}(T_\theta(\xi))\, d\theta,$$

by virtue of continuity of the maps $t \to T_t(\xi)$ and $x \to \dot{V}(x)$ and (c), that

$$V(T_t(\xi)) \leq V(\xi) < r \qquad \text{for all } t \geq 0.$$

Hence $T_t(\xi) \in E_r \subset N_\varepsilon(D)$ for all $t \geq 0$, and by Definition 4.5.15, the set D is stable. ■

For asymptotic stability we have the following result.

Theorem 4.5.18 (Asymptotic stability of sets). *Suppose the condi-*

tions of Theorem 4.5.17 *hold with* (c) *replaced by*
(c)′ $\dot{V}(x)<0$ *on* $G\backslash D$ *and* $\dot{V}(x)=0$ *on* D.
Then the set D is asymptotically stable.

Proof Since, under the given assumptions, D is stable it is only required to prove that D is an attractor. Since D is stable, for every $\varepsilon>0$, there exists a $\delta>0$, such that $\xi\in N_\delta(D)$ implies that $T_t(\xi)\in N_\varepsilon(D)\subset G$ for all $t\geqslant 0$. By virtue of (c)′, $t\to V(T_t(\xi))$ is non-increasing and $V(T_t(\xi))\geqslant 0$ for all $t\geqslant 0$. Hence $\lim_{t\to\infty} V(T_t(\xi))\equiv\eta_0$ exists. Since $T_t(\xi)$ is contained in a bounded set, it has a non-empty positive limit set $\Gamma^+(\xi)$ and $T_t(\xi)\to\Gamma^+(\xi)$ as $t\to\infty$. Hence by continuity of V and $t\to T_t(\xi)$, we have $V(\Gamma^+(\xi))=\eta_0$. That is, V is constant on $\Gamma^+(\xi)$ and hence $\dot{V}\equiv 0$ on $\Gamma^+(\xi)$ and by virtue of (c)′ $\Gamma^+(\xi)\subset D$ and consequently D is an attractor. Therefore, D is asymptotically stable. ■

Remark 4.5.19 If the assumptions of the above theorems hold for $G=\Omega$, then D is globally stable (globally asymptotically stable).

To illustrate the above results we present a few examples.

Example 4.5.20 (Satellite with dead zone). Consider the example of the spacecraft (Example 4.4.14, equation (4.4.15)) and suppose the applied torques T_1, T_2, T_3 are provided by control mechanisms having dead zones. That is,

$$T_i=-u_i(\xi)=-K_i d_\alpha(\xi),\qquad i=1,2,3$$

$$d_\alpha(\xi)=\begin{cases}1 & \xi>\alpha\\ 0 & |\xi|\leqslant\alpha\\ -1 & \xi<-\alpha\end{cases}\tag{4.5.18}$$

and $K_i>0$, $\alpha>0$.
Define

$$D\equiv\{(p,q,r)\colon |p|\leqslant\alpha,\ |q|\leqslant\alpha,\ |r|\leqslant\alpha\}.\tag{4.5.19}$$

Using the same V-function as given in Example 4.4.14 we have

$$\dot{V}\begin{cases}=0 & \text{for } (p,q,r)\in D\\ <0 & \text{for } (p,q,r)\in R^3\backslash D\end{cases}\tag{4.5.20}$$

Taking $G=\Omega=R^3$ one can readily check that the conditions (a)–(c) and (c)′ of Theorems 4.5.17 and 4.5.18 are satisfied. Hence, the set D is asymptotically stable. That is, $(p(t),q(t),r(t))\to D$ as $t\to\infty$. For fuel economy this control is preferable to bang-bang controls in certain applications where attitude of the satellite is immaterial.

Example 4.5.21 Consider Example 4.3.6, equation (4.3.23). We have seen that it has a limit cycle

$$(\xi_1(t), \xi_2(t)) = (\cos t, \sin t) \in D = \{(x_1, x_2): x_1^2 + x_2^2 = 1\}.$$

Using Theorem 4.5.18, we show that the limit cycle or equivalently the set D is asymptotically stable. Let

$$S = \{(x_1, x_2): (x_1)^2 + (x_2)^2 \leq 1\}$$

denote the unit disc on the plane and define

$$C_S(\xi_1, \xi_2) = \begin{cases} 1, & (\xi_1, \xi_2) \in S \\ 0, & \text{otherwise.} \end{cases}$$

For the Lyapunov function we take

$$V = \tfrac{1}{2}(\xi_1^2 + \xi_2^2 - 1)(1 - C_S(\xi_1, \xi_2)), \tag{4.5.21}$$

and note that conditions (a)–(c′) of Theorem 4.5.18 are satisfied:

(a) $V > 0$ outside S and $V = 0$ on S
(b) $d = 0 \leq V(\xi_1, \xi_2)$, $(\xi_1, \xi_2) \in R^2 \backslash S$
(c)′ $\dot{V} = 0$ on S and $\dot{V} < 0$ on $R^2 \backslash S$.

The condition (c)′ follows from the expression

$$\dot{V} = (\xi_1^2 + \xi_2^2)(1 - \xi_1^2 - \xi_2^2) < 0 \qquad \text{on } R^2 \backslash S.$$

Hence, by Theorem 4.5.18 the closed unit disc is asymptotically stable, that is, any trajectory starting from outside the disc will eventually enter the disc. In fact, the set $D = \partial S$ is asymptotically stable. Indeed taking

$$V = \begin{cases} \tfrac{1}{2}(1 - \xi_1^2 - \xi_2^2) & \text{for } (\xi_1, \xi_2) \in S \\ 0 & \text{elsewhere,} \end{cases} \tag{4.5.22}$$

one has $\dot{V} < 0$ on $S \backslash D$ and $\dot{V} = 0$ on D. Hence by the previous theorem the set D is asymptotically stable.

Example 4.5.22 (An ecological model). The system

$$\begin{aligned} \dot{x}_1 &= ax_1 - bx_1x_2 \\ \dot{x}_2 &= -cx_2 + dx_1x_2 \end{aligned} \qquad a, b, c, d > 0 \tag{4.5.23}$$

describes the temporal evolution of two species one of which preys upon the other, for example, x_1 for rabbits and x_2 for foxes. There are two equilibrium points $p_1 = (0, 0)$ and $p_2 = (x_1^0, x_2^0) = (c/d, a/b)$. The point p_1

is unstable and p_2 is a focus, that is, the trajectories spiral around p_2. The solution $(x_1(t), x_2(t))$, corresponding to any initial condition

$$(x_{10}, x_{20}) \in R_+^2 = [0, \infty) \times [0, \infty),$$

lies entirely in the first quadrant R_+^2. One can verify that the function

$$V(x_1, x_2) \equiv c((d/c)x_1 - \ln(d/c)x_1) + a((b/a)x_2 - \ln(b/a)x_2) \quad (4.5.24)$$

is a Lyapunov function for the system which is positive on R_+^2 and $V(x) \to \infty$ as $|x| \to \infty$ for $x \in R_+^2$. The time derivative of V along any trajectory vanishes,

$$\dot{V} = 0 \qquad \text{for all } (x_1, x_2) \in R_+^2. \quad (4.5.25)$$

Therefore, the system is conservative in the sense that for any given initial state $(x_{10}, x_{20}) \in R_+^2$,

$$V(x_1(t), x_2(t)) = V(x_{10}, x_{20}) = k_0 \qquad \text{for all } t \geq 0, \quad (4.5.26)$$

and describes a unique limit cycle in R_+^2. A few of these limit cycles corresponding to data $a = c = 5$, $b = d = 0.5$, are shown in figure 4.5.

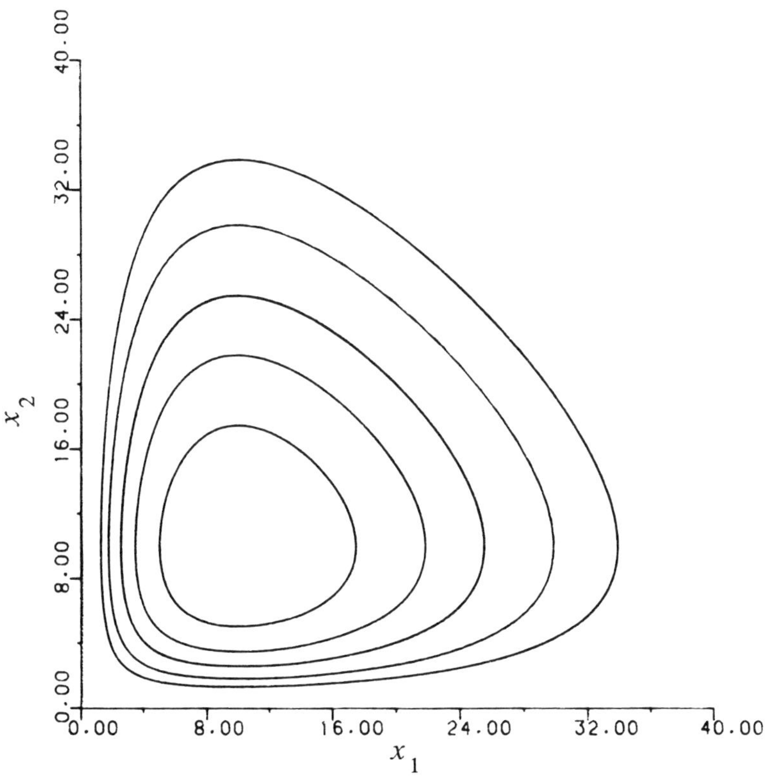

Figure 4.5 Stable limit cycles for system (4.5.23)

Each limit cycle $D = \{(x_1, x_2) \in R^2_+ : V(x_1, x_2) = k_0\}$ is stable but not asymptotically stable since $\dot{V} = 0$.

Example 4.5.23 Consider the electrical circuit of Example 4.5.12 with $L = C = 1$. It is clear from the expressions, $\dot{V} = -x_2 g(x_2)$ and $g(x_2) = \sin x_2$, that if x_2 initially lies in the open interval $(\pi, 2\pi)$, then both $\dot{V}$ and x_2 grow with time and x_2 tends to converge towards 2π, while, if it initially starts out from the open interval $(2\pi, 3\pi)$, both $\dot{V}$ and x_2 diminish with time and again x_2 tends to converge towards 2π. Hence there is a limit cycle in the neighbourhood of the circle of radius 2π. The limit cycle is locally asymptotically stable in the sense that, if the system is not perturbed too strongly, all its motions, originating in the neighbourhood of the limit cycle, will eventually converge to the limit cycle. In fact, one can easily check that there are infinitely many (locally) asymptotically stable limit cycles with radii $r_n = 2n\pi$, $n = 1, 2, \ldots$. Two of these limit cycles are shown in figure 4.6.

Remark 4.5.24 For further study concerning the domain of asymptotic stability the reader is referred to the specialized treatise of Zubov [150]. Here one would find several interesting results on this topic (Theorem 19, Theorem 22, etc.) including methods for detecting the boundary of the region of asymptotic stability.

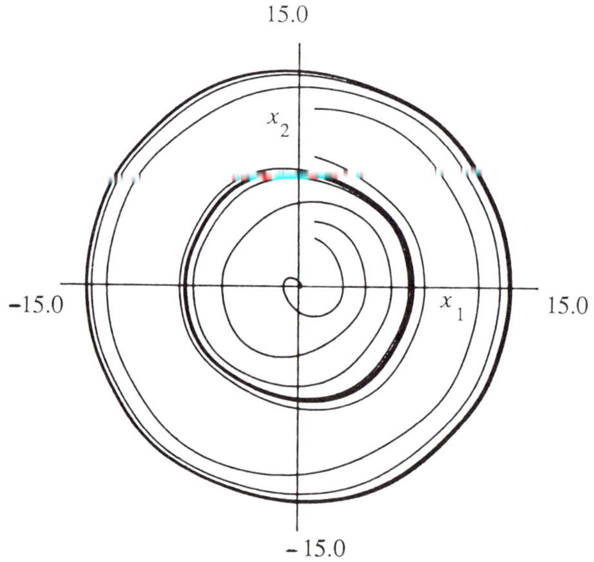

Figure 4.6 Stable limit cycles for system (4.5.12)

Strange attractors and chaos

In recent years the notion of strange attractors and chaos has drawn attention of many workers in the field. There are many deterministic (nonlinear) dynamical systems which are highly sensitive even to infinitesimal perturbations around their equilibria and generally exhibit chaotic or unpredictable motions. Examples of such systems are found in mechanics (buckling of beams), hydrodynamics (turbulence) and quantum dynamics (laser cavities, Josephson junctions) etc. The subject is at its early stage of development and may need several years before a coherent theory emerges.

Interested readers may refer to [151] and many references therein.

Exercises

4.2.P1 Show that the equation $dz/dt = A'z + zA$, $z(0) = \Gamma$ has the solution $z(t) = e^{tA'}\Gamma e^{tA}$, $t \geqslant 0$.

4.2.P2 Show that if A is a stability matrix,

$$\lim_{t\to\infty} z(t) = \lim_{t\to\infty} e^{tA'}\Gamma e^{tA} = 0.$$

4.2.P3 By use of the Lyapunov equation $A'Y + YA = -I$, show that the system $\dot{x} = Ax$, with

$$A = \begin{bmatrix} -3 & 2 \\ -1 & -1 \end{bmatrix},$$

is asymptotically stable.

4.2.P4 Prove that the solutions of the equations (see equation (4.2.24))

$$\dot{P}(t) = A(t)P(t) - P(t)B, \qquad P(0) = I$$

$$\dot{Q}(t) = -Q(t)A(t) + BQ(t), \qquad Q(0) = I$$

satisfy the property: $Q(t)P(t) = I$ for all $t \geqslant 0$.

4.2.P5 Verify that $X(t)$, as given in Example 4.2.5, is the fundamental solution of the equation $\dot{x}(t) = A(t)x(t)$ with $A(t)$ as given there.

4.2.P6

(a) Show that the eigenvalues of the matrix

$$A(t) = \begin{bmatrix} -2\cos^2 t & -1 - 2\sin t\cos t \\ 1 - 2\sin t\cos t & -2\sin^2 t \end{bmatrix}$$

are $\lambda_1(t) = \lambda_2(t) = -1$.

(b) Determine whether the periodic system $\dot{x} = A(t)x$ is stable or not. (*Hint*: Follow the procedure as in Example 4.2.5)

4.3.P7 Show that the nonlinear system

$$\begin{cases} \dot{x}_1 = ax_1 - bx_2 - \alpha x_1 x_2 \\ \dot{x}_2 = -cx_2 + dx_1 + \beta x_1 x_2 \end{cases} \qquad a, b, c, d, \alpha, \beta > 0$$

is locally asymptotically stable near the zero state if $c > a$ and $bd > ac$.

4.3.P8 Consider $b = d = 0$ in Exercise 4.3.P7.
(a) Find the rest states.
(b) Show that the rest state $(x_1^*, x_2^*) = (c/\beta, a/\alpha)$ is a focus.

4.3.P9
(a) Verify that $(\eta_1(t), \eta_2(t)) \equiv (\cos t, \sin t)$ is a periodic solution of the nonlinear system

$$\begin{cases} \dot{\eta}_1 = -\eta_2 - \eta_1(1 - \eta_1^2 - \eta_2^2) \\ \dot{\eta}_2 = \eta_1 - \eta_2(1 - \eta_1^2 - \eta_2^2) \end{cases}$$

(b) Find the corresponding (linear) variational equation around the periodic solution $(\cos t, \sin t)$ and show that the eigenvalues are $\lambda_1(t) = \lambda_2(t) = 1$.

4.3.P10 The dynamics of a synchronous generator connected to an infinite bus is given by

$$\begin{cases} \dot{\delta} = a_1 n & \delta \equiv \text{torque angle} \\ \dot{n} = a_2 - a_3 E \sin\delta - a_4 n & n \equiv \text{speed deviation} \\ \dot{E} = a_5 - a_6 E + a_7 \cos\delta & E \equiv \text{generator internal voltage} \\ & a_i > 0,\ i = 1, 2, \ldots, 7 \end{cases}$$

Find the rest states that are locally stable and asymptotically stable.

4.3.P11 (Econometric model) Consider the econometric model of Example 4.3.3 and verify that the function $f(k) \equiv \gamma k^\alpha$, $0 < \alpha < 1$, $\gamma > 0$ satisfies all the conditions (4.3.14) to qualify for a production function.

4.3.P12 (Fisheries model) An approximate fisheries model is governed by the system of equations

$$\dot{x} = \alpha(1 - x/k)x - xy, \qquad \alpha, k > 0$$
$$\dot{y} = \beta(px - c)y \qquad \beta, p, c > 0$$

where at time t, $x(t)$ and $y(t)$ represent the fish population (biomass) and level of fishing efforts respectively. The parameters α and k (carrying capacity) are biological factors, and β, p, c are economic factors ($p =$

price per unit biomass, c = cost per unit fishing effort, β = stiffness factor).

(a) Find the equilibria and determine their stability properties for small and large β.
(b) Does the system have a limit cycle?

4.4.P13 Find Lyapunov functions for the nonlinear systems

(a) $\ddot{\xi} + a\dot{\xi} + b\xi^3 = 0,\ a, b > 0$
(b) $\ddot{\xi} + a\,|\dot{\xi}|\,\dot{\xi} + b\xi = 0,\ a, b > 0$
(c) $\ddot{\xi} + a(\dot{\xi})^3 + g(\xi) = 0,\ a > 0,\ g(\xi) = \tan\xi,\ \text{sign}\,\xi,\ d_\alpha(\xi)$,

and determine whether they are asymptotically stable near the origin.

4.4.P14
(a) Show that the system

$$\ddot{\xi} + k_1\dot{\xi} + k_2(\dot{\xi})^3 + \xi = 0$$

is asymptotically stable near the origin for $k_1,\ k_2 > 0$.
(b) Determine its stability behaviour for the cases
(i) $k_1, k_2 < 0$, (ii) $k_1 > 0,\ k_2 < 0$, (iii) $k_1 < 0,\ k_2 > 0$.

4.4.P15 Show that Van der Pol's equation of Example 4.4.13 has a limit cycle which is asymptotically stable.

4.4.P16 Find the dynamic equilibrium of the system

$$\ddot{\theta} + a\dot{\theta} = b, \qquad a > 0, \quad b \text{ constant};$$

and using a Lyapunov function show that the equilibrium is asymptotically stable.

4.5.P17 Give an estimate of the domain of asymptotic stability of the zero state for the system

$$\ddot{\xi} + a\dot{\xi} + 2b\xi + 3\xi^2 = 0, \qquad a, b > 0.$$

4.5.P18 Prove the properties: (a)–(c) of the domain of attraction of a compact set K (see Definition 4.5.16).

4.5.P19 Consider Example 4.5.12. Find the domain of attraction of the limit cycles near $2k\pi$, $k = \pm 1, \pm 2, \ldots$.

4.5.P20 The attitude dynamics of a spherical satellite in geostationary orbit is approximately given by

$$\left.\begin{array}{lll} I\dot{p} + w_0 I r = T_1, & I\dot{q} = T_2, & I\dot{r} - w_0 I p = T_3 \\ \dot{\theta}_1 = p, & \dot{\theta}_2 = q, & \dot{\theta}_3 = r. \end{array}\right\}$$

Given that the controls are $T_1 = -k_1 \operatorname{sign} p - u_1(\theta_1)$, $T_2 = -k_2 \operatorname{sign} q - u_2(\theta_2)$, $T_3 = -k_3 \operatorname{sign} r - u_3(\theta_3)$ with k_i $(i = 1, 2, 3) > 0$, show that the system is asymptotically stable at the origin if the attitude feedback $u_i(\xi) = \xi$; what if $u_i(\xi)$ is an arbitrary function satisfying $u_i(\xi)\xi \geqslant 0$?

4.5.P21 Consider the system of 4.5.P20. Given that $u_i(\xi) = d_\alpha(\xi)$,

$$d_\alpha(\xi) = \begin{cases} 1, & \xi > \alpha \\ 0, & |\xi| \leqslant \alpha \\ -1, & \xi < -\alpha, \end{cases}$$

find the set D containing the origin that is asymptotically stable.

5

Observability, controllability and stabilizability

5.0 Introduction

The concepts of controllability and observability play significant roles in the theory of modern control. They are intimately connected with the theory of optimal control as well as with the questions of stabilizability of naturally unstable systems. In the previous chapter our objective was limited to the question as to whether or not a given system was stable. In this chapter we go beyond that and enquire into the possibility of providing feedback control laws that stabilize unstable systems. This is a problem of profound interest in engineering practice, since many of the physical systems encountered are either unstable or do not satisfy transient specifications and hence must be provided with stabilizing controls or compensators. We study these questions in some detail in the following sections.

In sections 5.1 and 5.2 we study the questions of observability and controllability for both linear time-invariant and time-varying systems. The questions of controllability with constraints on the control is considered in section 5.3. In section 5.4 stabilizability of linear systems using state feedback or using dynamic observers are considered. The questions of controllability and stabilizability for nonlinear systems are considered briefly in section 5.5. The nonlinear theory is rather incomplete and there is sufficient scope for further development.

5.1 Observability

Consider the linear system in R^n

$$\begin{aligned} \dot{x} &= A(t)x + B(t)u, \qquad t \geq 0 \\ x(0) &= x_0, \end{aligned} \tag{5.1.1}$$

with the control (input) u and the initial state x_0 given. The system is in operation during the time interval $[0, t_1]$. At some intermediate time, say $t_0^+ \in (0, t_1)$, it is subjected to an impulsive disturbance, as a consequence of which the system finds itself in state $x(t_0^+) = x(t_0) + \xi$ instead of $x(t_0)$, where $\xi \in R^n$. For example an aircraft in its flight path may be subjected to an unexpected wind gust leading to an abrupt change of state. The question is: is it possible to determine the impact ξ from the knowledge of the output history

$$y(t) = H(t)x(t), \qquad t \in [t_0, t_1]. \tag{5.1.2}$$

Clearly for $t > t_0$ the state is given by

$$x(t) = \Phi(t, t_0)(x(t_0) + \xi) + \int_{t_0}^{t} \Phi(t, \theta)B(\theta)u(\theta)\,d\theta, \tag{5.1.3}$$

where Φ is the transition operator corresponding to A. The output $y(t)$ is given by

$$\begin{aligned} y(t) &= \eta(t) + H(t)\Phi(t, t_0)\xi, \qquad t > t_0, \\ \eta(t) &\equiv H(t)\Phi(t, t_0)x(t_0) + \int_{t_0}^{t} H(t)\Phi(t, \theta)B(\theta)u(\theta)\,d\theta. \end{aligned} \tag{5.1.4}$$

Since x_0 is given, η is known and hence

$$\tilde{y}(t) \equiv y(t) - \eta(t) = H(t)\Phi(t, t_0)\xi, \qquad t > t_0$$

is available for observation. Thus the equivalent question is, given the history of $\tilde{y}$ over the time interval $(t_0, t_1]$, is it possible to determine ξ uniquely?

Definition 5.1.1 (Observability). The system (5.1.1) is said to be finite-time observable at time t_0 if there exists a finite time $t_1 > t_0$ such that the initial impact (state) ξ, occurring at t_0^+, can be uniquely determined from the history of the output $y(t)$, $t \in [t_0, t_1]$. If both $\xi \in R^n$ and $t_1 > t_0$ can take arbitrary values, then the system is said to be globally observable. In this case we shall simply call it observable.

5.1.1 Linear time-invariant systems

Consider the linear time-invariant observed system

$$\begin{cases} \dot{x} = Ax + Bu, & t \geq 0 \\ y = Hx, \end{cases} \tag{5.1.5}$$

with $A \in R(n \times n) = \mathscr{L}(R^n)$, $B \in R(n \times m)$ and $H \in R(r \times n)$.

Theorem 5.1.2 (Observability). *The system* (5.1.5) *is observable over the time interval* $[0, T]$ *if and only if the observability matrix* M_0, *defined by*

$$M_0 \equiv \int_0^T e^{tA'} H' H e^{tA} \, dt, \tag{5.1.6}$$

is positive definite (*i.e.* $M_0 > 0$).

Proof (Necessary condition): We prove that observability implies positivity of the matrix M_0. Since the system is observable, for any pair of distinct initial states x_{01}, $x_{02} \in R^n$, the outputs $y_1(t) \equiv y(t, x_{01})$ and $y_2(t) \equiv y(t, x_{02})$ are distinct, that is,

$$y_1(t) \not\equiv y_2(t) \qquad \text{for } t \in I = [0, T]. \tag{5.1.7}$$

Hence

$$\int_I |y_1(t) - y_2(t)|_r^2 \, dt = \int_I (y_1(t) - y_2(t), y_1(t) - y_2(t))_r \, dt > 0. \tag{5.1.8}$$

Substituting

$$y_i(t) = He^{tA} x_{0i} + \int_0^t He^{(t-\theta)A} Bu(\theta) \, d\theta, \qquad i = 1,2,$$

in (5.1.8) we have,

$$\begin{aligned} \int_I |y_1(t) - y_2(t)|^2 \, dt &= \int_I (He^{tA}(x_{01} - x_{02}), He^{tA}(x_{01} - x_{02}))_r \, dt \\ &= \int_I (e^{tA'} H' He^{tA}(x_{01} - x_{02}), x_{01} - x_{02})_n \, dt \\ &= (M_0 \xi, \xi) > 0, \end{aligned} \tag{5.1.9}$$

where $(\cdot, \cdot)_k$, $k = r, n$ denotes scalar products in R^k and $\xi \equiv x_{01} - x_{02}$. Since, by hypothesis, the system is observable over I, distinct initial states produce distinct outputs. Hence (5.1.9) holds for all $\xi \in R^n \backslash \{0\}$ and consequently M_0 is positive definite (written $M_0 > 0$).

(Sufficient condition): We show that if $M_0 > 0$ then the system is observable. We prove this by contradiction. Suppose $M_0 > 0$ but the system is not observable. In that case there exist initial states $x_0 \neq x_0'$ such that

$$y(t) \equiv y(t, x_0) = y(t, x_0') \equiv y'(t) \qquad \text{for all } t \in I.$$

Then for $\eta \equiv (x_0 - x_0') \neq 0$,

$$0 = \int_I |y(t) - y'(t)|^2 \, dt = (M_0 \eta, \eta). \tag{5.1.10}$$

This contradicts the hypothesis that $M_0 > 0$ and hence completes the proof. ■

As a consequence of the above theorem the reader can easily verify the following result.

Remark 5.1.3 Since $M_0 > 0$, it is invertible and hence the initial state that yields the output $\tilde{y}(t)$, $t \in I$, can be determined by the expression

$$x_0 = M_0^{-1}\left(\int_0^T e^{tA'}H'\tilde{y}(t)\,dt\right). \tag{5.1.11}$$

In general the positivity condition, $M_0 > 0$, is never used practically, since its verification requires a great deal of computation as suggested by the above expression (see also Exercise 5.3.P14). However, there is an equivalent characterization which offers an extremely simple procedure for numerical verification. This is the so-called rank condition.

Theorem 5.1.4 (Rank condition). *Let*

$$D \equiv \begin{bmatrix} H \\ HA \\ \vdots \\ HA^{n-1} \end{bmatrix} \in R(nr \times n) = \mathscr{L}(R^n, R^{nr}).$$

The (observed) system (5.1.5) *is observable if and only if the rows of D span* R^n, *or* $\bigcap_{k=0}^{n-1} \operatorname{Ker}(HA^k) = 0$, *or* Rank $(D') = n$.

Proof (Necessary condition): By Theorem 5.1.2, observability implies that $M_0 > 0$ which, in turn, means that

$$(M_0\xi, \xi) \equiv \int_0^T (He^{tA}\xi, He^{tA}\xi)\,dt = 0 \qquad \text{if and only if } \xi = 0.$$

In other words $He^{At}\xi \equiv 0$ for $t \in I = [0, T]$, if and only if $\xi = 0$. Equivalently

$$\left\{\frac{d^k}{dt^k}(He^{tA}\xi) \equiv 0, \qquad k = 0, 1, 2, \ldots, t \in I\right\}$$

if and only if $\xi = 0$, or

$$\{HA^k\xi = 0,\ k = 0, 1, 2, \ldots\} \qquad \text{if and only if } \xi = 0. \tag{5.1.12}$$

By the Cayley–Hamilton theorem, A^n is expressible as a linear combination of $\{A^k, k = 0, 1, 2, \ldots, n-1\}$. Hence (5.1.12) is equivalent to

$$\{HA^k\xi = 0, k = 0, 1, \ldots, n-1\} \qquad \text{if and only if } \xi = 0. \tag{5.1.13}$$

This means that $\{\text{Rows}\{HA^k\}, k = 0, 1, \ldots, n-1\} = \text{Rows}(D)$ has row space R^n, which is equivalent to $\bigcap_{k=0}^{n-1} \text{Ker}(HA^k) = \{0\}$. This condition is equivalent to the statement that the columns of D' span R^n, or $\text{Rank}\, D' = n$.

(Sufficient Condition): We prove that if $\text{Rank}\, D' = n$, then the system (5.1.5) is observable. Suppose not. Then there exists $\xi \neq 0$ such that $y(t) = He^{At}\xi \equiv 0$ for $t \in I$. Therefore $H\xi = 0$, $HA\xi = 0, \ldots, HA^{n-1}\xi = 0$. Since $\text{Rank}\,(D') = n$, ξ must equal zero. The contradiction proves the assertion. ■

Remark 5.1.5 According to the above theorem the observability of the system (5.1.5) is determined by verifying that

$$\text{Rank}\,(D') \equiv \text{Rank}\,(H', A'H', (A')^2H', \ldots, (A')^{n-1}H') = n. \tag{5.1.14}$$

It is clear from the above results that if $\text{Rank}\, D' = \ell < n$, then the system is not completely (globally) observable. Let $D_0 \equiv D_0(H, A)$ denote the subspace of R^n spanned by the columns (D'). Clearly the dimension of $(D_0) = \ell$, and the system is observable relative to D_0; that is, given x_{01}, $x_{02} \in D_0$ with $x_{01} \neq x_{02}$, $He^{At}x_{01} \not\equiv He^{At}x_{02}$, $t \in I$. Clearly if D_0 is a proper subspace of R^n we can write

$$R^n = D_0 \oplus D_0^\perp$$

where $D_0^\perp$ denotes the orthogonal complement of D_0.

Corollary 5.1.6 (Observability subspace)

(a) *$D_0^\perp$ is always invariant under A.*

(b) *If A is nonsingular then D_0 is also invariant under A.*

(c) *D_0 is invariant under A'.*

Proof Follows from the Cayley–Hamilton theorem and the definition of D_0. ■

Example 5.1.7 A spherical satellite is governed by the equation

$$\left.\begin{aligned} I\dot{p} + w_0 I r &= T_1 \\ I\dot{q} \qquad &= T_2 \\ I\dot{r} - w_0 I p &= T_3 \end{aligned}\right\} \cong \left\{\begin{aligned} \dot{p} &= -w_0 r + u_1 \\ \dot{q} &= u_2 \\ \dot{r} &= w_0 p + u_3, \end{aligned}\right. \qquad u_i = T_i/I, \tag{5.1.15}$$

where p, q, r are the angular velocities, I the moment of inertia, T_i ($i = 1, 2, 3$) are the torques and w_0 the spin velocity of the earth (in radians).

(Case i) $y = p,\ H = [1, 0, 0]$

(Case ii) $y = p + q,\ H = [1, 1, 0]$

(Case iii) $y = (q + r),\ H = [0, 1, 1]$.

The plant matrix is given by

$$A = \begin{bmatrix} 0 & 0 & -w_0 \\ 0 & 0 & 0 \\ w_0 & 0 & 0 \end{bmatrix}.$$

(Case i): $$H' = \begin{bmatrix} 1 \\ 0 \\ 0 \end{bmatrix}, \quad A'H' = \begin{bmatrix} 0 \\ 0 \\ -w_0 \end{bmatrix}, \quad (A')^2H' = \begin{bmatrix} -w_0^2 \\ 0 \\ 0 \end{bmatrix}.$$

Hence Rank $(H', A'H', A'^2H') = 2 < 3$, and the system is not observable.

(Case ii): $$H' = \begin{bmatrix} 1 \\ 1 \\ 0 \end{bmatrix}, \quad A'H' = \begin{bmatrix} 0 \\ 0 \\ -w_0 \end{bmatrix}, \quad (A')^2H' = \begin{bmatrix} -w_0^2 \\ 0 \\ 0 \end{bmatrix}.$$

Hence Rank $(H', A'H', A'^2H') = 3$ and the system is observable.
(Case iii): The system is observable.

Example 5.1.8 A system governed by an nth-order differential equation with scalar input and having constant coefficients with first component observed is always observable.

Consider

$$z^{(n)} + a_{n-1}z^{(n-1)} + \cdots + a_1z^{(1)} + a_0z = u \text{ (scalar)},$$

$$y = z.$$

Writing this in the canonical form, $\dot{x} = Ax + bu$, where

$$A = \begin{bmatrix} 0 & 1 & 0 & \cdots & 0 \\ 0 & 0 & 1 & \cdots & 0 \\ 0 & 0 & 0 & & 0 \\ \vdots & \vdots & \vdots & \cdots & \vdots \\ -a_0 & -a_1 & -a_2 & \cdots & -a_{n-1} \end{bmatrix}, \quad b = \begin{bmatrix} 0 \\ 0 \\ \vdots \\ 0 \\ 1 \end{bmatrix}, \quad x = \begin{bmatrix} x_1 \\ x_2 \\ \vdots \\ x_{n-1} \\ x_n \end{bmatrix} \tag{5.1.16}$$

where $x_i = z^{(i-1)}$ and $H = [1, 0, 0, \ldots, 0]$. One can easily verify that the rank of $\{H', A'H', \ldots, (A')^{n-1}H'\}$ is n and hence the system is observable.

5.1.2 Linear time-varying systems (observability)

Consider the observed time-varying system

$$\begin{cases} \dot{x} = A(t)x + B(t)u \\ y = H(t)x + G(t)u, \end{cases} \qquad t \geqslant 0, \tag{5.1.17}$$

where for each t, $A(t) \in R(n \times n)$, $B(t) \in R(n \times m)$, $H(t) \in R(r \times n)$ and $G(t) \in R(r \times m)$. We assume throughout this section, unless otherwise stated, that H, B, $G \in L_\infty$. The output is given by

$$y(t) = H(t)\Phi(t, 0)x_0 + \int_0^t H(t)\Phi(t, \theta)B(\theta)u(\theta)\,\mathrm{d}\theta + G(t)u \tag{5.1.18}$$

which is observed. Again, given the input u over the time interval $[t_0, t_1]$, we know

$$\tilde{y}(t) \equiv y(t) - \int_0^t H(t)\Phi(t, \theta)B(\theta)u(\theta)\,\mathrm{d}\theta - G(t)u(t), \qquad t \geqslant 0.$$

Hence $\tilde{y}(t) = H(t)\Phi(t, 0)x_0$ for $t \geqslant 0$. The question is, can we determine x_0 uniquely given the history of $\tilde{y}$.

Theorem 5.1.9 (Observability matrix). *The system* (5.1.17) *is observable at $t_0 \geqslant 0$ over the time interval $[t_0, t_1]$, $t_1 < \infty$, if and only if the observability matrix $M_0(t_0, t_1)$, given by*

$$M_0(t_0, t_1) \equiv \int_{t_0}^{t_1} \Phi'(t, t_0)H'(t)H(t)\Phi(t, t_0)\,\mathrm{d}t \tag{5.1.19}$$

is positive definite.

Proof The proof is identical to that of Theorem 5.1.2 where $\mathrm{e}^{A(t-\tau)}$ is replaced by $\Phi(t, \tau)$, $0 \leqslant \tau \leqslant t < \infty$. ■

Again one can easily check that if $\tilde{y}(t)$, $t \in [t_0, t_1]$, constitutes the observed data, the corresponding state at t_0 can be identified as

$$x_0 = M_0^{-1}(t_0, t_1)\left(\int_{t_0}^{t_1} \Phi'(t, t_0)H'(t)\tilde{y}(t)\,\mathrm{d}t\right). \tag{5.1.20}$$

Remark 5.1.10 Note that

$$M_0(t_0, t_1) > 0 \Leftrightarrow \text{Ker}\{H(t)\Phi(t, t_0), t \in [t_0, t_1]\} = \{0\}. \tag{5.1.21}$$

Hence this is an alternative characterization of observability.

Input identification

Fundamentally the question of observability treated above is equivalent to the question of unique identifiability of the initial state (impact) from the knowledge of the output history. Also an equally important question is: can we identify the input u from the knowledge of the initial state and the output history. Consider the system (5.1.17) for $t \geqslant t_0$ with $x(t_0) = x_0$ given and the output $y(t)$, $t \in [t_0, t_1]$ also given.

Defining

$$z(t) \equiv y(t) - H(t)\Phi(t, t_0)x_0, \qquad t \in [t_0, t_1], \tag{5.1.22}$$

we have

$$z(t) = \int_{t_0}^{t} H(t)\Phi(t, \theta)B(\theta)u(\theta)\, d\theta + G(t)u(t), \tag{5.1.23}$$

$t \in I = [t_0, t_1]$. Thus the question of observability or identifiability of u from the knowledge of the history of z is equivalent to the question of solvability of the Volterra integral equation (5.1.23) of the third kind. This is a much more difficult problem since G is not necessarily invertible. In case $G(t) \equiv 0$, we have an integral equation of the first kind. However, if z and H are differentiable, one can convert

$$z(t) = \int_{t_0}^{t} H(t)\Phi(t, \theta)B(\theta)u(\theta)\, d\theta \tag{5.1.24}$$

into

$$\dot{z}(t) = H(t)B(t)u(t) + \int_{t_0}^{t} \left(\frac{\partial}{\partial t} H(t)\Phi(t, \theta)\right)B(\theta)u(\theta)\, d\theta. \tag{5.1.25}$$

Again this is an integral equation of the third kind, and since HB is not necessarily invertible, there is no way to reduce it to the popular equation of the second kind.

In this situation one can develop an approximation theory for identification of the input u. Assuming H, B and G to have entries which are essentially bounded and measurable, one may consider finding a u^* that minimizes the functional

$$\ell(u) = \|z - Lu\|^2 \tag{5.1.26}$$

where L is the operator determined by

$$(Lu)(t) = G(t)u(t) + \int_{t_0}^{t} H(t)\Phi(t, \theta)B(\theta)u(\theta)\, d\theta, \tag{5.1.27}$$

$t \in I = [t_0, t_1]$. We can write

$$\begin{aligned} \ell(u) &= \|z - Lu\|^2_{L_2(I, R^r)} \\ &= \|z\|^2 - 2(L^*z, u) + (L^*Lu, u) \end{aligned} \tag{5.1.28}$$

where L^* is the adjoint of the operator L. We wish to find a u_0 such that $\ell(u_0) \leqslant \ell(u)$ for all u in the class of admissible input $\mathscr{U}$. The question of existence of a minimizing element u_0 will be discussed in chapter 6. At this point we simply note that if $\mathscr{U} \equiv L_2(I, R^m)$, the Gâteaux derivative ℓ' of ℓ evaluated at u_0 must vanish. That is,

$$\ell'(u_0) = -2L^*z + 2L^*Lu_0 = 0. \tag{5.1.29}$$

If (L^*L) is invertible in the Hilbert space $L_2(I, R^m)$, then

$$u_0 = (L^*L)^{-1}L^*z. \tag{5.1.30}$$

In general L^*L is not invertible and in this case one must construct an approximating sequence $\{u_n\}$ by using a suitable iterative technique, for example,

$$u_{n+1} = u_n - \varepsilon\ell'(u_n) \tag{5.1.31}$$

for a suitable $\varepsilon > 0$, possibly depending on n.

5.2 Controllability without control constraints

As mentioned in the introduction, the concept of controllability plays a fundamental role in control theory. In optimal control theory one usually starts with the assumption that the system is controllable and then considers the questions of existence of optimal controls and necessary conditions of optimality.

In this section we study the necessary and sufficient conditions of controllability of linear systems. Roughly speaking a system is said to be controllable if, by use of an admissible control policy, the system can be steered from the current state to a (desired) target state in finite time.

5.2.1 Linear time-invariant systems

Consider the system

$$\dot{x} = Ax + Bu, \qquad t \geqslant 0, \tag{5.2.1}$$

and suppose the class of admissible controls is given by $\mathcal{U} \equiv L_2^{\text{loc}}(0, \infty; R^m)$ or $L_\infty(0, \infty; R^m)$. Suppose a time interval $[0, T]$, and the initial and final (target) states x_0, $x_1 \in R^n$ are specified. Let $x(t, t_0, x_0, u)$ denote the solution of the problem (5.2.1) starting from the state x_0 at time $t_0 = 0$ corresponding to the control (input) u. The question is, does there exist a control $u \in \mathcal{U}$ and a time $t_1 \in [0, T]$ such that $x(t_1, t_0, x_0, u) = x_1$. That is,

$$x_1 = e^{t_1 A} x_0 + \int_0^{t_1} e^{(t_1 - \theta)A} Bu(\theta)\, d\theta. \tag{5.2.2}$$

Definition 5.2.1 (Controllability). The system (5.2.1) is said to be controllable with respect to the initial state x_0 and the final state x_1 if there exists an admissible control that transfers the system from the state x_0 to the state x_1 in finite time, that is, (5.2.2) is satisfied for some admissible u and $t_1 < \infty$.

Definition 5.2.2 (Global or complete controllability). The system is said to be globally (or completely) controllable if for each pair x_0, $x_1 \in R^n$ and $0 < t_1 < \infty$, there exists a control $u \in \mathcal{U}$, $u(t) = u(t, x_0, x_1, t_1)$, such that (5.2.2) is satisfied.

Theorem 5.2.3 (Global controllability). *The system* (5.2.1) *is completely controllable if and only if the controllability matrix* $M_c(t_1)$, *defined by*

$$M_c(t_1) = \int_0^{t_1} e^{(t_1 - \theta)A} BB' e^{(t_1 - \theta)A'}\, d\theta, \tag{5.2.3}$$

is positive definite. The control executing the transition is given by

$$u(t) \equiv B' e^{(t_1 - t)A'} M_c^{-1}(t_1)(x_1 - e^{t_1 A} x_0). \tag{5.2.4}$$

Proof (Necessary condition): In order that the system be globally controllable it is necessary and sufficient that, for each $z \in R^n$ and $0 < t_1 < \infty$,

$$L_{t_1}(u) \equiv \int_0^{t_1} e^{(t_1 - \tau)A} Bu(\tau)\, d\tau = z \tag{5.2.5}$$

for some $u \in \mathcal{U}$. This requires that range of the operator L_{t_1}, denoted *Range* (L_{t_1}), be equal to R^n; that is, Range $(L_{t_1}) = R^n$. This is equivalent to the statement that

$$(L_{t_1} u, \xi) = 0 \quad \text{for all } u \in \mathcal{U} \qquad \text{implies } \xi = 0,$$

or equivalently

$$(u, L_{t_1}^* \xi) = 0 \quad \text{for all } u \in \mathcal{U} \qquad \text{implies } \xi = 0. \tag{5.2.6}$$

Equivalently, $L_{t_1}^*\xi = 0$ implies $\xi = 0$ and hence

$$(L_{t_1}^*\xi,\ L_{t_1}^*\xi) = (L_{t_1}L_{t_1}^*\xi,\ \xi) = 0 \qquad \text{implies } \xi = 0. \tag{5.2.7}$$

The adjoint operator $L_{t_1}^*$ is easily identified as follows. For each $\eta \in R^n$,

$$\begin{aligned}(L_{t_1}u,\ \eta)_n &= \int_0^{t_1} (\mathrm{e}^{(t_1-\theta)A}Bu(\theta),\ \eta)_n\,\mathrm{d}\theta \\ &= \int_0^{t_1} (u(\theta),\ B'\mathrm{e}^{(t_1-\theta)A'}\eta)_m\,\mathrm{d}\theta \\ &= (u,\ L_{t_1}^*\eta)_{L_2}.\end{aligned} \tag{5.2.8}$$

Hence

$$(L_{t_1}^*\eta)(t) = B'\mathrm{e}^{(t_1-t)A'}\eta, \qquad t \in [0,\ t_1]. \tag{5.2.9}$$

Therefore,

$$\begin{aligned}L_{t_1}L_{t_1}^*\xi &= \int_0^{t_1} \mathrm{e}^{(t_1-\theta)A}B(L_{t_1}^*\xi)(\theta)\,\mathrm{d}\theta \\ &= \left(\int_0^{t_1} \mathrm{e}^{(t_1-\theta)A}BB'\mathrm{e}^{(t_1-\theta)A'}\,\mathrm{d}\theta\right)\xi = M_c(t_1)\xi.\end{aligned} \tag{5.2.10}$$

Thus it follows from (5.2.7) and (5.2.10) that

$$(M_c(t_1)\xi,\ \xi) = 0 \qquad \text{implies } \xi = 0. \tag{5.2.11}$$

Thus the controllability matrix $M_c(t_1)$ is positive definite, that is $M_c(t_1) > 0$.

(Sufficient condition): Given that $M_c(t_1) > 0$, we must show that the system is (globally) controllable. At the same time we shall verify (5.2.4). Suppose $M_c(t_1) > 0$ and $z \equiv (x_1 - \mathrm{e}^{t_1A}x_0) \in R^n$ is arbitrary. We must show that there is a control u_0 such that $z = L_{t_1}u_0$. Take u_0 such that

$$u_0(t) = B'\mathrm{e}^{(t_1-t)A'}(M_c(t_1))^{-1}z, \qquad t \in [0,\ t_1]. \tag{5.2.12}$$

Since $(M_c(t_1))^{-1}$ exists, the control u_0, as defined above, belongs to $L_2(0,\ t_1;\ R^m) \cap L_\infty(0,\ t_1;\ R^m)$ and hence $u_0 \in \mathcal{U}$. Clearly using this control we have,

$$\begin{aligned}L_{t_1}(u_0) &\equiv \int_0^{t_1} \mathrm{e}^{(t_1-\theta)A}Bu_0(\theta)\,\mathrm{d}\theta \\ &= \left(\int_0^{t_1} \mathrm{e}^{(t_1-\theta)A}BB'\mathrm{e}^{(t_1-\theta)A'}\,\mathrm{d}\theta\right)(M_c(t_1))^{-1}z = z.\end{aligned} \tag{5.2.13}$$

This completes the proof including (5.2.4). ■

Again, as stated in the case of observability, numerical verification of the positivity of the controllability matrix $M_c(t_1)$ is quite strenuous. Fortunately, for time-invariant linear systems, we have a simple algebraic criterion which is much easier to verify.

This is presented in the following result.

Theorem 5.2.4 (Algebraic criterion). *The following statements are equivalent*:

(a) *The system is (globally) controllable,*
(b) $M_c(t_1) > 0$ for $0 < t_1 < \infty$,
(c) $\operatorname{Ker} L_{t_1}^* = \{0\}$,
(d) $\operatorname{Rank}(B, AB, A^2B, \ldots, A^{n-1}B) = n$.

Proof That (a) is equivalent to (b) is a direct consequence of Theorem 5.2.3. In the course of the proof of the previous theorem (see (5.2.7)) we have seen that $M_c(t_1) > 0$ if and only if $L_{t_1}^*\xi = 0$ implies $\xi = 0$. That is, $\operatorname{Ker} L_{t_1}^* = \{0\}$. Hence (b) is equivalent to (c). We show that (d) follows from (c). $\operatorname{Ker} L_{t_1}^* = \{0\}$ is given, that is, $L_{t_1}^*\eta = 0$ implies $\eta = 0$, or equivalently

$$B' e^{(t_1 - t)A'} \eta = 0, \quad t \in [0, t_1], \qquad \text{implies } \eta = 0, \tag{5.2.14}$$

or

$$\left\{ \frac{d^k}{dt^k} (B' e^{(t_1 - t)A'} \eta) \Big|_{t = t_1} = 0, \; k = 0, 1, 2, \ldots \right\} \qquad \text{implies } \eta = 0. \tag{5.2.15}$$

By the Cayley–Hamilton theorem, it suffices to take $0 \leq k \leq n-1$, hence

$$\left\{ \frac{d^k}{dt^k} (B' e^{(t_1 - t)A'} \eta) \Big|_{t = t_1} = 0, \; 0 \leq k \leq n-1 \right\} \qquad \text{implies } \eta = 0,$$

or equivalently

$$\{B'\eta, B'A'\eta, \ldots, B'(A')^{n-1}\eta\} \equiv 0 \qquad \text{implies } \eta = 0.$$

Hence the rows of $\{B', B'A', \ldots, B'(A')^{n-1}\}$ span R^n, or equivalently there are n linearly independent row vectors in the above set, or the columns of $\{B, AB, \ldots, A^{n-1}B\}$ span R^n, and hence $\operatorname{Rank}\{B, AB, \ldots, A^{n-1}B\} = n$. Finally we show that (d) implies (a). Indeed we must show that if $\operatorname{Rank}\{B, AB, \ldots, A^{n-1}B\} = n$ then the system is controllable. Suppose the contrary. Then there exist $x_0, x_1 \in R^n$, $0 < t_1 < \infty$, such that

$$z \equiv (x_1 - e^{t_1 A} x_0) \notin \operatorname{Range} L_{t_1} \equiv R(L_{t_1}).$$

Since Range L_{t_1} is a linear subspace of R^n, $z \in R^{\perp}(L_{t_1})$ and hence $(z, L_{t_1}u) = (L_{t_1}^* z, u) = 0$ for all $u \in \mathcal{U}$ and consequently $(L_{t_1}^* z)(t) = 0$ for $t \in [0, t_1]$. Hence

$$B'z = B'A'z = \cdots = B'(A')^{n-1}z = 0$$

and, as a consequence, $(z, Bv) = 0$, $(z, ABv) = 0, \ldots, (z, A^{n-1}Bv) = 0$ for all $v \in R^n$, implying that Rank $(B, AB, \ldots, A^{n-1}B) \neq n$, which is a contradiction. This completes the proof. ■

Example 5.2.5 Consider the satellite example (Example 5.1.6) and suppose $T_i = \alpha_i u$, u a scalar control. Then $b = (\alpha_1, \alpha_2, \alpha_3)'$, $Ab = (-w_0\alpha_3, 0, w_0\alpha_1)'$ and $A^2b = (-w_0^2\alpha_1, 0, -w_0^2\alpha_3)'$. If $\alpha_2 \neq 0$ and either one of α_1 or $\alpha_3 \neq 0$, then Rank $(b, Ab, A^2b) = 3$ and the system is controllable. In case the control torques T_i, $i = 1, 2, 3$, are independent the system is always controllable.

Example 5.2.6 Consider the single-input system of Example 5.1.8:

$$z^{(n)} + a_{n-1}z^{(n-1)} + \cdots + a_1 z^{(1)} + a_0 z = u. \tag{5.2.16}$$

No matter what the coefficients are, this system is always controllable. This follows from the fact that the controllability matrix $(A^{n-1}b, A^{n-2}b, \ldots, Ab, b)$ is a square matrix and its determinant is 1 and hence its rank is n.

Recall that we used the symbol $D_0 = D_0(H, A)$ to denote the observable subspace of the observed system (5.1.5). Consider the controllability matrix

$$D(A, B) \equiv (B, AB, A^2B, \ldots, A^{n-1}B) \tag{5.2.17}$$

and let D_c denote the subspace of R^n spanned by the columns of the matrix D and call this the controllable subspace (justified later) of the controlled system $\dot{x} = Ax + Bu$.

Lemma 5.2.7 *The controllable subspace D_c is invariant under A, that is, $A(D_c) \subset D_c$.*

Proof Let $\xi \in D_c$; we show that $A\xi \in D_c$. Since $\xi \in D_c$ it must be given by some linear combination of the column vectors of $D(A, B)$. Hence

$$\xi = \sum_{k=0}^{n-1} A^k B w^k, \tag{5.2.18}$$

for some $\{w^k, k = 0, 1, \ldots, n-1\} \subset R^m$. Then

$$A\xi = \sum_{k=0}^{n-1} A^{k+1} B w^k. \tag{5.2.19}$$

By the Cayley–Hamilton theorem $A^n = \sum_{k=0}^{n-1} a_k A^k$ for a suitable family of constants $\{a_k, k = 0, 1, \ldots, n-1\}$. Hence

$$A^n B w^{n-1} = \sum_{k=0}^{n-1} a_k A^k B w^{n-1}, \tag{5.2.20}$$

and

$$A\xi = \sum_{k=0}^{n-1} A^k B v^k, \tag{5.2.21}$$

where

$$v^k = (a_k w^{n-1} + w^{k-1}), \qquad k = 0, 1, \ldots, n-1,$$

with the convention $w^{-1} = 0$. It follows from (5.2.21) that $A\xi$ is given by a suitable linear combination of the column vectors of $D(A, B)$. Hence $A\xi \in D_c$ whenever $\xi \in D_c$, proving $A(D_c) \subset D_c$. ■

As the name suggests we expect that the controllable subspace D_c is a subspace of the state space R^n in which the system is controllable. That is, any state in D_c can be reached from any other state in D_c by use of a suitable admissible control.

Let D_c be a proper subspace of R^n with

$$\text{dimension}(D_c) \equiv \dim(D_c) < n.$$

Theorem 5.2.8 (Controllability subspace). *The system $\dot{x} = Ax + Bu$, $u \in \mathcal{U}$, is controllable with respect to the (initial and target) states x_0, $x_1 \in R^n$ if and only if $y \equiv (\mathrm{e}^{-t_1 A} x_1 - x_0) \in D_c$ for some $t_1 < \infty$.*

Proof For the necessary condition, we must show that if the pair $\{x_0, x_1\}$ is controllable then necessarily $y \in D_c$. In other words, we must show that $R(L_{t_1}) \subset D_c$, where $R(L_{t_1})$ denotes the range of the operator L_{t_1} as defined by (5.2.5). By the Cayley–Hamilton theorem, for any $u \in \mathcal{U}$, we have

$$L_{t_1}(u) = \sum_{k=0}^{n-1} A^k B w^k \tag{5.2.22}$$

where

$$w^k \equiv \int_0^{t_1} \alpha_k(\tau) u(\tau)\, \mathrm{d}\tau, \qquad 0 \leqslant k \leqslant n-1, \tag{5.2.23}$$

with the functions $\{\alpha_k(t), t \in [0, t_1]\}$ bounded and given by the relation

$$\mathrm{e}^{-\tau A} = \sum_{m=0}^{\infty} \frac{(-\tau)^m}{m!} A^m = \sum_{k=0}^{n-1} \alpha_k(\tau) A^k. \tag{5.2.24}$$

Hence it follows from (5.2.22) that $L_{t_1}(u) \in D_c$ for each $u \in \mathcal{U}$ implying that $R(L_{t_1}) \subset D_c$.

For the sufficient condition we must show that if $y \in D_c$ then the system is controllable, or equivalently $D_c \subset R(L_{t_1})$. We prove this by showing that

$$R^{\perp}(L_{t_1}) \subset D_c^{\perp}. \tag{5.2.25}$$

Indeed let, $\eta \in R^{\perp}(L_{t_1})$, then $(w, \eta) = 0$ for all $w \in R(L_{t_1})$, or equivalently

$$\left(\int_0^{t_1} e^{-\tau A} B u(\tau)\, d\tau, \eta\right) = 0 \qquad \text{for all } u \in \mathcal{U},$$

or

$$\int_0^{t_1} (u(\tau), B' e^{-\tau A'} \eta)\, d\tau = 0 \qquad \text{for all } u \in \mathcal{U}. \tag{5.2.26}$$

Hence

$$B' e^{-\tau A'} \eta \equiv 0 \qquad \text{for all } \tau \in [0, t_1], \tag{5.2.27}$$

and consequently

$$(B' e^{-\tau A'} \eta, \xi) = 0 \qquad \text{for all } \tau \in [0, t_1] \text{ and } \xi \in R^m.$$

Therefore

$$(\eta, e^{-\tau A} B \xi) = 0 \qquad \text{for all } \tau \in [0, t_1] \text{ and } \xi \in R^m.$$

By differentiating the above expression and setting $\tau = 0$ we have

$$(\eta, B\xi) = 0, (\eta, AB\xi) = 0, \ldots, (\eta, A^{n-1}B\xi) = 0 \tag{5.2.28}$$

for all $\xi \in R^m$. But D_c is the linear subspace spanned by the columns of the controllability matrix $D(A, B) \equiv (B, AB, \ldots, A^{n-1}B)$. Hence (5.2.28) implies that $\eta \in D_c^{\perp}$. Thus we have shown that $\eta \in D_c^{\perp}$ whenever it is in $R^{\perp}(L_{t_1})$. Hence $R^{\perp}(L_{t_1}) \subset D_c^{\perp}$ and as a consequence $D_c \subset R(L_{t_1})$. Combining the above results we have $D_c = R(L_{t_1})$. This completes the proof. ■

Remark 5.2.9 A sufficient condition for $y = (e^{-t_1 A} x_1 - x_0)$ to belong to D_c is that both x_0 and $x_1 \in D_c$. In fact if $x_1 \in D_c$ then by the invariance (Lemma 5.2.7) $e^{-t_1 A} x_1 \in D_c$ and hence $y \in D_c$ provided $x_0 \in D_c$.

The significance of the above results is that every state in D_c can be reached from every other state in D_c. From the above result it is also clear that if the initial state $x_0 \in D_c$, then the whole trajectory

$$x(t, t_0, x_0) \equiv e^{tA} x_0 + \int_0^t e^{(t-\theta)A} B u(\theta)\, d\theta \in D_c$$

for all $t \geqslant 0$ and all $u \in \mathcal{U}$. In other words, the system can not leave D_c once it is there. Similarly one can verify that if $x_0(\neq 0) \in R^n \backslash D_c$, then $x(t, t_0, x_0) \in R^n \backslash D_c$ for all $t \geqslant 0$.

The preceding results lead to an interesting decomposition of the system $\dot{x} = Ax + Bu$. Suppose $\dim(D_c) = v < n$. Let us choose a new basis for the state space R^n so that the first v basis vectors form a basis for D_c. In terms of the new basis we have a new representation for all the matrices $\{x, A, B, u\}$ and they are denoted by $\{\tilde{x}, \tilde{A}, \tilde{B}, \tilde{u}\}$. This coordinate transformation can be realized by a nonsingular matrix Q so that $x = Q\tilde{x}$, $\tilde{A} = Q^{-1}AQ$, $\tilde{B} = Q^{-1}B$ and $\tilde{u} = u$. In this new basis the system has the representation

$$\dot{\tilde{x}} = \tilde{A}\tilde{x} + \tilde{B}\tilde{u} \tag{5.2.29}$$

where $\tilde{x} = \tilde{x}^1 + \tilde{x}^2$ with $\tilde{x}^1 \in D_c$ and $\tilde{x}^2 \in D_c^{\perp}$. Since the last $(n - v)$ entries of $\tilde{x}^1$ are all zero and the first v elements of $\tilde{x}^2$ are all zero, we can consider $\tilde{x}^1$ and $\tilde{x}^2$ as v and $n - v$ vectors and write $\tilde{x} = \binom{\tilde{x}^1}{\tilde{x}^2}$. In this form we can write our system (5.2.29) as

$$\frac{\mathrm{d}}{\mathrm{d}t}\begin{bmatrix}\tilde{x}^1\\ \tilde{x}^2\end{bmatrix} = \begin{bmatrix}\tilde{A}_{11} & \tilde{A}_{12}\\ \tilde{A}_{21} & \tilde{A}_{22}\end{bmatrix}\begin{bmatrix}\tilde{x}^1\\ \tilde{x}^2\end{bmatrix} + \begin{bmatrix}\tilde{B}_1\\ \tilde{B}_2\end{bmatrix}\tilde{u}, \tag{5.2.30}$$

where $\tilde{A}_{11} \in R(v \times v)$, $\tilde{A}_{12} \in R(v \times (n - v))$, $\tilde{A}_{21} \in R((n - v) \times v)$, $\tilde{A}_{22} \in R((n - v) \times (n - v))$, $\tilde{B}_1 \in R(v \times m)$, $\tilde{B}_2 \in R((n - v) \times m)$. Since the columns of B are elements of D_c, in the new representation $\tilde{B}_2 = 0$. Further, for

$$\tilde{x} = \begin{bmatrix}\tilde{x}^1\\ 0\end{bmatrix} \in D_c$$

we have

$$\xi \equiv \tilde{A}\tilde{x} = \begin{bmatrix}\tilde{A}_{11} & \tilde{A}_{12}\\ \tilde{A}_{21} & \tilde{A}_{22}\end{bmatrix}\begin{bmatrix}\tilde{x}^1\\ 0\end{bmatrix} = \begin{bmatrix}\tilde{A}_{11}\tilde{x}^1\\ \tilde{A}_{21}\tilde{x}^1\end{bmatrix}. \tag{5.2.31}$$

Since D_c is invariant under A it is also invariant under $\tilde{A}$ and hence ξ must necessarily belong to D_c. For arbitrary $\tilde{x}^1$, this is possible if and only if $\tilde{A}_{21} \equiv 0$. Hence the system (5.2.30) takes the form

$$\frac{\mathrm{d}}{\mathrm{d}t}\begin{bmatrix}\tilde{x}^1\\ \tilde{x}^2\end{bmatrix} = \begin{bmatrix}\tilde{A}_{11} & \tilde{A}_{12}\\ 0 & \tilde{A}_{22}\end{bmatrix}\begin{bmatrix}\tilde{x}^1\\ \tilde{x}^2\end{bmatrix} + \begin{bmatrix}\tilde{B}_1\\ 0\end{bmatrix}\tilde{u}, \tag{5.2.32}$$

or equivalently

$$\begin{aligned}\dot{\tilde{x}}^1 &= \tilde{A}_{11}\tilde{x}^1 + \tilde{A}_{12}\tilde{x}^2 + \tilde{B}_1\tilde{u}\\ \dot{\tilde{x}}^2 &= \tilde{A}_{22}\tilde{x}^2.\end{aligned} \tag{5.2.33}$$

It is clear from this representation that the system $\dot{\tilde{x}}^1 = \tilde{A}_{11}\tilde{x}^1 + \tilde{B}_1\tilde{u}$ is completely controllable in D_c while $\dot{\tilde{x}}^2 = \tilde{A}_{22}\tilde{x}^2$ is completely uncontrollable in its state space $D_c^\perp$. Thus we have proved the following result.

Corollary 5.2.10 *If the dimension of the controllable subspace of the system $\dot{x} = Ax + Bu$, $u \in \mathcal{U}$, denoted* $\dim(D_c) = \nu$, *is less than n, then the system has the decomposition* (5.2.32), *which can be realized by a non-singular transformation Q.*

We shall find this result useful in the study of stabilizability of linear systems by state feedback.

Another consequence of the previous results is given in the following corollary.

Corollary 5.2.11 *The system $\dot{x} = Ax + Bu$ is controllable if and only if the observed system $\dot{\xi} = -A'\xi$, $z = B'\xi$, is observable.*

This result can be briefly stated as: (A, B) is controllable iff $(B', -A')$ is observable. The proof is left as an exercise for the reader.

Consider the system (5.2.1) and suppose that the control u is given by a sum of two independent controls $u = u_c + u_0$, where u_c is to be provided by linear state feedback, $u_c = Kx$, $K \in R(m \times n)$, and u_0 is a control exercised by an exogenous agent, for example, a human operator. The question is, is the system

$$\dot{x} = (A + BK)x + Bu_0$$

controllable? The answer is given in the following result.

Theorem 5.2.12

(a) *The pair $(A + BK, B)$ is controllable if and only if (A, B) is controllable.*

(b) *For any $\lambda \in R$, the pair $(A + \lambda I, B)$ is controllable if and only if the pair (A, B) is controllable.*

Proof The result (a) follows from the facts that

(i) $$(B, (A + BK)B, (A + BK)^2B, \ldots, (A + BK)^{n-1}B) = (B, AB, A^2B, \ldots, A^{n-1}B)\Gamma, \quad (5.2.34)$$

where

$$\Gamma \equiv \begin{bmatrix} I & KB & KAB + (KB)^2 & \ldots \\ 0 & I & KB & \ldots \\ & & I & \ldots \\ & & & I \end{bmatrix}. \quad (5.2.35)$$

with $\Gamma \in R(nm \times nm)$ nonsingular and I is an $(m \times m)$ identity matrix, and

(ii) the rank of a rectangular matrix does not change if it is multiplied on the left or on the right by a non-singular matrix.

Result (b) follows from similar arguments. ■

System equivalence

Let $S(A, B)$ denote the linear controlled system $\dot{x} = Ax + Bu$ with admissible controls $\mathscr{U}$ fixed and let

$$\mathscr{L} \equiv \{S(A, B): A \in R(n \times n), B \in R(n \times m)\} \tag{5.2.36}$$

denote the class of all linear time-invariant controlled systems with a fixed common set of admissible controls $\mathscr{U}$. Clearly $\mathscr{L}$ is algebraically isomorphic to the space $R(n \times n) \times R(n \times m)$. On $\mathscr{L}$ we can introduce an equivalence relation through the controllability properties of its members. Let $S_1 \equiv S(A^{(1)}, B^{(1)})$ and $S_2 \equiv S(A^{(2)}, B^{(2)})$ be any two members of $\mathscr{L}$ and $D_c(S_1)$ and $D_c(S_2)$ the corresponding controllable subspaces $\subset R^n$. The system S_1 is said to be equivalent to S_2, written $S_1 \doteqdot S_2$, if and only if $D_c(S_1) = D_c(S_2)$ and the matrices $\tilde{A}_{22}^{(1)}$ and $\tilde{A}_{22}^{(2)}$ (see (5.2.32)) are either both stable or unstable. In other words two systems are equivalent if both have identical controllability properties. Similarly one can introduce observability equivalence, and combined observability–controllability equivalence, etc.

Another stronger concept of equivalence, based on system outputs, could be of substantial practical importance. Consider, more broadly, the linear observed–controlled system,

$$S(A, B, H, U) \equiv \begin{cases} \dot{x} = Ax + Bu, & u(t) \in U \quad \text{a.e.}, \quad t \geq 0 \\ z = Hx, & x(0) = x_0. \end{cases} \tag{5.2.37}$$

Let $\mathscr{K}(R^m)$ denote the space of compact subsets of R^m, and define $\mathscr{L} \equiv \{S(A, B, H, U): A \in R(n \times n),\ B \in R(n \times m),\ H \in R(r \times n)$, and $U = \mathscr{K}(R^m)\}$ to be the class of all time-invariant linear observed–controlled systems. Define

$$\mathscr{A}(t, A, B, U) \equiv \left\{ \xi \in R^n: \xi = e^{tA}x_0 + \int_0^t e^{(t-\theta)A} Bu(\theta)\, d\theta,\ u(t) \in U \text{ a.e.} \right\}, \tag{5.2.38}$$

to be the set of all attainable states at time t and

$$Z(t, A, B, H, U) \equiv H\mathscr{A}(t, A, B, U) \tag{5.2.39}$$

the set of all attainable outputs of the system $S(A, B, H, U)$ at time t. Since each $S \in \mathscr{L}$ corresponds to some

$$(A, B, H, U) \in R(n \times n) \times R(n \times m) \times R(r \times n) \times \mathscr{K}(R^m),$$

$\mathscr{L}$ is algebraically isomorphic to $R(n \times n) \times R(n \times m) \times R(r \times n) \times \mathscr{K}(R^m)$. If U is compact then $\mathscr{A}(t, A, B, U)$ is compact and hence $Z(t, A, B, H, U)$ is also compact. On the space $\mathscr{L}$ we can introduce a metric

$$\rho_{\mathscr{L}}(S_1, S_2) \equiv \sup\{\rho_H(Z(t, S_1), Z(t, S_2)), t \geq 0\} \tag{5.2.40}$$

where $S_1 \cong \{A^{(1)}, B^{(1)}, H^{(1)}, U^{(1)}\}$, $S_2 \cong \{A^{(2)}, B^{(2)}, H^{(2)}, U^{(2)}\}$ and ρ_H the Hausdorf metric (see Definition 6.1.1) on the space $\mathscr{K}(R^r)$. Hence $(\mathscr{L}, \rho_{\mathscr{L}})$ is a complete metric space. Clearly the metric (5.2.40) induces an equivalence relation on the class of controlled systems. We shall not pursue this subject further since, as yet, these concepts have not been introduced in the literature. However, the author expects that in future these concepts may prove to be important in system design.

5.2.2 Linear time-varying systems

Consider the system

$$\dot{x} = A(t)x + B(t)u, \qquad u \in \mathscr{U} = L_2^{\text{loc}} \tag{5.2.41}$$

and $B \in L_\infty(0, \infty; R(n \times m))$.

Definition 5.2.13 The system (5.2.41) is said to be completely controllable at time $t_0 \geq 0$ if, for every pair $x_0, x_1 \in R^n$, there exists a time $t_1 = t_1(x_0, x_1) < \infty$ and a control $u \in L_2(t_0, t_1; R^m)$, possibly depending on $\{t_0, t_1, x_0, x_1\}$, such that the corresponding solution $x(\cdot, u)$ satisfies the boundary conditions $x(t_0, u) = x_0$, $x(t_1, u) = x_1$.

For the given $x_0, x_1 \in R^n$ define

$$\eta \equiv x_1 - \Phi(t_1, t_0)x_0 \tag{5.2.42}$$

and the operator Γ by

$$\Gamma u \equiv \int_{t_0}^{t_1} \Phi(t_1, t)B(t)u(t)\, dt. \tag{5.2.43}$$

Then the question of controllability is equivalent to the question of solvability of the equation

$$\eta = \Gamma u \tag{5.2.44}$$

for $u \in L_2(t_0, t_1; R^m)$.

Theorem 5.2.14 *The system* (5.2.41) *is completely (or globally) controllable at time t_0 during the period $[t_0, t_1]$, $t_1 < \infty$, if and only if the controllability matrix $M_c(t_0, t_1)$, given by*

$$M_c(t_0, t_1) = \int_{t_0}^{t_1} \Phi(t_1, t)B(t)B'(t)\Phi'(t_1, t)\,\mathrm{d}t, \tag{5.2.45}$$

is positive definite (i.e. $M_c(t_0, t_1) > 0$).

Proof We prove that $M_c > 0$ implies controllability. Since $M_c > 0$ its inverse exists and we can take

$$u^0(t) = B'(t)\Phi'(t_1, t)M_c^{-1}(t_0, t_1)\eta, \qquad t \in [t_0, t_1]. \tag{5.2.46}$$

Then it is easy to verify that $\Gamma u^0 = \eta$ and hence the controllability follows. Next we show that controllability implies positivity of M_c. Controllability (global) implies that $\text{Range}\,\Gamma \equiv R(\Gamma) \equiv R^n$. Hence $(\Gamma u, \xi) = 0$ for all $u \in L_2(t_0, t_1, R^m)$ implies that $\xi = 0$. That is,

$$\int_{t_0}^{t_1} (\Phi(t_1, \theta)B(\theta)u(\theta), \xi)\,\mathrm{d}\theta$$

$$= \int_{t_0}^{t_1} (u(\theta), B'(\theta)\Phi'(t_1, \theta)\xi)\,\mathrm{d}\theta = (u, \Gamma^*\xi) = 0 \tag{5.2.47}$$

for all $u \in L_2(t_0, t_1; R^m)$ implies that $\xi = 0$. But

$$\int_{t_0}^{t_1} (u(\theta), B'(\theta)\Phi'(t_1, \theta)\xi)\,\mathrm{d}\theta = 0$$

for all $u \in L_2(t_0, t_1; R^m)$ implies that $B'(t)\Phi'(t_1, t)\xi = 0$ for almost all $t \in I = [t_0, t_1]$. Since the entries of B are in L_∞ and Φ is continuous and bounded, $B'(\cdot)\Phi'(t_1, \cdot)\xi \in L_2$ and since the integral must vanish for all $u \in L_2$ it also vanishes for the choice $u = B'\Phi'\xi$. Hence from (5.2.47) we have the equivalent statement,

$$\int_{t_0}^{t_1} |B'(\theta)\Phi'(t_1, \theta)\xi|_m^2\,\mathrm{d}\theta = \|\Gamma^*\xi\|_{L_2}^2 = 0$$

implies $\xi = 0$. This, in turn, is equivalent to

$$(M_c(t_0, t_1)\xi, \xi) = 0 \qquad \text{implies } \xi = 0. \tag{5.2.48}$$

Hence $M_c > 0$. ■

Some immediate consequences of this result are given in the following corollaries.

Corollary 5.2.15 *The following statements are equivalent*:

(a) *The system* (5.2.41) *is globally controllable at* t_0 *on* $I=[t_0, t_1]$,
(b) $M_c(t_0, t_1)>0$,
(c) $\operatorname{Ker}\Gamma^* = \operatorname{Ker}\{B'(t)\Phi'(t_1, t), t \in I\} = \{0\}$.

Corollary 5.2.16 *If the controllability matrix* $M_c(t_0, t_1)>0$, *then* $M_c(t_0', t_1')>0$ *also for any* $t_0' \leqslant t_0$ *and* $t_1' \geqslant t_1$.

This shows that if the system is globally controllable over an interval I, then it is also controllable over $J \supset I$.

Remark 5.2.17 $M_c(t_0, t_1)>0$, but $M_c(t_0, \tau)$ may fail to be positive definite if $\tau<t_1$, implying insufficient time for global controllability. Similarly $M_c(s, t_1)$ may fail to be positive definite if $s>t_0$, implying loss of power (too late for global controllability).

The results of this section are very useful in optimal control theory even though numerically inefficient for detecting controllability (see also Problem 5.3.P14).

It is interesting to note that the control given by the expression (5.2.46) not only makes the transition from the state x_0 to x_1 possible but also it does so with minimum energy. This is explained in the following result.

Corollary 5.2.18 *The control* $u^0(t)$ *defined by*

$$u^0(t) = B'(t)\Phi'(t_1, t)M_c^{-1}(t_0, t_1)(x_1 - \Phi(t_1, t_0)x_0) \tag{5.2.49}$$

transfers the state x_0 *to* x_1 *with minimum expenditure of energy, i.e.*

$$\int_{t_0}^{t_1} |u^0(t)|^2 \, dt \leqslant \int_{t_0}^{t_1} |u(t)|^2 \, dt \tag{5.2.50}$$

for all $u \in L_2(t_0, t_1; R^m)$ *which transfers* x_0 *to* x_1.

Proof Let $u(t)=u^0(t)+\tilde{u}(t)$ for some $\tilde{u} \in L_2(t_0, t_1; R^m)$. Since the control u transfers the initial state x_0 to the final state x_1, we have

$$\begin{aligned} x_1 &= \Phi(t_1, t_0)x_0 + \int_{t_0}^{t_1} \Phi(t_1, \tau)B(\tau)u(\tau)\, d\tau \\ &= \Phi(t_1, t_0)x_0 + \int_{t_0}^{t_1} \Phi(t_1, \tau)B(\tau)u^0(\tau)\, d\tau + \int_{t_0}^{t_1} \Phi(t_1, \tau)B(\tau)\tilde{u}(\tau)\, d\tau. \end{aligned}$$

Since u^0 also transfers x_0 to x_1, it is clear from the above expression that

$$\int_{t_0}^{t_1} \Phi(t_1, \tau)B(\tau)\tilde{u}(\tau)\,d\tau = 0. \tag{5.2.51}$$

Using (5.2.49) and (5.2.51) one can easily verify that

$$\int_{t_0}^{t_1} (u^0(t), \tilde{u}(t))\,dt = 0.$$

Consequently,

$$\begin{aligned}\int_{t_0}^{t_1} |u(t)|^2\,dt &= \int_{t_0}^{t_1} |u^0(t)|^2\,dt + \int_{t_0}^{t_1} |\tilde{u}(t)|^2\,dt + 2\int_{t_0}^{t_1} (u^0(t), \tilde{u}(t))\,dt \\ &= \int_{t_0}^{t_1} |u^0(t)|^2\,dt + \int_{t_0}^{t_1} |\tilde{u}(t)|^2\,dt \geq \int_{t_0}^{t_1} |u^0(t)|^2\,dt. \quad \blacksquare\end{aligned}$$

Remark 5.2.19 It is easy to verify that the minimum control energy required to transfer x_0 to x_1 using the control u^0 is given by

$$J^0 = \int_{t_0}^{t_1} |u^0(t)|^2\,dt = (M_c^{-1}(t_0, t_1)\eta, \eta),$$

where $\eta = x_1 - \Phi(t_1, t_0)x_0$.

Example 5.2.20 Consider the linear time-varying system

$$\begin{bmatrix}\dot{x}_1 \\ \dot{x}_2\end{bmatrix} = \begin{bmatrix}0 & 0 \\ 2t & 0\end{bmatrix}\begin{bmatrix}x_1 \\ x_2\end{bmatrix} + \begin{bmatrix}1 \\ 0\end{bmatrix}u(t), \qquad t \geq 0.$$

The state transition matrix for this system is given by

$$\Phi(t, \tau) = \begin{bmatrix}1 & 0 \\ t^2 - \tau^2 & 1\end{bmatrix}.$$

Thus one can verify that

$$M_c(0, t_1) = \begin{bmatrix}t_1 & \frac{2}{3}t_1^3 \\ \frac{2}{3}t_1^3 & \frac{8}{15}t_1^5\end{bmatrix},$$

which is positive definite. Hence this system is controllable. A control for transferring $x_0 = (0, 1)$ to $x_1 = (0, 0)$ is obtained using (5.2.49) and is given by

$$u(t) = \frac{45}{4t_1^5}(t^2 - \tfrac{1}{3}t_1^2).$$

Note that as the transition time t_1 is made smaller, the required amount of control force becomes larger, as intuition also suggests.

Corollary 5.2.21 *The system*

$$S: \dot{x} = A(t)x + B(t)u$$

is controllable if and only if the adjoint system

$$S_a: \begin{array}{l} \dot{\psi} = -A'(t)\psi \\ z = B'(t)\psi \end{array}$$

is observable.

Proof The proof follows from the simple observation that the controllability matrix of S and the observability matrix of S_a with clock reversed are identical. ■

5.3 Controllability under control constraints

In many practical problems there are constraints on the control magnitudes. In this section, we present some salient results on controllability with control constraints based on the work of several authors, notably [107]–[110], [138]. For infinite-dimensional systems, see [153].

Consider the system

$$S \begin{cases} \dot{x} = A(t)x + B(t)u \\ u \in \mathcal{U}, \end{cases} \tag{5.3.1}$$

where $A \in L_1^{\text{loc}}(0, \infty; R(n \times n))$, $B \in L_1^{\text{loc}}(0, \infty; R(n \times m))$ and $\mathcal{U} = M(U)$ is the space of admissible controls which consists of measurable functions of time with values in a non-empty set $U \subset R^m$. In the majority of applications U is a compact set.

For given t_0, $t_1 \in R$, $x_0 \in R^n$ and U a compact subset of R^m we define the attainable set,

$$\mathscr{A}(t_1, t_0, x_0, U) \equiv \left\{ \xi \in R^n: \xi = \Phi(t_1, t_0)x_0 + \int_{t_0}^{t_1} \Phi(t_1, \theta)B(\theta)u(\theta)\, \mathrm{d}\theta,\ u \in \mathcal{U} \right\}, \tag{5.3.2}$$

where Φ is the transition operator corresponding to A. This is the set of states that the system can attain given the initial state x_0, the time constraint $[t_0, t_1]$ and the control constraint U. Clearly if $U \subset V$ then

$\mathscr{A}(t_1, t_0, x_0, U) \subset \mathscr{A}(t_1, t_0, x_0, V)$, and if U is a non-empty compact subset of R^m, then $\mathscr{A}(t_1, t_0, x_0, U)$ is a compact subset of R^n. If U is fixed we shall write $\mathscr{A}(t_1, t_0, x_0)$ instead of $\mathscr{A}(t_1, t_0, x_0, U)$.

Definition 5.3.1 (U-null controllability) The system (5.3.1) is said to be U-null controllable at $\{t_0, x_0\}$ if, given the initial state x_0 and time t_0, there exists a control $u \in \mathscr{U} \equiv M(U)$ such that $x(t_0, u) = x_0$ and $x(t, u) = 0$ for some $t \in [t_0, \infty)$; in other words $0 \in \mathscr{A}(t, t_0, x_0, U)$ for some finite $t \geqslant t_0$.

Definition 5.3.2 (Local and global null controllability). The system (5.3.1) is said to be locally U-null controllable (or simply locally null controllable in case U is fixed) at t_0 if there exists an open neighbourhood N of the origin such that (5.3.1) is U-null controllable at (t_0, x_0) for every $x_0 \in N$. It is said to be globally U-null controllable at t_0 if $N = R^n$.

We define the set of null controllability as

$$\mathscr{C}(t_0, t_1, U) \equiv \left\{ x_0 \in R^n : \; -\Phi(t_1, t_0)x_0 = \int_{t_0}^{t_1} \Phi(t_1, t)B(t)u(t)\,\mathrm{d}t,\ u \in \mathscr{U} \right\}. \tag{5.3.3}$$

This gives us the set of initial states (given at time $t = t_0$) from which the system can be driven to the zero state during the time interval $[t_0, t_1]$ by application of admissible controls. The set $\mathscr{C}(t_0, U)$, given by

$$\mathscr{C}(t_0, U) \equiv \bigcup_{\infty > t_1 > t_0} \mathscr{C}(t_0, t_1, U), \tag{5.3.4}$$

is the domain of finite-time null controllability at t_0.

For time-invariant systems,

$$\mathscr{A}(t_1, t_0, x_0, U) = \mathscr{A}(t_1 - t_0, x_0, U) \equiv \mathscr{A}(\tau, x_0, U),$$

and

$$\mathscr{C}(t_0, t_1, U) = \mathscr{C}(t_1 - t_0, U) \equiv \mathscr{C}(\tau, U), \tag{5.3.5}$$

where $\tau = (t_1 - t_0)$ is the time difference. In this case there is an interesting relationship between the attainable sets and the domain of null controllability. Before we discuss this question, recall the definition of the domain of asymptotic stability of the origin. There we were interested in the domain of stability of uncontrolled systems. Here the uncontrolled system may be unstable but we can exercise controls to steer the system to the desired states. This gives rise to interest in the properties of the domain of null controllability. In case it is non-empty and contains the origin we are assured of the existence of control policies that steer the system to the origin, not merely asymptotically but in finite time. For a fixed non-empty set $U \subset R^m$ define

$$\mathscr{A} \equiv \bigcup_{\infty > \tau \geqslant 0} \mathscr{A}(\tau, 0, U),$$

and (5.3.6)

$$\mathscr{C} \equiv \bigcup_{\infty > \tau \geqslant 0} \mathscr{C}(\tau, U).$$

Lemma 5.3.3 ($\mathscr{C} = \mathscr{A}_-$). *Let U be a nonempty compact subset of R^m and consider the system*

$$S_-: \dot{y} = -Ay - Bu, \ u \in \mathscr{U}$$
$$y(0) = 0 \tag{5.3.7}$$

and let $\mathscr{A}_-$ denote the corresponding attainable set. Then the domain of null controllability of the system S (5.3.1), with $A(t) \equiv A$, $B(t) \equiv B$ constant, equals the attainable set of the system S_- (5.3.7), that is, $\mathscr{C} = \mathscr{A}_-$.

Proof For any $\tau \in (0, \infty)$ note that

$$\mathscr{C}(\tau, U) \equiv \left\{x_0 \in R^n : x_0 = -\int_0^\tau e^{-\theta A} Bu(\theta)\, \mathrm{d}\theta, \ u \in \mathscr{U}\right\} \tag{5.3.8}$$

and

$$\begin{aligned} \mathscr{A}_-(\tau, 0, U) &\equiv \left\{x_1 \in R^n : x_1 = -\int_0^\tau \mathrm{e}^{-(\tau-\theta)A} Bu(\theta)\, \mathrm{d}\theta, \ u \in \mathscr{U}\right\} \\ &= \left\{x_1 \in R^n : x_1 = -\int_0^\tau \mathrm{e}^{-\theta A} Bu(\tau - \theta)\, \mathrm{d}\theta, \ u \in \mathscr{U}\right\} \\ &= \left\{x_1 \in R^n : x_1 = -\int_0^\tau \mathrm{e}^{-\theta A} Bv(\theta)\, \mathrm{d}\theta, \ v \in \mathscr{U}\right\}. \end{aligned} \tag{5.3.9}$$

Hence

$$\mathscr{C}(\tau, U) = \mathscr{A}_-(\tau, 0, U)$$

for every $\tau \in (0, \infty)$ and consequently

$$\mathscr{C} \equiv \bigcup_{0 \leqslant \tau < \infty} \mathscr{C}(\tau, U) = \bigcup_{0 \leqslant \tau < \infty} \mathscr{A}_-(\tau, 0, U) \equiv \mathscr{A}_-. \quad \blacksquare \tag{5.3.10}$$

Theorem 5.3.4 (Local null controllability). *Consider the system* (5.3.1) *with $A(t) \equiv A, B(t) \equiv B$ constant, $u \in \mathscr{U} \equiv M(U)$, where U is a* *non-empty compact subset of R^m containing the origin in its interior. Then the domain of null controllability $\mathscr{C}$ is an open subset of R^n if and only if the system is controllable.*

Proof First note that the system $S: \dot{x} = Ax + Bu$, $u \in \mathscr{U}$ is controllable if and only if the system $S_-: \dot{y} = -Ay - Bu$, $u \in \mathscr{U}$, is controllable. This follows from the fact that $\text{Rank}(B, AB, A^2B, \ldots, A^{n-1}B) = \text{Rank}(-B, AB, -A^2B, \ldots, (-1)^n A^{n-1}B)$. Since $0 \in \text{Int } U$ (interior U), there exists a $\delta > 0$ such that $B_\delta \equiv \{\xi \in R^m: |\xi| < \delta\} \subset U$. Let $\mathscr{A}_-^\delta$ denote the set of attainability for the system S_- with U replaced by B_δ. Clearly $0 \in \mathscr{A}_-^\delta \subset \mathscr{A}_-$ and hence by the previous lemma, $\mathscr{A}_-^\delta \subset \mathscr{A}_- = \mathscr{C}$. Thus $\mathscr{C}$ contains a neighbourhood of the origin and since any point of $\mathscr{C}$ can be steered to the origin, we conclude that $\mathscr{C}$ is open in R^n. Next, we show that if $\mathscr{C}$ is open, then the system S is controllable. Since $\mathscr{C} = \mathscr{A}_-$ is open in R^n, each point $x_1 \in \mathscr{C}$ is attainable from the zero state by the system S_-. Hence

$$\mathscr{C} = \mathscr{A}_- = \left\{ x_1 \in R^n: x_1 = -\int_0^\tau e^{(\theta-\tau)A} Bu(\theta)\, d\theta \right.$$
$$\left. \text{for some } u \in \mathscr{U} \text{ and } 0 \leq \tau < \infty \right\}. \tag{5.3.11}$$

We prove controllability of S by proving that of S_-. Suppose S_- is not controllable then since $0 \in \text{Int } U$ there exists an $\eta \in R^n$ such that

$$(Bv, \eta) = (ABv, \eta) = (A^2Bv, \eta) = \cdots = (A^{n-1}Bv, \eta) = 0$$

for all $v \in U$. Then for any $x_1 \in \mathscr{C}$ we have, for some $\tau < \infty$,

$$\begin{aligned}(x_1, \eta) &= -\int_0^\tau (e^{(\theta-\tau)A} Bu(\theta), \eta)\, d\theta \\ &= -\sum_{k=0}^{n-1} \int_0^\tau \alpha_k(\theta - \tau)(A^k Bu(\theta), \eta)\, d\theta = 0. \end{aligned} \tag{5.3.12}$$

Thus $\mathscr{C}$ lies in a hyperplane normal to η. But this is impossible since $\mathscr{C}$ has interior points. Hence S_- and, consequently, S must be controllable. ■

Corollary 5.3.5 (Global null controllability). *Consider the system $S: \dot{x} = Ax + Bu$, $u \in \mathscr{U}$ with the control restraint set U satisfying the hypothesis of Theorem* 5.3.4. *Suppose S is controllable with $U = R^m$ and all the eigenvalues of A have negative real parts. Then the domain of null controllability $\mathscr{C} = R^n$, that is, the system is globally null controllable.*

Proof Let x_0 be an arbitrary element of R^n. Since all the eigenvalues of A have negative real parts, the state of the system, driven by control $u \equiv 0$, will asymptotically approach the origin. Hence as the state enters the domain of null controllability $\mathscr{C}$, one can choose a nonzero control that steers the system to the origin in finite time. Since $\mathscr{C}$ is an open neighbourhood of the origin, the time to reach $\mathscr{C}$ (with $u = 0$) is also finite. Using these facts one can conclude that there exists a control that steers the system from the state x_0 to the origin in finite time. Hence $x_0 \in \mathscr{C}$ and, since x_0 is an arbitrary element of R^n, $\mathscr{C} = R^n$. ■

Remark 5.3.6 The above result can be summarized by stating that if a linear time-invariant system is asymptotically stable (with respect to the origin) and locally null controllable, then it is globally null controllable. We will see later that in the time-varying case this result is not necessarily true.

We shall now consider linear time-varying systems and develop necessary and sufficient conditions for null controllability. From now on we shall assume that U is fixed and hence drop it from the argument of $\mathscr{A}(t_1, t_0, x_0, U) \equiv \mathscr{A}(t_1, t_0, x_0)$.

Consider the time-varying system S (equation (5.3.1)) and define for each $x_0 \in R^n$, $t_0 \in R$, $\eta \in R^n$ and $t \geq t_0$,

$$J(t, t_0, x_0, \eta) \equiv (\Phi(t, t_0)x_0, \eta) + \int_{t_0}^{t} H_U(B'(\theta)\Phi'(t, \theta)\eta)\, d\theta, \quad (5.3.13)$$

where

$$H_U(w) \equiv \sup\{(w, v), v \in U\}. \quad (5.3.14)$$

We shall prove that the time-varying system S is U-null controllable at (t_0, x_0) if and only if $J(t_1, t_0, x_0, \eta) \geq 0$ for all $\eta \in R^n$ for some $t_1 \in [t_0, \infty)$. For the proof of this result we shall need the following lemma.

Lemma 5.3.7 *The function $w \to H_U(w)$ from R^m to R is continuous if U is a compact subset of R^m.*

Proof Let $w_n \to w_0$ in R^m and let v_n and v_0 be suitable elements of U such that $H_U(w_n) = (v_n, w_n)$ and $H_U(w_0) = (v_0, w_0)$. Since U is compact such points exist. Then clearly

$$\begin{aligned} H_U(w_n) \geq (v_0, w_n) &= (v_0, w_n - w_0) + (v_0, w_0) \\ &= (v_0, w_n - w_0) + H_U(w_0) \end{aligned} \quad (5.3.15)$$

and

$$\begin{aligned} H_U(w_0) \geq (v_n, w_0) &= (v_n, w_0 - w_n) + (v_n, w_n) \\ &= (v_n, w_0 - w_n) + H_U(w_n). \end{aligned} \quad (5.3.16)$$

Since $w_n \to w_0$ and U is compact and hence $\sup_n |v_n| < \infty$, it follows from the above inequalities that

$$\underline{\lim}_n H_U(w_n) \geq H_U(w_0),$$

and

$$H_U(w_0) \geq \overline{\lim}^n H_U(w_n). \tag{5.3.17}$$

Combining these inequalities we have

$$H_U(w_0) \leq \underline{\lim}_n H_U(w_n) \leq \lim_n H_U(w_n) \leq \overline{\lim}^n H_U(w_n) \leq H_U(w_0),$$

and consequently

$$\lim_n H_U(w_n) = H_U(w_0). \tag{5.3.18}$$

This proves the continuity. ■

Theorem 5.3.8 (Null controllability). *Let U be a non-empty compact subset of R^m and Λ any subset of R^n with $0 \in \operatorname{Int} \Lambda$. Then the (time-varying) system S is U-null controllable at (t_0, x_0) if and only if*

$$\tilde{J}(t_1, t_0, x_0) \equiv \min\{J(t_1, t_0, x_0, \eta), \eta \in \Lambda\} = 0 \tag{5.3.19}$$

for some $t_1 \in [t_0, \infty)$.

Proof For simplicity of the proof we shall assume that U is also convex, though this is not essential as we shall see later. Under the convexity assumption it is easy to verify that the attainable set $\mathscr{A}(t_1, t_0, x_0)$ is also convex. It is also compact since U is compact. By definition, x_0 can be steered to the zero state at time t_1 if and only if $0 \in \mathscr{A}(t_1, t_0, x_0)$. We shall prove that this is equivalent to requiring that

$$\sup\{(\eta, a), a \in \mathscr{A}(t_1, t_0, x_0)\} \geq 0, \tag{5.3.20}$$

for any $\eta \in R^n$. Indeed if $0 \in \mathscr{A}(t_1, t_0, x_0)$ then the condition (5.3.20) is obviously satisfied. We must prove the converse. That is, if (5.3.20) is satisfied then $0 \in \mathscr{A}(t_1, t_0, x_0)$. We prove this by contradiction. Suppose $0 \notin \mathscr{A}(t_1, t_0, x_0)$ while (5.3.20) is satisfied. Since the set $\mathscr{A}$ is compact and convex there exists an $\varepsilon > 0$ such that 0 is at least at a distance ε from it. Hence one can find a hyperplane passing through the origin at a distance at least ε from the attainable set $\mathscr{A}(t_1, t_0, x_0)$. Then there exists an $\eta^* \in R^n$ which is normal to the hyperplane such that

$$(\eta^*, a) < -\varepsilon \qquad \text{for all } a \in \mathscr{A}(t_1, t_0, x_0). \tag{5.3.21}$$

But this contradicts (5.3.20) and hence $0 \in \mathscr{A}(t_1, t_0, x_0)$. Thus we have proved that the null controllability of (t_0, x_0) at time t_1 is equivalent to

(5.3.20). To prove (5.3.19), first we show that

$$\sup\{(\eta, a): a \in \mathcal{A}(t_1, t_0, x_0)\} = J(t_1, t_0, x_0, \eta), \tag{5.3.22}$$

where J is given by the expression (5.3.13). In view of the expression for the attainable set (5.3.2) and the definition for J, it suffices to show that

$$\sup\{\ell(u), u \in \mathcal{U}\} = \int_{t_0}^{t_1} H_U(B'(t)\Phi'(t_1, t)\eta)\, dt, \tag{5.3.23}$$

where

$$\ell(u) \equiv \int_{t_0}^{t_1} (u(t), B'(t)\Phi'(t_1, t)\eta)\, dt. \tag{5.3.24}$$

If the entries of B are merely locally integrable, then $u \to \ell(u)$ is weakstar continuous on $\mathcal{U}$, where $\mathcal{U}$ is a weakstar compact subset of L_∞. On the other hand, if the entries of B are essentially bounded or even locally square integrable, then $u \to \ell(u)$ is weakly continuous on $\mathcal{U}$, with $\mathcal{U}$ considered as a weakly compact subset of L_2. In any event ℓ attains its minimum and maximum on $\mathcal{U}$. Suppose ℓ attains its maximum at $u^* \in \mathcal{U}$. Then

$$\sup\{\ell(u), u \in \mathcal{U}\} = \int_{t_0}^{t_1} (u^*(\theta), B'(\theta)\Phi'(t_1, \theta)\eta)\, d\theta, \tag{5.3.25}$$

and hence, for the proof of (5.3.23), it suffices to show that

$$\int_{t_0}^{t_1} (u^*(\theta), B'(\theta)\Phi'(t_1, \theta)\eta)\, d\theta = \int_{t_0}^{t} H_U(B'(\theta)\Phi'(t_1, \theta)\eta)\, d\theta. \tag{5.3.26}$$

Note, in order that the integral in the right-hand side of the expression (5.3.26) is well defined, it is necessary that the function

$$t \to H_U(B'(t)\Phi'(t_1, t)\eta), \tag{5.3.27}$$

be measurable on $[t_0, t_1]$. But this follows from the facts that a continuous function of a measurable function is measurable and that, by Lemma 5.3.7, $H_U(\cdot)$ is continuous, and $t \to B'(t)\Phi(t_1, t)\eta$ is measurable on $[t_0, t_1]$. For the proof of (5.3.26) we use the obvious expression

$$\int_{t_0}^{t_1} (u^*(\theta), B'(\theta)\Phi'(t_1, \theta)\eta)\, d\theta \geq \int_{t_0}^{t_1} (u(\theta), B'(\theta)\Phi'(t_1, \theta)\eta)\, d\theta \tag{5.3.28}$$

which holds for all $u \in \mathcal{U}$. Let t be an arbitrary element of $I = (t_0, t_1)$ and define

$$I_n \equiv \{\theta \in R: t - 1/n \leq \theta \leq t + 1/n\} \cap I$$

and

$$u(t) \equiv \begin{cases} u^*(t), & t \in I \backslash I_n \\ v \text{ for } t \in I_n, & v \in U. \end{cases} \tag{5.3.29}$$

Clearly $u \in \mathscr{U}$. Using this u in (5.3.28) we obtain

$$\int_{I_n} (u^*(\theta), B'(\theta)\Phi'(t_1, \theta)\eta)\, \mathrm{d}\theta \geq \int_{I_n} (v, B'(\theta)\Phi'(t_1, \theta)\eta)\, \mathrm{d}\theta \tag{5.3.30}$$

for all $v \in U$ and all integers $n \geq 1$. Letting $\mu(I_n)$ denote the Lebesgue measure (length) of I_n, we have

$$\begin{aligned} &\frac{1}{\mu(I_n)} \int_{I_n} (u^*(\theta), B'(\theta)\Phi'(t_1, \theta)\eta)\, \mathrm{d}\theta \\ &\quad \geq \frac{1}{\mu(I_n)} \int_{I_n} (v, B'(\theta)\Phi'(t_1, \theta)\eta)\, \mathrm{d}\theta \end{aligned}$$

and letting $n \to \infty$ it follows from the above inequality that

$$(u^*(t), B'(t)\Phi'(t_1, t)\eta) \geq (v, B'(t)\Phi'(t_1, t)\eta) \tag{5.3.31}$$

for all $v \in U$. Hence

$$(u^*(t), B'(t)\Phi'(t_1, t)\eta) \geq H_U(B'(t)\Phi'(t_1, t)\eta),$$

but since $u^*(t) \in U$ a.e. on $[t_0, t_1]$

$$(u^*(t), B'(t)\Phi'(t_1, t)\eta) \leq H_U(B'(t)\Phi'(t_1, t)\eta) \qquad \text{a.e.}$$

and therefore, t being arbitrary,

$$(u^*(t), B'(t)\Phi'(t_1, t)\eta) = H_U(B'(t)\Phi'(t_1, t)\eta) \qquad \text{a.e.}$$

Integrating this over $[t_0, t_1]$ equation (5.3.26) follows and hence we have proved (5.3.23).

Thus we have proved the equality

$$\begin{aligned} &\sup\{(\eta, a), a \in \mathscr{A}(t_1, t_0, x_0)\} \\ &\quad = (\eta, \Phi(t_1, t_0)x_0) + \int_{t_0}^{t_1} H_U(B'(t)\Phi'(t_1, t)\eta)\, \mathrm{d}t \\ &\quad \equiv J(t_1, t_0, x_0, \eta). \end{aligned} \tag{5.3.32}$$

Hence it follows from (5.3.20) and (5.3.32) that, for the system to be null controllable at (t_0, x_0), it is necessary and sufficient that

$$J(t_1, t_0, x_0, \eta) \geq 0 \qquad \text{for all } \eta \in R^n. \tag{5.3.33}$$

Since $\eta \to J(t_1, t_0, x_0, \eta)$ is positively homogeneous in the sense that $J(t_1, t_0, x_0, \alpha\eta) = \alpha J(t_1, t_0, x_0, \eta)$, $\alpha \geq 0$, it follows from (5.3.33) that for any $\Lambda \subset R^n$ with $0 \in \operatorname{Int} \Lambda$

$$\tilde{J}(t_1, t_0, x_0) = \min\{J(t_1, t_0, x_0, \eta), \eta \in \Lambda\} = 0. \tag{5.3.34}$$

This completes the proof. ■

As a consequence of the above result we note that if (t_0, x_0) is not null controllable during the time period $[t_0, t_1]$, then $\tilde{J}(t, t_0, x_0) < 0$ for all $t \in [t_0, t_1]$. If $\tilde{J}(t, t_0, x_0) < 0$ for all finite t, then (t_0, x_0) is not finite-time null controllable. If τ is the first time after t_0 for which $\tilde{J}(\tau, t_0, x_0) = 0$, then $(\tau - t_0)$ is the minimum time of transition from the state x_0 to the zero state. Thus for $\Lambda = B_1 \equiv \{\xi \in R^n: |\xi| \leq 1\}$ the function $t \to \tilde{J}(t, t_0, x_0)$ starts out with value $-|x_0|$ at $t = t_0$ and increases with time and meets the t-axis in finite time if the system is finite-time null controllable. In fact for $\Lambda = B_1$, $x_0 \neq 0$,

$$\tilde{J}(t, t_0, x_0) \equiv \max\{-|x(t, t_0, x_0, u)|, u \in \mathcal{U}\} < 0 \tag{5.3.35}$$

if (t_0, x_0) is not null controllable at time t. This tells us nothing more than the basic fact that by minimizing the norm $|x(t, t_0, x_0, u)|$ by choosing controls u and repeating this until time t^* at which $\min\{|x(t^*, t_0, x_0, u)|, u \in \mathcal{U}\} = 0$ one can determine the desired control. We shall discuss this further while considering optimal controls.

Remark 5.3.9 Theorem 5.3.8 can be used to estimate the domain of null controllability. Given $t > t_0$, the domain of null controllability during the period $[t_0, t]$ is given by

$$\mathscr{C}_{t,t_0} = \{x_0 \in R^n: \tilde{J}(t, t_0, x_0) = 0\}. \tag{5.3.36}$$

Since $0 \in \mathscr{C}_{t,t_0}$ the set is non-empty and $t \to \mathscr{C}_{t,t_0}$ is a monotone non-contracting set-valued function on $t \geq t_0$, in the sense that $\mathscr{C}_{t_1,t_0} \subset \mathscr{C}_{t_2,t_0}$ for $t_1 < t_2$. We illustrate this result by an example.

Example 5.3.10 Consider the null controllability problem for the scalar system

$$\begin{cases} \dot{x} = ax + u(t)\exp(-pt), & t \geq 0, \\ u(t) \in U = [-1, 1]. \end{cases} \tag{5.3.37}$$

Use (5.3.13) to compute

$$J(t_1, 0, x_0, \eta) = \eta e^{at_1}\{x_0 + ((\operatorname{sign} \eta)/(a + p))(1 - e^{-(a+p)t_1})\}. \tag{5.3.38}$$

By virtue of (5.3.33), for null controllability we must have $J(t_1, 0, x_0, \eta) \geq 0$ for all $\eta \in \Lambda$, where Λ is any interval containing zero in its interior. Taking $\Lambda = [-1, 1]$, it is easy to verify that the domain of null controllability over $[0, t_1]$ is given by the intervals

$$(-1/(a+p))(1-e^{-(a+p)t_1}) < x_0 < (1/(a+p))(1-e^{-(a+p)t_1}) \qquad \text{for } a+p \neq 0$$

$$-t_1 < x_0 < t_1 \qquad \text{for } a+p=0$$

and for finite-time null controllability these are

$$(-1/(a+p)) < x_0 < (1/(a+p)) \qquad \text{for } a+p>0$$
$$-\infty < x_0 < \infty \qquad \text{for } a+p \leqslant 0. \tag{5.3.39}$$

Theorem 5.3.8 gives us necessary and sufficient conditions for point null controllability. In the following theorem we discuss global null controllability.

Theorem 5.3.11 (Global null controllability). *Suppose U is a non-empty compact subset of R^m with $0 \in U$. Then the system S* (5.3.1) *is globally U-null controllable at t_0 if and only if*

$$\int_{t_0}^{\infty} H_U(B'(t)\psi(t))\,dt = \infty \tag{5.3.40}$$

for all nontrivial solutions ψ of the adjoint equation

$$S_a\colon \dot{\psi} = -A'(t)\psi, \qquad t \geqslant t_0. \tag{5.3.41}$$

Proof (Only if): Suppose the system is globally U-null controllable at t_0. We show that this implies (5.3.40). We prove this by contradiction. Suppose there exists a nontrivial solution $\hat{\psi}$ of (5.3.41) such that

$$\int_{t_0}^{\infty} H_U(B'(t)\hat{\psi}(t))\,dt < \infty. \tag{5.3.42}$$

Then clearly there exists a number $0 < \beta < \infty$ such that

$$\int_{t_0}^{\infty} H_U(B'(t)\hat{\psi}(t))\,dt \leqslant \beta < \infty. \tag{5.3.43}$$

Define

$$\hat{x}_0 = -2\beta(\hat{\psi}(t_0)/|\hat{\psi}(t_0)|^2). \tag{5.3.44}$$

Since $\hat{\psi}$ is nontrivial, $\hat{\psi}(t_0) \neq 0$. We show that $\hat{x}_0$, taken as the initial state, cannot be steered to the origin by any admissible control. Recall the function

$$J(t, t_0, \hat{x}_0, \eta) \equiv (\hat{x}_0, \Phi'(t, t_0)\eta) + \int_{t_0}^{t} H_U(B'(\theta)\Phi'(t, \theta)\eta)\,d\theta. \tag{5.3.45}$$

Taking $\eta = \hat{\psi}(t)$ and using (5.3.43) we have

$$J(t, t_0, \hat{x}_0, \hat{\psi}(t)) = (\hat{x}_0, \hat{\psi}(t_0)) + \int_{t_0}^{t} H_U(B'(\theta)\hat{\psi}(\theta))\,\mathrm{d}\theta$$

$$= -2\beta + \int_{t_0}^{t} H_U(B'(\theta)\hat{\psi}(\theta))\,\mathrm{d}\theta \leq -\beta. \tag{5.3.46}$$

Hence $J(t, t_0, \hat{x}_0, \hat{\psi}(t)) < 0$ for all $t \geq 0$, and, by Theorem 5.3.8, $\hat{x}_0$ can not be driven to the zero state in finite time, contradicting the fact that the system is globally null controllable. Hence our hypothesis, that the integral (5.3.43) is finite for a nontrivial solution of the adjoint system, is false.

(If): We show that if (5.3.40) holds then, the system S is globally U-null controllable at t_0. Suppose for every nontrivial solution ψ of the adjoint problem S_a,

$$\int_{t_0}^{\infty} H_U(B'(t)\psi(t))\,\mathrm{d}t = \infty, \tag{5.3.47}$$

but the system is not globally null controllable. Then there exists $x_0 \neq 0$ which cannot be driven to the origin in finite time. Hence by virtue of the previous theorem there exists a sequence $t_k \geq t_0$ and $\eta_k \in R^n$ with $t_k \to \infty$ such that

$$J(t_k, t_0, x_0, \eta_k) < 0 \qquad \text{for all integers } k \geq 1. \tag{5.3.48}$$

Define

$$\psi_k = (\Phi'(t_k, t_0)\eta_k / |\Phi'(t_k, t_0)\eta_k|), \qquad k = 1, 2, \ldots. \tag{5.3.49}$$

Since $\eta_k \neq 0$ and Φ is invertible, $|\Phi'(t_k, t_0)\eta_k| \neq 0$ and hence the sequence $\{\psi_k\}$ is well defined with $|\psi_k| = 1$. Since the unit sphere in R^n is compact there exists a subsequence $\{\psi_{k_m}\} \subset \{\psi_k\}$ and a corresponding subsequence $\{t_{k_m}\} \subset \{t_k\}$ and a ψ_0 in the unit sphere in R^n such that

$$\begin{aligned} &\psi_{k_m} \to \psi_0 \\ &t_{k_m} \to \infty. \end{aligned} \tag{5.3.50}$$

To avoid the double subscript we shall relabel the subsequence $\{\psi_{k_m}, t_{k_m}\}$ as $\{\psi_k, t_k\}$ and assume that $\psi_k \to \psi_0$ and $t_k \to \infty$. Using ψ_k for the initial condition of the adjoint system we have, for all integers $k \geq 1$,

$$J(t_k, t_0, x_0, \alpha_k) = (x_0, \psi_k) + \int_{t_0}^{t_k} H_U(B'(\theta)\Phi'(t_0, \theta)\psi_k)\,\mathrm{d}\theta < 0, \tag{5.3.51}$$

where

$$\alpha_k = \Phi'(t_0, t_k)\psi_k.$$

Define

$$f_k(t) = \begin{cases} B'(t)\Phi'(t_0, t)\psi_k, & t_0 \leqslant t \leqslant t_k, \\ 0 & \text{elsewhere}, \end{cases} \tag{5.3.52}$$

and

$$f(t) = B'(t)\Phi'(t_0, t)\psi_0 = B'(t)\psi(t). \tag{5.3.53}$$

Clearly

$$0 \leqslant \int_{t_0}^{t_k} H_U(f_k(\theta))\, d\theta \leqslant |(x_0, \psi_k)| \leqslant |x_0|. \tag{5.3.54}$$

Since by Lemma 5.3.7, $w \to H_U(w)$ is continuous and $f_k(t) \to f(t)$ a.e. it follows from Fatou's lemma (Proposition 2.1.33) that

$$\int_{t_0}^{\infty} H_U(f(t))\, dt \leqslant \underline{\lim}_k \int_{t_0}^{t_k} H_U(f_k(t))\, dt$$

$$= \underline{\lim}_k \int_{t_0}^{\infty} H_U(f_k(t))\, dt \leqslant |x_0|. \tag{5.3.55}$$

Therefore

$$\int_{t_0}^{\infty} H_U(B'(t)\psi(t)) < \infty,$$

contradicting (5.3.40). Hence our hypothesis that the system is not globally null controllable is false. This completes the proof. ■

Remark 5.3.12 It is clear from the above result that if the control matrix $B(t)$, $t \geqslant t_0$, has bounded support then $\int_{t_0}^{\infty} H_U(B'(t)\psi(t))\, dt < \infty$ for nontrivial ψ and hence the system is not globally null controllable. The value of the integral is crucially dependent on the combined action of B' and Φ' and hence that of $\{A(t), B(t), t \geqslant t_0\}$.

An obvious consequence of the above theorem is given in the following remark.

Remark 5.3.13 The system is globally $B_1 \equiv \{\xi \in R^m: |\xi| \leqslant 1\}$ null controllable at t_0 if and only if

$$\int_{t_0}^{\infty} H_{B_1}(B'(t)\psi(t))\, dt = \int_{t_0}^{\infty} |B'(t)\psi(t)|\, dt = \infty \tag{5.3.56}$$

for all nontrivial solutions of the adjoint system.

It is also interesting to note the relationship between Theorem 5.3.11

and Corollary 5.2.15. For $t \geq t_0$, define

$$\Gamma_t u = \int_{t_0}^{t} \Phi(t, \theta) B(\theta) u(\theta) \, d\theta$$

and let Γ_t^* denote its adjoint.

Corollary 5.3.14 *If the system S* (5.3.1) *is globally U-null controllable at t_0, then*

$$\bigcap_{t \geq t_0} \text{Ker}\, \Gamma_t^* = \{0\}. \tag{5.3.57}$$

The proof is left as an exercise for the reader. In the case of a time-invariant system (5.3.57) implies Rank $(B, AB, \ldots, A^{n-1}B) = n$ (see Theorem 5.2.4). In case $U = R^m$, the condition (5.3.57) is a necessary and sufficient condition for global null controllability.

Yet another implication of Theorem 5.3.11 is given in the following corollary.

Corollary 5.3.15 *If U is any non-empty bounded subset of R^m with $0 \in \text{Int}\, U$ $(\neq \emptyset)$ and the system is globally U-null controllable, then for any ε-ball $B_\varepsilon \subset U$, the system is globally B_ε-null controllable.*

Proof Since U is a bounded subset of R^m, there exists a closed ball B_r containing U. Clearly the system is globally B_r-null controllable since it is so with respect to U. Consequently by Theorem 5.3.11,

$$\int_{t_0}^{\infty} H_{B_r}(B'(t)\psi(t)) \, dt = \infty, \tag{5.3.58}$$

for every nontrivial solution ψ of the adjoint system. Since $0 \in \text{Int}\, U$, there exists an ε-ball $B_\varepsilon \subset U$. Then

$$\begin{aligned}\int_{t_0}^{\infty} H_{B_\varepsilon}(B'(t)\psi(t)) \, dt &= \int_{t_0}^{\infty} \varepsilon \, |B'(t)\psi(t)| \, dt \\ &= \frac{\varepsilon}{r} \int_{t_0}^{\infty} r \, |B'(t)\psi(t)| \, dt = \left(\frac{\varepsilon}{r}\right) \int_{t_0}^{\infty} H_{B_r}(B'(t)\psi(t)) \, dt \\ &= \infty,\end{aligned}$$

due to (5.3.58). Hence by Theorem 5.3.11 the system is globally B_ε-null controllable. ■

For illustration we present two examples.

Example 5.3.16 Consider the system of Example 5.3.10 with $a = -1$, making the uncontrolled system asymptotically stable. We have seen that the system is locally null controllable with domain given by the open interval

$$(-1/(p-1)) < x_0 < (1/(p-1)) \quad \text{for } p > 1.$$

So one may suspect, according to Corollary 5.3.5, that the system is globally null controllable. But this is simply not true (see Remark 5.3.6). By Theorem 5.3.11, for global null controllability we must have

$$\int_0^\infty H_U(B'(t)\psi(t))\,\mathrm{d}t = \infty,$$

for nontrivial adjoint state ψ. The adjoint problem is $\dot{\psi} = -a\psi = \psi$. Here $U = [-1, +1]$, $B(t) = \mathrm{e}^{-pt}$, $\psi(t) = \mathrm{e}^{t}\psi_0$. Hence

$$H_U(B'(t)\psi(t)) = |\psi_0|\exp(-(p-1)t)$$

and

$$\int_0^\infty |\psi_0|\,\mathrm{e}^{-(p-1)t}\,\mathrm{d}t = (|\psi_0|/(p-1)) < \infty.$$

Hence by Theorem 5.3.11 the system is not globally null controllable, contradicting Corollary 5.3.5. This is a counter example constructed by Colmenares ([108], [109]) which shows that for time-varying systems, asymptotic stability and local null controllability do not guarantee global null controllability. The reader can verify this result directly without resorting to Theorem 5.3.11. Also note that for $p < 1$, the system is globally null controllable.

To illustrate Theorem 5.3.11 we present another example.

Example 5.3.17

$$\ddot{\xi} + \{1/(1+t)\}\dot{\xi} = u, \qquad U = [-1, +1]. \tag{5.3.59}$$

Writing $x_1 = \xi$, $x_2 = \dot{\xi}$, we have

$$A(t) = \begin{bmatrix} 0 & 1 \\ 0 & -1/(1+t) \end{bmatrix}, \qquad B = \begin{bmatrix} 0 \\ 1 \end{bmatrix}.$$

The transition operator Φ is given by

$$\Phi(t, \tau) = \begin{bmatrix} 1 & (1+\tau)\ln(1+t) - (1+\tau)\ln(1+\tau) \\ 0 & (1+\tau)/(1+t) \end{bmatrix}.$$

Hence for $0 \leqslant \tau \leqslant t_1$ and $\eta = (\eta_1, \eta_2)'$,

$$B'(\tau)\Phi'(t_1, \tau)\eta = \eta_1(1+\tau)\ln\{(1+t_1)/(1+\tau)\} + \eta_2(1+\tau)/(1+t_1)$$

and

$$H_U(B'(\tau)\Phi'(t_1, \tau)\eta) = (1+\tau)|\eta_1 \ln\{(1+t_1)/(1+\tau)\} + \eta_2\{1/(1+t_1)\}|.$$

Then one can find a sufficiently large number $t^* > 0$ and a constant $c = c(|\eta|) > 0$ such that

$$\int_0^{t_1} H_U(B'(\tau)\Phi'(t_1, \tau)\eta)\, d\tau \geqslant c(|\eta|)t_1^2 \qquad \text{for all } t_1 \geqslant t^* \text{ and } |\eta| \neq 0.$$

Therefore

$$\lim_{t_1 \to \infty} \int_0^{t_1} H_U(B'(\tau)\Phi'(t_1, \tau)\eta)\, d\tau = \infty$$

for every $\eta \neq 0$. Hence the system (5.3.59) is globally $U(=[-1, +1])$–null controllable at $t_0 = 0$.

In many problems dealing with socio-economic and ecological systems the controls take only positive (or negative) values with zero being an extreme point. In this regard the following results can be proved using Theorem 5.3.11.

Corollary 5.3.18 *Consider the time-invariant system*

$$\dot{x} = Ax + bu, \tag{5.3.60}$$

with $b \in R^n$, $u \in M(U)$, $U = [0, 1]$. *Then the system is globally null controllable (controllable to the origin) if and only if*

(a) Rank $(b, Ab, A^2b, \ldots, A^{n-1}b) = n$,
(b) Im eV$(A) \neq 0$ *and* Re eV$(A) \leqslant 0$ (Re eV$(A) \geqslant 0$),

where eV(A) *denotes the set of eigenvalues of* A *and* Im z (Re z) *denotes the imaginary part (real part) of* z.

Corollary 5.3.19 *The system* (5.3.60) *is globally controllable if and only if*

(a) Rank $(b, Ab, A^2b, \ldots, A^{n-1}b) = n$
(b) Im eV$(A) \neq 0$ *and* Re eV$(A) = 0$.

In case $U = [0, \infty)$, the condition Re eV$(A) = 0$ can be dispensed with.

Note that the conditions (b) of the above corollaries require that the system (5.3.60) be of even dimension (n even).

5.4 Stabilizability of linear systems with state feedback and observers

In this section we study the questions of stabilizability of linear systems by use of linear feedback control laws. In case complete information about system states is available, one can use state feedback laws. However, if only output information is available, one may use dynamic compensators called observers. We shall consider observers of the same dimension as the plant (called full-order observer) and also observers of reduced order equal to the dimension of the missing states. For linear time-invariant systems the theory of observers has been found to be practically useful, while for time-varying and nonlinear systems the results are not yet well developed.

Definition 5.4.1 (Stabilizability). The system in R^n,

$$\dot{x} = Ax + Bu, \tag{5.4.1}$$

is said to be stabilizable by a linear state feedback law if there exists a $(m \times n)$ matrix K such that, with $u = Kx$, the closed loop system $\dot{x} = (A + BK)x$ is stable, or equivalently $(A + BK)$ is a stability matrix.

The first result in this direction is given in the following theorem.

Theorem 5.4.2 (State feedback). *If the system* (5.4.1) *is controllable, then there exists an* $(m \times n)$ *matrix* K *such that the system with the feedback control* $u = Kx$ *is asymptotically stable, or equivalently* $(A + BK)$ *is a stability matrix.*

Proof We show that controllability implies the existence of a matrix K such that $(A + BK)$ is a stability matrix. By Corollary 5.2.11, (A, B) is controllable if and only if $(B', -A')$ is observable. The observability of $(B', -A')$ implies that the observability matrix,

$$M_\tau \equiv \int_0^\tau (\mathrm{e}^{-At}BB'\mathrm{e}^{-A't})\,\mathrm{d}t, \tag{5.4.2}$$

for any $\tau > 0$, is real symmetric and positive definite. Defining

$$K \equiv -B'M_\tau^{-1}, \tag{5.4.3}$$

we show that $(A + BK)$ is a stability matrix. Clearly

$$AM_\tau + M_\tau A' = -\int_0^\tau \frac{\mathrm{d}}{\mathrm{d}t}(\mathrm{e}^{-At}BB'\mathrm{e}^{-A't})\,\mathrm{d}t = BB' - \mathrm{e}^{-A\tau}BB'\mathrm{e}^{-A'\tau}.$$

Hence

$$(A - BB'M_\tau^{-1})M_\tau + M_\tau(A - BB'M_\tau^{-1})' = -\Gamma \tag{5.4.4}$$

where

$$\Gamma \equiv BB' + e^{-A\tau}BB'e^{-A'\tau}. \tag{5.4.5}$$

Using the definition of K in (5.4.4) we obtain

$$(A + BK)M_\tau + M_\tau(A + BK)' = -\Gamma. \tag{5.4.6}$$

We note that Γ is real symmetric and positive and (5.4.6) is the Lyapunov equation we have seen in section 4.2. Since M_τ is real, symmetric and positive definite it follows from Theorem 4.2.1 that $(A+BK)'$, and hence $(A + BK)$, is a stability matrix. The desired K is given by (5.4.3). This completes the proof. ■

It is clear from the above result that by choosing $K = -B'M_\tau^{-1}$ one can stabilize the unstable system $\dot{x} = Ax + Bu$. The computation of K involves several operations, such as matrix multiplications, integrations, and inversions. Obviously for large n, this is not very economic. There is an alternative procedure suggested by Bass [95] that reduces these computations substantially. This is given in the following result.

Theorem 5.4.3 (Stabilizer). *Consider the system* (5.4.1) *with* (A, B) *assumed controllable. Let*

$$\lambda_m \equiv \begin{cases} \sup\{|(A\xi, \xi)|, |\xi| \leq 1\} \\ \text{or} \\ \sup\left\{\sum_{j=1}^{n} |a_{ij}|\ i = 1, 2, \ldots, n\right\}. \end{cases} \tag{5.4.7}$$

Then $u = Kx$ *is a stabilizing control law where* $K = -B'Z^{-1}$ *and* Z *is the solution of the matrix equation*

$$(A + \lambda I)Z + Z(A + \lambda I)' = BB' \tag{5.4.8}$$

for any $\lambda > \lambda_m$.

Proof For $\lambda > \lambda_m$, $-(A + \lambda I)$ is a stability matrix. Indeed let x be any solution of the equation $\dot{x} = -(A + \lambda I)x$. Then $(\dot{x}, x) = -((A + \lambda I)x, x)$ and hence

$$\frac{d}{dt}|x(t)|^2 = -2((A + \lambda I)x(t), x(t)). \tag{5.4.9}$$

In view of (5.4.7), we have

$$\frac{d}{dt}|x(t)|^2 \leq -2(\lambda - \lambda_m)\,|x(t)|^2$$

and hence,

$$|x(t)|^2 \leq |x_0|^2 \exp(-2(\lambda - \lambda_m)t), \qquad t \geq 0. \tag{5.4.10}$$

Consequently, for $\lambda > \lambda_m$, $-(A + \lambda I)$ is a stability matrix. Therefore by Theorem 4.2.1, the matrix equation

$$-(A + \lambda I)Z - Z(A + \lambda I)' = -BB' \tag{5.4.11}$$

has a real, symmetric, and positive definite solution Z. Clearly

$$(A + \lambda I - BB'Z^{-1})Z + Z(A + \lambda I - BB'Z^{-1})' = -BB' \tag{5.4.12}$$

and hence, again by Theorem 4.2.1, $(A + \lambda I - BB'Z^{-1})'$ and therefore $(A + \lambda I - BB'Z^{-1})$ is a stability matrix. We know, by Theorem 5.2.12(b), (A, B) is controllable if and only if $(A + \lambda I, B)$ is controllable for every $\lambda \in R$. Since by assumption (A, B) is controllable, $(A + \lambda I, B)$ is controllable and hence $(A + \lambda I, B)$ is stabilizable with feedback control $u = Kx = -B'Z^{-1}x$. Since $\lambda > \lambda_m > 0$, the eigenvalues of $(A - BB'Z^{-1})$ are at a distance λ to the left of the eigenvalues of $(A + \lambda I - BB'Z^{-1})$. Hence $(A - BB'Z^{-1})$ is also a stability matrix and consequently the law, $u = Kx$, also stabilizes the system $\dot{x} = Ax + Bu$. This completes the proof. ■

Remark 5.4.4 The valuable point in the method given by Theorem 5.4.3 is that it requires very little computation compared to that given by Theorem 5.4.1. Further one can freely choose λ ($>\lambda_m$) to shift the closed-loop eigenvalues *en masse*, which is certainly a useful feature in design problems. For illustration we consider the following example.

Example 5.4.5 Consider the ecological system

$$\begin{aligned} \dot{F} &= 6F - 2W \\ \dot{W} &= -W + 3F - u \end{aligned} \tag{5.4.13}$$

where F represents the population of prey and W that of predator. The control u represents removal rate of W. Here

$$A = \begin{bmatrix} 6 & -2 \\ 3 & -1 \end{bmatrix}, \qquad B = \begin{bmatrix} 0 \\ -1 \end{bmatrix}.$$

One can easily verify that the uncontrolled system is unstable but (A, B) is controllable with $\det(B, AB) \neq 0$. It is easy to verify that $0 < \lambda_m < 6$ and hence taking $\lambda = 8$, we have

$$(A + \lambda I) = \begin{bmatrix} 14 & -2 \\ 3 & 7 \end{bmatrix}, \qquad BB' = \begin{bmatrix} 0 & 0 \\ 0 & 1 \end{bmatrix}.$$

For the solution of the equation (5.4.8), write

$$Z = \begin{bmatrix} z_1 & z_2 \\ z_2 & z_3 \end{bmatrix},$$

giving

$$\begin{aligned} 14z_1 - 2z_2 &= 0 \\ 6z_2 + 14z_3 &= 1 \\ 3z_1 + 21z_2 - 2z_3 &= 0. \end{aligned} \tag{5.4.14}$$

Solving these equations we have $z_1 = (1/1092)$, $z_2 = (7/1092)$, $z_3 = (75/1092)$,

$$Z^{-1} = 42\begin{bmatrix} 75 & -7 \\ -7 & 1 \end{bmatrix}$$

and $K = -B'Z^{-1} = -42[7, -1]$. Hence

$$BK = -42\begin{bmatrix} 0 & 0 \\ -7 & 1 \end{bmatrix} \quad \text{and} \quad \tilde{A} = (A + BK) = \begin{bmatrix} 6 & -2 \\ 297 & -43 \end{bmatrix}.$$

The eigenvalues of $\tilde{A}$ are $\lambda_{1,2} = -(37/2) \pm (1/2)\sqrt{65} < 0$. Hence the closed-loop matrix $(A + BK)$ is a stability matrix.

We have seen in Theorem 5.4.2 that if the system $\dot{x} = Ax + Bu$ is controllable, then it is also stabilizable. But the converse is not true as evidenced by the following result.

Theorem 5.4.6 (Non stabilizability). *Let the controllable subspace D_c of the system $\dot{x} = Ax + Bu$ be a proper subspace of R^n with* $\dim(D_c) = \nu < n$. *Then the system is stabilizable if and only if $\tilde{A}_{22}$ of* (5.2.30) *is a stability matrix.*

Proof If $\dim(D_c) = \nu \leqslant n$, then the system has the decomposition (5.2.33) where the pair $(\tilde{A}_{11}, \tilde{B}_1)$ is controllable and hence stabilizable. Considering the feedback control $\tilde{u} = K_1\tilde{x}^1 + K_2\tilde{x}^2$, the decomposed system (5.2.32) takes the form

$$\frac{d}{dt}\begin{bmatrix} \tilde{x}^1 \\ \tilde{x}^2 \end{bmatrix} = \begin{bmatrix} \tilde{A}_{11} + \tilde{B}_1K_1 & \tilde{A}_{12} + \tilde{B}_1K_2 \\ 0 & \tilde{A}_{22} \end{bmatrix}\begin{bmatrix} \tilde{x}^1 \\ \tilde{x}^2 \end{bmatrix}. \tag{5.4.15}$$

The eigenvalues of the closed-loop system depend on those of $\tilde{A}_{11} + \tilde{B}_1K_1$ and $\tilde{A}_{22}$ only. By virtue of controllability of $(\tilde{A}_{11}, \tilde{B}_1)$ we can choose K_1 to make $(\tilde{A}_{11} + \tilde{B}_1K_1)$ a stability matrix. Hence the system is stabilizable if and only if $\tilde{A}_{22}$ is a stability matrix. On the other hand, if $\tilde{A}_{22}$ is not a stability matrix, then there is no linear state feedback stabilizer. ■

We have seen in Examples 5.1.8 and 5.2.6 that the single input system

$$z^{(n)} + a_{n-1}z^{(n-1)} + \cdots + a_1 z^{(1)} + a_0 z = u \tag{5.4.16}$$

has the canonical representation $\dot{x} = Ax + bu$, where A and b are as given in Example 5.1.8. Since this system is always controllable whatever the coefficients $\{a_0, a_1, \ldots, a_{n-1}\}$ may be, it is also stabilizable. Indeed let $u = k'x = (k, x)$ where k is an n-vector. Then

$$\dot{x} = Ax + bu = (A + bk')x, \tag{5.4.17}$$

where the matrix $(A + bk')$ is given by

$$\tilde{A} = (A + bk') = \begin{bmatrix} 0 & 1 & 0 & 0 \\ 0 & 0 & 1 & 0 \\ \vdots & \vdots & \vdots & \vdots \\ 0 & 0 & 0 & 1 \\ -a_0 + k_1 & -a_1 + k_2 & -a_2 + k_3 & -a_{n-1} + k_n \end{bmatrix} \tag{5.4.18}$$

Hence, by appropriate choice of the vector k, all the coefficients of the system (5.4.17) and hence that of (5.4.16) can be modified to desired values. In other words all the eigenvalues of the closed-loop system (5.4.17) can be modified as required. This shows that if we can transform an arbitrary system, $\dot{x} = Ax + Bu$, into the form given by Example 5.1.8, then we can place the eigenvalues of the closed-loop system $(A + BK)$ anywhere in the complex plane by choosing K appropriately. We shall see that this can be always done provided the system is controllable.

Theorem 5.4.7 (Eigenvalue placement). *Consider the system with single input*

$$\dot{x} = Ax + bu \tag{5.4.19}$$

and suppose (A, b) is controllable. Then, for any given set of desired eigenvalues $\{\lambda_1, \lambda_2, \ldots, \lambda_n\}$, $\lambda_i \in \mathbb{C}$, with complex roots appearing in complex conjugate pairs, there exists a feedback law $u = k'x$ such that the eigenvalues of the closed-loop system matrix $(A + bk')$ coincide with the given set $\{\lambda_1, \lambda_2, \ldots, \lambda_n\}$.

Proof According to the previous discussions all that is required to prove the theorem is that there exists a vector k such that $(A + bk')$ has the form given by (5.4.18). Controllability implies that

$$D \equiv (A^{n-1}b, A^{n-2}b, \ldots, Ab, b) \tag{5.4.20}$$

is a non-singular (square) matrix. Consider first the coordinate transformation $x = D\xi$ transforming (5.4.19) into

$$\dot{\xi} = D^{-1}AD\xi + D^{-1}bu = A_1\xi + b_1u$$

with (5.4.21)

$$A_1 = D^{-1}AD \qquad \text{and} \qquad b_1 = D^{-1}b.$$

Note that

$$I = D^{-1}D = D^{-1}(A^{n-1}b, A^{n-2}b, \ldots, Ab, b)$$
$$= (e_1, e_2, \ldots, e_n) \tag{5.4.22}$$

where e_i denotes the unit vector with ith entry equal to 1 and the other entries zero. This shows that

$$D^{-1}A^{n-1}b = e_1, \quad D^{-1}A^{n-2}b = e_2, \ldots, D^{-1}Ab = e_{n-1}, \quad D^{-1}b = e_n. \tag{5.4.23}$$

Then

$$AD = (A^nb, A^{n-1}b, A^{n-2}b, \ldots, A^2b, Ab)$$

and by (5.4.23)

$$D^{-1}AD = (D^{-1}A^nb, e_1, e_2, \ldots, e_{n-2}, e_{n-1}). \tag{5.4.24}$$

Let

$$\det(\lambda I - A) = |\lambda I - A| = \lambda^n + \alpha_{n-1}\lambda^{n-1} + \alpha_{n-2}\lambda^{n-2} + \cdots + \alpha_1\lambda + \alpha_0, \tag{5.4.25}$$

then by the Cayley–Hamilton theorem, we have

$$A^n = -\sum_{i=0}^{n-1} \alpha_i A^i \tag{5.4.26}$$

and

$$D^{-1}A^nb = -\sum_{i=0}^{n-1} \alpha_i D^{-1}A^ib = -\sum_{i=0}^{n-1} \alpha_i e_{n-i}. \tag{5.4.27}$$

Using (5.4.27) in (5.4.24), we obtain

$$A_1 = D^{-1}AD = \left(-\sum_{i=0}^{n-1} \alpha_i e_{n-i}, e_1, e_2, \ldots, e_{n-2}, e_{n-1}\right)$$
$$= \begin{bmatrix} -\alpha_{n-1} & 1 & 0 & 0 & \cdots & 0 \\ -\alpha_{n-2} & 0 & 1 & 0 & \cdots & 0 \\ -\alpha_{n-3} & 0 & 0 & 1 & \cdots & 0 \\ \vdots & \vdots & \vdots & \vdots & & \vdots \\ -\alpha_1 & 0 & 0 & 0 & \cdots & 1 \\ -\alpha_0 & 0 & 0 & 0 & \cdots & 0 \end{bmatrix}. \tag{5.4.28}$$

Note also that

$$b_1 = D^{-1}b = e_n. \tag{5.4.29}$$

The matrix A_1 is quite close to the desired form (5.4.18). We need one more transformation to reduce it to the form of (5.4.18). All that is required is an exchange of position between the first column and last row. This can be done by a nonsingular transformation given by

$$E = [E_1, E_2, \ldots, E_n] \tag{5.4.30}$$

where each E_i is an n-vector of the form

$$E_i = [0, 0, \ldots, 1, \alpha_{n-1}, \ldots, \alpha_{i+1}, \alpha_i]', \qquad 1 \leqslant i \leqslant n-1 \tag{5.4.31}$$

where 1 appears at the ith position, $E_n = e_n$, and $\{\alpha_i\}$ are the coefficients of the polynomial (5.4.25). Using the non-singular matrix E we can transform (5.4.21) into the desired canonical form. Define

$$\xi = E\eta, \tag{5.4.32}$$

then

$$\begin{aligned} \dot{\eta} &= (E^{-1}A_1E)\eta + E^{-1}b_1u \\ &= A_2\eta + b_2u. \end{aligned} \tag{5.4.33}$$

We show that (5.4.33) gives us the desired form (5.4.18). Using (5.4.28) and (5.4.31) one can easily verify that

$$A_1E = [-\alpha_0E_n, E_1 - \alpha_1E_n, E_2 - \alpha_2E_n, \ldots, E_{n-1} - \alpha_{n-1}E_n],$$

and hence

$$A_2 \equiv E^{-1}A_1E = [-\alpha_0e_n, e_1 - \alpha_1e_n, e_2 - \alpha_2e_n, \ldots, e_{n-1} - \alpha_{n-1}e_n], \tag{5.4.34}$$

which is precisely the canonical form we have been looking for. For the matrix b_2, note that $b_1 = E_n = e_n$ and hence

$$b_2 \equiv E^{-1}b_1 = e_n. \tag{5.4.35}$$

Hence the transformed system

$$\dot{\eta} = A_2\eta + b_2u \tag{5.4.36}$$

is in the control canonical form (5.4.17) and (5.4.18). By taking $u = \ell'\eta$, where $\ell \in R^n$, we have

$$\dot{\eta} = (A_2 + b_2\ell')\eta \tag{5.4.37}$$

where

$$(A_2 + b_2\ell') = \begin{bmatrix} 0 & 1 & 0 & \cdots & 0 \\ 0 & 0 & 1 & \cdots & 0 \\ \vdots & \vdots & \vdots & & \vdots \\ 0 & 0 & 0 & & 1 \\ -\alpha_0 + \ell_1 & -\alpha_1 + \ell_2 & -\alpha_2 + \ell_3 & \cdots & -\alpha_{n-1} + \ell_n \end{bmatrix} \tag{5.4.38}$$

Clearly by choosing the vector ℓ we can realize any set of n specified eigenvalues for the closed-loop system (5.4.37). Since the eigenvalues are invariant under similarity transformation, the eigenvalues of $(A_2 + b_2\ell')$ coincide with those of the original system matrix $(A + bk')$. Indeed

$$DE(A_2 + b_2\ell')E^{-1}D^{-1} = A + b(\ell'E^{-1}D^{-1}) = A + bk', \tag{5.4.39}$$

where

$$k' = \ell'E^{-1}D^{-1} = \ell'(DE)^{-1}. \tag{5.4.40}$$

Therefore, once ℓ is chosen to obtain the desired set of eigenvalues, one can use (5.4.40) to obtain k. This completes the proof. ■

Remark 5.4.8 Let

$$P_{A_2+b_2\ell'}(\lambda) = \lambda^n + (\alpha_{n-1} - \ell_n)\lambda^{n-1} + (\alpha_{n-2} - \ell_{n-1})\lambda^{n-2} + \cdots + (\alpha_1 - \ell_2)\lambda + (\alpha_0 - \ell_1). \tag{5.4.41}$$

If $\{\lambda_i^0\}$ is the desired set of eigenvalues, then

$$P_{\text{des}}(\lambda) = (\lambda - \lambda_1^0)(\lambda - \lambda_2^0) \cdots (\lambda - \lambda_n^0) = \lambda^n + \beta_{n-1}\lambda^{n-1} + \beta_{n-2}\lambda^{n-2} + \cdots + \beta_1\lambda + \beta_0. \tag{5.4.42}$$

Equating the coefficients of like powers of λ we obtain

$$\ell_i = (\alpha_{i-1} - \beta_{i-1}), \qquad i = 1, 2, \ldots, n. \tag{5.4.43}$$

Using this in (5.4.40) one obtains the desired k.

For illustration let us consider the following example.

Example 5.4.9

$$\frac{d}{dt}\begin{bmatrix} x_1 \\ x_2 \\ x_3 \end{bmatrix} = \begin{bmatrix} 0 & 1 & 0 \\ 0 & 0 & 0 \\ -1 & 0 & 0 \end{bmatrix}\begin{bmatrix} x_1 \\ x_2 \\ x_3 \end{bmatrix} + \begin{bmatrix} 0 \\ 1 \\ 0 \end{bmatrix} u. \tag{5.4.44}$$

The characteristic polynomial $P_A(\lambda) = |\lambda I - A| = \lambda^3$, hence $\alpha_2 = \alpha_1 = \alpha_0 = 0$, and $\lambda = 0$ is an eigenvalue of multiplicity 3. Suppose the desired eigenvalues are $\lambda_1 = -1 - i$, $\lambda_2 = -1 + i$, $\lambda_3 = -1$, and we wish to find k such that the closed-loop system matrix $(A + bk')$ has the specified eigenvalues. We use the procedure given by the previous theorem. Set

$$P_{\text{des}}(\lambda) = (\lambda - \lambda_1)(\lambda - \lambda_2)(\lambda - \lambda_3) = \lambda^3 + 3\lambda^2 + 4\lambda + 2, \tag{5.4.45}$$

hence $\beta_0 = 2$, $\beta_1 = 4$, $\beta_2 = 3$ and by (5.4.43), $\ell_1 = -2$, $\ell_2 = -4$, $\ell = -3$. We use the vector $\ell = (-2, -4, -3)'$ in the expression (5.4.40) to obtain

$$k = ((DE)^{-1})'\ell = (E^{-1}D^{-1})'\ell.$$

We compute the matrices D and E using the expressions (5.4.20) and (5.4.30) respectively. Hence

$$k = (-4, -3, 2)'. \tag{5.4.46}$$

The matrix k, given above, yields the desired eigenvalues for $(A + bk') \equiv \tilde{A}$ with

$$\tilde{A} = \begin{bmatrix} 0 & 1 & 0 \\ -4 & -3 & 2 \\ -1 & 0 & 0 \end{bmatrix}.$$

The eigenvalue placement (or the spectral assignment) property also holds for multiple input systems. We state the following result, whose proof is based on similar arguments to those detailed in Theorem 5.4.7.

Theorem 5.4.10 (Eigenvalue placement). *Consider the system $\dot{x} = Ax + Bu$, where $B \in R(n \times m)$, $m \geq 1$, and suppose (A, B) is controllable. Then there exists $K \in R(m \times n)$ such that the eigenvalues of $(A + BK)$ coincide with any specified set of eigenvalues.*

Further results on this topic can be found in [117].

Dynamic compensator (Full-order observer)

Consider the observed system

$$\begin{cases} \dot{x} = Ax + Bu \\ z = Hx \\ u = Kz, \end{cases} \tag{5.4.47}$$

with closed-loop system $\dot{x} = (A + BKH)x$. Even though (A, B) is controllable and (H, A) observable, in general there exists *no* stabilizing K. The problem arises due to missing information regarding the state x. In this situation one provides an estimate of the missing state through what is known as the Luenberger observer [129]. This requires an analogue simulation of a dynamic system, similar to the original one given by

$$\dot{y} = Ay + BKy + Le, \tag{5.4.48}$$

where K is chosen so as to make $(A + BK)$ a stabilizing matrix, and $e = (z - Hy)$ is the error vector, a measure of the deviation of the simulated output Hy from that of the actual physical system. Then the control input of the physical plant is provided by $u = Ky$, where y is the simulated output which is always available. As a result we obtain the following system of equations,

$$\begin{aligned} \dot{x} &= Ax + BKy \\ \dot{y} &= Ay + BKy + L(z - Hy), \end{aligned} \tag{5.4.49}$$

or

$$\begin{aligned} \dot{x} &= Ax + BKy \\ \dot{y} &= (A + BK - LH)y + LHx. \end{aligned} \tag{5.4.50}$$

Writing this in the matrix partitioned form we have

$$\frac{\mathrm{d}}{\mathrm{d}t}\begin{bmatrix} x \\ y \end{bmatrix} = \begin{bmatrix} A & BK \\ LH & A + BK - LH \end{bmatrix}\begin{bmatrix} x \\ y \end{bmatrix} \equiv \tilde{A}\begin{bmatrix} x \\ y \end{bmatrix}, \tag{5.4.51}$$

which is a differential system of order $2n$. If by appropriate choice of K and L we can stabilize this system, then we have, in effect, provided a stabilizing control law for the original system. This is given in the following result.

Theorem 5.4.11 (Observer design). *If (A, B) is stabilizable and (H, A) observable (equivalently $(A', -H')$ controllable), then there exist matrices K and L such that the observed system*

$$\dot{x} = Ax + Bu, \qquad z = Hx, \tag{5.4.52}$$

with dynamic compensator (or observer),

$$\dot{y} = (A + BK)y + L(z - Hy), \qquad u = Ky, \tag{5.4.53}$$

is stabilizable.

Proof Consider the $(2n \times 2n)$ matrix

$$J = \begin{bmatrix} I & 0 \\ I & -I \end{bmatrix}, \tag{5.4.54}$$

where I is the $(n \times n)$ identity matrix. Note that $J^{-1} = J$, or $J \cdot J = I_{2n \times 2n}$.

Define

$$\begin{bmatrix} x \\ y \end{bmatrix} = J \begin{bmatrix} \xi \\ \eta \end{bmatrix}. \tag{5.4.55}$$

Substituting this in (5.4.51) we have

$$\begin{aligned} \frac{\mathrm{d}}{\mathrm{d}t}\begin{bmatrix} \xi \\ \eta \end{bmatrix} &= J^{-1}\tilde{A}J\begin{bmatrix} \xi \\ \eta \end{bmatrix} = J\tilde{A}J\begin{bmatrix} \xi \\ \eta \end{bmatrix} \\ &= \begin{bmatrix} A+BK & -BK \\ 0 & A-LH \end{bmatrix}\begin{bmatrix} \xi \\ \eta \end{bmatrix}. \end{aligned} \tag{5.4.56}$$

Since (A, B) is stabilizable there exists a $K \in R(m \times n)$ such that $(A + BK)$ is a stability matrix. Similarly, (H, A) being observable, $(A', -H')$ is controllable and hence there exists an $L \in R(n \times r)$ such that $(A' - H'L')$ is a stability matrix. Hence $(A - LH)$ is a stability matrix. Thus for appropriate choice of K and L, which exists, $J\tilde{A}J$ is a stability matrix. In other words K and L can be chosen to stabilize system (5.4.56), giving

$$\lim_{t\to\infty}\begin{bmatrix} \xi(t) \\ \eta(t) \end{bmatrix} = \begin{bmatrix} 0 \\ 0 \end{bmatrix},$$

and hence

$$\lim_{t\to\infty}\begin{bmatrix} x(t) \\ y(t) \end{bmatrix} = \lim_{t\to\infty} J\begin{bmatrix} \xi(t) \\ \eta(t) \end{bmatrix} = \begin{bmatrix} 0 \\ 0 \end{bmatrix}.$$

This follows from the fact that the eigenvalues of the matrix in (5.4.56) are entirely determined by those of $(A + BK)$ and $(A - LH)$ which we can choose. Further, under similarity transformation, eigenvalues being invariant, $\tilde{A}$ of (5.4.51) have the same eigenvalues as those of $J\tilde{A}J$. This completes the proof. ■

Remark 5.4.12 Note that the plant–observer system (5.4.52) and (5.4.53) has the asymptotic tracking property in the sense that $\lim_{t\to\infty} y(t) = \lim_{t\to\infty} x(t)$. This is left as an exercise for the reader. For faster tracking it is desirable to place the eigenvalues of $(A - LH)$ to the left of those of $(A + BK)$.

For illustration, consider the following example.

Example 5.4.13 $\ddot{x} = u$, $z = x_1$. Writing this in state-space language, we have

$$\frac{\mathrm{d}}{\mathrm{d}t}\begin{bmatrix} x_1 \\ x_2 \end{bmatrix} = \begin{bmatrix} 0 & 1 \\ 0 & 0 \end{bmatrix}\begin{bmatrix} x_1 \\ x_2 \end{bmatrix} + \begin{bmatrix} 0 \\ 1 \end{bmatrix} u = Ax + Bu$$

$$z = [1, 0]\begin{bmatrix} x_1 \\ x_2 \end{bmatrix} = Hx.$$

The system is controllable and observable but not stabilizable by output feedback $u = Kz$. So we may use observers. According to Theorem 5.4.11 we introduce the observer and choose the matrices K and L so as to make $(A + BK)$ and $(A - LH)$ stability matrices. Since (A, B) is controllable, according to Theorem 5.4.10, we can assign any set of eigenvalues for $(A + BK)$. Letting the desired eigenvalues to be -1, -2, we have $k_1 = -2$, $k_2 = -3$. Similarly assuming -3, -4 to be the desired eigenvalues for $A - LH$, we have $\ell_1 = 7$, $\ell_2 = 12$. The closed-loop plant observer dynamics is given by

$$\frac{\mathrm{d}}{\mathrm{d}t}\begin{bmatrix} x_1 \\ x_2 \\ y_1 \\ y_2 \end{bmatrix} = \begin{bmatrix} 0 & 1 & 0 & 0 \\ 0 & 0 & -2 & -3 \\ 7 & 0 & -7 & 1 \\ 12 & 0 & -14 & -3 \end{bmatrix}\begin{bmatrix} x_1 \\ x_2 \\ y_1 \\ y_2 \end{bmatrix}$$

with the eigenvalues -1, -2, -3, -4 as expected.

Dynamic compensator (Reduced-order observer)

The dimension of observers considered in the previous section equals that of the plant. Such an observer was called a full-order observer. In practice it is unnecessary and uneconomic to design a full-order observer if the plant can be stabilized by a reduced-order observer. Since the state is partially observed through the output $z = Hx$, it suffices to estimate the missing components of the state variable and design a reduced-order compensator.

We shall see below that almost any system can serve as an observer. Consider the uncontrolled observed system

$$\begin{aligned} \dot{x} &= Ax \\ z &= Hx, \qquad H \in R(r \times n), \end{aligned} \tag{5.4.57}$$

and the laboratory (auxiliary) system

$$\dot{y} = Fy + Rz = Fy + RHx = Fy + Gx, \tag{5.4.58}$$

with $F \in R((n-r) \times (n-r))$ and $G \in R((n-r) \times n)$ where Gx acts as an input to the observer. The matrix G depends on the output structure of the plant and the input structure of the observer as reflected in the expression $G = RH$. Clearly y carries some information about the state x possibly complementary to that contained in the output z. The information contained in $[z \ y]'$ is then exploited to regulate the controlled system, as we shall see shortly.

Let T be the solution of the matrix equation

$$FT - TA + G = 0 \tag{5.4.59}$$

and define

$$\xi = y - Tx. \tag{5.4.60}$$

Then one can easily verify that $\dot{\xi} = F\xi$, and hence

$$\xi(t) = e^{Ft}\xi_0 = e^{Ft}(y(0) - Tx(0)). \tag{5.4.61}$$

Therefore, if F is chosen to be a stability matrix from $R((n-r) \times (n-r))$, then, whatever be the initial state, $\lim_{t\to\infty} \xi(t) = 0$, or equivalently

$$\lim_{t\to\infty} y(t) = \lim_{t\to\infty} Tx(t). \tag{5.4.62}$$

Thus the observer state y asymptotically approaches the transformed plant state Tx. This fact remains valid also for the controlled system $\dot{x} = Ax + Bu$, provided the observer is redefined as $\dot{y} = Fy + Gx + TBu$. Thus for the combined plant–observer system

$$\begin{aligned} \dot{x} &= Ax + Bu, \ z = Hx \\ \dot{y} &= Fy + Gx + TBu, \end{aligned} \tag{5.4.63}$$

we have

$$y(t) = Tx(t) + e^{Ft}(y(0) - Tx(0)). \tag{5.4.64}$$

Note that system (5.4.63) is of dimension $(2n - r)$ while the corresponding system for the full-order observer (5.4.49) is of dimension $2n$. By choosing F to be a stability matrix with real parts of its eigenvalues being negative and far to the left of those of A, we can consider $y \cong Tx$ and hence

$$\begin{bmatrix} z \\ y \end{bmatrix} \cong \begin{bmatrix} H \\ T \end{bmatrix} x. \tag{5.4.65}$$

If the $(n \times n)$ matrix

$$\begin{bmatrix} H \\ T \end{bmatrix}$$

is invertible, we can write

$$x \cong \begin{bmatrix} H \\ T \end{bmatrix}^{-1} \begin{bmatrix} z \\ y \end{bmatrix} \tag{5.4.66}$$

at least in the limit as $t \to \infty$. Therefore we can consider

$$\begin{bmatrix} H \\ T \end{bmatrix}^{-1} \begin{bmatrix} z \\ y \end{bmatrix}$$

as an estimate of the true state x and use this estimate as the input to the feedback controller

$$u = K \begin{bmatrix} H \\ T \end{bmatrix}^{-1} \begin{bmatrix} z \\ y \end{bmatrix}. \tag{5.4.67}$$

Defining

$$\begin{bmatrix} H \\ T \end{bmatrix}^{-1} = (\Gamma_1, \Gamma_2) \tag{5.4.68}$$

with $\Gamma_1 \in R(n \times r)$, $\Gamma_2 \in R(n \times (n-r))$, we can rewrite the plant–observer system (5.4.63) as the closed loop $(2n-r)$-dimensional system

$$\begin{aligned} \dot{x} &= (A + BK\Gamma_1 H)x + BK\Gamma_2 y \\ \dot{y} &= (F + TBK\Gamma_2)y + (G + TBK\Gamma_1 H)x, \end{aligned} \tag{5.4.69}$$

or

$$\frac{\mathrm{d}}{\mathrm{d}t}\begin{bmatrix} x \\ y \end{bmatrix} = \begin{bmatrix} A + BK\Gamma_1 H & BK\Gamma_2 \\ G + TBK\Gamma_1 H & F + TBK\Gamma_2 \end{bmatrix} \begin{bmatrix} x \\ y \end{bmatrix}. \tag{5.4.70}$$

Using the non-singular coordinate transformation

$$\Lambda = \begin{bmatrix} I_n & 0_{n \times n-r} \\ -T & I_{n-r} \end{bmatrix} \tag{5.4.71}$$

and introducing the new variables

$$\begin{bmatrix} \xi \\ \eta \end{bmatrix} = \Lambda \begin{bmatrix} x \\ y \end{bmatrix}, \tag{5.4.72}$$

we can rewrite (5.4.70) as

$$\frac{\mathrm{d}}{\mathrm{d}t}\begin{bmatrix} \xi \\ \eta \end{bmatrix} = \begin{bmatrix} A + BK(\Gamma_1 H + \Gamma_2 T) & BK\Gamma_2 \\ FT - TA + G & F \end{bmatrix} \begin{bmatrix} \xi \\ \eta \end{bmatrix}. \tag{5.4.73}$$

Using (5.4.59) and (5.4.68) this reduces to

$$\frac{\mathrm{d}}{\mathrm{d}t}\begin{bmatrix} \xi \\ \eta \end{bmatrix} = \begin{bmatrix} A + BK & BK\Gamma_2 \\ 0 & F \end{bmatrix} \begin{bmatrix} \xi \\ \eta \end{bmatrix}. \tag{5.4.74}$$

We have seen that if (A, B) is stabilizable we can find a matrix K such that $(A+BK)$ is a stability matrix. Hence by choosing F to be a stability matrix, we can make the system (5.4.74) and hence the system (5.4.69) asymptotically stable.

The basic theory for reduced-order observers, as described above, is due to Luenberger [129]. A simple design procedure, based on plant decomposition, is due to Gopinath [121]. We shall discuss this procedure after we have proved the following simple fact.

Lemma 5.4.14 *Consider the observed system*

$$\begin{aligned} \dot{x} &= Ax, \qquad x \in R^n \\ z &= Hx, \qquad z \in R^r, \quad 1 \leqslant r \leqslant n, \end{aligned} \tag{5.4.75}$$

and suppose $\mathrm{Rank}(H) = r$. *Then the system admits a decomposition of the form*

$$\begin{aligned} \frac{\mathrm{d}}{\mathrm{d}t}\begin{bmatrix} \tilde{x}^1 \\ \tilde{x}^2 \end{bmatrix} &= \begin{bmatrix} \tilde{A}_{11} & \tilde{A}_{12} \\ \tilde{A}_{21} & \tilde{A}_{22} \end{bmatrix}\begin{bmatrix} \tilde{x}^1 \\ \tilde{x}^2 \end{bmatrix}, \\ \tilde{z} &= \tilde{x}^1. \end{aligned} \tag{5.4.76}$$

where the observed variable is the vector of the first r components of $\tilde{x}$ denoted $\tilde{x}^1$. Then (H, A) is observable if and only if $(\tilde{A}_{12}, \tilde{A}_{22})$ is observable.

Proof Since $\mathrm{Rank}(H) = r$, there exists a non-singular $(n \times n)$ matrix P such that $\tilde{H} \equiv HP^{-1} = (I_r, 0)$ where I_r is an r-dimensional identity matrix and 0 is an $r \times (n-r)$ zero matrix. Define $\tilde{x} = Px$. Then

$$\dot{\tilde{x}} = P\dot{x} = PAx = PAP^{-1}\tilde{x} = \tilde{A}\tilde{x},$$

and $\tilde{z} = z = HP^{-1}\tilde{x}$. Hence we have the decomposition (5.4.76). Here $\tilde{A}_{11} \in R(r \times r)$, $\tilde{A}_{12} \in R(r \times (n-r))$, $\tilde{A}_{21} \in R((n-r) \times r)$ and $\tilde{A}_{22} \in R((n-r) \times (n-r))$. We prove that (H, A) is observable if and only if $(\tilde{A}_{12}, \tilde{A}_{22})$ is observable. Suppose (H, A) is observable but $(\tilde{A}_{12}, \tilde{A}_{22})$ is not. Then there exists a nontrivial solution $\tilde{x}^2$ of

$$\dot{\tilde{x}}^2 = \tilde{A}_{22}\tilde{x}^2$$

satisfying

$$\tilde{A}_{12}\tilde{x}^2 \equiv 0. \tag{5.4.77}$$

Then clearly $\begin{bmatrix} 0 \\ \tilde{x}^2 \end{bmatrix}$ is a nontrivial solution of

$$\frac{\mathrm{d}}{\mathrm{d}t}\begin{bmatrix} \tilde{x}^1 \\ \tilde{x}^2 \end{bmatrix} = \begin{bmatrix} \tilde{A}_{11} & \tilde{A}_{12} \\ \tilde{A}_{21} & \tilde{A}_{22} \end{bmatrix}\begin{bmatrix} \tilde{x}^1 \\ \tilde{x}^2 \end{bmatrix} \tag{5.4.78}$$

with

$$\tilde{z} = z = \tilde{H}\tilde{x} = \tilde{x}^1 \equiv 0. \tag{5.4.79}$$

Since (H, A) and hence $(\tilde{H}, \tilde{A})$ is observable this is a contradiction. For the converse, suppose $(\tilde{A}_{12}, \tilde{A}_{22})$ is observable but (H, A) and $(\tilde{H}, \tilde{A})$ is not. Then (5.4.78) has a nontrivial solution

$$\begin{bmatrix} 0 \\ \tilde{x}^2 \end{bmatrix}$$

satisfying (5.4.79). As a consequence it follows from (5.4.78) that

$$\dot{\tilde{x}}^2 = \tilde{A}_{22}\tilde{x}^2$$
$$\tilde{z} = \tilde{A}_{12}\tilde{x}^2 = \dot{\tilde{x}}^1 - \tilde{A}_{11}\tilde{x}^1 = 0$$

contradicting the hypothesis that $(\tilde{A}_{12}, \tilde{A}_{22})$ is observable. ■

Using this lemma we can now prove the following theorem.

Theorem 5.4.15 (Reduced-order observer, design). *If (H, A) is observable and (A, B) stabilizable, then there exists an $(n-r)$th-order observer for the system*

$$\begin{aligned} \dot{x} &= Ax + Bu \\ z &= Hx. \end{aligned} \tag{5.4.80}$$

Proof By virtue of Lemma 5.4.14, we may assume, without loss of generality, that $H = [I_r, 0]$. Then we can rewrite (5.4.80) in the form

$$\begin{aligned} \begin{bmatrix} \dot{x}^1 \\ \dot{x}^2 \end{bmatrix} &= \begin{bmatrix} A_{11} & A_{12} \\ A_{21} & A_{22} \end{bmatrix} \begin{bmatrix} x^1 \\ x^2 \end{bmatrix} + \begin{bmatrix} B^1 u \\ B^2 u \end{bmatrix}, \\ z &= x^1. \end{aligned} \tag{5.4.81}$$

Here x^1 is available as the observed (measured) parameter and B^1u is also a measurable parameter and hence

$$A_{12}x^2 = (\dot{x}^1 - A_{11}x^1 - B^1u) \tag{5.4.82}$$

is a measurable parameter which is an indicator of the unobserved state x^2 of the system $\dot{x}^2 = A_{22}x^2 + (A_{21}x^1 + B^2u)$. Thus we have the observed system

$$\begin{aligned} \dot{x}^2 &= A_{22}x^2 + (A_{21}x^1 + B^2u) \qquad \text{dimension } (n-r), \\ \tilde{z} &= A_{12}x^2, \qquad \text{dimension } r. \end{aligned} \tag{5.4.83}$$

For this reduced system we can design a full-order observer by simply identifying the variables corresponding to the full-order observer of (5.4.52) and (5.4.53) as follows:

$$\begin{cases} x \to x^2 \\ A \to A_{22} \\ Bu \to A_{21}x^1 + B^2u \\ H \to A_{12}. \end{cases} \tag{5.4.84}$$

Accordingly we have

$$\begin{aligned} \dot{y} &= (A_{22} - LA_{12})y + LA_{12}x^2 + (A_{21}x^1 + B^2u) \\ &= (A_{22} - LA_{12})y + L(\dot{x}^1 - A_{11}x^1 - B^1u) + (A_{21}x^1 + B^2u) \\ &= (A_{22} - LA_{12})y + (A_{21} - LA_{11})x^1 + (B^2 - LB^1)u + L\dot{x}^1. \end{aligned} \tag{5.4.85}$$

Introducing $y \equiv \tilde{y} + Lx^1$ we can eliminate the last term containing the derivative of x^1. This gives

$$\begin{aligned} \dot{\tilde{y}} = (A_{22} - LA_{12})\tilde{y} + [A_{21} - LA_{11} + (A_{22} - LA_{12})L]x^1 \\ + (B^2 - LB^1)u. \end{aligned} \tag{5.4.86}$$

Comparing (5.4.86) with the observer dynamics of (5.4.63), we have

$$\begin{aligned} &F \equiv (A_{22} - LA_{12}) \in R((n-r) \times (n-r)), \\ &G \equiv [A_{21} - LA_{11} + (A_{22} - LA_{12})L, 0] \in R((n-r) \times n), \end{aligned} \tag{5.4.87}$$

where $0 \in R((n-r) \times (n-r))$. For the operator T, we have

$$TBu \equiv (B^2 - LB^1)u = [-L, I]\begin{bmatrix} B^1 \\ B^2 \end{bmatrix} u$$

and hence

$$T = [-L, I] \in R((n-r) \times n) \tag{5.4.88}$$

with $-L \in R((n-r) \times r)$ and $I \in R((n-r) \times (n-r))$. Since (A_{12}, A_{22}) is observable, there exists a matrix L such that $F \equiv (A_{22} - LA_{12})$ is a stability matrix. In order to complete the proof of existence of an $(n-r)$-dimensional observer for the system (5.4.80) all that remains to verify is that the matrices F, G and T satisfy the identity (5.4.59). Indeed

$$\begin{aligned} FT - TA &= (A_{22} - LA_{12})[-L, I] - [-L, I]\begin{bmatrix} A_{11} & A_{12} \\ A_{21} & A_{22} \end{bmatrix} \\ &= -[A_{21} - LA_{11} + (A_{22} - LA_{12})L, 0] = -G. \end{aligned} \tag{5.4.89}$$

Since (A, B) is stabilizable, there exists a K such that $(A + BK)$ is a stability matrix. Hence the system (5.4.80) and the observer (5.4.86) constitute a complete $(2n - r)$-dimensional plant–observer system which is asymptotically stable with

$$u = K\begin{bmatrix} H \\ T \end{bmatrix}^{-1}\begin{bmatrix} z \\ \tilde{y} \end{bmatrix}.$$

Thus we have not only proved the existence of a $(n - r)$-dimensional observer but we have also given its construction through (5.4.86). This completes the proof. ■

Corollary 5.4.16 (Closed-loop plant–observer). *Under the assumptions of Theorem* 5.4.15, *the closed-loop plant–observer system is given by the* $(2n - r)$*-dimensional system*

$$\frac{\mathrm{d}}{\mathrm{d}t}\begin{bmatrix} x \\ \tilde{y} \end{bmatrix} = \begin{bmatrix} A + BK\Gamma_1 H & BK\Gamma_2 \\ G + (B^2 - LB^1)K\Gamma_1 H & A_{22} - LA_{12} + (B^2 - LB^1)K\Gamma_2 \end{bmatrix}\begin{bmatrix} x \\ \tilde{y} \end{bmatrix}. \tag{5.4.90}$$

Proof The proof follows from (5.4.80) and (5.4.86) by setting

$$u = K\begin{bmatrix} H \\ T \end{bmatrix}^{-1}\begin{bmatrix} z \\ \tilde{y} \end{bmatrix} = K[\Gamma_1, \Gamma_2]\begin{bmatrix} Hx \\ \tilde{y} \end{bmatrix}. \quad ■ \tag{5.4.91}$$

For illustration we consider the following example.

Example 5.4.17 It is required to design a reduced-order observer for the second order system $\ddot{\xi} + 2\dot{\xi} - 3\xi = u$, $z = \dot{\xi}$. The corresponding state equation is

$$\begin{aligned} \frac{\mathrm{d}}{\mathrm{d}t}\begin{bmatrix} x_1 \\ x_2 \end{bmatrix} &= \begin{bmatrix} 0 & 1 \\ 3 & -2 \end{bmatrix}\begin{bmatrix} x_1 \\ x_2 \end{bmatrix} + \begin{bmatrix} 0 \\ 1 \end{bmatrix}u \\ z &= [0\ 1]\begin{bmatrix} x_1 \\ x_2 \end{bmatrix} = x_2. \end{aligned} \tag{5.4.92}$$

Note that the uncontrolled system is unstable, (A, B) is controllable, and (H, A) observable but the system is not stabilizable by output feedback. We wish to design a reduced order observer. We can use Corollary 5.4.16, specifically equation (5.4.90) to write the complete plant–observer dynamics provided we use the transformation

$$P = \begin{bmatrix} 0 & 1 \\ 1 & 0 \end{bmatrix}$$

to write (5.4.92) into the form (5.4.81). Using this transformation we have,

$$\tilde{H} = HP^{-1} = [1, 0]$$

$$\tilde{A} = PAP^{-1} = \begin{bmatrix} -2 & 3 \\ 1 & 0 \end{bmatrix} = \begin{bmatrix} \tilde{A}_{11} & \tilde{A}_{12} \\ \tilde{A}_{21} & \tilde{A}_{22} \end{bmatrix}$$

$$\tilde{B} = PB = \begin{bmatrix} 1 \\ 0 \end{bmatrix} = \begin{bmatrix} \tilde{B}^1 \\ \tilde{B}^2 \end{bmatrix}.$$

The characteristic equation corresponding to $(\tilde{A} + \tilde{B}K)$ is given by $\lambda^2 + (2 - k_1)\lambda - (3 + k_2) = 0$. For specified eigenvalues say $\lambda_1 = -1$ and $\lambda_2 = -2$, we must have $k_1 = -1$ and $k_2 = -5$ giving $K = [-1, -5]$. For Γ_1 and Γ_2 we have

$$\begin{bmatrix} H \\ T \end{bmatrix}^{-1} = (\Gamma_1, \Gamma_2).$$

$\tilde{H} = [1, 0]$ and from (5.4.88) $T = [-L, I] = [-\ell, 1]$.
Hence

$$\begin{bmatrix} \tilde{H} \\ T \end{bmatrix}^{-1} = \begin{bmatrix} 1 & 0 \\ -\ell & 1 \end{bmatrix}^{-1} = \begin{bmatrix} 1 & 0 \\ \ell & 1 \end{bmatrix}, \quad \text{and} \quad \Gamma_1 = \begin{bmatrix} 1 \\ \ell \end{bmatrix} \quad \text{and} \quad \Gamma_2 = \begin{bmatrix} 0 \\ 1 \end{bmatrix}.$$

Therefore

$$\tilde{B}K\Gamma_1\tilde{H} = \begin{bmatrix} -(1 + 5\ell) & 0 \\ 0 & 0 \end{bmatrix}, \tag{5.4.93}$$

and

$$\tilde{B}K\Gamma_2 = \begin{bmatrix} -5 \\ 0 \end{bmatrix}. \tag{5.4.94}$$

From (5.4.89)

$$G = [\tilde{A}_{21} - \ell\tilde{A}_{11} + (\tilde{A}_{22} - \ell\tilde{A}_{12})\ell, 0]$$

$$= [1 + 2\ell - 3\ell^2, 0] \tag{5.4.95}$$

$$(\tilde{B}^2 - \ell\tilde{B}^1)K\Gamma_1\tilde{H} = [\ell(1 + 5\ell), 0], \tag{5.4.96}$$

and hence

$$G + (\tilde{B}^2 - L\tilde{B}^1)K\Gamma_1\tilde{H} = [1 + 3\ell + 2\ell^2, 0]. \tag{5.4.97}$$

Similarly we have

$$\tilde{A}_{22} - L\tilde{A}_{12} = -3\ell, \qquad (\tilde{B}^2 - L\tilde{B}^1) = -\ell$$

$$(\tilde{B}^2 - L\tilde{B}^1)K\Gamma_2 = 5\ell,$$

and hence

$$(\tilde{A}_{22} - L\tilde{A}_{12}) + (\tilde{B}^2 - L\tilde{B}^1)K\Gamma_2 = 2\ell. \tag{5.4.98}$$

Using (5.4.93)–(5.4.98) in the closed-loop plant–observer system (5.4.90), we have

$$\frac{\mathrm{d}}{\mathrm{d}t}\begin{bmatrix} \tilde{x}_1 \\ \tilde{x}_2 \\ \tilde{y} \end{bmatrix} = \begin{bmatrix} -(3+5\ell) & 3 & -5 \\ 1 & 0 & 0 \\ 1+3\ell+2\ell^2 & 0 & 2\ell \end{bmatrix}\begin{bmatrix} \tilde{x}_1 \\ \tilde{x}_2 \\ \tilde{y} \end{bmatrix}.$$

In order for the observer to have eigenvalue, -3, say, it is necessary that the characteristic equation, $|\lambda I - (\tilde{A}_{22} - L\tilde{A}_{12})| = 0$, has the root -3. Thus $(\lambda + 3\ell) = 0$, and hence, for $\ell = 1$, the specified eigenvalue is realized. The plant–observer eigenvalues are $\lambda = -1, \ -2, \ -3$ as expected.

5.5 Controllability of nonlinear systems

Theories of controllability, observability, and stabilizability of nonlinear systems are not so well developed as those for linear systems. The subject is of current interest and is in the process of slow but continuous progress. Results presently available are not so popular as linear theory, the reason being the difficulty in verifying the assumptions used. For the sake of completeness we shall present here some basic results. Readers interested in research in this field may consult current literature and relevant references listed.

Consider the nonlinear time-invariant system in R^n

$$\dot{x} = f(x, u), \qquad u(t) \in U \subset R^m, \tag{5.5.1}$$

where U is a nonempty subset of R^m, and $\mathcal{U}$ the class of admissible controls which are measurable functions on $R_0 = [0, \infty)$ with values in U.

We wish to find sufficient conditions for global null controllability of the system (5.5.1). For this we shall first consider the question of local null controllability (see Definition 5.3.2). Throughout, we shall assume that f is C^1 in all the variables, and define

$$\begin{aligned} f_x(0, 0) &= \left(\frac{\partial f}{\partial x}\right)(0, 0) = A \\ f_u(0, 0) &= \left(\frac{\partial f}{\partial u}\right)(0, 0) = B. \end{aligned} \tag{5.5.2}$$

Lemma 5.5.1 (Local null controllability). *Suppose* $0 \in \operatorname{Int} U$ *and*

$f \in C^1$ with $f(0, 0) = 0$. Then the system (5.5.1) *is locally null controllable if (A, B) is controllable, that is,* Rank $D(A, B) = n$.

The result is intuitively clear since the system is nearly linear close to the origin. However, its rigorous proof is based on the implicit function theorem and continuous dependence of solutions on initial data. We leave it as an exercise for the reader.

Lemma 5.5.2 *Suppose there exists a function $\sigma: R^n \to U$ which is C^1 and there exists a Lyapunov function V for the system $\dot{x} = f(x, \sigma(x))$ such that*

(a) $V(x) \geq 0$, $V(x) = 0$ *iff* $x = 0$,
(b) $\lim_{|x| \to \infty} V(x) = \infty$, *and*
(c) $\sum_{i=1}^{n} f_i(x, \sigma(x)) \frac{\partial V}{\partial x_i} < 0$ *for* $x \neq 0$.

Then the system (5.5.1) *with $u(t) = \sigma(x(t))$ is globally asymptotically stable (with respect to the origin).*

Proof Follows directly from Theorem 4.4.9. ■

With the help of the above results we can prove global null controllability of the system (5.5.1).

Theorem 5.5.3 (Global null controllability). *Consider the system* (5.5.1) *and suppose*

(a) *f is C^1 in all the variables,*
(b) $f(0, 0) = 0$,
(c) $0 \in \text{Int } U$,
(d) *there exists a $\sigma: R^n \to U$ with $\sigma \in C^1$ satisfying the hypotheses of Lemma* 5.5.2,
(e) *(A, B) is controllable, where $A = f_x(0, 0)$, $B = f_u(0, 0)$.*

Then the system is globally null controllable.

Proof Since, by Lemma 5.5.2, the system is globally asymptotically stable with respect to the origin, for any $x_0 \in R^n$ there exists a finite time $t_0 > 0$ such that the solution $x(t, x_0)$ of the system

$$\dot{x} = f(x, \sigma(x)), \qquad x(0) = x_0$$

arrives in the domain N of local controllability at t_0. Then, by Lemma 5.5.1, we can find a control $u^*(t)$, $t \geq t_0$, with $u^*(t) \in U$ a.e., and a finite time $t_1 \geq t_0$ such that $x^*(t_1) = 0$ where x^* is the solution of the initial-value problem

$$\dot{x}^*(t) = f(x^*(t), u^*(t)), \qquad t \geq t_0$$
$$x^*(t_0) = x(t_0, x_0).$$

This completes the proof. ■

Now we consider nonlinear time-varying systems of the form

$$\begin{aligned} &\dot{x} = f(t, x, u), \qquad t \geq 0, \\ &u \in \mathcal{U}. \end{aligned} \tag{5.5.3}$$

In Theorem 5.5.6 we present sufficient conditions for global asymptotic stability of the system (5.5.3) and in Theorem 5.5.7 we prove global null controllability. For this we need the following preliminary results which are essentially due to Chukwu [107].

Lemma 5.5.4 *Let $f: R_0 \times R^n \times R^m \to R^n$ such that f and $f_x \equiv \partial f / \partial x$ are continuous in (t, x, u). Suppose there is a constant $M > -\infty$ such that the characteristic roots ν_k, $k = 1, 2, \ldots, n$, of the matrix $\frac{1}{2}(f_x + f'_x)$ satisfy the inequality $\nu_k \leq M$, $k = 1, 2, \ldots, n$, for some $u = u^* \in \mathcal{U}$ uniformly in $x \in R^n$ and $t \in R_0$. Then the scalar product*

$$G \equiv (f(t, x + \xi, u^*(t)) - f(t, x, u^*(t)), \xi) \tag{5.5.4}$$

satisfies

$$G \leq M\, |\xi|^2 \tag{5.5.5}$$

for all $t \in R_0$, $x \in R^n$.

Proof By the mean value theorem

$$f(t, x + \xi, u^*(t)) = f(t, x, u^*(t)) + \int_0^1 f_x(t, x + \theta\xi, u^*(t)) \cdot \xi \, d\theta. \tag{5.5.6}$$

Hence

$$\begin{aligned} G &= \int_0^1 (f_x(t, x + \theta\xi, u^*(t))\xi, \xi) \, d\theta \\ &= \frac{1}{2} \int_0^1 ((f_x + f'_x)(t, x + \theta\xi, u^*(t))\xi, \xi) \, d\theta. \end{aligned} \tag{5.5.7}$$

Since by assumption the eigenvalues $\{\nu_k\}$ of $\frac{1}{2}(f_x + f'_x)$ are uniformly bounded by M we have $G \leqslant M\,|\xi|^2$ for all $x \in R^n$ and $t \in R_0$. This proves the assertion. ■

For convenience of notation we shall write $f_x(t, x, u^*(t)) \equiv J$.

Lemma 5.5.5 *Suppose f satisfies the assumptions of Lemma 5.5.4 and that for the system (5.5.3) there exists a control $u^* \in \mathcal{U}$ and a symmetric positive definite matrix Y such that the eigenvalues $\lambda_i(t, x, u^*)$, $i = 1, 2, \ldots, n$, of the matrix $K = \frac{1}{2}(YJ + J'Y)$ satisfy*

$$\lambda_i \leqslant -\delta < 0, \qquad i = 1, 2, \ldots, n \tag{5.5.8}$$

for all $(t, x) \in R_0 \times R^n$ where δ is a constant. Then, for each finite $T \geqslant 0$, and $\rho \in [1, 2]$, each solution of (5.5.3) with $u = u^*$ satisfies the inequality

$$|x(t)|^\rho \leqslant e^{-\lambda t}\{C_1(T)e^{-\lambda t} + C_2 g(t)\}, \qquad t \geqslant T, \tag{5.5.9}$$

where

$$\begin{aligned} \lambda &= (\rho\delta/2\alpha) \\ C_1(T) &= 2^\rho(\alpha/\alpha')^{\rho/2}|x(T)|^\rho\, e^{2\lambda T} \\ C_2 &= (2\alpha(\rho-1)/\delta\rho)^{\rho-1} \cdot (2\,\|Y\|/\alpha')^\rho \\ g(t) &\equiv \int_T^t |f(\theta, 0, u^*(\theta))|^\rho e^{\lambda\theta}\, d\theta \end{aligned} \tag{5.5.10}$$

and α and α' are the largest and the smallest eigenvalues of Y respectively.

Proof Let $u^* \in \mathcal{U}$ and x the corresponding solution of the system (5.5.3) with initial state $x(0) = x_0 \in R^n$. For convenience of notation write $f(t, x(t), u^*(t)) = f(x)$. Since Y is positive definite the quadratic form

$$V(t) \equiv (Yx(t), x(t)) \tag{5.5.11}$$

is positive and satisfies the inequality

$$\alpha'\,|x(t)|^2 \leqslant V(t) \leqslant \alpha\,|x(t)|^2 \tag{5.5.12}$$

for all $t \in R_0$. Computing the time derivative of V and using the symmetry of Y we have

$$\dot V(t) = 2(Y(f(x) - f(0)), x) + 2(Yf(0), x). \tag{5.5.13}$$

Using the mean value theorem for the first term in (5.5.13) and the assumption (5.5.8) we obtain

$$\begin{aligned} \tfrac{1}{2}\dot V(t) &= \frac{1}{2}\int_0^1 ((Yf_x(\theta x) + f_x'(\theta x)Y)x, x)\, d\theta + (Yf(0), x) \\ &\leqslant -\delta\,|x(t)|^2 + (Yf(0), x). \end{aligned} \tag{5.5.14}$$

Using (5.5.12) into (5.5.14) one can deduce that

$$\dot{V}(t) + 2\delta_0 V(t) \leq \delta_1 |f(0)| V^{1/2}(t) \tag{5.5.15}$$

with $\delta_0 = (\delta/\alpha)$ and $\delta_1 = (2\,\|Y\|)/(\alpha')^{1/2}$. Hence we have

$$\frac{\mathrm{d}}{\mathrm{d}t}(V^{1/2}(t)\mathrm{e}^{\delta_0 t}) \leq (\delta_1/2)\,|f(0)|\,\mathrm{e}^{\delta_0 t}, \qquad t \geq 0, \tag{5.5.16}$$

where we recall that $f(0) \equiv f(t, 0, u^*(t)) \equiv f^*(t)$. Integrating (5.5.16) over the interval $[T, t]$, $t \geq T \geq 0$, we obtain

$$V^{1/2}(t)\mathrm{e}^{\delta_0 t} \leq V^{1/2}(T)\mathrm{e}^{\delta_0 T} + (\delta_1/2)\int_T^t |f^*(\theta)|\,\mathrm{e}^{\delta_0\theta}\,\mathrm{d}\theta. \tag{5.5.17}$$

Hence for any $\rho \in [1, 2]$,

$$V^{\rho/2}(t)\mathrm{e}^{\rho\delta_0 t} \leq 2^\rho V^{\rho/2}(T)\,\mathrm{e}^{\rho\delta_0 T} + (\delta_1)^\rho\left(\int_T^t |f^*(\theta)|\,\mathrm{e}^{\delta_0\theta}\,\mathrm{d}\theta\right)^\rho. \tag{5.5.18}$$

By Hölder's inequality (see chapter 2)

$$\int_T^t |f^*(\theta)|\,\mathrm{e}^{\delta_0\theta}\,\mathrm{d}\theta \leq \left(\int_T^t |f^*(\theta)|^\rho\,\mathrm{e}^{(\rho\delta_0/2)\theta}\,\mathrm{d}\theta\right)^{1/\rho}\left(\int_T^t \mathrm{e}^{(\rho'\delta_0/2)\theta}\,\mathrm{d}\theta\right)^{1/\rho'} \tag{5.5.19}$$

where $(1/\rho) + (1/\rho') = 1$. Using (5.5.19) into (5.5.18) one can easily check that

$$V^{\rho/2}(t) \leq \mathrm{e}^{-(\rho\delta_0/2)t}\left\{2^\rho V^{\rho/2}(T)\mathrm{e}^{\rho\delta_0 T}\mathrm{e}^{-(\rho\delta_0/2)t} + \delta_1^\rho\left(\frac{2(\rho-1)}{\delta_0\rho}\right)^{\rho-1} g(t)\right\}. \tag{5.5.20}$$

where

$$g(t) = \int_T^t |f^*(\theta)|^\rho \mathrm{e}^{(\rho\delta_0\theta/2)}\,\mathrm{d}\theta.$$

Using the relation (5.5.12) once again, the results (5.5.9) and (5.5.10) follow from (5.5.20). This completes the proof. ∎

With the help of the above results we shall first prove that the system (5.5.3) is globally asymptotically stable.

Theorem 5.5.6 (Global asymptotic stability). *Suppose the assumptions of Lemma 5.5.5 hold and that there exists an $r > 0$ such that*

$$\lim_{t\to\infty}\int_t^{t+r} |f^*(\theta)|^\rho\,\mathrm{d}\theta = 0. \tag{5.5.21}$$

Then the system (5.5.3) with $u = u^*$ is globally asymptotically stable.

Proof Partition the interval $[T, t]$ into $(m+1)$ subintervals,

$$T < t - mr < t - (m-1)r < \cdots < t - r < t,$$

giving

$$[T, t] = [T, t - mr] \cup \left\{ \bigcup_{j=1}^{m} I_j \right\} \tag{5.5.22}$$

where

$$I_j \equiv [t - jr, t - (j-1)r], \qquad j = 1, 2, \ldots, m,$$

and

$$m = \left[\frac{t-T}{r} \right] \equiv \text{largest integer} < \left(\frac{t-T}{r} \right),$$

with the convention that for $m = 0$ the second factor in (5.5.22) is empty. Then we can write

$$\begin{aligned} g(t) &= \int_T^t |f^*(\theta)|^\rho e^{\lambda\theta} \, d\theta \\ &= \left\{ \int_T^{t-mr} |f^*(\theta)|^\rho e^{\lambda\theta} \, d\theta + \sum_{j=1}^{m} \int_{t-jr}^{t-(j-1)r} |f^*(\theta)|^\rho e^{\lambda\theta} \, d\theta \right\}. \end{aligned} \tag{5.5.23}$$

Define

$$\gamma_T \equiv \sup_{t \geq T} \int_t^{t+r} |f^*(\theta)|^\rho \, d\theta \geq 0. \tag{5.5.24}$$

Using (5.5.24) into (5.5.23) we obtain the estimate

$$g(t) \leq e^{\lambda t}(1 - e^{-\lambda r})^{-1} \gamma_T, \qquad t \geq T. \tag{5.5.25}$$

By virtue of Lemma 5.5.5 it follows from (5.5.9) and (5.5.25) that

$$|x(t)|^\rho \leq C_1(T) e^{-2\lambda t} + C_2(1 - e^{-\lambda r})^{-1} \gamma_T, \tag{5.5.26}$$

for all $t \geq T$. By virtue of assumption (5.5.21),

$$\lim_{T \to \infty} \gamma_T = 0. \tag{5.5.27}$$

Hence, for any $\varepsilon > 0$, there exists a finite $T^* > 0$ such that

$$\gamma_T < (\varepsilon(1 - e^{-\lambda r})/C_2) \qquad \text{for all } T \geq T^*.$$

Therefore, for all $t \geq T^*$,

$$|x(t)|^\rho \leq C_1(T^*) e^{-2\lambda t} + \varepsilon, \tag{5.5.28}$$

and since $C_1(T^*) < \infty$ for $T^* < \infty$, we have

$$\overline{\lim} \, |x(t)|^\rho \leq \varepsilon. \tag{5.5.29}$$

Since $\varepsilon > 0$ is arbitrary, it follows from (5.5.29) that $\overline{\lim} |x(t)|^p = 0$ and hence $\lim_{t \to \infty} x(t) = 0$. Since this is true for any initial state $x_0 \in R^n$, we conclude that the system (5.5.3), with $u = u^*$, is globally asymptotically stable. ■

Using the above result we can now prove global finite-time null controllability. Let

$$\begin{cases} f(t, 0, 0) = 0 \\ f_x(t, 0, 0) \equiv A(t) \\ f_u(t, 0, 0) \equiv B(t) \end{cases} \tag{5.5.30}$$

and recall the definition of the support function $H_U(\cdot)$ as given in Lemma 5.3.7.

Theorem 5.5.7 (Global null controllability). *Consider the system* (5.5.3) *with the control restraint set U a compact subset of R^m with $0 \in U$. Suppose f satisfies the assumptions of Theorem* 5.5.6 *and, further, for every finite $T > 0$,*

$$\int_T^\infty H_U(B'(t)\psi(t))\,\mathrm{d}t = \infty \tag{5.5.31}$$

for every nontrivial solution ψ of the adjoint system

$$\dot{\psi} = -A'(t)\psi. \tag{5.5.32}$$

Then the system (5.5.3) *is globally finite-time null controllable.*

Proof Under the assumption (5.5.31) it follows from Theorem 5.3.11 that the linearized system

$$\dot{\xi} = A(t)\xi + B(t)u, \qquad u(t) \in U \quad \text{a.e.} \tag{5.5.33}$$

is globally finite-time null controllable. Hence, f being a C^1 function, the nonlinear system is locally finite-time null controllable. By Theorem 5.5.6, there exists an admissible control u^* such that the system (5.5.3), with $u = u^*$, is globally asymptotically stable with respect to the zero state. Combining these results the conclusion of the theorem follows. This completes the proof. ■

Exercises

5.1.P1 Verify whether or not the following systems are observable

(a) $\ddot{\theta} = u, \quad y = \theta$ (b) $\ddot{\theta} + u, \quad y = \dot{\theta}$

(c) $J\ddot{\theta} + B\dot{\theta} = u, \quad y = \theta$ (d) $J\ddot{\theta} + B\dot{\theta} = u, \quad y = \dot{\theta}.$

5.1.P2 Verify that the time-varying system of Example 5.2.20 is observable with $H = [0, 1]$.

5.1.P3 Prove the statements of Corollary 5.1.6.

5.2.P4 Show that the systems of Problem 5.1.P1 are controllable. Obtain a control law for the system 5.1.P1(a) which transfers the initial state $(1, 0)'$ to the origin $(0, 0)'$ in $t_1 = 1$ second.

5.2.P5 Consider the dynamics of a controlled oscillator

$$\begin{bmatrix} \dot{x}_1 \\ \dot{x}_2 \end{bmatrix} = \begin{bmatrix} 0 & w \\ -w & 0 \end{bmatrix} \begin{bmatrix} x_1 \\ x_2 \end{bmatrix} + \begin{bmatrix} b_1 \\ b_2 \end{bmatrix} u.$$

Find the conditions on b_1 and b_2 for which the system is (a) controllable, (b) not controllable.

5.2.P6 The dynamics of a voltage controlled d.c. servo-system is given by the following set of equations:

$$J\frac{\mathrm{d}^2\theta}{\mathrm{d}t^2} + B\frac{\mathrm{d}\theta}{\mathrm{d}t} + K\theta = C_1 i$$

$$L\frac{\mathrm{d}i}{\mathrm{d}t} + Ri + C_2\frac{\mathrm{d}\theta}{\mathrm{d}t} = v$$

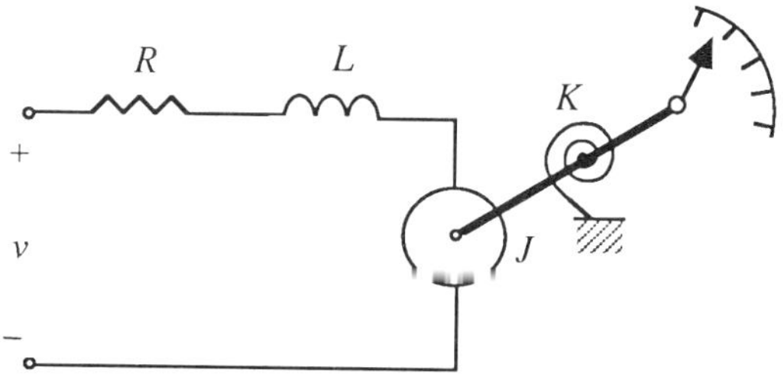

(a) Verify controllability of the system.
(b) Is the system observable if only θ is measured (observed)?

5.2.P7 Consider the system as given in Problem 5.2.P6 and suppose that $\theta(t)$, $t \in [0, t_1] \equiv I$, has been recorded as the output data. Assuming that the initial state is known, show that the input signal $u(t)$, $t \in I$, is given by the solution u of the linear Volterra integral equation (of the second kind)

$$u(t) = \xi(t) + \int_0^t K(t, \tau)u(\tau)\,\mathrm{d}\tau.$$

Give the expressions for the free term ξ and the kernel K in terms of the output data $y = \theta$ and the system parameters.

5.2.P8 Verify Remark 5.2.19, and prove Corollary 5.2.11.

5.3.P9 Show that the second-order system $\ddot{\theta} = u$, $u(t) \in U = [-1, +1]$, is globally null controllable.

5.3.P10 Show that the system

$$\begin{aligned} \dot{x}_1 &= x_2 \\ \dot{x}_2 &= -x_2 + u, \end{aligned} \qquad u(t) \in U = [-1, +1]$$

is globally null controllable.

5.3.P11 Prove Corollaries 5.3.18 and 5.3.19 (*Hint*: Use Theorem 5.3.11).

5.3.P12 Verify that the oscillator

$$\ddot{\xi} + w^2 \xi = u, \qquad u(t) \in U = [0, 1]$$

is globally null controllable.

5.3.P13 Verify that the system

$$\begin{bmatrix} \dot{x}_1 \\ \dot{x}_2 \end{bmatrix} = \begin{bmatrix} -1 & e^{2t} \\ 0 & -1 \end{bmatrix} \begin{bmatrix} x_1 \\ x_2 \end{bmatrix} + \begin{bmatrix} 0 \\ 1 \end{bmatrix} u,$$

is globally null controllable for $U = [-1, +1]$, while for $U = [0\ 1]$ it is not (*Hint*: Use Theorem 5.3.11).

5.3.P14 Verify that the observability matrix M_0 (equation (5.1.19)) and the controllability matrix M_c (equation (5.2.45)) satisfy the following differential equations

$$\text{(a)} \quad \begin{cases} \dfrac{\partial M_0}{\partial t}(t_0, t_1) = -A'(t)M_0(t, t_1) - M_0(t, t_1)A(t) - H'(t)H(t) \\ M_0(t_1, t_1) = 0, \qquad t \leqslant t_1. \end{cases}$$

$$\text{(b)} \quad \begin{cases} \dfrac{\partial M_c}{\partial t}(t_0, t) = A(t)M_c(t_0, t) + M_c(t_0, t)A'(t) + B(t)B'(t) \\ M_c(t_0, t_0) = 0, \qquad t \geqslant t_0. \end{cases}$$

Note that numerical verification of observability and controllability by use of equations (5.1.19) and (5.2.45) involve rather lengthy computations (evaluation of state-transition operator, matrix multiplications, integrations etc). This is conveniently avoided by solving the differential equations (a) and (b). The reader is encouraged to try some examples using a computer.

5.4.P15 For small deflections, the equation of an inverted pendulum is given by $\ddot{\xi} - \alpha\xi = u$, $\alpha > 0$. Using Theorem 5.4.2, find the stabilizing state feedback matrix K.

5.4.P16 Repeat Problem 5.4.P15 using Theorem 5.4.3 and comment on its advantage.

5.4.P17 Assuming only one group of delayed neutron emitters, the reactor dynamics (1.2.65) reduces to the following equation:

$$\frac{dN}{dt} = \frac{N}{\ell}(u - \beta) + \lambda R$$

$$\frac{dR}{dt} = (\beta\mu/\ell)N - \lambda R.$$

(a) Show that the system attains an equilibrium provided

$$u_0 = (1 - \mu)\beta \qquad \text{and} \qquad (R_0/N_0) = (\beta\mu/\lambda\ell).$$

Since the thermal power generated is proportional to neutron flux density, for a desired power level, N_0 and hence R_0 is fixed.

(b) Linearize the system around the equilibrium (N_0, R_0, u_0) and verify that

$$\frac{d}{dt}\begin{bmatrix} n \\ r \end{bmatrix} = \begin{bmatrix} -\beta\mu/\ell & \lambda \\ \beta\mu/\ell & -\lambda \end{bmatrix}\begin{bmatrix} n \\ r \end{bmatrix} + \begin{bmatrix} N_0/\ell \\ 0 \end{bmatrix} u$$

where n, r represent the variations of N and R from N_0 and R_0 respectively.

(c) Assuming that only neutron flux density is observed, design (i) a full-order observer (ii) a reduced-order observer.

5.4.P18 Design (a) a full order observer and (b) a reduced order observer for the following systems:

(i) $$\frac{d}{dt}\begin{bmatrix} x_1 \\ x_2 \end{bmatrix} = \begin{bmatrix} 0 & 1 \\ 1 & 0 \end{bmatrix}\begin{bmatrix} x_1 \\ x_2 \end{bmatrix} + \begin{bmatrix} 0 \\ 1 \end{bmatrix} u$$

$$y = x_1$$

(ii) $$\dddot{\theta} = u, \quad y = \theta.$$

5.5.P19 Verify the inequality (5.5.20).

5.5.P20 Show that the time-varying nonlinear system

$$\begin{aligned} \dot{x}_1 &= x_2 \\ \dot{x}_2 &= -x_1^3 - tx_2 + u \end{aligned} \qquad t \geqslant 0$$

with $U = [-1, +1]$ is globally null controllable. (*Hint*: Use $V = x_1^4 + 2x_2^2$ as the Lyapunov function for $u = 0$.)

5.5.P21 Show that the nuclear reactor of example (5.4.P17) is globally null controllable for $U = [-\delta, +\delta]$, where $\delta > \mu - \beta$.

6

Optimal control

6.0 Introduction

There has been an enormous development of optimal control theory over the last three decades. The richness of the subject, measured by its mathematical contents blending both pure and applied mathematics and real world engineering applications, has attracted mathematicians, engineers, economists, scientists and many others. It is now a matured field covering extensive ground in systems governed by ordinary differential equations (Pontryagin *et al.* [25], Cesari [4]), delay differential equations (Oğuztöreli [22], Warga [34]), stochastic differential equations (Fleming and Rishel [42], Krylov [49], Wonham [52], Ahmed [37]), partial differential equations and abstract evolution equations (Lions [58], Butkovsky [55], Ahmed and Teo [53], Teo and Wu [59]).

In this chapter we consider systems governed by ordinary differential equations, leaving stochastic systems for the following chapter. Obviously we can cover only a small amount of ground, and therefore we limit our attention only to those results which we can now consider classical and have found extensive applications. We shall present the LaSalle bang-bang principle, the Pontryagin minimum principle, the Bellman principle of optimality, and briefly discuss the recent thrust in the so called viscosity solution of the Hamilton–Jacobi–Bellman equation which seems to have great potential towards a sound theory for optimal feedback controls. We shall also discuss the question of existence of optimal controls, and conclude the chapter with a section on numerical methods for optimal control and system identification.

6.1 Time-optimal control of linear systems

Consider the linear time-varying system

$$\dot{x} = A(t)x + B(t)u, \qquad t \geqslant t_0 \geqslant 0, \tag{6.1.1}$$

and let U be a compact subset of R^m and $\mathscr{U}$ the class of all measurable functions $\{u\}$ from $[t_0, \infty)$ to U. We call $\mathscr{U}$ the class of admissible controls. Let x_0 and x_1 be any two arbitrary elements of R^n. The problem is to find a control $u^* \in \mathscr{U}$ that transfers the system from the initial state x_0 to the target state x_1 in minimum time. This is the time-optimal control problem; and apparently it was LaSalle who first gave a complete solution to this problem.

Recall that the solution of the initial-value problem (6.1.1) is given by

$$x(t, u) = X(t)X^{-1}(t_0)x_0 + \int_{t_0}^{t} X(t)X^{-1}(\tau)B(\tau)u(\tau)\,d\tau, \tag{6.1.2}$$

where X is the fundamental solution of the problem

$$\begin{aligned} &\dot{X}(t) = A(t)X(t), \\ &X(t_0) = I \qquad \text{(identity matrix)}, \end{aligned} \tag{6.1.3}$$

and $u \in \mathscr{U}$. For each $t \in R_0 = \{t \geq t_0\}$, define the sets

$$\mathscr{A}(t) \equiv \{\xi \in R^n : \xi = x(t, u),\ u \in \mathscr{U}\} \tag{6.1.4}$$

and

$$\mathscr{R}(t) = \left\{\xi \in R^n : \xi = \int_{t_0}^{t} Y(\tau)u(\tau)\,d\tau,\ u \in \mathscr{U}\right\}, \tag{6.1.5}$$

where $Y(t) \equiv X^{-1}(t)B(t)$, $t \in R_0$. The set $\mathscr{A}(t)$ is called the attainable set, and $\mathscr{R}(t)$ the reachable set. Clearly the two set-valued functions are related through the following expressions:

$$\left.\begin{aligned} &\mathscr{A}(t) = X(t)x_0 + X(t)\mathscr{R}(t) \\ &\mathscr{R}(t) = X^{-1}(t)\mathscr{A}(t) - x_0 \end{aligned}\right\} \quad t \in R_0. \tag{6.1.6}$$

In general the time-optimal control problem is meaningful only if the system is controllable, that is, for some $\bar{t} \in R_0$, the target

$$x_1 \in \mathscr{A}(\bar{t}),$$

or equivalently

$$X^{-1}(\bar{t})x_1 - x_0 \in \mathscr{R}(\bar{t}). \tag{6.1.7}$$

Thus if the system is controllable, that is, $Y(t)\eta \not\equiv 0$ for $\eta \neq 0$, then there exists a $\bar{t} \in R_0$, and $\bar{u} \in \mathscr{U}$ such that $x(\bar{t}, \bar{u}) = x_1$, or equivalently

$$X^{-1}(\bar{t})x_1 - x_0 = \int_{t_0}^{\bar{t}} Y(\theta)\bar{u}(\theta)\,d\theta. \tag{6.1.8}$$

In certain problems, for example, spacecraft rendezvous, a robot arm picking up a moving object, etc, the target itself is in motion, and may be described by a continuous trajectory $z(t) \in R^n$, $t \in R_0$. In this case it is required that we have

$$x(\tilde{t}, \tilde{u}) = z(\tilde{t}),$$

or equivalently

$$X^{-1}(\tilde{t})z(\tilde{t}) - x_0 = \int_{t_0}^{\tilde{t}} Y(\theta)\tilde{u}(\theta)\,\mathrm{d}\theta. \tag{6.1.9}$$

Our problem is to find from among all such admissible controls the one that minimizes the transition time $(\tilde{t} - t_0)$. Throughout, we shall use the notation

$$y(t, u) = \int_{t_0}^{t} Y(\theta)u(\theta)\,\mathrm{d}\theta, \qquad u \in \mathcal{U}. \tag{6.1.10}$$

6.1.1 The LaSalle bang-bang principle

Throughout this section we assume that the system is controllable. Thus we must focus our attention on the following questions:

(Q1) If the system is controllable, does there exist a time optimal control, or equivalently a $t^* \geqslant t_0$ such that $z(t^*) \in \mathcal{A}(t^*)$ and $z(t) \notin \mathcal{A}(t)$ for any $t \in [t_0, t^*)$?

(Q2) If such a t^* exists, how can we find the corresponding control?

In certain situations the optimal control turns out to be bang-bang, which is easy to implement. This raises the following question:

(Q3) If an optimal control exists, does it belong to the class of bang-bang controls

$$\mathcal{U}_b \equiv \{u \in \mathcal{U}: u(t) \in \partial U \equiv \text{boundary of } U \text{ a.e.}\}? \tag{6.1.11}$$

We shall consider these questions after we have presented some basic properties of the sets of attainability $\mathcal{A}(t)$ and reachability $\mathcal{R}(t)$. Recall the Definitions 3.1.6 and 3.1.4 for convex and compact sets respectively.

The concept of a distance between two sets plays an important role in control theory. Let $CC(R^n)$ denote the class of all compact convex subsets of R^n. For $G \in CC(R^n)$, define the distance of a point x from G by

$$\rho(x, G) = \rho(G, x) = \inf\{|x - y|, y \in G\}. \tag{6.1.12}$$

Note that the infimum exists and is attained in G, and that convexity is not essential (Proposition 2.1.43).

Definition 6.1.1 (Hausdorff metric). Let G_1, $G_2 \in CC(R^n)$, and define

$$\rho_H(G_1, G_2) = \tfrac{1}{2}\left\{\sup_{x \in G_2} \rho(G_1, x) + \sup_{x \in G_1} \rho(x, G_2)\right\} \tag{6.1.13}$$

Since $x \to \rho(x, G)$ is continuous and $G_i (i = 1,2)$ is compact, the supremum and hence ρ_H are well defined. The function ρ_H is called the Hausdorff metric, and it satisfies all the axioms of a metric space. Thus $CC(R^n)$ is a metric space with the metric ρ_{H}.

In the following lemma we collect some of the basic properties of the sets of attainability and reachability. We assume throughout, unless mentioned otherwise, that the entries of the matrices $A(t)$ and $B(t)$ are locally Lebesgue integrable.

Lemma 6.1.2 *Suppose U is a closed bounded convex subset of R^m. Then*

(a) *for each $t \in R_0$, $\mathcal{A}(t)$ and $\mathcal{R}(t)$ are convex and compact subsets of R^n,*
(b) *$\mathcal{R}(t)$ is a non-contracting set, in the sense that $\mathcal{R}(t_1) \subseteq \mathcal{R}(t_2)$ for all $t_2 \geq t_1$ provided $0 \in U$,*
(c) *$\mathcal{R}(t)$ is continuous in the Hausdorff metric ρ_H,*
(d) *$\mathcal{R}(t) = \mathcal{R}_b(t) \equiv \{\xi \in R^n: \xi = \int_{t_0}^t Y(\theta)u(\theta)\, \mathrm{d}\theta,\ u \in \mathcal{U}_b\}$ for each $t \in R_0$.*

Proof (a) The convexity follows from that of U and the linearity of the system. For compactness one must show that for each $t \in R_0$, $\mathcal{A}(t)$ (resp. $\mathcal{R}(t)$) is closed and bounded. Boundedness of U implies that of $\mathcal{A}(t)$ (resp. $\mathcal{R}(t)$); closure and convexity of U implies the closure of $\mathcal{U}$ in the w^*-topology, that is, the topology of convergence of $f_n \to f$ in the sense that $\int g(t) f_n(t)\, \mathrm{d}t \to \int g(t) f(t)\, \mathrm{d}t$ for each $g \in L_1$. Hence the conclusion follows from linearity.

(b) For $t_0 \leq t_1 \leq t_2$,

$$\mathcal{R}(t_2) = \mathcal{R}(t_1) + \left\{\int_{t_1}^{t_2} Y(\theta)u(\theta)\, \mathrm{d}\theta, \quad u(\theta) \in U \text{ a.e.}\right\}$$

and hence the result follows by setting $u(t) \equiv 0$ on $[t_1, t_2]$.

(c) Since U is bounded, there exists a finite number M such that

$$\sup\{|\eta|, \eta \in U\} = M.$$

Hence for $t_1, t_2 \in R_0$,

$$\rho_H(\mathcal{R}(t_2), \mathcal{R}(t_1)) = \sup\left\{\left|\int_{t_1}^{t_2} Y(\theta)u(\theta)\,\mathrm{d}\theta\right|, u \in \mathcal{U}\right\}$$

$$\leqslant M \int_{t_1}^{t_2} \|Y(\theta)\|\,\mathrm{d}\theta, \tag{6.1.14}$$

and the continuity follows from local integrability of Y.

(d) This is essentially the bang-bang principle due to LaSalle, which we shall consider later in this section. ■

Remark 6.1.3 Lemma 6.1.2 is also valid for a variable control constraint set $U = U(t)$, $t \in R_0$.

In fact the result (d) is a consequence of a more fundamental result due to Auman. Let $\mathcal{P}(R^n)$ denote the set of all subsets of R^n called the power set, and suppose $F: R_0 \to \mathcal{P}(R^n)$ such that for each $t \in R_0$, $F(t) \neq \varnothing$ (non-empty). Define the function g given by

$$g(t) \equiv \sup\{|\eta|: \eta \in F(t)\}. \tag{6.1.15}$$

If g is locally integrable, then F is said to be a locally integrable set-valued function, and one can define

$$\int_I F(t)\,\mathrm{d}t = \left\{\int_I f(t)\,\mathrm{d}t,\ f \text{ measurable}, f(t) \in F(t) \text{ a.e.}\right\}.$$

The property (d) is the consequence of a fundamental result that states

$$\int_I \overline{\mathrm{Co}}\, F(t)\,\mathrm{d}t = \int_I F(t)\,\mathrm{d}t, \tag{6.1.16}$$

where $\overline{\mathrm{Co}}\, F \equiv$ closure of the convex hull of F. Thus, even if F is not convex valued, its integral is, and hence

$$\left\{\int f(t)\,\mathrm{d}t, f(t) \in U \text{ a.e.}\right\} = \left\{\int f(t)\,\mathrm{d}t, f(t) \in \partial U \text{ a.e.}\right\},$$

whenever U is convex. We shall not use this result in this book. For further discussion on this topic the reader is referred to Hermes and LaSalle [8].

We now return to the time-optimal control problem. Assuming controllability, we prove the existence of time-optimal controls. More precisely we have the following result.

Theorem 6.1.4 (Existence of optimal controls). *Suppose the set U satisfies the assumptions of Lemma* 6.1.1. *Then, if there is a control $\tilde{u} \in \mathcal{U}$*

and $\bar{t} \in R_0$ *such that*

$$\tilde{z}(\bar{t}) \equiv X^{-1}(\bar{t})z(\bar{t}) - x_0 \in \mathcal{R}(\bar{t}),$$

there is a (time)-optimal control.

Proof Define $T \equiv \{t \in [t_0, \bar{t}]: \tilde{z}(t) \in \mathcal{R}(t)\}$, and note that it is a non-empty set since $\bar{t} \in T$. We prove that there exists a $u^* \in \mathcal{U}$ such that $y(t^*, u^*) = \tilde{z}(t^*)$, where $t^* = \inf T$. If T contains only one point it is necessarily $\bar{t}$ and there is nothing to prove. Hence let $\{t_n\} \in T$. Without loss of generality we may assume that $\{t_n\}$ is a decreasing sequence converging to t^*. Clearly $\tilde{z}(t_n) \in \mathcal{R}(t_n)$ and hence there exists a sequence of controls $\{u_n\} \in \mathcal{U}$ such that $\tilde{z}(t_n) = y(t_n, u_n)$, where by definition

$$y(t_n, u_n) = \int_{t_0}^{t_n} Y(\theta)u_n(\theta)\,\mathrm{d}\theta.$$

We must show that there exists a $u^* \in \mathcal{U}$ such that $\tilde{z}(t^*) = y(t^*, u^*)$. Clearly we can write

$$\begin{aligned}\tilde{z}(t^*) - y(t^*, u_n) &= \tilde{z}(t^*) - \tilde{z}(t_n) + \tilde{z}(t_n) - y(t^*, u_n)\\ &= \tilde{z}(t^*) - \tilde{z}(t_n) + \int_{t^*}^{t_n} Y(\theta)u_n(\theta)\,\mathrm{d}\theta\end{aligned}$$

and hence, by the boundedness of U, we have

$$|\tilde{z}(t^*) - y(t^*, u_n)| \leq |\tilde{z}(t^*) - \tilde{z}(t_n)| + M\int_{t^*}^{t_n} \|Y(\theta)\|\,\mathrm{d}\theta,$$

where $M = \sup\{|\eta|, \eta \in U\}$. Since $\tilde{z}$ is continuous and $\|Y(t)\|$ is locally integrable, it follows from the above inequality that, as $n \to \infty$, $y(t^*, u_n) \to \tilde{z}(t^*)$. But $y(t^*, u_n) \in \mathcal{R}(t^*)$ for all n, and $\mathcal{R}(t^*)$ is a closed set by Lemma 6.1.2 and hence the limit $\tilde{z}(t^*) \in \mathcal{R}(t^*)$. In other words there exists an $u^* \in \mathcal{U}$ such that $\tilde{z}(t^*) = y(t^*, u^*) = \int_{t_0}^{t^*} Y(\theta)u^*(\theta)\,\mathrm{d}\theta$. This completes the proof. ■

Remark 6.1.5 The optimal control u^* need not have any bearing on the existence or non-existence of a limit of the sequence $\{u_n\}$.

Remark 6.1.6 Theorem 6.1.4 also holds for the variable control constraint set $U(t)$ satisfying convexity and compactness for each $t \geq t_0$.

The questions (Q2) and (Q3) are related to the necessary conditions for optimality, that is, conditions that characterize optimal controls leading to their numerical evaluation. Consider

$$y(t, u) \equiv \int_{t_0}^{t} Y(\theta)u(\theta)\,\mathrm{d}\theta, \qquad u \in \mathcal{U}, \quad t \in R_0.$$

Clearly y is absolutely continuous and hence its derivative exists a.e. and is given by $\dot{y}(t, u) = Y(t)u(t)$, $y(t_0, u) = 0$. Let $v \in R^n$ with $|v| = 1$, and consider the expression

$$(\dot{y}, v) = (Y(t)u, v) = (Y'(t)v, u). \tag{6.1.17}$$

Clearly for maximum rate of change of y, and hence the state x, in the direction v, we must choose u to maximize the functional

$$f(t, u) \equiv (Y'(t)v, u) \tag{6.1.18}$$

for each $t \in R_0$. The maximizing element depends on the geometry of the set U. If U is a unit ball in R^m, furnished with the usual Euclidean norm $|\cdot|$, then u must be taken as

$$u^*(t) = Y'(t)v/|Y'(t)v|, \tag{6.1.19}$$

except where $Y'(t)v = 0$ and is set arbitrarily if otherwise. If U is the unit cube $= \{u \in R^m\colon |u_i| \leq 1,\ i = 1, 2, \ldots, m\}$, then

$$u^*(t) = \operatorname{sign} Y'(t)v, \tag{6.1.20}$$

where $\operatorname{sign} z = (\operatorname{sign} z_1, \operatorname{sign} z_2, \ldots, \operatorname{sign} z_m)'$, and if U is a polyhedron, given by the intersection of a finite number of half spaces determined by a finite number of hyperplanes, then $u^*(t)$ takes values from the vertices $V(U)$ of U.

A precise result is stated in the following lemma.

Lemma 6.1.7 *Suppose U is the unit cube. Then $y(t^*, u^*) \in \partial\mathcal{R}(t^*)$ if and only if*

$$u^*(t) = \operatorname{sign}(Y'(t)v), \qquad t \in R_0, \tag{6.1.21}$$

for some nonzero unit vector $v \in R^n$.

Proof For the sufficient condition, suppose that (6.1.21) hold. Then we must show that $y(t^*, u^*) = \int_{t_0}^{t^*} Y(t)u^*(t)\,\mathrm{d}t$ is a boundary point of $\mathcal{R}(t^*)$. By virtue of (6.1.21)

$$(u^*(t), Y'(t)v) \geq (u(t), Y'(t)v) \qquad \text{for all } u \in \mathcal{U}, \quad t \in [t_0, t^*],$$

and hence

$$\int_{t_0}^{t^*} (u^*(t), Y'(t)v)\,\mathrm{d}t \geq \int_{t_0}^{t^*} (u(t), Y'(t)v)\,\mathrm{d}t \qquad \text{for all } u \in \mathcal{U}.$$

Therefore

$$(y(t^*, u^*), v) \geq (y(t^*, u), v) \qquad \text{for all } u \in \mathcal{U},$$

or equivalently

$$(\xi, v) \leqslant (y(t^*, u^*), v) \qquad \text{for all } \xi \in \mathcal{R}(t^*),$$

which means that $\mathcal{R}(t^*)$ lies on the 'left' of the hyperplane

$$H(v) \equiv \{\xi \in R^n: (\xi, v) = (y(t^*, u^*), v)\}$$

passing through the point $y^* = y(t^*, u^*)$. Hence $y^* \in \partial\mathcal{R}(t^*)$.

For the necessary condition, we must show that if $y^* \in \partial\mathcal{R}(t^*)$, then the control u that takes $y(t, u)$ to y^* at $t = t^*$ must have the form $u(t) = \operatorname{sign}(Y'(t)\eta)$ for some nonzero unit vector $\eta \in R^n$. Since $\mathcal{R}(t^*)$ is convex and $y^* \in \partial\mathcal{R}(t^*)$, there exists a hyperplane $H(\eta)$, with η denoting the outward normal to the hyperplane, such that $\mathcal{R}(t^*)$ is on the left of $H(\eta)$, that is

$$(\xi, \eta) \leqslant (y^*, \eta) \qquad \text{for all } \xi \in \mathcal{R}(t^*). \tag{6.1.22}$$

Since $y^* \in \partial\mathcal{R}(t^*)$, there exists a $u^{**} \in \mathcal{U}$ such that

$$y^* = \int_{t_0}^{t^*} Y(\theta)u^{**}(\theta)\, d\theta.$$

Hence it follows from the inequality (6.1.22) that

$$\int_{t_0}^{t^*} (Y(\theta)u(\theta), \eta)\, d\theta \leqslant \int_{t_0}^{t^*} (Y(\theta)u^{**}(\theta), \eta)\, d\theta,$$

or equivalently

$$\int_{t_0}^{t^*} (u(\theta), Y'(\theta)\eta)\, d\theta \leqslant \int_{t_0}^{t^*} (u^{**}(\theta), Y'(\theta)\eta)\, d\theta \tag{6.1.23}$$

for all $u \in \mathcal{U}$. For an arbitrary $t \in [t_0, t^*]$, a Lebesgue density point, define

$$I_\varepsilon \equiv \{\theta \in [t_0, t^*]: |t - \theta| \leqslant \varepsilon\} \qquad \text{for } \varepsilon > 0,$$

and take

$$u(\theta) = \begin{cases} u^{**}(\theta) & \text{if } \theta \in [t_0, t^*] \backslash I_\varepsilon, \\ v & \text{if } \theta \in I_\varepsilon, \end{cases}$$

for any $v \in U$. Then it follows from (6.1.23) that

$$\int_{I_\varepsilon} (v, Y'(\theta)\eta)\, d\theta \leqslant \int_{I_\varepsilon} (u^{**}(\theta), Y'(\theta)\eta)\, d\theta$$

for all $v \in U$ and $\varepsilon > 0$. Dividing this expression by the measure (length) $\mu(I_\varepsilon)$ and letting $\varepsilon \downarrow 0$, we obtain

$$(v, Y'(t)\eta) \leqslant (u^{**}(t), Y'(t)\eta) \tag{6.1.24}$$

for all $v \in U$. Hence it follows from our previous discussion that $u^{**}(t) = \operatorname{sign}(Y'(t)\eta)$. Since t is an arbitrary Lebesgue density point, this completes the proof. ■

With the help of the above lemma, we can prove the following necessary condition of optimality.

Theorem 6.1.8 (Necessary conditions of optimality). *If u^* is an optimal control with minimum transition time $(t^* - t_0)$, then u^* is of the form*

$$u^*(t) = \operatorname{sign}(Y'(t)v), \qquad t \in [t_0, t^*] \tag{6.1.25}$$

for some nonzero vector $v \in R^n$.

Proof If u^* is the optimal control with transition time $(t^* - t_0)$, then

$$y(t^*, u^*) \equiv \int_{t_0}^{t^*} Y(\theta)u^*(\theta)\, d\theta = \tilde{z}(t^*). \tag{6.1.26}$$

We show that $\tilde{z}(t^*)$, and hence $y(t^*, u^*)$, is a boundary point of $\mathcal{R}(t^*)$, because then, by virtue of Lemma 6.1.7, we can conclude that there exists a $v \neq 0$ (unit vector) such that u^* has the form (6.1.25). We prove this by contradiction. Suppose $\tilde{z}(t^*) \notin \partial\mathcal{R}(t^*)$, then the only other possibility is that $\tilde{z}(t^*) \in \operatorname{Int} \mathcal{R}(t^*) \equiv$ interior of $\mathcal{R}(t^*)$. Then there exists an $\varepsilon > 0$ such that the ε-neighbourhood of $\tilde{z}(t^*)$, denoted by $N_\varepsilon(\tilde{z}(t^*))$, is contained in $\operatorname{Int} \mathcal{R}(t^*)$. Since, by Lemma 6.1.2, $\mathcal{R}(t)$ is noncontracting, there exists a $\tau < t^*$ such that $N_\varepsilon(\tilde{z}(t^*)) \subset \mathcal{R}(\tau)$ and $\tilde{z}(t^*) \in \mathcal{R}(\theta)$ for all $\theta \geq \tau$. Then the continuity of $t \to \tilde{z}(t)$ implies that there exists a $\tau^* \in [\tau, t^*)$ such that $\tilde{z}(\tau^*) \in \mathcal{R}(\tau^*)$. Hence there exists a control $\tilde{u} \in \mathcal{U}$ such that $y(\tau^*, \tilde{u}) = \tilde{z}(\tau^*)$. This contradicts the optimality of the pair $\{u^*, t^*\}$. Therefore $\tilde{z}(t^*) = y(t^*, u^*) \in \partial\mathcal{R}(t^*)$, and hence, by the previous lemma, u^* has the form (6.1.25). This completes the proof. ■

In case $(Y'(t)v) = 0$ on a set of positive measure in $[t_0, t^*]$, the necessary condition (Theorem 6.1.8) does not provide any information on the structure of the control during that period of time. Hence the concepts of 'normality' and 'properness' were introduced. Let $b^i(t)$ denote the ith column of the matrix $B(t)$, and $v(\neq 0) \in R^n$, and define the sets

$$I_i(v) \equiv \{t \in [t_0, t^*]\colon ((X(t))^{-1}b^i(t), v) = 0, \text{ and } b^i(t) \neq 0\}$$

and

$$K_i(v) \equiv \{t \in [t_0, t^*]\colon ((X(t))^{-1}b^i(t), v) = 0\}. \tag{6.1.27}$$

The system $\dot{x} = A(t)x + B(t)u$, $t \geq t_0$, is said to be essentially normal (resp. normal) if the (Lebesgue) measure $\mu(I_i(v)) = 0$ (resp. $\mu(K_i(v)) = 0$) for all $i = 1, 2, \ldots, m$, and $v \neq 0$. Clearly $I_i(v) \subset K_i(v)$ and the system is essentially normal if it is normal. The system is said to be proper if the condition $Y'(t)v \equiv 0$, on any set of positive measure, implies $v = 0$. Any system that is essentially normal is also proper. If the system is essentially normal, then for each $i = 1, 2, \ldots, m$,

$$u_i^*(t) = \begin{cases} \operatorname{sign}(Y'(t)v)_i & \text{if } b_i(t) \neq 0, \\ \text{arbitrary} & \text{if } b_i(t) = 0, \end{cases} \tag{6.1.28}$$

and if it is normal then

$$u_i^*(t) = \operatorname{sign}(Y'(t)v)_i. \tag{6.1.29}$$

Thus whenever the system is essentially normal (resp. normal) the optimal control, if it exists, is given by the above expressions and it is essentially bang-bang (resp. bang-bang), and hence it is essentially unique (resp. unique). The uniqueness follows from the fact that if u^1 and u^2 are two optimal controls, then $t_1^* = t_2^*$ and both are bang-bang and $u = \frac{1}{2}(u^1 + u^2)$ is also optimal and bang-bang and therefore $u^1 = u^2$. Note that if the system is essentially normal, then the reachable set $\mathcal{R}(t)$ is expanding and strictly convex, in the sense that the support plane $H(v)$ through any boundary point $y^* \in \partial\mathcal{R}(t)$ contains no other point of $\mathcal{R}(t)$ other than y^*, that is, $\partial\mathcal{R}(t)$ has no flat component. In general a system is proper if and only if $\mathcal{R}(t)$ is expanding, that is $\mathcal{R}(t_1) \subset \operatorname{Int} \mathcal{R}(t_2)$ whenever $t_1 < t_2$.

The distinctions between normal and proper systems can be best appreciated through linear time-invariant systems. The linear time-invariant system, $\dot{x} = Ax + Bu$, is normal if and only if $(e^{-tA}b^i, v) \neq 0$ a.e. on $[t_0, t^*]$ for all $v \neq 0$ and each $i = 1, 2, \ldots, m$. This is true if and only if the vectors $\{b^i, Ab^i, \ldots, A^{n-1}b^i\}$ are linearly independent for each $i = 1, 2, \ldots, m$. On the other hand the system is proper if and only if $Y'(t)v = (B'e^{-tA'}v) \equiv 0$ on a set of positive measure implies $v = 0$, or equivalently $\operatorname{Rank}(B, AB, \ldots, A^{n-1}B) = n$, or (A, B) is controllable. Thus a time-invariant system is normal if and only if

$$\operatorname{Rank}(b^i, Ab^i, \ldots, A^{n-1}b^i) = n \qquad \text{for } i = 1, 2, \ldots, m, \tag{6.1.30}$$

while the system is proper if and only if

$$\operatorname{Rank}(B, AB, \ldots, A^{n-1}B) = n. \tag{6.1.31}$$

Thus normality is a much stronger condition than properness. However, for a single input system of the form $\dot{x} = Ax + bu$, $b \in R^n$, u scalar, normality and properness are equivalent.

In many situations, the state $x(t)$, $t \geq t_0$, represents the error process which is to be driven to zero in minimum time. This requires that $-x_0 \in \mathcal{R}(t^*)$ with t^* being the minimum time. In this case the necessary condition is also sufficient.

Theorem 6.1.9 (Necessary and sufficient condition). *If the system is proper and if $u^* \in \mathcal{U}$ drives the system from the state x_0 to the zero state, then u^* is optimal if and only if $u^*(t) = \text{sign}(Y'(t)v)$ for some $v \neq 0$.*

For a more exhaustive study of time-optimal controls for linear systems the reader is referred to Hermes and LaSalle [8]. Before we conclude this section we show the relationship between the LaSalle bang-bang principle and the celebrated Pontryagin maximum principle, which is to be treated in detail in the following section. By virtue of the necessary condition, if u^* is optimal, then for all $v \in U$

$$(v, Y'(t)v) \leq (u^*(t), Y'(t)v) \quad \text{a.e. on } I = [t_0, t^*],$$

or equivalently

$$(v, B'(t)\psi(t)) \leq (u^*(t), B'(t)\psi(t)) \qquad \text{a.e. on } I,$$

where

$$\psi(t) \equiv (X(t^*)X^{-1}(t))'\eta = (\Phi(t^*, t))'\eta, \qquad t_0 \leq t \leq t^*,$$

is the solution of the adjoint differential equation,

$$\frac{d\psi}{dt} = -A'(t)\psi, \qquad t \in I,$$

$$\psi(t^*) = \eta = (X'(t^*))^{-1}v.$$

Defining

$$H(t, x, \psi, u) \equiv (A(t)x, \psi) + (B(t)u, \psi),$$

one can rewrite the necessary conditions for optimality in the form

$$H(t, x^*(t), \psi^*(t), v) \leq H(t, x^*(t), \psi^*(t), u^*(t)) \quad \text{a.e. on } I \quad \text{for all } v \in U,$$

$$\dot{x}^*(t) = H_\psi(t, x^*(t), \psi^*(t), u^*(t)), \quad x^*(t_0) = x_0,$$

$$\dot{\psi}^*(t) = -H_x(t, x^*(t), \psi^*(t), u^*(t)), \qquad \psi^*(t^*) = \eta.$$

This is a special case of Pontryagin's necessary conditions of optimality, which are treated in the next section.

Example 6.1.10 Considering the satellite dynamics of Example 5.1.6, and using the equations (6.1.30) and (6.1.31), one can easily verify that the system is proper but not normal.

Example 6.1.11 Consider the system $\ddot{\xi} + \xi = u$, $|u| \leqslant 1$ to be steered to the zero state in minimum time. The system is normal and hence proper. The optimal control has the form $u^* = \text{sign}(\sin(t + r))$ for some $r \in R$. Since the system is normal any control of this form, which steers the system to the zero state, is optimal.

Example 6.1.12 Consider the system (d.c. motor) $\ddot{\theta} + k\dot{\theta} = u$, $|u| \leqslant \beta$, k, $\beta > 0$. The system is equivalent to $\dot{x}_1 = x_2$, $\dot{x}_2 = -kx_2 + u$. Suppose it is required to transfer the system from an arbitrary state to the manifold $M = \{(x_1, x_2): x_2 = N\}$ where $|N| < \beta/k$ (= the maximum attainable speed). By direct integration one can verify that the phase plane trajectories are given by

$$x_1 = c_1 + \frac{1}{k^2}\{(\beta - kx_2) - \beta \ln(\beta - kx_2)\} \qquad \text{for } u = +\beta,$$

$$x_1 = c_1 - \frac{1}{k_2}\{(\beta + kx_2) - \beta \ln(\beta + kx_2)\} \qquad \text{for } u = -\beta.$$

The optimal feedback control which tends to maintain the speed x_2 near the desired speed N is given by $u^* = \text{sign}(N - x_2)$. This is shown in a phase plane diagram (see figure 6.1).

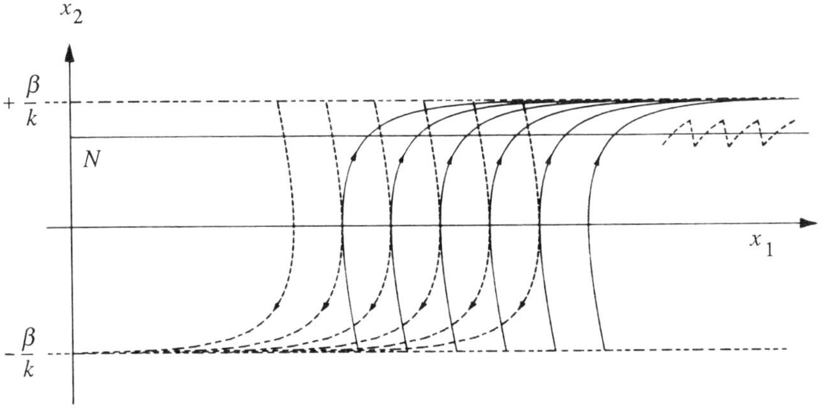

Figure 6.1

6.2 Calculus of variations and optimal control

6.2.1 The Euler–Lagrange equation

The basic problem of the calculus of variation is to find a function $x \in AC(I, R^n)$($\equiv$the space of absolutely continuous functions on $I = [t_1, t_2]$ to R^n) that imparts a minimum (or maximum) to the functional

$$\varphi(x) = \int_{t_1}^{t_2} \ell(t, x(t), \dot{x}(t))\, \mathrm{d}t, \tag{6.2.1}$$

subject to the constraints: $x(t_1) = x_1$, $x(t_2) = x_2$, where $\ell(t, x, y)$ is a possibly non-negative C^1 function from $I \times R^n \times R^n$ to R. In classical mechanics the function ℓ may represent the sum of potential and kinetic energies. By the principle of least action the system tends to follow the path (trajectory) that minimizes the functional φ. In fact the principle of least action applies to physics, biology, and even to social systems.

The classical calculus of variation deals with two basic questions: (a) the question of existence of an

$$x^0 \in X \equiv \{x \in AC(I, R^n): x(t_1) = x_1,\ x(t_2) = x_2\}$$

such that $\varphi(x^0) \leqslant \varphi(x)$ for all $x \in X$, and (b) the necessary conditions of optimality, that is, conditions that characterize the minimizing elements potentially leading to their determination.

It is known that Tonelli was the pioneer in proving the existence of a solution to the basic problem. He proved the existence of a solution in X under the assumptions that ℓ is C^2 and satisfies the following coercivity and convexity conditions:

(a) (coercivity) there exist $\alpha \in R$, $\beta > 0$ such that

$$\ell(t, x, y) \geqslant \alpha + \beta\, |y|^2 \qquad \text{for all } (t, x, y) \in I \times R^n \times R^n,$$

(b) (convexity) for each $(t, x) \in I \times R^n$, $y \to \ell(t, x, y)$ is convex.

The convexity condition implies that ℓ_{yy} is a positive semidefinite matrix. For general results on this topic the reader is referred to the excellent book of Cesari [4]. We shall also consider some existence questions for control problems later. Here we wish to consider only the second question, and derive the basic necessary condition of optimality giving the Euler–Lagrange equation.

The derivation is fairly straightforward. Let $x^0 \in X$ impart a minimum to the functional φ, that is, $\varphi(x^0) \leqslant \varphi(x)$ for all $x \in X$. Let x be an arbitrary element of X and $\varepsilon \in R$, and consider the function

$$\eta(\varepsilon) \equiv \varphi(x^0 + \varepsilon(x - x^0)). \tag{6.2.2}$$

Clearly η attains its minimum at $\varepsilon = 0$ and hence $\eta'(\varepsilon)|_{\varepsilon=0} = 0$. Differentiating η with respect to ε and setting $\varepsilon = 0$, we have

$$\eta'(0) = \int_{t_1}^{t_2} \{(\ell_x(t, x^0, \dot{x}^0), x - x^0) + (\ell_y(t, x^0, \dot{x}_0), \dot{x} - \dot{x}^0)\}\, \mathrm{d}t. \tag{6.2.3}$$

Integrating by parts and recalling that $x(t_1) = x^0(t_1)$, $x(t_2) = x^0(t_2)$ for $x, x^0 \in X$ and $\eta'(0) = 0$, we have

$$\eta'(0) = \int_{t_1}^{t_2} \left(\left((\ell_x(t, x^0, \dot{x}^0) - \frac{\mathrm{d}}{\mathrm{d}t}\ell_y(t, x^0, \dot{x}^0)\right), x - x^0\right) \mathrm{d}t = 0 \tag{6.2.4}$$

for all $x \in X$. The reader can easily verify that if for $h \in L_1$, $\int_{t_1}^{t_2} (h(t), w(t))\, \mathrm{d}t = 0$ for all $w \in AC(I, R^n)$ satisfying $w(t_1) = 0$ and $w(t_2) = 0$, then $h(t) = 0$ a.e. on I. Hence we obtain the system of Euler–Lagrange equations,

$$\begin{aligned} &\frac{\mathrm{d}}{\mathrm{d}t}(\ell_{\dot{x}}) = \ell_x, \qquad \text{a.e. on } I, \\ &x(t_1) = x_1, \qquad x(t_2) = x_2. \end{aligned} \tag{6.2.5}$$

This usually gives a system of n second-order or $2n$ first-order ordinary differential equations with the boundary conditions as stated above. In principle, by solving this two-point boundary-value problem one obtains x^0.

The Euler–Lagrange approach applies equally well to variational problems involving functions of many variables, for example

$$\varphi(v) = \int_{t_1}^{t_2} \int_{\Omega} \ell(t, \xi, v(t, \xi), v_t, D^1 v, D^2 v)\, \mathrm{d}\xi\, \mathrm{d}t, \tag{6.2.6}$$

where $v(t, \xi) = v(t, \xi_1, \xi_2, \ldots, \xi_n) \equiv p$ is a function of $(n+1)$ variables, and

$$v_t = \frac{\partial v}{\partial t} \equiv q,$$

$$D^1 v = \left\{D_i v = \frac{\partial v}{\partial \xi_i}, i = 1, 2, \ldots, n\right\} \equiv r \equiv (r_1, r_2, \ldots, r_n),$$

$$\begin{aligned} D^2 v &= \left\{D_{ij} v = \frac{\partial^2 v}{\partial \xi_i \partial \xi_j}, i, j = 1, 2, \ldots, n\right\} \\ &\equiv s \equiv \{s_{ij}, i, j = 1, 2, \ldots, n\}, \end{aligned}$$

and Ω is an open bounded subset of R^n. The function $\ell = \ell(t, \xi, p, q, r, s)$ is defined from $I \times \Omega \times R \times R \times R^n \times R^{n^2}$ to R, and is C^1 in all the variables. The necessary condition for the minimum is given by the Euler–Lagrange equation,

$$\frac{\partial}{\partial t}(\ell_q) + \sum_{i=1}^{n} \frac{\partial}{\partial \xi_i}(\ell_{r_i}) - \sum_{i,j=1}^{n} \frac{\partial^2}{\partial \xi_i \partial \xi_j}(\ell_{s_{ij}}) = \ell_p. \tag{6.2.7}$$

This gives a partial differential equation for v, second order in t and fourth order in ξ.

For illustration, we consider the free vibration of a string of length L stretched along the x-axis with $y = v(t, x)$ representing its vertical displacement from the rest state at the position x at time t. Using ρ to denote the mass density (per unit length), the kinetic energy is given by

$$\mathrm{KE} = \frac{1}{2}\int_0^L \rho(v_t)^2\,\mathrm{d}x. \tag{6.2.8}$$

The potential energy is a function of the stress developed due to elongation of the string as it vibrates. An element of length Δx stretches to

$$\Delta s = ((v(t, x+\Delta x) - v(t, x))^2 + (\Delta x)^2)^{1/2}.$$

Hence

$$\mathrm{d}s = (1 + (v_x)^2)^{1/2}\,\mathrm{d}x, \tag{6.2.9}$$

and the deformation is given by

$$\mathrm{d}s - \mathrm{d}x = \{(1 + (v_x)^2)^{1/2} - 1\}\,\mathrm{d}x, \tag{6.2.10}$$

which approximates to $\mathrm{d}s - \mathrm{d}x \cong \frac{1}{2}(v_x)^2\,\mathrm{d}x$ for small deflection. Assuming that the potential energy (PE) is directly proportional to the deformation with proportionality coefficient $k(x)$, we have

$$\mathrm{PE} = \frac{1}{2}\int_0^L k(x)(v_x)^2\,\mathrm{d}x. \tag{6.2.11}$$

Hence the Lagrangian is given by

$$\mathrm{KE} - \mathrm{PE} = \frac{1}{2}\int_0^L \{\rho(v_t)^2 - k(v_x)^2\}\,\mathrm{d}x$$

$$= \int_0^L \ell(t, x; v_t, v_x)\,\mathrm{d}x, \tag{6.2.12}$$

and it follows from the Euler–Lagrange equation,

$$\frac{\partial}{\partial t}(\ell_q) + \frac{\partial}{\partial x}(\ell_r) = \ell_p,$$

that

$$\frac{\partial}{\partial t}(\rho v_t) - \frac{\partial}{\partial x}(k v_x) = 0. \tag{6.2.13}$$

If ρ, k are constant, this equation reduces to

$$\rho \frac{\partial^2 v}{\partial t^2} - k \frac{\partial^2 v}{\partial x^2} = 0. \tag{6.2.14}$$

This is the homogeneous string equation which is solved for v under appropriate initial and boundary conditions. If there is an external force $f(t, x)$ acting vertically on the string, then the work done on an element dx is $\rho f v \, dx$. Hence the Lagrangian is modified, giving

$$\mathrm{KE} - \mathrm{PE} = \int_0^L \{\tfrac{1}{2}\rho(v_t)^2 + \rho v f - \tfrac{1}{2}k(v_x)^2\} \, dx, \tag{6.2.15}$$

and we obtain the equation of forced vibration

$$\frac{\partial}{\partial t}(\rho v_t) - \frac{\partial}{\partial x}(k v_x) = \rho f, \tag{6.2.16}$$

which reduces to

$$\rho \frac{\partial^2 v}{\partial t^2} - k \frac{\partial^2 v}{\partial x^2} = \rho f \tag{6.2.17}$$

for constant coefficients. This equation is also used in the study of propagation of sound waves.

For a thin rod subject to external force f, we have

$$\ell(t, \xi, v, v_t, D^1 v, D^2 v) = \tfrac{1}{2}\rho(v_t)^2 - \tfrac{1}{2}k(v_{\xi\xi})^2 + \rho f v, \tag{6.2.18}$$

where $\tfrac{1}{2}k(v_{\xi\xi})^2$ denotes the potential energy due to curvature (neglecting elongation). Hence using equation (6.2.7) we obtain the dynamics of vibration of a rod,

$$\frac{\partial}{\partial t}(\rho v_t) - \frac{\partial^2}{\partial \xi^2}\left(-k \frac{\partial^2 v}{\partial \xi^2}\right) = \rho f, \tag{6.2.19}$$

which reduces to

$$\rho \frac{\partial^2 v}{\partial t^2} + k \frac{\partial^4 v}{\partial \xi^4} = \rho f \tag{6.2.20}$$

if the coefficients are constant. This equation is solved for v with appropriate initial and boundary conditions. Equation (6.2.19) or (6.2.20) is also used in the study of vibration of beams, bridges, and buildings etc. [70].

In fact the calculus of variations has many interesting applications in physics and mechanics, and is often used to construct mathematical models for physical systems.

The Euler–Lagrange approach also applies to variational problems with side constraints such as:

(a) differential constraints:

$$f_i(t, x, \dot{x}) = 0, \qquad i = 1, 2, \ldots, m, \quad m < n \tag{6.2.21}$$

(b) isoperimetric constraints:

$$\int_{t_1}^{t_2} g(t, x, \dot{x})\, \mathrm{d}t = z_i, \qquad i = 1, 2, \ldots, m, \quad m < n. \tag{6.2.22}$$

Since these problems are covered by control theory, we shall not discuss them separately. Our interest here is in the variational problems arising in optimal control theory. The relationship between the classical calculus of variations and the optimal control theory can be illustrated through the following simple example.

Suppose the functional φ, given by equation (6.2.1), is to be minimized subject to the additional constraint that $|\dot{x}(t)| \leqslant \alpha$ a.e. on I. Let B_α denote the closed ball of radius α centred at the origin in R^n, and consider the problem:

$$\min_{u \in \mathcal{U}} J(u) = \min_{u \in \mathcal{U}} \int_{t_1}^{t_2} \ell(t, x, u)\, \mathrm{d}t, \tag{6.2.23}$$

subject to the constraints, $x(t_1) = x_1$, $x(t_2) = x_2$, and

$$\dot{x} = u, \quad \text{and} \quad u \in \mathcal{U} \equiv \{u \text{ measurable: } u(t) \in B_\alpha \text{ a.e.}\}.$$

This is the simplest control problem. In the spirit of the calculus of variations, one introduces some undetermined multipliers, called Lagrange multipliers, $\psi \in AC(I, R^n)$, and $\lambda \geqslant 0$, a constant, to take care of the constraints, giving

$$\tilde{J}(x, u) \equiv \int_{t_1}^{t_2} \{\lambda \ell(t, x, u) + (\psi, \dot{x} - u)\}\, \mathrm{d}t \equiv \int_{t_1}^{t_2} \tilde{\ell}\, \mathrm{d}t. \tag{6.2.24}$$

Assuming the existence of a solution u, such that $u(t) \in \operatorname{Int} B_\alpha$ for $t \in I$, one can write the Euler–Lagrange equations, giving

$$\left.\begin{aligned} \frac{\mathrm{d}}{\mathrm{d}t}(\tilde{\ell}_{\dot{x}}) &= \tilde{\ell}_x \\ \frac{\mathrm{d}}{\mathrm{d}t}(\tilde{\ell}_{\dot{\psi}}) &= \tilde{\ell}_\psi \\ \frac{\mathrm{d}}{\mathrm{d}t}(\tilde{\ell}_{\dot{u}}) &= \tilde{\ell}_u \end{aligned}\right\} \Rightarrow \left\{\begin{aligned} \dot{\psi} &= \lambda \ell_x \\ \dot{x} &= u \\ \psi &= \lambda \ell_u. \end{aligned}\right. \tag{6.2.25}$$

One immediate objection to the above formalism is that B_α is a compact set and there is no guarantee that the optimal u would lie in the interior of B_α for all $t \in I$. Therefore arbitrary variation in the classical manner leading to the Euler–Lagrange equations is not permissible. Hence these equations are not always valid. Another objection comes from the fact that ℓ need not be differentiable in u.

In the late 1950s and early 1960s a group of Soviet mathematicians, L.S. Pontryagin, V.G. Boltyanskii, R.V. Gamkrelidze and E.F. Mishchenko, succeeded in overcoming this difficulty. They developed the necessary conditions of optimality (known as the Pontryagin maximum principle) which are close to those of the Euler–Lagrange equations, by use of a non-standard perturbation technique that automatically guarantees admissibility of variations, removing at the same time the differentiability requirement as mentioned above.

The variation introduced by Pontryagin *et al.* can be explained as follows. Let U be a compact control constraint set, and $\mathcal{U}$ the class of measurable functions with values $u(t) \in U$ a.e. Let u denote an extremal control, and suppose it is replaced on certain arbitrary disjoint subsets $\{I_i\} \subset I$ by $\{v_i\}$ taken from U, giving the perturbed control

$$\tilde{u}(t) = \begin{cases} u(t) & \text{if } t \in I \Big\backslash \bigcup_{i=1}^{s} I_i, \\ v_i & \text{if } t \in I_i, \end{cases} \qquad i = 1, 2, \ldots, s. \tag{6.2.26}$$

Hence

$$\delta u(t) = \sum_{i=1}^{s} \chi_{I_i}(t)(v_i - u(t)),$$

and

$$\tilde{u} = u + \delta u.$$

Since one is dealing with measurable controls both u and $\tilde{u}$ belong to $\mathcal{U}$ and hence δu is admissible.

6.2.2 Pontryagin's minimum principle and transversality conditions

The basic problems of optimal control dealing with the necessary conditions of optimality can be described as follows. Consider the system

$$\dot{x} = f(t, x, u), \qquad t \in [t_1, t_2] \equiv I,$$

with initial state $x(t_1) = x_1 \in R^n$. The problem is to find a control $u \in \mathcal{U}$

that minimizes the cost functional

$$J(u)=\int_{t_1}^{t_2} \ell(t, x(t), u(t))\,\mathrm{d}t, \qquad (\mathrm{P_L} \equiv \text{Lagrange problem})$$

or

$$J(u)=g(t_2, x(t_2)), \qquad (\mathrm{P_M} \equiv \text{Meyer problem})$$

or (6.2.27)

$$J(u)=\int_{t_1}^{t_2} \ell(t, x(t), u(t))\,\mathrm{d}t+\varphi(x(t_2)), \qquad (\mathrm{P_B} \equiv \text{Bolza problem}).$$

Both the Lagrange and Bolza problems can be reduced to that of Meyer. Indeed, define

$$x_{n+1}(t)=\int_{t_1}^{t} \ell(\theta, x(\theta), u(\theta))\,\mathrm{d}\theta, \qquad t \in I,$$

$$\tilde{f}=(f', \ell)' \in R^{n+1},$$

$$\tilde{x}=(x', x_{n+1})' \in R^{n+1},$$

and

$$g(t_2, \tilde{x}(t_2))=x_{n+1}(t_2).$$

Then the problem $\mathrm{P_L}$ reduces to

$$(\mathrm{P_L})': \qquad \begin{cases} \dot{\tilde{x}}=\tilde{f}(t, \tilde{x}, u) \\ \tilde{x}(t_1)=(x_1', 0)' \\ \min\{J(u) \equiv g(t_2, \tilde{x}(t_2)) \equiv x_{n+1}(t_2)\} \end{cases} \tag{6.2.28}$$

and $\mathrm{P_B}$ reduces to

$$(\mathrm{P_B})': \qquad \begin{cases} \dot{\tilde{x}}=\tilde{f}(t, \tilde{x}, u) \\ \tilde{x}(t_1)=(x_1', 0)' \\ \min\{J(u) \equiv g(t_2, \tilde{x}(t_2)) \equiv x_{n+1}(t_2)+\varphi(x(t_2))\}. \end{cases} \tag{6.2.29}$$

In fact the reader can verify that all the above problems are equivalent.

Pontryagin problem ($\mathrm{P_0}$)

Consider the (possibly appended) system $\dot{\tilde{x}}=\tilde{f}(t, \tilde{x}, u)$ with fixed initial time t_1 and free terminal time $t_2 \geqslant t_1$. The problem is to transfer the system from a given state, say $\tilde{x}_1$, to a moving target described by a continuous curve (trajectory) $z(t)$, $t \geqslant t_1$, in R^n or a set-valued target $S(t)$, $t \geqslant t_1$, continuous in the Hausdorff metric, at some time $t_2 \geqslant t_1$ while minimizing the cost $J(u)=\tilde{g}(t_2, \tilde{x}(t_2))$. Define

$$\mathcal{M}_2 \equiv \{(t, z(t)): t \geqslant t_1\} \quad \text{or} \quad \{(t, y): y \in S(t), t \geqslant t_1\},$$

and introduce the time coordinate as an additional component in the state variable, giving

$$\begin{aligned} &\dot{x}_0(t)=1, \qquad x_0(t_1)=t_1, \\ &\dot{\tilde{x}}(t)=\tilde{f}(t, \tilde{x}(t), u(t)), \qquad \tilde{x}(t_1)=\tilde{x}_1. \end{aligned} \tag{6.2.30}$$

Defining $x=(x_0, \tilde{x}')'$, $f=(1, \tilde{f}')'$, we have the following equivalent control problem. Find a control $u \in \mathcal{U}$ that transfers the system

$$\dot{x}=f(x, u), \qquad t \geqslant t_1,$$

from the initial state x_1 (in the phase space) to the desired manifold $\mathcal{M}_2$ at some time t_2 while minimizing the cost functional,

$$J(u)=g(x(t_2))=\tilde{g}(t_2, \tilde{x}(t_2)). \tag{6.2.31}$$

Note, if $\tilde{g}(t_2, \tilde{x}(t_2))=t_2$, we have the problem of hitting the target $z(t)$ or $S(t)$ in minimum time. This is the time-optimal control problem. In case $z(t) \equiv \tilde{x}_2$, we have a fixed target to be reached in minimum time. In certain problems both the starting time t_1 and the initial state $\tilde{x}(t_1)=\tilde{x}_1$ may be chosen from a given (set) manifold $\mathcal{M}_1 \subset R^{n+1}$, and in this case the problem is to find a control that transfers the system from $\mathcal{M}_1$ to $\mathcal{M}_2$ while minimizing the cost functional

$$J(u)=\tilde{g}(t_1, \tilde{x}(t_1), t_2, \tilde{x}(t_2)).$$

General control problem

Thus we have the following general control problem. Consider the system

$$\dot{x}=f(t, x, u), \tag{6.2.32}$$

with the cost functional

$$J(x, u)=g(t_1, x(t_1), t_2, x(t_2)). \tag{6.2.33}$$

Let $B \subset R \times R^n \times R \times R^n \equiv R^{2(n+1)}$ be a closed set, and define the set of admissible pairs $\{(x, u)\}$ by

$$\begin{aligned} G \equiv \{(x, u): x \in AC, u \in \mathcal{U}, \dot{x}=f(t, x, u) \text{ a.e., and } e(x) \\ \equiv (t_1, x(t_1), t_2, x(t_2)) \in B\}, \end{aligned} \tag{6.2.34}$$

where AC denotes the class of all absolutely continuous functions with values $x(t) \in R^n$, and $\mathcal{U}$ denotes the class of measurable functions with values $u(t) \in U$ a compact subset of R^m or a compact non-empty set-valued map $U(t) \subset R^m$. The basic control problem is to find a pair

$(x, u) \in G$ such that

$$J(x, u) \leq J(\tilde{x}, \tilde{u}) \qquad (6.2.35)$$

for all $(\tilde{x}, \tilde{u}) \in G$. A pair satisfying this property is called *optimal.* The following theorem gives the necessary conditions that an optimal pair must satisfy. This is the most important result of this section. For its proof we follow the variational arguments of Cesari [4]. The original proof due to Pontryagin *et al.* can be found in [25].

Theorem 6.2.1 (Necessary conditions of optimality). *Suppose f is measurable in t, C^1 in x and continuous in u and, for each (t, x), $f(t, x, U(t)) \equiv Q(t, x)$ is convex, g is C^1 in all the variables, and B is a closed subset of $R^{2(n+1)}$. Define the Hamiltonian H on $R \times R^n \times R^n \times R^m \to R$ by $H(t, x, \psi, u) \equiv \langle \psi, f(t, x, u)\rangle$. If the pair $(x, u) \in G$ is optimal with initial and terminal times $\{t_1, t_2\}$, then there exist multipliers $\psi \in AC(I, R^n)$, $I = [t_1, t_2]$, $\lambda_0 \geq 0$, not all simultaneously zero, such that*

(a) $\int_I H(t, x(t), \psi(t), u(t))\, dt \leq \int_I H(t, x(t), \psi(t), w(t))\, dt$ *for all* $w \in \mathcal{U}$, *or equivalently*

$$H(t, x(t), \psi(t), u(t)) \leq H(t, x(t), \psi(t), v) \quad \text{for all } v \in U(t), \text{ a.e. on } I, \qquad (6.2.36)(a)$$

(b) $\dot{x}(t) = f(t, x(t), u(t))$ *a.e. on* I, $x(t_1) = x_1$, $x(t_2) = x_2$, (6.2.36)(b)

(c) $\dot{\psi}(t) = -f_x'(t, x(t), u(t))\psi(t)$ *a.e. on* I, (6.2.36)(c)

and

(d)
$$(\lambda_0 g_{t_1} - M(t_1))\, dt_1 + (\lambda_0 g_{t_2} + M(t_2))\, dt_2 + \langle \lambda_0 g_{x_1} + \psi(t_1), \xi_1\rangle + \langle \lambda_0 g_{x_2} - \psi(t_2), \xi_2\rangle = 0, \qquad (6.2.36)(d)$$

for all $(dt_1, \xi_1, dt_2, \xi_2) \in T_B(t_1, x_1, t_2, x_2)$, *where* $T_B(t_1, x_1, t_2, x_2)$ *is the tangent plane to* B *at the point* $(t_1, x_1, t_2, x_2) \in B$ *and* $M(t) \equiv H(t, x(t), \psi(t), u(t))$, $t \in I$.

Proof For simplicity we assume t_1, t_2 to be fixed. Let (x^0, u^0) be the optimal pair and $\Lambda \equiv [0, 1]$, and consider the C^1 functions $X_1(\theta)$, $X_2(\theta)$: Λ to R^n satisfying $X_1(0) = x^0(t_1) = x_1^0$, $X_2(0) = x^0(t_2) = x_2^0$, and $(t_1, X_1(\theta), t_2, X_2(\theta)) \in B$ for $\theta \in \Lambda$. As θ describes Λ, $(t_1, X_1(\theta), t_2, X_2(\theta))$ describes a neighbourhood of (t_1, x_1^0, t_2, x_2^0) relative to B and further, $dX_1(\theta)/d\theta|_{\theta=0} \equiv X_1'(0)$ is tangent to the curve $C_1 \equiv \{X_1(\theta), \theta \in \Lambda\}$ at $\theta = 0$, and $X_2'(0)$ is tangent to the curve $C_2 \equiv \{X_2(\theta), \theta \in \Lambda\}$ at $\theta = 0$. Let $z \equiv (\tau_1, \xi_1, \tau_2, \xi_2) \in R^{2(n+1)}$ denote any vector tangent to the set B at the point $P_0 \equiv (t_1, x_1^0, t_2, x_2^0) \in B$. Since t_1, t_2 are assumed to be fixed $z = (0, \xi_1, 0, \xi_2)$ and, for C^1 curves $X_1(\theta)$, $X_2(\theta)$, $\theta \in \Lambda$, we can identify ξ_1 with $X_1'(0)$ and ξ_2 with $X_2'(0)$. Define

$$x(t_1, \theta) \equiv X_1(\theta), \quad x(t_2, \theta) \equiv X_2(\theta), \qquad \theta \in \Lambda,$$

such that

$$x(t_1, 0) = x_1^0, \qquad x(t_2, 0) = x_2^0.$$

Let V denote the set of vectors $\{v\} = \{(c, u, z)\}$ satisfying $c \geq 0$, $u \in \mathcal{U}$, and $z = (0, \xi_1, 0, \xi_2)$ any vector tangent to B at the point P_0. Consider the system

$$\dot{x}(t, \theta) = (1 - c\theta)f(t, x(t, \theta), u^0(t)) + c\theta f(t, x(t, \theta), u(t)) \equiv q_\theta(t, x)$$
$$x(t_1, \theta) = X_1(\theta), \qquad \theta \in \Lambda, \quad t_1 \leq t \leq t_2. \tag{6.2.37}$$

Since $Q(t, x) \equiv f(t, x, U(t))$ is convex, $q_\theta(t, x) \in Q(t, x)$ for all u^0, $u \in \mathcal{U}$ and $0 \leq c\theta \leq 1$. Hence $x(t, \theta)$, $\theta \in \Lambda$, is an admissible trajectory implying the existence of an $\{u(t, \theta), t \in I\} \in \mathcal{U}$ such that

$$\dot{x}(t, \theta) = f(t, x(t, \theta), u(t, \theta)) \quad \text{a.e. on } I. \tag{6.2.38}$$

We can extend the definition of the curves $X_1(\theta)$, $X_2(\theta)$ from Λ to $[-1, 1]$ without requiring that $(t_1, X_1(\theta), t_2, X_2(\theta))$ lie in B for $\theta \in [-1, 0)$. Differentiating the equation (6.2.37) with respect to θ, we have

$$\begin{aligned}\frac{\mathrm{d}}{\mathrm{d}t}\left(\frac{\partial x}{\partial \theta}(t, \theta)\right) = \Big\{ & -cf(t, x(t, \theta), u^0(t)) \\ & + (1 - c\theta)f_x(t, x(t, \theta), u^0(t))\frac{\partial x}{\partial \theta}(t, \theta) \\ & + cf(t, x(t, \theta), u(t)) \\ & + c\theta f_x(t, x(t, \theta), u(t))\frac{\partial x}{\partial \theta}(t, \theta)\Big\}\end{aligned} \tag{6.2.39}$$

$$\frac{\partial x}{\partial \theta}(t_1, 0) = X_1'(\theta), \qquad \theta \in \Lambda, \quad t \in I.$$

From the continuous dependence of solutions on parameters (see chapter 3), it follows that

$$\lim_{\theta \to 0} x(t, \theta) = x(t, 0) = x^0(t) \qquad \text{uniformly in } t \in I.$$

Letting $\theta \to 0$ in the equation (6.2.39) and using y to denote the limit of $(\partial x/\partial \theta)$, we obtain

$$\frac{\mathrm{d}y}{\mathrm{d}t} = f_x(t, x^0(t), u^0(t))y + c(f(t, x^0(t), u(t)) - f(t, x^0(t), u^0(t))) \tag{6.2.40}$$

$$y(t_1) = \xi_1.$$

This is a linear system called the variational equation, and has the standard form (see chapter 2)

$$\frac{dy}{dt} = A(t)y + h(t), \qquad t \in I, \qquad y(t_1) = \xi_1, \tag{6.2.41}$$

where

$$A(t) \equiv f_x(t, x^0(t), u^0(t)), \qquad h(t) \equiv c(f(t, x^0(t), u(t)) - f(t, x^0(t), u^0(t))). \tag{6.2.42}$$

For each variation $v = (c, u, z) \in V$, with $z = (0, \xi_1, 0, \xi_2)$, consider the $(n+1)$ vector,

$$\tilde{Y}(v) \equiv (Y^0(v), Y^1(v), \ldots, Y^n(v)) = (Y^0(v), Y(v)), \tag{6.2.43}$$

defined by

$$Y^0(v) \equiv g_{x_1} \cdot \xi_1 + g_{x_2} \cdot \xi_2 \equiv (g_{x_1}, \xi_1) + (g_{x_2}, \xi_2), \qquad Y(v) \equiv y(t_2, v) - \xi_2, \tag{6.2.44}$$

where the gradients g_{x_1} and g_{x_2} are evaluated at the point (t_1, x_1^0, t_2, x_2^0), with $x_1^0 = x^0(t_1)$ and $x_2^0 = x^0(t_2)$. Define the set

$$K \equiv \{\eta \in R^{n+1}: \eta = \tilde{Y}(v), v \in V\}. \tag{6.2.45}$$

It is easy to verify that for each pair $\eta_1, \eta_2 \in K$ and $\alpha_1, \alpha_2 \geqslant 0$, $\alpha_1\eta_1 + \alpha_2\eta_2 \in K$. Hence K is a convex cone with vertex at the origin $0 \in R^{n+1}$. Further one can show that the point $\eta^* \equiv (-1, 0, 0, \ldots, 0)' \in R^{n+1}$ is not an interior point of K. This is proved by showing that, on the contrary, there would exist a pair $(x, u) \in G$ such that $J(x, u) \leqslant J(x^0, u^0)$ contradicting the optimality of (x^0, u^0). Hence there exists a hyperplane $H(\tilde{\lambda}) \subset R^{n+1}$ with $\tilde{\lambda}$ a unit vector, $|\tilde{\lambda}| = 1$, normal to $H(\tilde{\lambda})$ such that $\tilde{\lambda} = (\lambda_0, \lambda)$ and

$$(\tilde{\lambda}, \eta^*) \leqslant 0 \leqslant (\tilde{\lambda}, \tilde{Y}(v)) \qquad \text{for all } v \in V. \tag{6.2.46}$$

That is, the cone K lies on the right of the hyperplane $H(\tilde{\lambda})$. Consequently

$$\lambda_0(g_{x_1} \cdot \xi_1 + g_{x_2} \cdot \xi_2) + (\lambda, y(t_2, v) - \xi_2) \geqslant 0 \tag{6.2.47}$$

for all $v \in V$. Denoting $g_{x_1} \cdot \xi_1 + g_{x_2} \cdot \xi_2$ by dg, we can state that if $dg \not\equiv 0$ then $\lambda \not\equiv 0$. Suppose $\lambda = 0$ but dg is not. Then it follows from (6.2.46) and (6.2.47) that K is contained in the half space $\{\eta \in R^{n+1}: (\tilde{\lambda}, \eta) \geqslant 0\} = \{\eta \in R^{n+1}: \lambda_0\eta_0 \geqslant 0\}$. Since $\lambda_0 \geqslant 0$ and $\tilde{\lambda} \not\equiv 0$, $\lambda_0 > 0$ and $K \subset \{\eta \in R^{n+1}: \eta_0 \geqslant 0\}$. Hence $Y^0(v) \geqslant 0$ for all $v \in V$, and consequently, evaluating g_{x_i}

at the point $e(x^0) = (t_1, x_1^0, t_2, x_2^0) = P_0$, we have

$$Y^0(v) = dg(e(x^0)) = g_{x_1}(e(x^0)) \cdot \xi_1 + g_{x_2}(e(x^0)) \cdot \xi_2 \geqslant 0 \tag{6.2.48}$$

for all ξ_1, ξ_2 such that $v = (c, u, z) \in V$ for $z = (0, \xi_1, 0, \xi_2)$. Since the inequality (6.2.48) changes its sign if both ξ_1 and ξ_2 change their signs, this must be an equality, that is, $dg(e(x^0)) = 0$. This is a contradiction and hence $\lambda \not\equiv 0$ whenever $dg \not\equiv 0$.

Consider the adjoint system

$$\begin{aligned} \frac{d\psi}{dt} &= -A'(t)\psi, \qquad t \in I = [t_1, t_2), \\ \psi(t_2) &= \lambda. \end{aligned} \tag{6.2.49}$$

Since the elements of A are integrable, this equation has a unique solution $\psi^0 \in AC(I, R^n)$. Multiplying the variational equation (6.2.41) by ψ^0, integrating by parts and using the adjoint equation (6.2.49), we obtain

$$(\psi^0(t_2), y(t_2)) - (\psi^0(t_1), y(t_1)) = \int_{t_1}^{t_2} (\psi^0(\tau), h(\tau))\, d\tau. \tag{6.2.50}$$

Hence it follows from (6.2.41)–(6.2.42) and the definition of Hamiltonian that

$$\begin{aligned} (\lambda, y(t_2) - \xi_2) &\equiv (\psi^0(t_1), \xi_1) - (\psi^0(t_2), \xi_2) \\ &\quad + c \int_{t_1}^{t_2} [H(t, x^0, \psi^0, u) - H(t, x^0, \psi^0, u^0)]\, dt. \end{aligned} \tag{6.2.51}$$

Substituting (6.2.51) into (6.2.47), we obtain

$$\begin{aligned} &(\lambda_0 g_{x_1} + \psi^0(t_1), \xi_1) + (\lambda_0 g_{x_2} - \psi^0(t_2), \xi_2) \\ &\quad + c \int_{t_1}^{t_2} [H(t, x^0, \psi^0, u) - H(t, x^0, \psi^0, u^0)]\, dt \geqslant 0. \end{aligned} \tag{6.2.52}$$

This inequality must hold for all $v = (c, u, z) \in V$, and hence, for $c = 0$, we have

$$(\lambda_0 g_{x_1} + \psi^0(t_1), \xi_1) + (\lambda_0 g_{x_2} - \psi^0(t_2), \xi_2) \geqslant 0 \tag{6.2.53}$$

for all $v \in V$. Since B is assumed to have a tangent plane $T_B(P_0)$ at $P_0 = (t_1, x_1^0, t_2, x_2^0)$, the inequality reduces to an equality, that is,

$$(\lambda_0 g_{x_1} + \psi^0(t_1), \xi_1) + (\lambda_0 g_{x_2} - \psi^0(t_2), \xi_2) = 0. \tag{6.2.54}$$

This is the transversality condition in case t_1, t_2 are fixed. Thus in general the transversality condition (6.2.36)(d) holds. Further, we have also proved the existence of multipliers $\lambda_0 \geqslant 0$ and a $\psi \in AC(I, R^n)$ that satisfies the adjoint equation (6.2.36)(c). This verifies (b)–(d). It remains to verify (a).

Since (6.2.52) holds for all $v \in V$, we have, for $z = (0, \xi_1, 0, \xi_2) = (0, 0, 0, 0)$, and $c > 0$,

$$\int_{t_1}^{t_2} H(t, x^0(t), \psi^0(t), u(t))\, dt \geq \int_{t_1}^{t_2} H(t, x^0(t), \psi^0(t), u^0(t))\, dt \quad (6.2.55)$$

for all $u \in \mathcal{U}$. This is the integral form of the minimum principle. For the proof of the pointwise minimum principle, we follow the following arguments. Let $t' \in I$ be such that the integrands in (6.2.55) are uniquely defined and $N_\varepsilon(t')$ denote the intersection of I with the ε-neighbourhood of t', and suppose $N_\varepsilon(t') \to \{t'\}$ as $\varepsilon \to 0$. Define, for any $v \in U(t')$,

$$u(t) = \begin{cases} u^0(t) & \text{if } t \in I \backslash N_\varepsilon(t') \\ v & \text{if } t \in N_\varepsilon(t'). \end{cases}$$

Then, letting $\mu(N_\varepsilon(t'))$ denote the Lebesgue measure of the set $N_\varepsilon(t')$, it follows from (6.2.55) that

$$\frac{1}{\mu(N_\varepsilon(t'))} \int_{N_\varepsilon(t')} [H(t, x^0(t), \psi^0(t), v) - H(t, x^0(t), \psi^0(t), u^0(t))]\, dt \geq 0,$$

and hence letting $\varepsilon \to 0$, we obtain

$$H(t', x^0(t'), \psi^0(t'), v) \geq H(t', x^0(t'), \psi^0(t'), u^0(t')) \quad (6.2.56)$$

for all $v \in U(t')$. Since, for measurable functions, almost all points of I are Lebesgue density points and t' is arbitrary, the result follows. In other words

$$\inf_{v \in U(t)} H(t, x^0(t), \psi^0(t), v) = H(t, x^0(t), \psi^0(t), u^0(t)) \text{ a.e. on } I. \quad (6.2.57)$$

This completes the proof. ■

Remark 6.2.2 In case the control restraint set $U(t)$ is a constant compact set U, the convexity property for the orientor (or velocity) field $Q(t, x) \equiv f(t, x, U)$ is not necessary.

Alternative proof of Theorem 6.2.1

In this proof one uses the Lagrange multiplier rule. Indeed, introducing the multipliers $\lambda_0 \geq 0$, and $\psi \in AC$, one applies the first variation to the extended cost functional

$$J_e(t_1, t_2, x, u) \equiv \lambda_0 g(t_1, x(t_1), t_2, x(t_2)) + \int_{t_1}^{t_2} \langle \psi(t), \dot{x}(t) - f(t, x(t), u(t)) \rangle \, dt,$$

where $\langle \cdot, \cdot \rangle$ denotes the scalar product in R^n. Assume $\{t_1, t_2, x^0, u^0\}$ to be optimal with the end point $e(x^0) = (t_1, x^0(t_1), t_2, x^0(t_2)) \equiv P_0 \in B$, and let $T_B(P_0)$ denote the tangent plane to B passing through the point P_0. Introduce the boundary perturbation $(dt_1, \delta x_1, dt_2, \delta x_2) \in T_B(P_0)$ so that $(t_1 + dt_1, x_1^0 + \delta x_1, t_2 + dt_2, x_2^0 + \delta x_2) \in B$ near P_0, and the control perturbation $\delta u = \chi_\sigma(t)(w - u^0(t))$, $w \in U$, where χ_σ is the indicator function of any measurable set $\sigma \subset I = [t_1, t_2]$, giving $u = u^0 + \delta u$, and let $x = x^0 + \delta x$ denote the corresponding trajectory. Considering the inequality

$$J_e(x^0, u^0) \leqslant J_e(x, u) \tag{6.2.58}$$

and carrying out some lengthy but straightforward computations, one arrives at

$$\begin{aligned} 0 \leqslant & (\lambda_0 g_{t_1} - M(t_1)) \, dt_1 \\ & + (\lambda_0 g_{t_2} + M(t_2)) \, dt_2 + (\lambda_0 g_{x_1} + \psi(t_1)) \cdot \delta x(t_1) \\ & + (\lambda_0 g_{x_2} - \psi(t_2)) \cdot \delta x(t_2) \\ & + \int_{t_1}^{t_2} \langle f_x'(t, x^0, u^0)\psi^0 + \dot{\psi}^0, \delta x \rangle \, dt + o(dt_i, \delta x(t_i)). \end{aligned}$$

This leads to the transversality condition (6.2.36)(d) and the adjoint equation (6.2.36)(c). This transversality condition gives the necessary end conditions for the optimal pair (x^0, u^0). Then considering (t_1, x_1, t_2, x_2) fixed and using (6.2.58), we obtain

$$\begin{aligned} & \int_{t_1}^{t_2} H(t, x^0(t), \psi^0(t), u^0(t)) \, dt \\ & \qquad \leqslant \int_{t_1}^{t_2} H(t, x^0(t), \psi^0(t), u^0(t) + \delta u(t)) \, dt + T_2 + \int_{t_1}^{t_2} o(\delta x) \, dt \end{aligned}$$

where

$$T_2 = \int_{t_1}^{t_2} \langle (f_x'(t, x^0(t), u^0(t) + \delta u(t)) - f_x'(t, x^0(t), u^0(t)))\psi^0(t), \delta x \rangle \, dt.$$

Two important facts are used in the proof:

(a) Continuous dependence of solutions on controls given in the form

$$\begin{aligned} \|x - x^0\| = \|\delta x\| \leqslant \beta\rho(u, u^0) = \beta\mu\{t: u(t) \neq u^0(t)\} \\ = \beta\mu(\sigma), \end{aligned}$$

where the set of admissible controls $\mathcal{U}$ is considered to be a complete metric space with respect to the metric ρ, and μ is the Lebesgue measure (see chapter 3). This allows one to estimate T_2 by

$$|T_2| \leq c\mu(\sigma) \int_\sigma \|f'_x(t, x^0, u) - f'_x(t, x^0, u^0)\| \, dt$$

where c depends only on β and $\|\psi^0\|$. Hence for integrable $f'_x(t, x^0(t), u(t))$, $u \in \mathcal{U}$, we have

$$\lim_{\mu(\sigma)\to 0} \left\{ \frac{T_2}{\mu(\sigma)} \right\} = 0.$$

(b) The function x^0 is extended beyond the interval $I = [t_1, t_2]$ by $x^0(t) = x^0(t_i)$ $(i = 1, 2)$, for t sufficiently near t_i outside I. ■

It may be instructive for the reader to carry out the above formalism in detail.

Remark 6.2.3 The only objection to the alternative proof lies in the lack of concern for the existence of the multipliers.

Some special cases

We discuss here some special but important cases of the general Theorem 6.2.1.

Corollary 6.2.4 (Terminal control problem). *Suppose the terminal times t_1, t_2 and the initial state x_1 are fixed leaving only x_2 free. Then an optimal pair $(x, u) \in G$ must satisfy the necessary conditions* (a)–(c) *with* $\psi(t_2) = g_{x_2}$.

Proof Since $\{t_1, t_2, x_1\}$ are fixed, the transversality condition (6.2.36)(d) reduces to $(\lambda_0 g_{x_2} - \psi(t_2), \xi_2) = 0$. As the terminal state x_2 is free, the equation must hold for all $\xi_2 \in R^n$. Hence $\psi(t_2) = \lambda_0 g_{x_2}$, and since $\lambda_0 > 0$ we can take $\lambda_0 = 1$. ■

Corollary 6.2.5 (Time optimal control). *Suppose the initial time t_1 and the initial and final* (*target*) *state $\{x_1, x_2\}$ are fixed and $g = (t_2 - t_1)$. Then a time optimal control and hence the corresponding optimal pair $(x, u) \in G$ must satisfy the necessary conditions* (a)–(c) *and the terminal condition*

(d)′: $$M(t_2) + 1 = (\psi(t_2), \dot{x}(t_2)) + 1$$
$$= (\psi(t_2), f(t_2, x_2, u(t_2))) + 1 = 0. \quad (6.2.59)$$

Proof Since $\{t_1, x_1, x_2\}$ are fixed, it follows from the transversality condition (6.2.36)(d) that all the components except the second disappear, giving $M(t_2) = -\lambda_0$. Since $\lambda_0 > 0$, we can take $\lambda_0 = 1$. Hence the result follows. ■

Consider the system

$$\dot{z} = f(t, z, u), \qquad z(t) \in R^s, \quad t \in [t_1, t_2] = I,$$

with the cost functional,

$$J(u) = \int_{t_1}^{t_2} \ell(t, z(t), u(t))\, dt. \tag{6.2.60}$$

The problem is to find a control $u \in \mathcal{U}$ that transfers the system starting at time t_1 from a smooth manifold $M_1 (\subset R^s)$ of dimension $\leq s$ to a smooth manifold $M_2 (\subset R^s)$ of dimension $\leq s$ at time t_2 (free) while minimizing the cost functional (6.2.60). Define the Hamiltonian

$$H(t, z, \psi, u) \equiv \langle \psi, f(t, z, u) \rangle + \ell(t, z, u). \tag{6.2.61}$$

Corollary 6.2.6 (Pontryagin's minimum principle). *In order that the pair (z, u) be optimal, it is necessary that there exists a $\psi \in AC(I, R^s)$ such that*

(a) $H(t, z(t), \psi(t), u(t)) \leq H(t, z(t), \psi(t), v)$ *for all* $v \in U$ *a.e. on* I,
(b) $\dot{z} = H_\psi, \ \dot{\psi} = -H_z$ (6.2.62)
(c) $\psi(t_1) \perp \mathcal{T}_{M_1}(z(t_1)), \ \psi(t_2) \perp \mathcal{T}_{M_2}(z(t_2))$
where $x \perp y$ *means* $(x, y) = 0$, $z_i \in M_i$, *and* $\mathcal{T}_{M_i}(z_i)$ *is the tangent plane to* M_i *passing through the point* z_i *of* M_i $(i = 1, 2)$.
(d) *If* $t_2 (\geq t_1)$ *is free,* $M(t_2) = H(t_2, z(t_2), \psi(t_2), u(t_2)) = 0$.

Proof The proof is based on specializing the general problem of Theorem 6.2.1 to that of the present one. Define

$$z_{s+1}(t) = \int_{t_1}^{t} \ell(\theta, z(\theta), u(\theta))\, d\theta \tag{6.2.63}$$

giving the system in R^{s+1},

$$\left.\begin{aligned} \dot{z}(t) &= f(t, z(t), u(t)) \\ \dot{z}_{s+1}(t) &= \ell(t, z(t), u(t)) \end{aligned}\right\} \cong \dot{\tilde{z}} = \tilde{f}(t, \tilde{z}, u). \tag{6.2.64}$$

Assume for the moment that t_2 is also fixed, and identify

$$\begin{aligned} &x \equiv \tilde{z} \\ &g(t_1, x_1, t_2, x_2) \equiv g(x_2) \equiv z_{s+1} \\ &\tilde{H} = H + \ell \cdot \tilde{\psi}_{s+1} \\ &\tilde{\psi} = \{\psi, \tilde{\psi}_{s+1}\}. \end{aligned} \tag{6.2.65}$$

Since both t_1 and t_2 are fixed, $z_{s+1}(t_1)=0$, and $z_{s+1}(t_2)$ is arbitrary, we take $B \equiv \{0\} \times \mathcal{M}_1 \times \{0\} \times \mathcal{M}_2$, where $\mathcal{M}_1 = M_1 \times \{0\}$ and $\mathcal{M}_2 = M_2 \times R$. Note that $\tilde{f}$ is independent of z_{s+1}. Hence it follows from (c) of Theorem 6.2.1 that $d\tilde{\psi}_{s+1}(t)/dt=0$, and consequently $\tilde{\psi}_{s+1}(t)$ is constant and we can take it equal to one. Therefore, (a) and (b) of (6.2.62) follows from (a)–(c) of Theorem 6.2.1.

For the transversality condition, since t_1 and t_2 are fixed, the first two terms of (d) (Theorem 6.2.1) disappear, leaving

$$(\lambda_0 g_{x_1} + \tilde{\psi}(t_1),\ \xi_1) + (\lambda_0 g_{x_2} - \tilde{\psi}(t_2),\ \xi_2) = 0 \tag{6.2.66}$$

for all $(\xi_1, \xi_2) \in T_B(t_1, x_1, t_2, x_2)$. Since the manifolds M_1 and M_2 are independent, we have

$$\begin{aligned} &(\lambda_0 g_{x_1} + \tilde{\psi}(t_1),\ \xi_1) = 0 \qquad \text{for all } \xi_1 \in \mathcal{T}_{\mathcal{M}_1}(x_1) \\ &(\lambda_0 g_{x_2} - \tilde{\psi}(t_2),\ \xi_2) = 0 \qquad \text{for all } \xi_2 \in \mathcal{T}_{\mathcal{M}_2}(x_2). \end{aligned} \tag{6.2.67}$$

By virtue of the facts that g is independent of x_1 and the last element of the vector ξ_1 is zero, it follows from the first equation of (6.2.67) that

$$(\psi(t_1),\ \xi_1) = 0 \qquad \text{for all } \xi_1 \in \mathcal{T}_{M_1}(z_1) \tag{6.2.68}$$

that is, $\psi(t_1) \perp \mathcal{T}_{M_1}(z_1)$. The $(s+1)$ vector g_{x_2} has all its entries zero except the last one which equals 1. The last entry of the vector $\tilde{\psi}(t_2)$ is $\tilde{\psi}_{s+1}(t_2)$ which we know is constant. Taking $\tilde{\psi}_{s+1}(t_2) = \lambda_0 = 1$, we have $\psi(t_2) \perp \mathcal{T}_{M_2}(z_2)$. This proves (a)–(c). Now freeing t_2 and recalling that g is independent of t_2, we have (from (6.2.36)(d)) $M(t_2)=0$. This completes the proof. ■

Remark 6.2.7 Using the minimum principle (6.2.62)(a), one can find u as a function of $\{t, z, \psi\}$. Substituting the u so obtained in (6.2.62)(b), one must solve the two-point boundary-value problem subject to the boundary conditions (6.2.62)(c). If the manifolds M_1 and M_2 are given respectively by k_1 and k_2 algebraic equations, the condition (6.2.62)(c) provides $(s-k_1)$ and $(s-k_2)$ algebraic equations relating $z(t_i)$ and $\psi(t_i)$, $i=1,2$. Thus we have $2s$ equations (b) with $(2s)$ boundary conditions which may be solved to determine the optimal (in fact extremal) control. Computational aspects will be discussed further in section 6.4.

Theorem 6.2.8 (Constancy of the Hamiltonian). *If the system is time invariant, then along the optimal trajectory $x(t)$, $t \in I$,*

$$M(x(t),\ \psi(t)) = \textit{constant}, \qquad t \in I, \tag{6.2.69}$$

and if it is time varying, then $M(t, x(t), \psi(t))$ is absolutely continuous and *is given by*

$$M(t, x(t), \psi(t)) = \tilde{c} - \int_t^{t_2} \langle \psi(\theta), f_\theta(\theta, x(\theta), u(\theta)) \rangle \, d\theta \tag{6.2.70}$$

where $\tilde{c}$ is a constant, f_t is the partial of f with respect to t.

Proof Consider the time-invariant case. Since f is C^1 in x and continuous in u, H_ψ and H_x are continuous in (x, ψ, u). Hence for a compact set U, H is locally Lipschitz in both ψ and x, that is,

$$|H(x, \psi, u) - H(x', \psi', u)| \leq K_r\{|x - x'| + |\psi - \psi'|\}$$

for all $x, \psi, x', \psi' \in B_r = \{\xi \in R^n: |\xi| \leq r\}$ uniformly with respect to $u \in U$ for some constant $0 \leq K_r < \infty$. Then $M(x, \psi) \equiv \inf\{H(x, \psi, v), v \in U\}$ is also locally Lipschitz. Indeed suppose $x, \psi, x', \psi' \in B_r$, and let v and $v' \in U$ such that $M(x, \psi) = H(x, \psi, v)$ and $M(x', \psi') = H(x', \psi', v')$, then

$$\begin{aligned} M(x, \psi) = H(x, \psi, v) &= (H(x, \psi, v) - H(x', \psi', v)) + H(x', \psi', v) \\ &\geq -K_r\varepsilon + H(x', \psi', v) \\ &\geq -K_r\varepsilon + M(x', \psi') \end{aligned}$$

for all $x, \psi, x', \psi' \in B_r$ satisfying $|x - x'| + |\psi - \psi'| < \varepsilon$. Hence

$$M(x, \psi) - M(x', \psi') \geq -K_r\varepsilon. \tag{6.2.71}$$

Similarly by reversing the roles of $\{x, \psi\}$ with those of $\{x', \psi'\}$, we have

$$M(x', \psi') - M(x, \psi) \geq -K_r\varepsilon. \tag{6.2.72}$$

Hence

$$|M(x, \psi) - M(x', \psi')| \leq K_r\{|x - x'| + |\psi - \psi'|\}, \tag{6.2.73}$$

proving local Lipschitz continuity. Since the solution trajectories $x^0 \equiv \{x^0(t), t \in I\}$, $\psi^0 \equiv \{\psi^0(t), t \in I\}$ are in $AC(I, R^n)$ and M is locally Lipschitz, it is clear that $t \to M(x^0(t), \psi^0(t)) \equiv M(t)$ is absolutely continuous and hence differentiable a.e. on I. We show that its derivative vanishes a.e. Let $t^* \in (t_1, t_2)$ where M is differentiable and $t_2 > t > t^*$, and denote $u^0(t^*)$ by u^*; then

$$\begin{aligned} &\frac{M(x^0(t), \psi^0(t)) - M(x^0(t^*), \psi^0(t^*))}{t - t^*} \\ &\qquad = \frac{H(x^0(t), \psi^0(t), u^0(t)) - H(x^0(t^*), \psi^0(t^*), u^*)}{t - t^*} \\ &\qquad \leq \frac{H(x^0(t), \psi^0(t), u^*) - H(x^0(t^*), \psi^0(t^*), u^*)}{t - t^*}. \end{aligned}$$

Hence letting $t \downarrow t^*$, we have

$$\dot{M}(t^*) \leq \left\{ \frac{\mathrm{d}}{\mathrm{d}t} H(x^0(t), \psi^0(t), u^*) \right\} \bigg|_{t=t^*} = \{(H_x, \dot{x}^0) + (H_\psi, \dot{\psi}^0)\} \bigg|_{t=t^*} = 0.$$

Similarly, by taking $t_1 < t < t^*$, one can verify that $\dot{M}(t^*) \geq 0$. This shows that $\dot{M}(t^*) = 0$. Since $M \in AC$, it is differentiable a.e. and hence $\dot{M}(t) = 0$ a.e. on I proving constancy of the Hamiltonian along the optimal trajectory.

For the time-varying case, define

$$\dot{x}_0(t) = 1, \qquad x_0(t_1) = t_1. \tag{6.2.74}$$

Appending this to the original system equations (6.2.32), one obtains the system

$$\dot{\tilde{x}} = \tilde{f}(\tilde{x}, u), \qquad \tilde{x}(t_1) = \begin{bmatrix} t_1 \\ x_1 \end{bmatrix} = \begin{bmatrix} x_0(t_1) \\ x(t_1) \end{bmatrix}$$

where

$$\tilde{f} = \begin{bmatrix} 1 \\ f \end{bmatrix}.$$

This is an autonomous system. Introducing the Hamiltonian, $\tilde{H} = \psi_0 + H$, we have the canonical equations,

$$\left.\begin{aligned} \frac{\mathrm{d}x_i}{\mathrm{d}t} &= \tilde{H}_{\psi_i} \\ \frac{\mathrm{d}\psi_i}{\mathrm{d}t} &= -\tilde{H}_{x_i} \end{aligned}\right. \qquad i = 0, 1, 2, \ldots n.$$

Along the optimal trajectory, $\tilde{M} \equiv \inf \tilde{H}$ must be constant. Hence

$$\psi_0^0(t) + \langle \psi^0(t), f(t, x^0(t), u^0(t)) \rangle = \text{constant}.$$

On the other hand

$$\frac{\mathrm{d}}{\mathrm{d}t} \psi_0^0 = -\tilde{H}_{x_0} = -\langle \psi^0, f_t \rangle = -H_t,$$

and

$$\psi_0^0(t) = \psi_0^0(t_2) + \int_t^{t_2} \langle \psi^0(\theta), f_\theta(\theta, x^0(\theta), u^0(\theta)) \rangle \,\mathrm{d}\theta.$$

Since the first term in $\tilde{H}$ does not contain u, it follows from the above that

$$M(t, x^0(t), \psi^0(t)) = \text{constant} - \int_t^{t_2} \langle \psi^0(\theta), f_\theta(\theta, x^0(\theta), u^0(\theta))\rangle \, d\theta \tag{6.2.75}$$

for all $t \in I$. This completes the proof. ■

To handle nonsmooth data Clarke [6] introduced the notion of generalized gradient given by

$$\partial_x f(x) = \left\{ \xi \in R^n : (\xi, y) \leq \limsup_{\substack{\varepsilon \downarrow 0 \\ \eta \to x}} \left(\frac{f(\eta + \varepsilon y) - f(\eta)}{\varepsilon} \right) \text{ for all } y \in R^n \right\},$$

where f is any Lipschitz function from R^n to R. Using this notion he generalized Pontryagin's necessary conditions of optimality. For example, using the minimum principle (a) of Corollary (6.2.3) and defining

$$M(t, x, \psi) \equiv \inf\{H(t, x, \psi, u), u \in U(t)\},$$

one can replace (6.2.62)(b) by the equation

$$\dot{x} = M_\psi, \qquad -\dot{\psi} = M_x. \tag{6.2.62)(b)$'$}$$

In case M is not differentiable in the classical sense, one uses the generalized gradient to replace (6.2.62)(b)′ by

$$\dot{x} \in \partial_\psi M, \qquad -\dot{\psi} \in \partial_x M, \tag{6.2.62)(b)$''$}$$

yielding a generalized minimum principle. In fact, this concept can be carried onto systems governed by differential inclusions

$$\begin{cases} \dot{x} \in F(t, x), & x(t_1) = x_1, \quad t \in [t_1, t_2] \\ \min \varphi(x(t_2)), & \end{cases}$$

giving the necessary conditions of optimality

$$\begin{pmatrix} \dot{x} \\ -\dot{\psi} \end{pmatrix} \in \partial M(t, x, \psi), \qquad x(t_1) = x_1, \quad \psi(t_2) = \varphi_x(x(t_2))$$

where

$$M(t, x, \psi) \equiv \inf\{(\psi, \xi) : \xi \in F(t, x)\}.$$

Example 6.2.9 (Attitude control). Under the assumption of small angular displacements, the attitude dynamics of a rigid-body satellite is given by three sets of decoupled second-order differential equations of the form

$$\ddot{\varphi} + a\varphi = u. \tag{6.2.76}$$

For this system we have state equation $\dot{\eta}_1 = \eta_2$, $\dot{\eta}_2 = -a\eta_1 + u$. We want to find a control u that steers the system from any initially perturbed state (η_{10}, η_{20}) to the desired final state $(0, 0)$ while minimizing the cost of fuel $J(u) = \int_0^{t_2} |u| \, d\theta$ subject to the thruster limitation $|u(t)| \leq 1$. Introducing a third variable $\eta_3(t) = \int_0^t |u| \, d\theta$, we have

$$\dot{\eta}_1 = \eta_2, \qquad \dot{\eta}_2 = -a\eta_1 + u, \qquad \dot{\eta}_3 = |u|, \qquad g(t_1, x_1, t_2, x_2) = \eta_3(t_2).$$

Here the Hamiltonian is

$$H = \psi_1\eta_2 + \psi_2(-a\eta_1 + u) + \psi_3 |u|,$$

and the adjoint equations are

$$\dot{\psi}_1 = -H_{\eta_1} = a\psi_2, \qquad \dot{\psi}_2 = -H_{\eta_2} = \psi_1, \qquad \dot{\psi}_3 = -H_{\eta_3} = 0,$$

with $\psi_3(t) = c$ (constant), and the optimal control has the form

$$u = -\,\mathrm{dz}\left(\frac{\psi_2}{c}\right),$$

where

$$\mathrm{dz}(\xi) = \begin{cases} 1 & \xi \geq 1, \\ 0 & |\xi| < 1, \\ -1 & \xi \leq -1, \end{cases}$$

and

$$M(\eta, \psi) = \psi_1\eta_2 - a\eta_1\psi_2 + c\left\{\left|\mathrm{dz}\left(\frac{\psi_2}{c}\right)\right| - \left(\frac{\psi_2}{c}\right)\mathrm{dz}\left(\frac{\psi_2}{c}\right)\right\}.$$

Since t_2 is free and g is independent of t_2, it follows from the transversality condition that (a) $M(t_2) = 0$ and (b) $\lambda_0 = c \geq 0$. Hence

$$0 = M(t_2) = M(\eta(t_2), \psi(t_2)) = c\left\{\left|\mathrm{dz}\left(\frac{\psi_2(t_2)}{c}\right)\right| - \left(\frac{\psi_2(t_2)}{c}\right)\mathrm{dz}\left(\frac{\psi_2(t_2)}{c}\right)\right\},$$

and, therefore, $\psi_2(t_2) = c$, or $-c$, or 0. The value $\psi_2(t_2) = 0$ is inadmissible, since $u = 0$ produces elliptic trajectories around the origin in R^2, and hence the origin cannot be reached. In other words $\psi_2(t_2) = c$, or $-c$, indicating that during the last part of the trajectory $u = -1$ or $+1$.

Example 6.2.10 (Firm economy). The firm economy can be approximately described by the equation

$$\dot{K} = I - \mu K, \tag{6.2.77}$$

where K denotes capital goods or machines used in the production of certain homogeneous consumable goods, I is the investment (rate) in purchase of capital goods, μ is the capital depreciation rate. The production rate q is given by

$$q = F(K, L),$$

where F is the production function (discussed in chapter 1) which depends on capital and labour employed. Letting p denote the unit price $p = p(q)$, the gross revenue earned is

$$R_g = pq. \tag{6.2.78}$$

In earning this revenue there are expenditures, such as (a) $WL \equiv$ wages paid to the labourers, (b) $\delta I \equiv$ cost of borrowing investment capital from banks, and (c) $C(I) \equiv$ adjustment cost (installation of machines, training of manpower etc). Hence the gross expenditure is

$$E_g = WL + \delta I + C(I). \tag{6.2.79}$$

Hence the total discounted profit is

$$J(L, I) \equiv \int_0^{\infty} \mathrm{e}^{-rt}\{pq - WL - \delta I - C(I)\}\, \mathrm{d}t. \tag{6.2.80}$$

The firm wishes to maximize its profit by choosing an appropriate policy of hiring (firing) and investment (disinvestment) subject to the constraint (6.2.77).

Here the Hamiltonian is

$$H = (I - \mu K)\psi - \mathrm{e}^{-rt}(pq - WL - \delta I - C(I)). \tag{6.2.81}$$

Since both hiring and firing, investment and disinvestment are permissible, one can minimize H by setting

$$\frac{\partial H}{\partial L} = 0, \qquad \frac{\partial H}{\partial I} = 0. \tag{6.2.82}$$

This gives

$$\psi = \mathrm{e}^{-rt}(\delta + C'(I)), \tag{6.2.83)(a}$$

$$W = \left(p + q\frac{\partial p}{\partial q}\right)\frac{\partial q}{\partial L} \equiv M_r \frac{\partial F}{\partial L}, \tag{6.2.83)(b}$$

where $M_r \equiv (p + q(\partial p/\partial q))$ is called the marginal revenue. The adjoint equation is given by

$$\frac{\mathrm{d}}{\mathrm{d}t}\psi = -H_K = \mu\psi + \mathrm{e}^{-rt}M_r\frac{\partial F}{\partial K}. \tag{6.2.84}$$

In economics one is interested in the equilibrium, if one exists. With this end in view, differentiating (6.2.83)(a) and equating with (6.2.84), one obtains the following set of equations:

$$\dot{K} = I - \mu K,$$

$$\dot{I} = \frac{1}{C''(I)}\left\{(r+\mu)(\delta + C'(I)) - W\frac{F_K}{F_L}\right\}, \tag{6.2.85}$$

$$W = M_r F_L.$$

In equilibrium the optimum level of capital stock and investment are given by the solution of the following set of algebraic equations:

$$I = \mu K, \qquad (6.2.86)(a)$$

$$C'(I) = -\delta + \left(\frac{W}{r+\mu}\right)\frac{F_K}{F_L}, \qquad (6.2.86)(b)$$

$$W = M_r F_L. \qquad (6.2.86)(c)$$

Note the properties of the functions $p(q)$ and $C(I)$. The price p is a non-increasing function of the production rate q giving $\partial p/\partial q \leq 0$. The adjustment cost $C(I)$ has the following properties: (a) $C(0) = 0$, (b) $C'(I) > 0$ for $I > 0$, and (c) $C''(I) > 0$ for $I > 0$. Using the Lagrange formula, one can express

$$C'(I) = C'(0) + \left(\int_0^1 C''(\theta I)\,\mathrm{d}\theta\right)I = C'(0) + \alpha(I)I,$$

where, by the property (c), $\alpha(I) > 0$. Using this in (6.2.86)(b), one has

$$I = \frac{1}{\alpha(I)}\left\{-(\delta + C'(0)) + \left(\frac{W}{r+\mu}\right)\frac{F_K}{F_L}\right\}. \tag{6.2.87}$$

If the price p is insensitive to production rate q (true in the case of a small firm), or if it is of the form

$$p = p_0\left(1 + \frac{\beta}{q}\right), \qquad p_0, \beta > 0, \tag{6.2.88}$$

then M_r is constant, say p_0. If the adjustment cost C is quadratic or nearly quadratic, then $\alpha(I) \cong \alpha_0$, a constant. Let (I^*, K^*, L^*) denote any non-negative solution of the equation (6.2.86). Then, by virtue of the fact that $F_K(K, L^*)$ is a decreasing function of K, it follows from (6.2.87) that I is a decreasing function of K and the two equations (6.2.86)(a) and (6.2.87) intersect at (I^*, K^*). The flow of the (I, K) pair, for fixed L^*, is illustrated in a phase plane diagram (see figure 6.2). It is clear that the larger $r + \mu$, the smaller is the equilibrium level (I^*, K^*). For details see the reference [135].

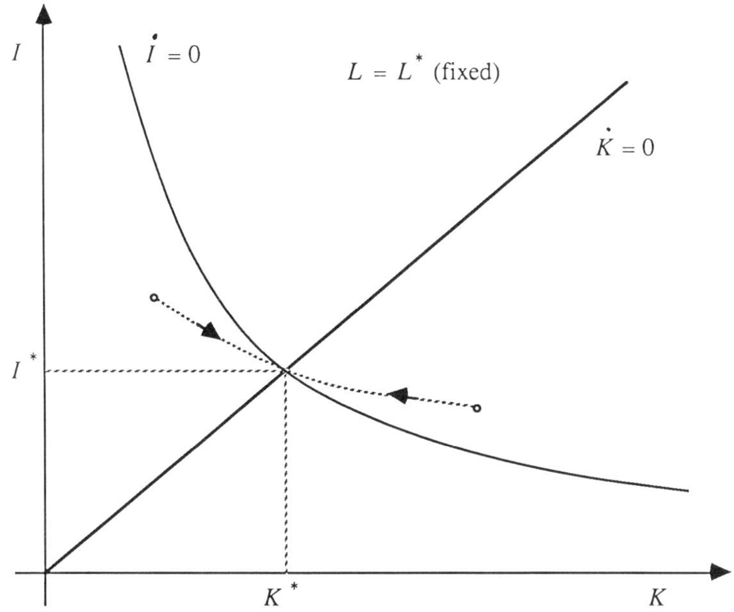

Figure 6.2

Example 6.2.11 (Pest control). Consider the pest population dynamics (1.2.73). Suppose the loss due to crop damage and the cost of administering pesticides are linearly proportional to the population level N and the application rate u respectively. Then, for α_i, β_i positive, the functional

$$J(u)=\int_{t_1}^{t_2}\left\{\sum_{i=1}^{n}\alpha_i N_i+\sum_{k=1}^{m}\beta_k u_k\right\}\mathrm{d}t, \tag{6.2.89}$$

representing the aggregated loss over the season $[t_1, t_2]$ arising from crop damage and cost of pesticides, must be minimized to maximize yield. Writing the Hamiltonian,

$$H(N,\psi,u)=\sum_{i=1}^{n}\left(a_i+\sum_{j=1}^{n}b_{ij}N_j+\sum_{k=1}^{m}c_{ik}u_k\right)N_i\psi_i+\sum_{i=1}^{n}\alpha_i N_i+\sum_{k=1}^{m}\beta_k u_k, \tag{6.2.90}$$

and assuming that the pesticide application rate u is limited (due to other reasons), $0\leqslant u_k(t)\leqslant r_k$ (r_k constant), one can easily deduce that the optimal control has the form

$$u_k=\left(\frac{r_k}{2}\right)\left\{1-\operatorname{sign}\left(\sum_{i=1}^{n}c_{ik}N_i\psi_i+\beta_k\right)\right\}. \tag{6.2.91}$$

The adjoint state ψ is given by the solution of the following system of equations

$$\frac{\mathrm{d}\psi_i}{\mathrm{d}t} = -g_i(t, N, u)\psi_i - \alpha_i, \qquad i = 1, 2, \ldots, n, \tag{6.2.92}$$

where

$$g_i = a_i + \sum_{j \neq i} b_{ij} N_j + 2b_{ii} N_i + \sum_{k=1}^{m} c_{ik} u_k. \tag{6.2.93}$$

By virtue of the transversality condition, $\psi(t_2) = 0$, and, hence, one can easily verify that

$$\psi_i(t) = \alpha_i \int_t^{t_2} \mathrm{d}\theta \exp\left(\int_t^{\theta} g_i \, \mathrm{d}\tau\right), \qquad i = 1, 2, \ldots, n. \tag{6.2.94}$$

Even without solving these equations one can make some interesting observations on the basis of expressions (6.2.91) and (6.2.94).

(1) If the kth pesticide has no effect on any of the pest population N, then $c_{ik} = 0$ $(i = 1, 2, \ldots n)$, and β_k being positive, we have $u_k = 0$. This is to be expected.
(2) If $N(t) \equiv 0$ on any subinterval $I \subset [t_1, t_2]$, then $u(t) \equiv 0$ on I. Again this is natural.
(3) Since by (6.2.94) $\psi_i(t) > 0$, for large crop damage factor α_i, $\psi_i(t)$ takes large values and hence $\sum_{i=1}^{n} c_{ik} N_i \psi_i$ has large negative values possibly leading to

$$\sum_{i=1}^{n} c_{ik} N_i \psi_i + \beta_k < 0. \tag{6.2.95}$$

This dictates, through (6.2.91), maximum application rate of the kth chemical, i.e. $u_k(t) = r_k$. For details see reference [83].

Control theory has found applications in many different fields of engineering, economics, and operations research. Currently, techniques of optimal control theory are also being applied in the control of robotic manipulators [147]–[149].

6.2.3 Existence of optimal controls

Throughout the previous section, we tacitly assumed the existence of optimal controls. In the absence of existence, the necessary conditions are meaningless since they attempt to characterize entities that may not exist in the first place. The question of existence is, therefore, very important and fortunately has been extensively studied. In this brief monograph we can only include a cursory introduction to the subject.

First note that a Lagrange problem, with the cost integrand $\ell > 0$, can be converted into a time-optimal control problem. Define

$$\tau(t) = \int_{t_1}^{t} \ell(\theta, x(\theta), u(\theta))\, d\theta,$$

with

$$\dot{x}(t) = \tilde{f}(t, x(t), u(t)), \qquad t \in [t_1, t_2],$$
$$x(t_1) = x_1.$$

Defining $f = (\tilde{f}/\ell)$, we arrive at the following equation

$$S: \quad \begin{cases} \dfrac{dx}{d\tau} = f(\tau, x(\tau), u(\tau)), & \tau \geqslant 0, \\ x(0) = x_1. \end{cases}$$

Thus if we prove the existence of any time-optimal control for this problem, we have proved the existence of an optimal control for the Lagrange problem. In general, define $I = [0, T]$, $T < \infty$, and let $t \to U(t)$ be a set-valued map from I to $K(R^n)$($\equiv$ the space of nonempty compact subsets of R^n). For each $t \in I$, $x \in R^n$, define

$$R(t, x) = \{\xi \in R^n: \xi = f(t, x, v), v \in U(t)\}$$
$$= f(t, x, U(t)). \tag{6.2.96}$$

The set-valued map $R(t, x)$ determines the set of possible directions of motion admitted by the system starting from the state x at time t. It is known as the velocity field or orientor field, and the control system may be described by the differential inclusion

$$S_0: \quad \begin{cases} \dot{x}(t) \in R(t, x(t)) & \text{a.e. on } I, \\ x(0) = x_1, \end{cases} \tag{6.2.97}$$

called the contingent equation.

Under the assumption of convexity, we shall be able to prove that the two systems, S and S_0, are equivalent in the sense that every solution of S is a solution of S_0, and conversely. We shall express this concept symbolically by writing $S \cong S_0$. For this we need a selection theorem or implicit function theorem.

Let $t \to \Gamma(t)$ denote any (non-empty) set-valued function from I to $K(R^s)$ (s = finite positive integer). The set-valued map Γ, with values $\Gamma(t) \in K(R^s)$, is said to be measurable if for every open set $E \subset R^s$, the set $\{t \in I: \Gamma(t) \cap E \neq \varnothing\}$ is a measurable subset of I. Given a measurable set-valued map Γ on I, one of the fundamental questions is, does there exist a measurable function $\xi: I \to R^s$ so that $\xi(t) \in \Gamma(t)$ a.e. on I. The set

$$S(\Gamma) \equiv \{\xi \text{ measurable functions from } I \text{ to } R^s: \xi(t) \in \Gamma(t) \text{ a.e.}\}$$

is called the set of measurable selections of Γ. In general, there is no guarantee that $S(\Gamma)$ is non-empty. Results concerning this are called selection theorems or implicit function theorems. One of the classical results in this direction is popularly known as the Filippov implicit function lemma.

Lemma 6.2.12 (Filippov implicit function lemma). *Let $(I, \mathscr{B}(I), \mu)$ denote any μ-measurable space, and $t \to G(t)$ a measurable set-valued map from I to K (R^s) (possibly continuous in the Hausdorff metric ρ_H) and the functions f: $R^s \to R^r$ is continuous and $h: I \to R^r$ is measurable such that $h(t) \in f(G(t))$ a.e. Then there exists a measurable function g: $I \to R^s$ such that $g(t) \in G(t)$ a.e. and $h(t) = f(g(t))$ a.e.*

Equivalently $S(D) \neq \varnothing$ (nonempty) where, for $t \in I$,

$$D(t) \equiv \{\xi \in G(t): h(t) = f(\xi)\}.$$

The content of the above lemma is illustrated by figure 6.3.

For many more general results in this direction the reader is referred to Cesari [4], Ahmed and Teo [53] and Young [35]. For our purpose the above result is sufficient.

Theorem 6.2.13 (Equivalence). *If the orientor field $R(t, x)$ is convex for each $(t, x) \in I \times R^n$ and f is continuous in all the variables, then $S \cong S_0$.*

Proof Let $x = \{x(t), t \in I\}$ be a solution of S. Then, by definition, there exists $u \in \mathscr{U}$ such that $\dot{x}(t) = f(t, x(t), u(t))$ a.e. and $x(0) = x_1$. Since

$$f(t, \xi, u(t)) \in f(t, \xi, U(t)) = R(t, \xi) \qquad \text{for all } (t, \xi) \in I \times R^n,$$

we have $\dot{x}(t) \in R(t, x(t))$ a.e. Conversely, suppose x is a solution of the

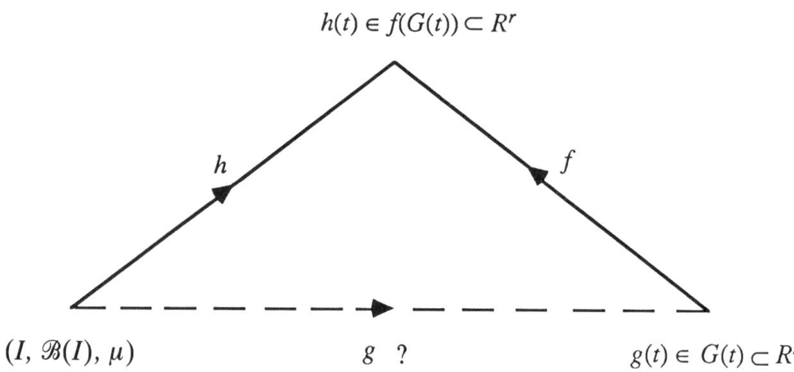

Figure 6.3

system S_0, that is, $x(0)=x_1$ and $\dot{x}(t)\in R(t, x(t))$ a.e. Define, for each $t\in I$, the set

$$G(t)\equiv\{\zeta\in I\times R^n\times R^m\colon \zeta=(t, x(t), v), v\in U(t)\}.$$

Since $U(t)$ is a compact set-valued map, $G(t)$ is also a compact set-valued map with values $G(t)\in K(I\times R^n\times R^m)$. Since $x\in AC$, $\dot{x}$ is Lebesgue measurable from I to R^n, further $G\colon I\to K(I\times R^n\times R^m)$ is measurable, and $\dot{x}(t)\in f(G(t))$ a.e. Hence from Lemma 6.2.12, it follows that there exists a measurable $g\colon I\to I\times R^n\times R^m$ such that $g(t)\in G(t)$ a.e. and $\dot{x}(t)=f(g(t))$ a.e. Consequently there exists a $u\in\mathcal{U}$ such that $g(t)=(t, x(t), u(t))\in G(t)$ a.e. and $\dot{x}(t)=f(t, x(t), u(t))$ a.e. with $x(0)=x_1$. This proves that x is a solution of S whenever it is a solution of S_0. This ends the proof. ■

Remark 6.2.14 Continuity of f in t is not essential; it suffices if f is merely measurable in t and continuous in the other variables.

For the proof of existence of optimal controls, we must establish certain closure and, possibly, compactness properties of admissible trajectories:

$$X=\{x\in AC\colon \dot{x}(t)\in R(t, x(t)) \text{ a.e.}, x(0)=x_1\}, \tag{6.2.98}$$

and attainable sets

$$\mathcal{A}(t)=\{\xi\in R^n\colon \xi=x(t) \text{ for some } x\in X\}, \tag{6.2.99}$$

$t\in I=[0, T]$. Let ρ denote any of the equivalent metrics on R^n. For any subset $G\subset R^n$, we shall use $G^\varepsilon\equiv\{\xi\in R^n\colon \rho(\xi, G)\leqslant\varepsilon\}$ to denote the closed ε-neighbourhood of G. The concept of upper semicontinuity of set-valued maps plays an essential role in the proof of existence of optimal controls.

Definition 6.2.15 (Upper semicontinuity). The orientor field $R(t, x)$, $(t, x)\in I\times R^n$, is said to be upper semicontinuous at $(t, x)\in I\times R^n$ with respect to set inclusion, if, for every $\varepsilon>0$, there exists a $\delta>0$, such that, whenever $(t', x')\in N_\delta(t, x)\equiv\delta$-neighbourhood of (t, x), $R(t', x')\subset R^\varepsilon(t, x)$. The field $R(\cdot, \cdot)$ is said to be upper semicontinuous with respect to set inclusion if it is so on its entire domain of definition.

Theorem 6.2.16 (Closure of X). *Suppose $t\to U(t)$ is a measurable set-valued map from I to $K(R^m)$, there exists a non-negative $\eta\in L_1^{\text{loc}}$ such that f satisfies the growth condition*

$$|f(t, x, u)|\leqslant\eta(t)[1+|x|] \tag{6.2.100}$$

for all $u \in U(t)$, $t \in I$, *and that, for each* $(t, x) \in I \times R^n$, *the velocity field* $R(t, x)$ *is a closed convex subset of* R^n *and upper semicontinuous with respect to set inclusion. Then* X *is a compact subset of* $C(I, R^n)$.

Proof Under the growth condition (6.2.100), the set X is a bounded subset of $C(I, R^n)$. We show that it is equicontinuous. Let x_u denote the solution of $\dot{x} = f(t, x, u)$ for $u \in \mathcal{U} \equiv \{u \text{ measurable}, u(t) \in U(t) \text{ a.e.}\}$ with $x_u(0) = x_1$. Then it follows from (6.2.100) that

$$|x_u(t+h) - x_u(t)| \leq \int_t^{t+h} \eta(\theta)[1 + |x_u(\theta)|]\, d\theta$$

for $h > 0$. Since $\sup\{\|x\|, x \in X\} \equiv b < \infty$, it follows from the above inequality that

$$|x_u(t+h) - x_u(t)| \leq (1+b) \int_t^{t+h} \eta(\theta)\, d\theta \tag{6.2.101}$$

independently of $u \in \mathcal{U}$. Hence, η being an element of L_1^{loc}, we have

$$\lim_{h \downarrow 0} \left\{ \sup_{u \in \mathcal{U}} \left[\sup_{t, t+h \in I} |x_u(t+h) - x_u(t)| \right] \right\} = 0.$$

The same property holds for $h < 0$ and $h \uparrow 0$. Hence by Proposition 2.1.37 we conclude that X is a conditionally compact subset of $C(I, R^n)$. Since a conditionally compact set is compact if it is closed, we must show that X is closed. Let $\{x_n\} \in X$ and suppose $x_n \to x$ in $C(I, R^n)$. We show that $x \in X$. Since $x_n \in X$, it follows from the definition of X that $\dot{x}_n(t) \in R(t, x_n(t))$ a.e. on I and $x_n(0) = x_1$ for all n. Further, due to (6.2.100), $x \in AC(I, R^n)$ and hence it is differentiable a.e. Let $t^* \in I$ at which $\dot{x}(t^*)$ is defined. Clearly for $n \geq 1$, and $t^* + h \in I$,

$$\frac{x_n(t^* + h) - x_n(t^*)}{h} = \int_0^1 \dot{x}_n(t^* + \theta h)\, d\theta. \tag{6.2.102}$$

Since $x_n(t) \to x(t)$ uniformly on I and $R(t, \xi)$ is upper semicontinuous with respect to inclusion, for every $\varepsilon > 0$, there exists $n_1 = n(\varepsilon)$ such that

$$R(t^*, x_n(t^*)) \subset R^{\varepsilon}(t^*, x(t^*))$$

for all $n \geq n_1$. Again due to upper semicontinuity of $R(t, \xi)$ and continuity of $\{x_n\}$, we have

$$R(t^* + \tau, x_n(t^* + \tau)) \subset R^{\varepsilon}(t^*, x_n(t^*))$$

for $|\tau| < \delta = \delta(\varepsilon)$. Hence, choosing h so that $|h| < \delta$, we have

$$\dot{x}_n(t^* + \theta h) \in R(t^* + \theta h, x_n(t^* + \theta h)) \subset R^{\varepsilon}(t^*, x_n(t^*)) \subset R^{2\varepsilon}(t^*, x(t^*)), \tag{6.2.103}$$

for all $0 \leq \theta \leq 1$ and $n \geq n_1$. By virtue of convexity of the orientor field, we may conclude from (6.2.103) that

$$\int_0^1 \dot{x}_n(t^* + \theta h)\, d\theta \in R^{2\varepsilon}(t^*, x(t^*)) \tag{6.2.104}$$

for all $n \geq n_1$ and $|h| < \delta$. Now letting $n \to \infty$, it follows from (6.2.102) and (6.2.104) and the fact that $x_n \to x$, that

$$\frac{x(t^* + h) - x(t^*)}{h} \in R^{2\varepsilon}(t^*, x(t^*))$$

for all $|h| < \delta$. Letting $h \to 0$, we have

$$\dot{x}(t^*) \in R^{2\varepsilon}(t^*, x(t^*))$$

for any $\varepsilon > 0$. Since $R(t^*, x(t^*))$ is a closed convex set, it follows from the above inclusion relation that $\dot{x}(t^*) \in R(t^*, x(t^*))$. The function x being absolutely continuous, $\dot{x}(t)$ is defined a.e. and hence the inclusion holds for almost all $t^* \in I$. Therefore the set X contains x and all such limit points implying its closure and hence compactness. This completes the proof. ■

Corollary 6.2.17 (Compactness of $\mathscr{A}(t)$). *For each $t \in I$, the attainable set $\mathscr{A}(t)$ is a compact subset of R^n.*

Proof Follows easily from Theorem 6.2.16. The reader is encouraged to carry out the details. ■

Now we consider some control problems.

Corollary 6.2.18 (Meyer's problem). *The problem*

$$\dot{x} = f(t, x, u), \qquad x(t_1) = x_1$$

$$J(u) = g(x(t_2)) = \textit{minimize}$$

with g continuous and bounded on bounded sets, has a solution.

Proof Since g is continuous, and, by the previous lemma $\mathscr{A}(t_2)$ is compact, g attains its minimum (also maximum) on $\mathscr{A}(t_2)$. Hence there exists $x^* \in X$ such that $g(x^*(t_2)) \leq g(x(t_2))$ for all $x \in X$ and consequently a control $u^* \in \mathscr{U}$ so that $J(u^*) \leq J(u)$ for all $u \in \mathscr{U}$. ■

Corollary 6.2.19 (Time optimal control). *If the system is controllable from $x_1 \in R^n$ to the moving target $\{z(t), t \in I\} \in C(I, R^n)$ at some time*

$t_2 \in I$, then, under the given assumptions, there exists a time-optimal control.

Proof Let $I_0 \equiv \{t \in [t_1, t_2]:\ z(t) \in \mathcal{A}(t)\}$ and $t^* = \inf I_0$. Since, by assumption, the system is controllable, I_0 is non-empty. Let $\{\tau_n\} \in I_0$ be a decreasing sequence such that $\tau_n \downarrow t^*$ where $t^* \in I = [t_1, t_2]$. Corresponding to $\{\tau_n\}$, there exists a sequence $\{x_n\} \in X$ such that $x_n(\tau_n) = z(\tau_n)$. Clearly

$$|x_n(t^*) - z(t^*)| \leq |x_n(t^*) - x_n(\tau_n)| + |z(\tau_n) - z(t^*)|.$$

Since $\{x_n\}$ is equicontinuous, z is continuous and $\tau_n \to t^*$, it follows from the above inequality that

$$\lim_{n\to\infty} x_n(t^*) = z(t^*).$$

But $x_n \in X$, and hence $x_n(t^*) \in \mathcal{A}(t^*)$ for all n, and therefore, by virtue of compactness of $\mathcal{A}(t^*)$, its limit $z(t^*) \in \mathcal{A}(t^*)$ and hence $t^* \in I_0$. This implies existence of a time optimal control. ■

Nonconvex problems

It follows from the above results that the convexity condition plays a crucial role in existence theory. In fact, we shall give examples for which there is no optimal control, and this happens precisely due to lack of convexity.

In such a situation one has two options:

(a) Enlarge the class of controls and prove the existence of an optimal control in the larger class and investigate if this can be approximated in some sense by one of the admissible controls.
(b) Introduce stronger topology on the control space (in other words impose additional constraints such as smoothness) so that nonconvexity does not matter.

In the first option, replace

$$\mathcal{U} \equiv \{u \text{ measurable, } u(t) \in U \text{ (compact) a.e.}\},$$

by

$$\mathcal{U}_r \equiv \{\mu \text{ measure-valued functions with support } U, \text{ that is}$$

$$\mu_t(U) = 1 \text{ for all } t \in I \text{ and } t \to \int_U g(u)\mu_t(\mathrm{d}u) \text{ is measurable for all } g \in C(U)\}.$$

In this case the system equation is driven by measure valued controls, like

$$\dot{x}(t) = \int_U f(t, x(t), u)\mu_t(\mathrm{d}u) \tag{6.2.105}$$

for some $\mu \in \mathcal{U}_r$. Under certain assumptions one can prove that this is equivalent to convexifying the orientor field $R(t, x)$ to

$$R_r(t, x) = \mathrm{Cl}\,\mathrm{Co}\, R(t, x). \tag{6.2.106}$$

Hence the system (6.2.105) is equivalent to the contingent equation

$$S_r: \quad \begin{cases} \dot{x}(t) \in R_r(t, x(t)) & \text{a.e.} \\ x(t_1) = x_1. \end{cases} \tag{6.2.107}$$

Let

$$X_r \equiv \{x \in AC(I, R^n): x(t_1) = x_1, \dot{x}(t) \in R_r(t, x(t)) \text{ a.e.}\}$$

denote the set of relaxed trajectories and, for each $t \in I$,

$$\mathcal{A}_r(t) \equiv \{\xi \in R^n\colon \xi = x(t) \text{ for some } x \in X_r\}$$

the corresponding attainable set.

Since the orientor field $R_r(t, x)$ of the relaxed system (6.2.107) is closed and convex, the results of Theorem 6.2.16 and Corollaries 6.2.17–6.2.19 hold with measure-valued controls taking the place of ordinary controls. These controls are called relaxed controls or chattering controls. The chattering controls were originally introduced by Gamkrelidze [7], and subsequently generalized and extensively studied by Warga [34], Young [35], Cesari [4], Ahmed and Teo [53], Teo and Wu [59], Ahmed [72], [73]. The usefulness of the generalized controls can be appreciated through the following results.

Theorem 6.2.20 ($\bar{X} = X_r$). *The set X is dense in X_r, that is, for every $\varepsilon > 0$ and $x \in X_r$, there exists an $x_\varepsilon \in X$ such that $\|x - x_\varepsilon\| < \varepsilon$.*

Corollary 6.2.21 ($\bar{\mathcal{A}}(t) = \mathcal{A}_r(t)$). *For each $t \in I$, $\mathcal{A}(t)$ is dense in $\mathcal{A}_r(t)$.*

These results allow us to determine the effectiveness of any suboptimal solution when there is no optimal solution of the original (unrelaxed) problem. In the second option there are several possibilities.

(a) Introduce constraints on the rate of change of controls. Consider the system

$$\begin{aligned} &\dot{x} = f(t, x, u) \\ &\dot{u}(t) \in \Gamma(t), \qquad t \in I \end{aligned} \tag{6.2.108}$$

where Γ is a measurable set-valued map with values $\Gamma(t)$ which are compact convex subsets of R^m. Then introducing a new variable y, we can write the system (6.2.108) as

$$\dot{z} = \tilde{f}(t, z, w) \qquad \text{in } R^{n+m},$$

$$w \in \mathcal{U}_c \equiv \{w \text{ measurable: } w(t) \in \Gamma(t) \text{ a.e.}\}$$

where

$$z = \begin{bmatrix} x \\ y \end{bmatrix}, \qquad \tilde{f} = \begin{bmatrix} f \\ w \end{bmatrix} = \begin{bmatrix} f(t, x, y) \\ w \end{bmatrix} \equiv \begin{bmatrix} f(t, z) \\ w \end{bmatrix}.$$

Define for $t \in I$ and $z \in R^{n+m}$, the orientor field

$$\tilde{R}(t, z) \equiv \{\xi \in R^{n+m}: \xi = \tilde{f}(t, z, w), w \in \Gamma(t)\}.$$

Clearly $\tilde{R}$ is a closed convex set-valued map, and we have the contingent equation

$$\dot{z}(t) \in \tilde{R}(t, z(t)) \quad \text{a.e. on } I,$$

$$z(t_1) = z_1 = \begin{bmatrix} x_1 \\ 0 \end{bmatrix}$$

for which the results given in this section remains valid.

(b) In certain infinite horizon problems one may use the space of band-limited (vector-valued) functions for controls and prove general existence theorems without convexity requirements (Ahmed [73]). Existence theory and necessary conditions of optimality for time-lag systems were given by Oğuztöreli [22], Ahmed [72], Georganas and Ahmed [120].

As stated earlier, in the absence of convexity there may not exist any optimal control. This is illustrated in the following examples due to Hermes and LaSalle [8].

Example 6.2.22 Consider the minimum time control problem of the system $\dot{\xi}_1 = (1 - (\xi_2)^2)u^2$, $\dot{\xi}_2 = u$ from state $0 \in R^2$ to $(1, 0) \in R^2$ during the interval $[0, 1]$ with control constraint $U = [-1, +1]$. Here the orientor field is not convex and we show that there is no optimal control. Consider the sequence of controls $\{u_n\}$ which alternately takes values $+1$ and -1 on $2n$ equal subintervals whose union is $I = [0, 1]$. The corresponding trajectories are

$$\xi_{1,n}(t) = \int_0^t (1 - (\xi_{2,n}(\theta))^2)(u^n(\theta))^2 \, d\theta = \int_0^t (1 - (\xi_{2,n}(\theta))^2) \, d\theta$$

$$\xi_{2,n}(t) = \int_0^t u_n(\theta) \, d\theta.$$

It is easy to verify that $\xi_{2,n}(t) \to 0$ uniformly on I and hence $(\xi_{1,n}(1), \xi_{2,n}(1)) \to (1, 0)$ as $n \to \infty$. But there is no control $u^* \in \mathcal{U} = \{u$ measurable, $u(t) \in U\}$ for which $(\xi_1^*(1), \xi_2^*(1)) = (1, 0)$, since, if there were any such u^*, then

$$\xi_2^*(t) = \int_0^t u^*(\theta)\,d\theta \not\equiv 0, \qquad \xi_2^*(1) = 0$$

and

$$\xi_1^*(t) = \int_0^t (1 - (\xi_2^*(\theta))^2)(u^*(\theta))^2\,d\theta \leq \int_0^t (1 - (\xi_2^*(\theta))^2)\,d\theta < 1$$

for all $t \in I$ and $\xi_1^*(1) < 1$ leading to a contradiction. Hence $(1, 0)$ is a limit point of the attainable set $\mathcal{A}(1)$ but $(1, 0) \notin \mathcal{A}(1)$.

However there is a relaxed control $\mu_t(du) = \frac{1}{2}(\delta_t(u-1) + \delta_t(u+1))$, $t \in [0, 1]$ for which $(1, 0) \in \mathcal{A}_r(1)$.

Example 6.2.23 The same conclusion holds for the time-optimal control problem: $\dot{\xi}_1 = 1 - (\xi_2)^2$, $\xi_1(0) = 0$, $\dot{\xi}_2 = u$, $\xi_2(0) = 0$, $t \in [0, 1]$, $U = \{-1, +1\}$ (2 points), desired target $(1, 0)$. Here U is not convex and hence the orientor field is nonconvex.

6.3 Bellman's principle of optimality and feedback controls

The necessary conditions of optimality given in the previous section provide methods for obtaining optimal open-loop controls. In many control problems such as economics, environment, management, etc. this is quite satisfactory. However, in many engineering problems open-loop controls, requiring external agencies or human operators, are considered unrealistic. In such problems it is desirable and sometimes indispensable to use feedback control laws, that is, a controller that monitors the system state and accordingly exercises control actions to achieve certain measures of performance. This is the topic of this section.

6.3.1 Bellman's equations for feedback controls

The most suitable technique for designing optimal feedback control laws is the principle of dynamic programming (principle of optimality) due to Bellman. Using this principle one obtains a first-order (hyperbolic) partial differential equation whose solution provides the optimal cost or the 'value' function and hence the feedback control law. In this section we shall derive the Bellman equation (also called the Hamilton–Jacobi–Bellman equation) and a simple verification theorem. We consider the questions of existence of solutions of the HJB equations in the section on nonlinear regulators. Consider the system

$$\begin{aligned} &\dot{y} = f(\tau, y, u), \qquad t < \tau \leqslant T \\ &y(t) = x, \qquad x \in R^n, \quad t \in I = [0, T], \quad T < \infty, \end{aligned} \tag{6.3.1}$$

and let $y_{t,x}(\tau)$ denote the solution starting from state x at time $\tau = t$. Let

$$J(t, x, u) \equiv \int_t^T \ell(\theta, y_{t,x}(\theta), u(\theta))\, d\theta + V_0(y_{t,x}(T)) \tag{6.3.2}$$

denote the cost of the control policy u over the interval $[t, T]$, where

$$u \in \mathcal{U} \equiv \{u \text{ measurable on } I \text{ to } R^m \text{ with values } u(t) \in U\}.$$

In general the control set U is a closed subset of R^m, and in particular a compact set or all of R^m. Our problem is to find a feedback control law $\eta: I \times R^n \to U$ such that the closed-loop system

$$\dot{y} = f(\tau, y, \eta(\tau, y)), \quad \tau \in I \tag{6.3.3}$$

is optimal, in the sense that, for

$$u^*(\tau) \equiv \eta(\tau, y_{t,x}(\tau)), \qquad t \leqslant \tau \leqslant T, \tag{6.3.4}$$

$$J(t, x, u^*) \leqslant J(t, x, u) \tag{6.3.5}$$

for all $u \in \mathcal{U}$ and $(t, x) \in I \times R^n$. In general η is a Borel measurable function from $I \times R^n \to U$. Define

$$V(t, x) \equiv \inf\{J(t, x, u), u \in \mathcal{U}\} \tag{6.3.6}$$

for $(t, x) \in I \times R^n$. The function V is called the value function. We wish to develop a suitable technique for determining V and hence the corresponding control law.

Principle of optimality (dynamic programming)

Let $\{u^*, y^*\}$ be an optimal pair over the interval $I = [0, T]$, and let $I_1 = [t_1, T]$, $0 < t_1 < T$, be an arbitrary subinterval. Then u^* restricted to I_1 is also optimal for the system

$$\begin{aligned} &\dot{z} = f(\tau, z, u), \qquad \tau \in I_1 \\ &z(t_1) = x_1, \end{aligned} \tag{6.3.7}$$

starting from the state $x_1 = y^*(t_1)$. This is the essence of Bellman's famous principle of optimality. We can state this in terms of the pay-off or value function V in the following result.

Lemma 6.3.1 (Principle of optimality). *For every $(t, x) \in I \times R^n$, and s, such that $t < s \leqslant T$,*

$$V(t, x) = \inf_{u \in \mathcal{U}} \left\{ \int_t^s \ell(\theta, y_{t,x}(\theta), u(\theta))\, d\theta + V(s, y_{t,x}(s)) \right\}. \tag{6.3.8}$$

This expression tells us that once the die is cast all is over, the player must make his future moves so as to minimize his losses.

Bellman's equations

Theorem 6.3.2 The value function V satisfies the first-order (hyperbolic) partial differential equation,

$$\begin{aligned} &\frac{\partial V}{\partial t} + \inf_{\xi \in U} \{(f(t, x, \xi), DV) + \ell(t, x, \xi)\} = 0, \qquad (t, x) \in I \times R^n, \\ &V(T, x) = V_0(x), \qquad x \in R^n. \end{aligned} \tag{6.3.9}$$

where $DV = V_x = \text{grad } V$.

Proof The proof essentially follows from the principle of optimality (6.3.8) (Lemma 6.3.1). Take $s = t + \Delta t$, then

$$y_{t,x}(t + \Delta t) = x + f(t, x, \xi)\Delta t + o(\Delta t), \qquad \xi \in U,$$

where $\lim_{\Delta t \to 0} (o(\Delta t)/\Delta t) = 0$. Then equation (6.3.8) can be written as

$$\begin{aligned} V(t, x) &= \inf_{\xi \in U} \{\ell(t, x, \xi)\Delta t + V(t + \Delta t, x + f(t, x, \xi)\Delta t + o(\Delta t))\} \\ &= \inf_{\xi \in U} \{\ell(t, x, \xi)\Delta t + V(t + \Delta t, x) \\ &\quad + ((DV)(t + \Delta t, x), f(t, x, \xi))\Delta t + o(\Delta t)\}. \end{aligned} \tag{6.3.10}$$

Hence

$$\frac{V(t, x) - V(t + \Delta t, x)}{\Delta t} = \inf_{\xi \in U} \left\{ \ell(t, x, \xi) + ((DV)(t + \Delta t, x), f(t, x, \xi)) + \frac{o(\Delta t)}{\Delta t} \right\}$$

and letting $\Delta t \to 0$, we obtain the first equation of (6.3.9). The second equation of (6.3.9) follows from (6.3.2) and (6.3.8) upon letting $t \to T$. ■

In dealing with infinite horizon problems one may introduce a discounted cost functional as follows. Let $I=[0, T)$, $0 \leqslant T \leqslant \infty$, and let c: $I \times R^n \times U \rightarrow [0, \infty]$ be a continuous uniformly bounded function, and define

$$J(t, x, u) \equiv \left\{ \int_t^T \ell(\theta, y_{t,x}(\theta), u(\theta)) \right.$$

$$\times \left[\exp\left(-\int_t^\theta c(\tau, y_{t,x}(\tau), u(\tau))\, \mathrm{d}\tau \right) \right] \mathrm{d}\theta$$

$$\left. + V_0(y_{t,x}(T)) \exp - \int_t^T c(\tau, y_{t,x}(\tau), u(\tau))\, \mathrm{d}\tau \right\} \qquad (6.3.11)$$

and

$$V(t, x) = \inf\{J(t, x, u),\ u \in \mathcal{U}\}.$$

In case $T=\infty$, we set $V_0 \equiv 0$ and assume that

$$\inf\{c(t, x, u),\ (t, x, u) \in [0, \infty) \times R^n \times U\} = \lambda > 0.$$

Theorem 6.3.3 (Discounted value). *Consider the system* (6.3.1) *with the cost functional* (6.3.11). *Then the corresponding value function* V *satisfies the following differential equation,*

$$\frac{\partial V}{\partial t} + \inf_{\xi \in U} \{\ell(t, x, \xi) + (DV, f(t, x, \xi))$$

$$- c(t, x, \xi) V(t, x)\} = 0, \qquad (t, x) \in I \times R^n, \quad (6.3.12)$$

$$V(T, x) = V_0(x), \qquad x \in R^n.$$

Proof It follows from the definition of V and (6.3.11) and the principle of optimality that

$$V(t, x) = \inf_{u \in \mathcal{U}} \left\{ \int_t^{t+\Delta t} \ell \exp\left(-\int_t^\theta c\, \mathrm{d}\tau \right) \mathrm{d}\theta \right.$$

$$\left. + V(t+\Delta t, y_{t,x}(t+\Delta t)) \left(\exp\left(-\int_t^{t+\Delta t} c\, \mathrm{d}\tau \right) \right) \right\}$$

where

$$\ell = \ell(\theta, y_{t,x}(\theta), u(\theta)),$$

$$c = c(\tau, y_{t,x}(\tau), u(\tau)).$$

For Δt sufficiently small, we have

$$\begin{aligned} V(t, x) &= \inf_{\xi \in U} \{\ell(t, x, \xi)\Delta t \\ &\quad + (1 - \Delta t\, c(t, x, \xi))V(t + \Delta t, y_{t,x}(t + \Delta t)) + o(\Delta t)\} \\ &= \inf_{\xi \in U} \{\ell(t, x, \xi)\Delta t + V(t + \Delta t, x) \\ &\quad + (DV(t + \Delta t, x), f(t, x, \xi))\Delta t \\ &\quad - c(t, x, \xi)V(t + \Delta t, x)\Delta t + o(\Delta t)\}. \end{aligned}$$

Hence

$$\left(\frac{V(t, x) - V(t + \Delta t, x)}{\Delta t}\right) = \inf_{\xi \in U}\left\{\ell(t, x, \xi) + (DV(t + \Delta t, x), f(t, x, \xi)) - c(t, x, \xi)V(t + \Delta t, x) + \frac{o(\Delta t)}{(\Delta t)}\right\},$$

and letting $\Delta t \to 0$, one obtains the first equation of (6.3.12). The equation, $V(T, x) = V_0(x)$, follows from (6.3.11) and the definition of V. ■

Let us define the Hamiltonians,

$$H(t, x, p) \equiv \inf_{\xi \in U} \{\ell(t, x, \xi) + (f(t, x, \xi), p)\} \tag{6.3.13}$$

and

$$H(t, x, q, p) \equiv \inf_{\xi \in U} \{\ell(t, x, \xi) + (f(t, x, \xi), p) - c(t, x, \xi) \cdot q\} \tag{6.3.14}$$

for $(t, x) \in I \times R^n$, $p \in R^n$, $q \in R$. Note that these functions are well defined whenever U is compact and f, ℓ, c are continuous. Using these Hamiltonians we can write the equations (6.3.9) and (6.3.12) in the compact form

$$\begin{aligned} &\frac{\partial V}{\partial t} + H(t, x, DV) = 0, \qquad (t, x) \in [0, T) \times R^n \\ &V(T, x) = V_0(x), \qquad x \in R^n \end{aligned} \tag{6.3.15}$$

and

$$\begin{aligned} &\frac{\partial V}{\partial t} + H(t, x, V, DV) = 0, \qquad (t, x) \in [0, T) \times R^n \\ &V(T, x) = V_0(x), \qquad x \in R^n. \end{aligned} \tag{6.3.16}$$

These equations are also called Hamilton–Jacobi–Bellman equations (HJB).

The question as to whether or not the solutions of these equations determine optimal controls is partially answered in the following simple verification theorem.

Theorem 6.3.4 (Verification theorem). *Suppose the equation* (6.3.15) *has a solution* $V \in C^1(Q_T)$, $Q_T \equiv [0, T) \times R^n$, *and suppose there exists a continuous function* η *on* Q_T *to* U, *such that*

$$H(t, x, p) = \ell(t, x, \eta(t, x)) + (f(t, x, \eta(t, x)), p)$$

for all $(t, x) \in Q_T$ *and* $p \in R^n$. *Then, for each* $(t, x) \in Q_T$, *the feedback control,*

$$u^*_{t,x}(s) \equiv \eta(s, y_{t,x}(s)), \qquad t \leqslant s \leqslant T,$$

or equivalently, the closed-loop system

$$\frac{dy}{ds} = f(s, y_{t,x}(s), \eta(s, y_{t,x}(s))), \qquad t \leqslant s \leqslant T, \quad t \in I$$

$$y_{t,x}(t) = x,$$

is optimal, in the sense that

$$V(t, x) = \inf J(t, x, u) = J(t, x, u^*_{t,x})$$

for all $(t, x) \in Q_T$.

Proof Consider the function $V(s, y_{t,x}(s))$ for $t \leqslant s \leqslant T$, and $(t, x) \in Q_T$. Since V is C^1, we have

$$\frac{d}{ds} V(s, y_{t,x}(s)) = \frac{\partial V}{\partial s} + ((DV)(s, y_{t,x}(s)), f(s, y_{t,x}(s), \eta(s, y_{t,x}(s)))).$$

Integrating this over the interval $[t, T]$, and recalling that V satisfies (6.3.15), we obtain

$$V(T, y_{t,x}(T)) - V(t, x) = -\int_t^T \ell(s, y_{t,x}(s), \eta(s, y_{t,x}(s)))\, ds.$$

Hence

$$V(t, x) = \int_t^T \ell(s, y_{t,x}(s), \eta(s, y_{t,x}(s)))\, ds + V(T, y_{t,x}(T))$$
$$= J(t, x, u^*_{t,x}).$$

This shows that the feedback control law η is optimal. ■

A similar result holds for the system (6.3.16).

Remark 6.3.5 In case V is sufficiently smooth, one can verify the equivalence between Bellman's equation and Pontryagin's minimum principle by defining $\psi(t) = DV(t, x(t))$ (along the optimal trajectory).

Unfortunately the equations (6.3.15) and (6.3.16), in general, do not have such smooth (C^1) solutions. We shall discuss this further in the section dealing with nonlinear regulators.

Control problems with state constraints

In some practical situations, the control system is automatically shut off (possibly for safety reasons as in nuclear reactors) as soon as it reaches the boundary of a certain specified region in the state space. Consider the system

$$\frac{dy}{dt} = f(t, y, u), \qquad t \in [0, T].$$

Let Ω be an open bounded connected subset of R^n with smooth boundary $\partial\Omega$. Let $x \in \Omega$ and let $\tau_x(u)$ denote the first time y hits the boundary $\partial\Omega$ and let χ_E denote the indicator function of any set E. Consider the cost function,

$$J(u) = \int_0^{T \wedge \tau_x(u)} \ell(\theta, y(\theta), u(\theta))\, d\theta + \chi_{T<\tau_x(u)} V_0(y(T)) + \chi_{\tau_x(u) \le T}\, \varphi(\tau_x(u), y(\tau_x(u))),$$

where $V_0 \in C(\Omega)$, $\varphi \in C(I \times \partial\Omega)$ and ℓ is a continuous bounded function on $I \times \bar{\Omega} \times U$. For $(t, x) \in I \times \Omega$, define the value function,

$$V(t, x) = \inf\left\{ \int_t^{T \wedge \tau_x(u)} \ell(\theta, y_{t,x}(\theta), u(\theta))\, d\theta + \chi_{T<\tau_x(u)} V_0(y_{t,x}(T)) + \chi_{\tau_x(u) \le T}\, \varphi(\tau_x(u), y_{t,x}(\tau_x(u))) \right\} \tag{6.3.17}$$

where $y_{t,x}(\cdot)$ is the solution starting from $(t, x) \in I \times \Omega$ depending on u.

Again by use of the principle of optimality we can prove the following result.

Theorem 6.3.6 (Value function on bounded domain). *The value function V, given by* (6.3.17), *satisfies the HJB equation with the Dirichlet* *boundary condition*:

$$\frac{\partial V}{\partial t} + H(t, x, DV) = 0 \qquad \text{in } [0, T) \times \Omega,$$
$$V(t, x) = \varphi(t, x) \qquad \text{in } [0, T] \times \partial\Omega, \tag{6.3.17)'$$
$$V(T, x) = V_0(x) \qquad \text{in } \Omega.$$

Remark 6.3.7 Note that if the cost integrand $\ell \geqslant 0$ and the equation (6.3.15) has a C^1-solution V, then V is a Lyapunov function for the closed-loop system with feedback control η (see Theorem 6.3.4). Further, if $\ell(t, 0, 0) = 0$, $f(t, 0, 0) = 0$ and $\ell(t, x, u) > 0$ otherwise, then the closed-loop system is asymptotically stable with respect to the zero state. This means that the optimal feedback control is also a stabilizing control.

Before closing this section we consider the stationary problem. For infinite horizon problems, involving time-invariant systems, stopped in a bounded domain Ω, we may define the value function as

$$V(x) \equiv \inf\left\{\int_0^{\tau_x(u)} \ell(y_x(\theta), u(\theta))\exp\left(-\int_0^{\theta} c(y_x(\tau), u(\tau))\,\mathrm{d}\tau\right)\mathrm{d}\theta + \varphi(y_x(\tau_x(u)))\left(\exp\left(-\int_0^{\tau_x(u)} c(y_x(\tau), u(\tau))\,\mathrm{d}\tau\right)\right)\right\} \quad \text{for } x \in \Omega. \tag{6.3.18}$$

Then by the principle of optimality, one can easily prove the following result.

Theorem 6.3.8 (∞-Horizon problem). *The value function V given by* (6.3.18) *satisfies the nonlinear Dirichlet boundary-value problem*

$$H(x, V, DV) = 0 \qquad \text{on } \Omega,$$
$$V(x) = \varphi(x) \qquad \text{on } \partial\Omega, \tag{6.3.18)'$$

where

$$H(x, q, p) \equiv \inf_{\xi \in U} \{(f(x, \xi), p) + \ell(x, \xi) - c(x, \xi)q\}.$$

6.3.2 Linear systems with quadratic cost functions

One of the most popular topics that has received widespread application over the last two decades is linear quadratic regulator theory (LQR). The subject is popular because of its simplicity and because the theory provides an optimal feedback control law which is linear and hence easily implementable.

Consider the system

$$\frac{dy}{dt} = A(t)y + B(t)u, \qquad t \in I = [0, T], \tag{6.3.19)(a}$$

with output

$$z(t) = H(t)y, \qquad t \in I, \tag{6.3.19)(b}$$

where $A \in C(I, R^{n\times n})$, $B \in C(I, R^{n\times m})$, $H \in C(I, R^{r\times n})$. Let $z_d(t)$, $t \in I$ denote the desired output trajectory and $\tilde{z}_d$ the desired terminal output. We wish to design a linear optimal feedback control law that minimizes the cost functional $J(u)$, given by

$$J(u) = \frac{1}{2}\int_I \{(Q(t)(H(t)y - z_d(t)), H(t)y - z_d(t)) + (R(t)u, u)\}\, dt$$
$$+ \tfrac{1}{2}(S(\tilde{H}y(T) - \tilde{z}_d), \tilde{H}y(T) - \tilde{z}_d), \tag{6.3.20}$$

where $Q(t)$, $R(t)$, $t \in I$, and S are suitable real symmetric positive matrices and $\tilde{H} \in R^{r\times n}$. We shall call this problem the linear quadratic regulator problem (LQR).

We apply Theorem 6.3.2 to find the feedback control law. In the notation of the theorem, we have

$$f(t, x, \xi) = A(t)x + B(t)\xi$$
$$\ell(t, x, \xi) = \tfrac{1}{2}\{(Q(t)(H(t)x - z_d(t)), H(t)x - z_d(t)) + (R(t)\xi, \xi)\}$$
$$V_0(x) \equiv \tfrac{1}{2}(S(\tilde{H}x - \tilde{z}_d), \tilde{H}x - \tilde{z}_d)$$
$$U \equiv R^m,$$

and

$$H(t, x, p) = \inf_{\xi\in R^m} \{(A(t)x + B(t)\xi, p)$$
$$+ \tfrac{1}{2}[(Q(t)(H(t)x - z_d(t)), H(t)x - z_d(t)) + (R(t)\xi, \xi)]\}. \tag{6.3.21}$$

Hence $\inf_{\xi\in R^m}\{(B(t)\xi, p) + \tfrac{1}{2}(R(t)\xi, \xi)\}$ determines the desired control law. Minimizing this, one obtains

$$\xi^* = -(R(t))^{-1}B'(t)p. \tag{6.3.22}$$

Substituting this in the Hamiltonian and suppressing t for the moment, we have

$$H(t, x, p) = (Ax, p) - (BR^{-1}B'p, p) + \tfrac{1}{2}(BR^{-1}B'p, p)$$
$$+ \tfrac{1}{2}(Q(Hx - z_d), Hx - z_d)$$
$$= (Ax, p) - \tfrac{1}{2}(BR^{-1}B'p, p) + \tfrac{1}{2}(Q(Hx - z_d), Hx - z_d). \tag{6.3.23}$$

Hence the Hamilton–Jacobi–Bellman equation (6.3.15) takes the form

$$\frac{\partial V}{\partial t} + (Ax, DV) - \tfrac{1}{2}(BR^{-1}B'DV, DV) + \tfrac{1}{2}(Q(Hx - z_d), Hx - z_d) = 0, \qquad (t, x) \in Q_T \quad (6.3.24)$$

$$V(T, x) = \tfrac{1}{2}(S(\tilde{H}x - \tilde{z}_d), \tilde{H}x - \tilde{z}_d), \qquad x \in R^n.$$

Anticipating that the value function V is quadratic in x, we choose

$$V(t, x) = \tfrac{1}{2}(K(t)x, x) + (\alpha(t), x) + \beta(t). \tag{6.3.25}$$

Substituting (6.3.25) into (6.3.24), we obtain

$$\begin{aligned} &\tfrac{1}{2}(\dot{K}x, x) + (\dot{\alpha}, x) + \dot{\beta} + (A'\alpha, x) + \tfrac{1}{2}(KAx, x) + \tfrac{1}{2}(A'Kx, x) \\ &\quad - \tfrac{1}{2}\left(BR^{-1}B'\frac{K+K'}{2}x, \frac{K+K'}{2}x\right) - \left(BR^{-1}B'\frac{K+K'}{2}x, \alpha\right) \\ &\quad - \tfrac{1}{2}\left(BR^{-1}B'\alpha, \alpha\right) + \tfrac{1}{2}(H'QHx, x) \\ &\quad - (H'Qz_d, x) + \tfrac{1}{2}(Qz_d, z_d) = 0 \end{aligned} \tag{6.3.26}$$

for all $(t, x) \in Q_T = (0, T] \times R^n$. Since (6.3.26) holds for all $x \in R^n$, and $t \in I$, equating individual homogeneous terms to zero, we have

$$(\dot{K}x, x) + (KAx, x) + (A'Kx, x) - \left(BR^{-1}B'\frac{K+K'}{2}x, \frac{K+K'}{2}x\right) + (H'QHx, x) = 0, \qquad (t, x) \in Q_T, \quad (6.3.27)(a)$$

$$(\dot{\alpha}, x) + (A'\alpha, x) - \left(BR^{-1}B'\frac{K+K'}{2}\alpha, x\right) - (H'Qz_d, x) = 0, \qquad (t, x) \in Q_T, \quad (6.3.27)(b)$$

$$\dot{\beta} - \tfrac{1}{2}(BR^{-1}B'\alpha, \alpha) + \tfrac{1}{2}(Qz_d, z_d) = 0, \qquad t \in I, \quad (6.3.27)(c)$$

and

$$\begin{aligned} V(T, x) &= \tfrac{1}{2}(K(T)x, x) + (\alpha(T), x) + \beta(T) \\ &= \tfrac{1}{2}(\tilde{H}'S\tilde{H}x, x) - (\tilde{H}'S\tilde{z}_d, x) + \tfrac{1}{2}(S\tilde{z}_d, \tilde{z}_d). \end{aligned} \tag{6.3.28}$$

Hence

$$\begin{cases} \dot{K} + A'K + KA - \left(\dfrac{K+K'}{2}\right)BR^{-1}B'\left(\dfrac{K+K'}{2}\right) + H'QH = 0 \\ K(T) = \tilde{H}'S\tilde{H}, \end{cases} \tag{6.3.29}$$

$$\begin{cases} \dot{\alpha} + A'\alpha - BR^{-1}B'\left(\dfrac{K+K'}{2}\right)\alpha - H'Qz_d = 0 \\ \alpha(T) = -\tilde{H}'S\tilde{z}_d, \end{cases} \tag{6.3.30}$$

and

$$\begin{cases} \dot{\beta} - \frac{1}{2}(BR^{-1}B'\alpha, \alpha) + \frac{1}{2}(Qz_d, z_d) = 0 \\ \beta(T) = \frac{1}{2}(S\tilde{z}_d, \tilde{z}_d). \end{cases} \tag{6.3.31}$$

Since R, Q and S are assumed to be symmetric, it is clear from (6.3.29) that K is also symmetric, and hence (6.3.29), (6.3.30) and (6.3.31) can be rewritten as

$$\begin{cases} \dot{K} + A'K + KA - KBR^{-1}B'K + H'QH = 0, & t \in [0, T) \\ K(T) = \tilde{H}'S\tilde{H}, \end{cases} \tag{6.3.29$'$}$$

$$\begin{cases} \dot{\alpha} + A'\alpha - BR^{-1}B'K\alpha - H'Qz_d = 0 \\ \alpha(T) = -\tilde{H}'S\tilde{z}_d \end{cases} \tag{6.3.30$'$}$$

$$\begin{cases} \dot{\beta} - \frac{1}{2}(BR^{-1}B'\alpha, \alpha) + \frac{1}{2}(Qz_d, z_d) = 0 \\ \beta(T) = \frac{1}{2}(S\tilde{z}_d, \tilde{z}_d) \end{cases} \tag{6.3.31$'$}$$

Remark 6.3.9 In case $z_d(t) \equiv 0$ and $\tilde{z}_d = 0$, the equation (6.3.30)$'$ reduces to a (linear) homogeneous equation with $\alpha(T) = 0$. Hence $\alpha(t) \equiv 0$, and it follows from (6.3.31)$'$ that $\beta(t) \equiv 0$. This leaves us with equation (6.3.29)$'$ only, the value function V reduces to

$$V(t, x) = \tfrac{1}{2}(K(t)x, x), \tag{6.3.32}$$

and the feedback control law is given by

$$u(t, x) = -R^{-1}B'Kx. \tag{6.3.33}$$

Remark 6.3.10 In case $z_d(t) \equiv 0$, $\tilde{z}_d = 0$ and $H(t) \equiv \tilde{H} \equiv I_n$ ($n \times n$ identity matrix) for $t \in I$, we obtain the classical Riccati differential equation

$$\begin{aligned} &\dot{K} + A'K + KA - KBR^{-1}B'K + Q = 0 \\ &K(T) = S, \end{aligned} \tag{6.3.34}$$

with the same value function and control law as in (6.3.32) and (6.3.33) respectively. Note that this equation can be also derived from Pontryagin's minimum principle (see Exercise 6.3.P24).

In view of the above discussion, we have the following result.

Theorem 6.3.11 (Existence, uniqueness). *Suppose that the matrix-valued functions A, B, H, Q, R are continuous in t on I; $Q(t)$, $t \in I$, and S are real symmetric and positive semidefinite; and $R(t)$, $t \in I$, is real symmetric and positive definite, $\tilde{z}_d \in R^r$, $z_d \in C(I, R^r)$ and $\tilde{H} \in R^{r \times n}$. Then the problem (LQR) has a unique solution given by the equations*

(6.3.29)′–(6.3.31)′ *with optimal control*

$$u_0(t, x) = -R^{-1}(t)B'(t)(K(t)x + \alpha(t)), \qquad t \in I, \quad x \in R^n, \tag{6.3.35}$$

and optimal cost

$$J(u_0) = V(0, x) = \tfrac{1}{2}(K(0)x, x) + (\alpha(0), x) + \beta(0), \qquad x \in R^n, \tag{6.3.36}$$

where $K(t)$, $t \in I$, *is positive semidefinite.*

Proof All that remains to complete the proof is to verify that the equations (6.3.29)′–(6.3.31)′ have unique solutions, and that $(K(t)\xi, \xi) \geqslant 0$, $t \in I$. Since equation (6.3.30)′ is linear and β is the integral of known functions once α is given, we have to prove only the existence of a unique solution of equation (6.3.29)′ and the positivity. For convenience, by reversing the flow of time, we can write (6.3.29)′ as

$$\begin{aligned} \dot{P} &= A'P + PA - P(BR^{-1}B')P + H'QH, \qquad t \in (0, T], \\ P(0) &= \tilde{H}'S\tilde{H} \equiv P_0, \end{aligned} \tag{6.3.37}$$

where we have redefined $M(t)$ for $M(T-t)$ with M being any of the matrices A, B, R, H, and Q. The equation (6.3.37) can be written as

$$\begin{aligned} \dot{P} &= F(t, P), \qquad t > 0, \\ P(0) &= P_0, \end{aligned} \tag{6.3.38}$$

where F represents the function on the right-hand side of (6.3.37). Since F is continuous in t and P, it follows from the basic existence theorem of chapter 3 that there exists a maximal time interval $[0, T')$, on which the equation has at least one absolutely continuous solution with the possibility of finite explosion time, that is,

$$\lim_{t \to T'} \|P(t)\| = \infty. \tag{6.3.39}$$

Let $T \in [0, T')$ such that $\|P(t)\| < \infty$ for all $t \in [0, T]$. First, we show that P is unique, and then show that $P(t) \geqslant 0$ and that it is defined on any time interval $I = [0, T]$ for $T < \infty$. We prove uniqueness by contradiction. Suppose there are two solutions P_1 and P_2. Then defining $E = (P_2 - P_1)$, one can easily verify that E satisfies the equation

$$\begin{aligned} \dot{E} &= (A' - P_1\Gamma)E - E(A - \Gamma P_2), \qquad t \in [0, T] \\ E(0) &= 0, \qquad \text{where } \Gamma = BR^{-1}B'. \end{aligned} \tag{6.3.40}$$

Since this is a homogeneous linear differential equation in E with zero initial condition, it follows that $E(t) \equiv 0$ on $[0, T]$. Hence (6.3.37), or equivalently (6.3.29)′, has a unique solution on the maximal interval of existence.

Now to complete the proof we show that $P(t) \geqslant 0$ (i.e. $(P(t)\xi, \xi) \geqslant 0$ for all $\xi \in R^n$, $t \geqslant 0$), and that for any finite T, $P(t)$ is dominated by a positive definite matrix-valued function on $[0, T]$. For convenience, we write equation (6.3.37) as

$$\begin{aligned} &\dot{P} = A'P + PA - P\Gamma P + \tilde{Q}, \qquad t \in (0, T], \\ &P(0) = P_0 \end{aligned} \tag{6.3.41}$$

where $\Gamma = BR^{-1}B'$, $\tilde{Q} = H'QH$, $P_0 = \tilde{H}'S\tilde{H}$. First we show that $P(t)$ is dominated from above by a positive and bounded (in norm) matrix-valued function. Let $\Phi(t, \tau)$, $0 \leqslant \tau \leqslant t \leqslant T$, denote the transition operator corresponding to A'. Then we can write an integral equation for P given by

$$\begin{aligned} P(t) = {} & \Phi(t, 0)P_0\Phi'(t, 0) \\ & - \int_0^t \Phi(t, \theta)P(\theta)\Gamma(\theta)P(\theta)\Phi'(t, \theta)\,\mathrm{d}\theta \\ & + \int_0^t \Phi(t, \theta)\tilde{Q}(\theta)\Phi'(t, \theta)\,\mathrm{d}\theta. \end{aligned}$$

Since $P_0 \geqslant 0$, $\tilde{Q}(t) \geqslant 0$, and $\Gamma(t) \geqslant 0$ for all $t \geqslant 0$, it is clear from the above expression that

$$P(t) \leqslant \Phi(t, 0)P_0\Phi'(t, 0) + \int_0^t \Phi(t, \theta)\tilde{Q}(\theta)\Phi'(t, \theta)\,\mathrm{d}\theta \tag{6.3.42}$$

for all $t \geqslant 0$ and, since Φ is bounded in norm on any finite time interval, it follows from this that $P(t)$ is dominated from above by a positive and bounded matrix-valued function. It remains to show that $P(t) \geqslant 0$. Clearly, we can write the equation (6.3.41) in the form

$$\begin{aligned} &\dot{P} = (A' - P\Gamma)P + P(A - \Gamma P) + P\Gamma P + \tilde{Q} \\ &P(0) = P_0. \end{aligned}$$

Letting Ψ denote the transition operator corresponding to $(A' - P\Gamma)$, we have

$$P(t) = \Psi(t, 0)P_0\Psi'(t, 0) + \int_0^t \Psi(t, \theta)(P\Gamma P + \tilde{Q})(\theta)\Psi'(t, \theta)\,\mathrm{d}\theta.$$

Since $P\Gamma P + \tilde{Q} \geqslant 0$, independently of whether P is positive or not, and $P_0 \geqslant 0$, it follows from the above expression that $P(t) \geqslant 0$. Thus we have

$$0 \leqslant P(t) \leqslant \Phi(t, 0)P_0\Phi'(t, 0) + \int_0^t \Phi(t, \theta)\tilde{Q}(\theta)\Phi'(t, \theta)\,\mathrm{d}\theta.$$

Since $P_0 \geqslant 0$, and $\tilde{Q}(t) \geqslant 0$ for all $t \geqslant 0$, and continuous, it follows from the above inequalities that, for each finite T, there exists a positive number $N_T < \infty$ such that

$$0 \leq \sup_{t \in [0, T]} (P(t)\xi, \xi) \leq N_T |\xi|^2.$$

Hence (6.3.41) or equivalently (6.3.29)′ has a unique positive solution $P(t)$, or equivalently $K(t)$, on the interval $[0, T]$ for any finite T. This completes the proof. ■

As a corollary of the above theorem, we have the following stability result.

Corollary 6.3.12 (Optimality to stability). *Consider the linear quadratic regulator problem* (LQR) *with* $H(t) \equiv \tilde{H} \equiv I_n$, *and* $z_d(t) = \tilde{z}_d \equiv 0$ *for* $t \geq 0$. *If* $Q(t) > 0$ *for all* $t \geq 0$, *then the closed-loop system with the feedback control* $u_0(t, x) = -R^{-1}(t)B'(t)K(t)x$ *is asymptotically stable with respect to the zero state.*

Proof The system (6.3.19)(a) with the feedback control $u_0(t, x) = -R^{-1}(t)B'(t)K(t)x$ takes the form

$$\dot{y} = (A - BR^{-1}B'K)y. \tag{6.3.43}$$

Then the value function $V(t, x) = \frac{1}{2}(K(t)x, x)$ is a Lyapunov function for the closed-loop system (6.3.43). Indeed, one can easily verify that along any solution of (6.3.43), we have

$$\frac{\mathrm{d}}{\mathrm{d}t} V(t, y(t)) = \tfrac{1}{2}((\dot{K} + KA + A'K - 2KBR^{-1}B'K)y, y)$$

$$= -\tfrac{1}{2}((Q + KBR^{-1}B'K)y, y).$$

Therefore, if $Q(t) > 0$ for $t \geq 0$, then we have $\mathrm{d}V(t, y(t))/\mathrm{d}t < 0$ along any solution y of (6.3.43). Hence by Lyapunov's theory (Theorem 4.4.8), the system is asymptotically stable. ■

Stationary control laws

In the case of time-invariant systems with constant A, B, Q, R, one expects that a steady state solution, if one exists, of the equation (6.3.34) may provide a stationary control law $u = -R^{-1}B'Kx$ which is obviously convenient for synthesis. This raises the question of existence of a positive stabilizing solution of the associated algebraic Riccati equation,

$$A'K + KA - KBR^{-1}B'K + Q = 0 \qquad \text{(ARE)}$$

so that the control law

$$u = -R^{-1}B'Kx$$

makes the closed-loop system

$$\dot{x} = (A - BR^{-1}B'K)x$$

asymptotically stable. This is a classical problem and has been extensively studied in the literature [see 132, 128, 118, 119]. We present here some typical results. Let $\mathrm{eV}(T)$ denote the set of eigenvalues of the matrix T.

Theorem 6.3.13 *The equation (ARE) has a real symmetric stabilizing solution K, that is,* $\mathrm{Re}\,\mathrm{eV}(A - BR^{-1}B'K) < 0$, *if and only if*

(a) *(A, B) is stabilizable, and*
(b) $\mathrm{Re}\,\mathrm{eV}(N) \neq 0$, *where*

$$N = \begin{bmatrix} A & BR^{-1}B' \\ Q & -A' \end{bmatrix}.$$

Proof For proof see [132]. ■

From Theorem 6.3.13, one obtains the following results.

Corollary 6.3.14 *Suppose $Q = H'H$. Then the equation (ARE) has a real symmetric solution K satisfying* $\mathrm{Re}\,\mathrm{eV}(A - BR^{-1}B'K) < 0$, *if and only if*

(a) *(A, B) is stabilizable*
(b) *all eigenvalues of A satisfying* $\mathrm{Re}\,\mathrm{eV}(A) \neq 0$ *are observable modes of (H, A).*

Corollary 6.3.15 *Suppose $Q = H'H$ and* $\mathrm{Re}\,\mathrm{eV}(A) \neq 0$, *then the equation (ARE) has a real symmetric stabilizing solution K if and only if (A, B) is stabilizable.*

Remark 6.3.16 (Computation of K). The solution of the algebraic Riccati equation (ARE) can be obtained by use of the Newton–Raphson technique with a starting value $K_0 = K(0)$ where $K(t)$, $0 \leq t \leq T$, is the solution of the Riccati differential equation (6.3.34). This approach is likely to improve the convergence of the Newton–Raphson iteration appreciably.

6.3.3 Nonlinear regulators

There is no elegant theory for nonlinear regulators comparable to that of linear quadratic regulators. It appears from section 6.3.1 that the theory of nonlinear regulators must be intimately connected with the Hamilton–Jacobi–Bellman equations (6.3.15), (6.3.16) and (6.3.17)′. In fact optimal feedback regulators can be constructed once the value function (pay-off function) V is known.

Indeed, if the functions f, ℓ, and c of equations (6.3.9) and (6.3.12) are assumed to be continuous in all their arguments and the control set U is assumed to be compact, then the infima in (6.3.13) and (6.3.14) can be replaced by minima. Since U is compact, these minima are attained in U. Let

$$v\colon [0, T] \times R^n \times R^n \to U$$

denote any measurable function such that

$$H(t, x, p) = \ell(t, x, v(t, x, p)) + (f(t, x, v(t, x, p)), p) \tag{6.3.44}$$

for all $(t, x, p) \in I \times R^n \times R^n$, and let

$$\mu\colon I \times R^n \times R \times R^n \to U$$

be a similar function such that

$$H(t, x, q, p) = \ell(t, x, \mu(t, x, q, p)) + (f(t, x, \mu(t, x, q, p)), p) - c(t, x, \mu(t, x, q, p))q \tag{6.3.45}$$

for all $(t, x, q, p) \in I \times R^n \times R \times R^n$.

If V denotes the solution of equation (6.3.15) or (6.3.16), then the feedback control laws are given by

$$\eta(t, x) = \begin{cases} v(t, x, (DV)(t, x)) & \text{in case of (6.3.15),} \\ \mu(t, x, V(t, x), (DV)(t, x)) & \text{in case of (6.3.16),} \end{cases} \tag{6.3.46}$$

provided $V \in C^1$. Thus if we can solve the equations (6.3.15), (6.3.16) and (6.3.17)′, we can construct the control laws through the relations (6.3.46). Hence the problem of optimal synthesis of regulators for nonlinear systems is equivalent to the problem of constructing suitable solutions of the equations (6.3.15), (6.3.16) and (6.3.17)′. This raises certain fundamental questions:

(a) Does equation 6.3.15 ((6.3.16), (6.3.17)′) have a solution?
(b) If a solution exists, is it unique?
(c) Given the existence and uniqueness, how regular (smooth) is the control law (6.3.46), and what are the topological properties of the sets of discontinuities of DV and hence that of η, and what are their significance in control synthesis?
(d) If there is a unique solution, can we find an efficient method for computing it for $n \geq 3$.

Only in recent years have answers to questions (a) and (b) begun to appear [19], [111]–[114], [139], [140]. Answers to questions (c) and (d) remain largely unexplored except for a very few trivial examples which can be studied by other means.

Here we shall briefly consider questions (a) and (c), and refer the reader to recent literature for further study.

Viscosity solutions

It has been known for a long time that the first-order nonlinear hyperbolic equations of the form (6.3.15), (6.3.16) and (6.3.17)′ do not, in general, possess classical solutions, that is, solutions V which are C^1 in both t and x and satisfy the equations everywhere on $I \times \Omega$ and $\lim_{t \to T} V(t, x) = V_0(x)$ for $X \in \Omega$.

In recent years a new notion of solution, called the viscosity solution, has been introduced. In this type of solution V is not required to be a C^1 function on $Q = I \times \Omega$ or $I \times R^n$. We shall present only a brief outline of the concept. First note that by reversal of the flow of time, $t \to T - t$, any of the equations (6.3.15)–(6.3.17)′ can be converted into initial-value (or initial–boundary-value) problems. For example, by redefining the Hamiltonian, we can write (6.3.16) in the form

$$\frac{\partial V}{\partial t} + H(t, x, V, DV) = 0, \qquad (t, x) \in (0, T] \times R^n = Q, \tag{6.3.47}$$
$$V(0, x) = V_0(x), \qquad x \in R^n.$$

In the case of infinite horizon problems for time-invariant systems, we obtain a stationary Hamilton–Jacobi equation

$$H(x, V, DV) = 0, \qquad x \in R^n. \tag{6.3.48}$$

Defining

$$(t, x) \equiv y, \; DV \equiv D_y V \equiv (D_t V, D_{x_1} V, \ldots, D_{x_n} V) \equiv (p_0, p_1, \ldots, p_n) \equiv p,$$

we can write all of these equations in a unified notation,

$$F(y, V, DV) = 0, \qquad y \in Q, \tag{6.3.49}$$

with appropriate side conditions (initial and boundary conditions) where

$$F(y, q, p) = p_0 + H(y, q, p_1, p_2, \ldots, p_n).$$

We shall introduce the notion of viscosity solution using this canonical form.

Let $C(Q)$ denote the space of real-valued continuous functions on Q with the topology of uniform convergence on compact subsets of Q.

Before formally introducing the definition of a viscosity solution, let us assume that $V \in C^1(Q)$ satisfies (6.3.49) and let $\varphi \in C_0^1(Q)(\equiv C^1$ functions on Q with compact supports), and suppose φV assumes a local maximum (or minimum) at y_0. Then $D(\varphi V)(y_0) = 0$, and hence $DV(y_0) = -(D\varphi(y_0)/\varphi(y_0))V(y_0)$. Similarly, for $k \in R$, $\varphi \in C_0^1(Q)$, if $\varphi(V-k)$ attains a local maximum (or minimum) at y_0, then

$$DV(y_0) = -\frac{D\varphi(y_0)}{\varphi(y_0)}(V(y_0)-k). \tag{6.3.50}$$

Hence, for every such $\varphi \in C_0^1(Q)$, $k \in R$ and $y_0 \in Q$, we can replace $DV(y_0)$ in (6.3.49) by the expression (6.3.50), giving

$$F\left(y_0, V(y_0), -\frac{D\varphi(y_0)}{\varphi(y_0)}(V(y_0)-k)\right) = 0. \tag{6.3.51}$$

In fact this substitution can be extended to $V \in C(Q)$. Let $V \in C(Q)$, $k \in R$ and $\varphi \in \mathscr{D}^+(Q) \equiv \{\varphi \in C_0^\infty(Q)\colon\ \varphi > 0\}$, and suppose $\varphi(V-k)$ attains a local positive maximum at $y_0 \in Q$. Let V^ε be a C^1 solution of the equation

$$-\varepsilon\Delta V + F(y, V, DV) = 0, \qquad y \in Q, \tag{6.3.52}$$

where Δ is the Laplacian $\Delta\psi \equiv \sum \partial^2\psi/\partial x_i^2$, and suppose $V^\varepsilon \to V$ uniformly on compact subsets of Q as $\varepsilon \downarrow 0$. Let $\varphi(V^\varepsilon - k)$ assume a local maximum at $y_\varepsilon \in Q$ and $y_\varepsilon \to y_0$ as $\varepsilon \downarrow 0$. Then computing $\Delta(\varphi(V^\varepsilon - k))$, one can easily verify that

$$\begin{aligned} -\varepsilon\Delta(\varphi(V^\varepsilon - k)) = {} & -\varepsilon(\Delta\varphi)(V^\varepsilon - k) \\ & - \varphi F(y, V^\varepsilon(y), DV^\varepsilon(y)) - 2\varepsilon(D\varphi, DV^\varepsilon). \end{aligned} \tag{6.3.53}$$

Evaluating (6.3.53) at y_ε and noting that $D\varphi(V^\varepsilon - k) = 0$ and $\Delta(\varphi(V^\varepsilon - k)) < 0$ at $y = y_\varepsilon$, it follows from (6.3.53) that

$$\begin{aligned} & F\left(y_\varepsilon, V^\varepsilon(y_\varepsilon), -\frac{D\varphi(y_\varepsilon)}{\varphi(y_\varepsilon)}(V^\varepsilon(y_\varepsilon)-k)\right) \\ & \quad \leqslant -\varepsilon\frac{\Delta\varphi(y_\varepsilon)}{\varphi(y_\varepsilon)}(V^\varepsilon(y_\varepsilon)-k) + 2\varepsilon\frac{|D\varphi(y_\varepsilon)|^2}{(\varphi(y_\varepsilon))^2}(V^\varepsilon(y_\varepsilon)-k). \end{aligned} \tag{6.3.54}$$

Since $V^\varepsilon \to V$ uniformly on compact subsets of Q and F is uniformly continuous on compact subsets of Q (by assumption), letting $\varepsilon \downarrow 0$, we obtain

$$F\left(y_0, V(y_0), -\frac{D\varphi(y_0)}{\varphi(y_0)}(V(y_0)-k)\right) \leqslant 0. \tag{6.3.55}$$

Similarly for any $k \in R$, $\varphi \in \mathscr{D}^+(Q)$, if $\varphi(V-k)$ possesses a negative minimum at $y_0 \in Q$, then

$$F\left(y_0, V(y_0), -\frac{D\varphi(y_0)}{\varphi(y_0)}(V(y_0)-k)\right) \geq 0. \tag{6.3.56}$$

Thus we have arrived at the formal definition of a viscosity solution.

Definition 6.3.17 (Viscosity solution). A function $V \in C(Q)$ is said to be a viscosity subsolution (resp. supersolution) of the equation

$$F(y, V, DV) = 0, \qquad y \in Q,$$

if, for every $\varphi \in \mathscr{D}^+(Q)$ and $k \in R$, for which there exists a $y_0 \in Q$ at which $\varphi(V-k)$ assumes a local positive maximum (resp. local negative minimum),

$$F\left(y_0, V(y_0), -\frac{D\varphi(y_0)}{\varphi(y_0)}(V(y_0)-k)\right) \leq 0 \qquad (\text{resp. } \geq 0). \tag{6.3.57}$$

V is said to be a viscosity solution if it is both a viscosity subsolution and supersolution.

It is clear from the above discussion that the name, viscosity solution, has been derived from the procedure used for the proof of existence of solutions of the equation $F(y, V, DV) = 0$. Introduction of the viscosity term, $-\varepsilon \Delta V$, in equation (6.3.52) turns a strongly nonlinear hyperbolic system into a semilinear parabolic system. Usually the equation (6.3.52) is much more well behaved than the original system (i.e. with $\varepsilon = 0$). In engineering terms, a nonviscous fluid is turned into a viscous one which has smoother flow. It also has a probabilistic interpretation in terms of the Markov diffusion process.

It is clear from the above discussion that the following result holds.

Theorem 6.3.18 (Viscosity–classical). *If $V \in C^1(Q)$ is a classical solution of* (6.3.49), *then it is also a viscosity solution, and conversely if $V \in C(Q)$ is a viscosity solution, then for every*

$$y_0 \in d(V) \equiv \{y \in Q\colon DV(y) \text{ exists}\}$$

$$F(y_0, V(y_0), DV(y_0)) = 0.$$

There are other convenient characterizations of viscosity solutions. For $y_0 \in Q$, define

$$D^+V(y_0) \equiv \left\{p \in R^m\colon \limsup_{y \to y_0} \left[\frac{V(y) - V(y_0) - (p, y - y_0)}{|y - y_0|}\right] \leq 0\right\}$$

and

$$D^-V(y_0) \equiv \left\{p \in R^m: \liminf_{y \to y_0} \left[\frac{V(y) - V(y_0) - (p, y - y_0)}{|y - y_0|}\right] \geq 0\right\}. \tag{6.3.58}$$

The reader can easily verify that, whenever these sets are non-empty, they are closed and convex. For the function $V(y) = |y|$, $y \in R$, we have $D^+V(0) = \varnothing$ (empty set) and $D^-V(0) = [-1, +1]$. If a function V is differentiable then $D^+V(y_0) = D^-V(y_0) = \{DV(y_0)\}$.

One can define viscosity solutions in terms of these sets.

Definition 6.3.19 (Viscosity solution characterized). A function $V \in C(Q)$ is a viscosity subsolution (respectively *supersolution*) of $F(y, V(y), DV(y)) = 0$ if and only if $F(y, V(y), p) \leq 0$ (resp. ≥ 0) for all $p \in D^+V(y)$ (resp. $D^-V(y)$) and $y \in Q$. The function V is a viscosity solution if it is both a viscosity subsolution and supersolution.

Definitions 6.3.17 and 6.3.19 are equivalent. The proof is elementary and is left as an exercise for the reader. For other related results the reader may refer to P.L. Lions [19]. Another characterization of a viscosity solution is given in the following theorem.

Theorem 6.3.20 (Viscosity solution characterized). *A function $V \in C(Q)$ is a viscosity subsolution (respectively supersolution) of $F(y, V(y), DV(y)) = 0$, $y \in Q$, if and only if for each $\varphi \in C^1(Q)$, $F(y_0, V(y_0), D\varphi(y_0)) \leq 0$ (resp. ≥ 0) at each point $y_0 \in Q$ at which $(V - \varphi)$ attains its local maximum (resp. minimum).*

Proof If: We consider only the subsolution; the proof is identical for supersolution. Let $\varphi \in C^1(Q)$, and $y_0 \in Q$ be a point of local maximum of $(V - \varphi)$ so that $F(y_0, V(y_0), D\varphi(y_0)) \leq 0$. For any $y \in Q$ in the neighbourhood of y_0,

$$\varphi(y) = \varphi(y_0) + (D\varphi(y_0), y - y_0) + o(|y - y_0|).$$

Since $V(y) - \varphi(y) \leq V(y_0) - \varphi(y_0)$ for y sufficiently near y_0, we have

$$V(y) \leq V(y_0) + \varphi(y) - \varphi(y_0) = V(y_0) + (D\varphi(y_0), y - y_0) + o(|y - y_0|).$$

Hence by definition of the set $D^+V(y_0)$, we have $D\varphi(y_0) \in D^+V(y_0)$. Since φ is an arbitrary C^1 function, we conclude that

$$F(y_0, V(y_0), p) \leq 0 \qquad \text{for all } p \in D^+V(y_0).$$

Only if: Suppose $F(y_0, V(y_0), p) \leqslant 0$ for all $p \in D^+V(y_0)$, and $y_0 \in Q$. Take any $p \in D^+V(y_0)$, and define the C^1-function

$$\varphi(y) \equiv V(y_0) + (p, y - y_0), \qquad y \in Q.$$

Then $V(y) - \varphi(y) = V(y) - V(y_0) - (p, y - y_0)$ and, since $p \in D^+V(y_0)$, we have $V(y) - \varphi(y) \leqslant 0$ for $y \in Q$ and y sufficiently near y_0 and $V(y_0) - \varphi(y_0) = 0$. This shows that y_0 is a point of local maximum of $(V - \varphi)$ and $D\varphi(y_0) = p$, and hence $F(y_0, V(y_0), D\varphi(y_0)) \leqslant 0$. ■

Clearly the above result provides yet another definition for viscosity solution which is found more convenient in certain situations. For example, the following theorem is easily proved using this definition.

Theorem 6.3.21 (Viscosity solutions as limits of classical solutions). *Let $F_\varepsilon \in C(Q \times R \times R^m)$, $\varepsilon > 0$, and suppose $F_\varepsilon \to F$ as $\varepsilon \to 0$, uniformly on compact subsets of $Q \times R \times R^m$. Let $V^\varepsilon \in C^1(Q)$ be a classical solution of*

$$-\varepsilon \Delta V^\varepsilon + F_\varepsilon(y, V^\varepsilon, DV^\varepsilon) = 0, \qquad y \in Q,$$

and suppose $V^\varepsilon \to V$ uniformly on compact subsets of Q. Then V is a viscosity solution of

$$F(y, V(y), DV(y)) = 0, \qquad y \in Q.$$

Proof By Theorem 6.3.20, it suffices to prove that, for each $\varphi \in C^1(Q)$,

$$F(y_0, V(y_0), D\varphi(y_0)) \leqslant 0 \qquad (\text{resp.} \geqslant 0)$$

at each $y_0 \in Q$ where $V - \varphi$ assumes its local maximum (minimum). Let $\eta \in C^1(Q)$ be such that $0 \leqslant \eta < 1$ for all $y \in Q \backslash \{y_0\}$ and $\eta(y_0) = 1$. First we prove for $\varphi \in C^2(Q)$ and suppose that $(V - \varphi)$ attains its local maximum at y_0; then clearly $(V - \varphi + \eta)$ has a strict local maximum at y_0. Hence, for $\varepsilon > 0$ sufficiently small, $(V^\varepsilon - (\varphi - \eta))$ attains its local maximum at, say y_ε, and $y_\varepsilon \to y_0$ as $\varepsilon \to 0$. Clearly

$$-\varepsilon \Delta(V^\varepsilon - (\varphi - \eta)) = -\varepsilon \Delta V^\varepsilon + \varepsilon \Delta(\varphi - \eta),$$

and hence

$$-\varepsilon \Delta(V^\varepsilon - (\varphi - \eta)) + F_\varepsilon(y, V^\varepsilon(y), DV^\varepsilon(y)) = \varepsilon \Delta(\varphi - \eta).$$

Clearly at $y = y_\varepsilon$, $D(V^\varepsilon - (\varphi - \eta)) = 0$ and $\Delta(V^\varepsilon - (\varphi - \eta)) \leqslant 0$ (resp. $\geqslant 0$), and we have

$$F_\varepsilon(y_\varepsilon, V^\varepsilon(y_\varepsilon), DV^\varepsilon(y_\varepsilon)) \leqslant \varepsilon \Delta(\varphi - \eta)(y_\varepsilon) \ (\text{resp.} \geqslant \varepsilon \Delta(\varphi - \eta)(y_\varepsilon)).$$

Since $F_\varepsilon \to F$ uniformly on compacts (compact subsets) of $Q \times R \times R^m$, and $V^\varepsilon \to V$ uniformly on compacts of Q, and $y_\varepsilon \to y_0$, and $D\eta(y_0) = 0$, letting $\varepsilon \downarrow 0$, we obtain

$$F(y_0, V(y_0), D\varphi(y_0)) \leqslant 0 \qquad (\text{resp. } \geqslant 0)$$

This is true for every $\varphi \in C^2$. For $\varphi \in C^1$, we choose a sequence $\varphi_n \in C^2$ such that $\varphi_n \to \varphi$ uniformly on Q. Let $\{y_n\}$ be a point in Q at which $V - \varphi_n$ attains its local maximum and $y_n \to y_0$. Clearly

$$F(y_n, V(y_n), D\varphi_n(y_n)) \leqslant 0 \qquad (\text{resp.} \geqslant 0),$$

and letting $n \to \infty$, we arrive at the conclusion. ■

The questions of existence and uniqueness of viscosity solutions are important, though it is beyond the scope of this book. However, it may be useful for some readers to know the outline of the procedure.

The function F is replaced by a sequence of smoother functions F_ε, $\varepsilon > 0$, such that $F_\varepsilon \to F$ uniformly on compacts of the set $Q \times R \times R^m$. Then one proves that the regularized problem,

$$-\varepsilon \Delta V + F_\varepsilon(y, V(y), DV(y)) = 0, \qquad y \in Q, \tag{6.3.59}$$

with appropriate side conditions, has a solution

$$V^\varepsilon \in C^2 \text{ or } W^{1,\infty}_{\text{loc}} \equiv \{\varphi\colon \|\varphi\|_{L^{\text{loc}}_\infty} + \|D\varphi\|_{L^{\text{loc}}_\infty} < \infty\},$$

and that there exists a constant β, independent of $\varepsilon > 0$, such that

$$\|V^\varepsilon\|_{L^{\text{loc}}_\infty} + \|DV^\varepsilon\|_{L^{\text{loc}}_\infty} \leqslant \beta. \tag{6.3.60}$$

From this one can conclude that there exists a function $V \in C(Q)$ such that $V^\varepsilon \to V$ uniformly on compacts of Q. Then one proves that V is a viscosity solution of the problem $F(y, V(y), DV(y)) = 0$, $y \in Q$, as given in the previous theorem. The crucial parts in the proof consist of (a) the question of existence of solution of the regularized problem (6.3.59), and (b) the uniform estimates (6.3.60). For these the reader is referred to the book of Ladyženskaja *et al.* [57]. We shall give only a brief outline for the following Cauchy problem:

$$\begin{aligned} &\frac{\partial V}{\partial t} + H(t, x, V, DV) = 0, \qquad (t, x) \in Q = (0, T] \times R^n = Q \\ &V(0, x) = V_0(x), \qquad x \in R^n. \end{aligned} \tag{6.3.61}$$

Let $X = BUC(Q)$ denote the Banach space of all bounded uniformly continuous functions on Q. Introducing the viscosity term $-\varepsilon \Delta V$, one considers the linearized problem

$$\frac{\partial V}{\partial t} - \varepsilon \Delta V = -H(t, x, W, DW), \qquad (t, x) \in Q,$$
$$V(0, x) = V_0(x), \qquad x \in R^n, \tag{6.3.62}$$

for any $W \in X$ such that $DW \in X$ and $H \in X$. Then, under reasonably mild assumptions on H (after regularization if necessary) one can prove that the problem (6.3.62) has a unique solution $V \in X$, and that V is continuously dependent on W. Let G denote the map that takes W to V, giving $V = GW$. One then proves that G is continuous and maps bounded subsets of X into compact sets, thereby concluding that G is a compact map. At this point one uses the Schauder fixed-point theorem (chapter 3) or more general fixed-point theorems (the Leray–Schauder degree principle [66]) to prove the existence of a fixed point of G, that is, an element $V_\varepsilon \in X$ such that $V_\varepsilon = GV_\varepsilon$. At the same time one obtains the estimate

$$\|V_\varepsilon\|_{L_\infty^{\text{loc}}} + \|DV_\varepsilon\|_{L_\infty^{\text{loc}}} \leq \beta$$

for some finite $\beta > 0$, independent of ε. This allows one to prove that V_ε converges to some V uniformly on compact subsets of Q. Then by use of Theorem 6.3.21, one concludes the existence of a $V^* \in X$ which is a viscosity solution of (6.3.61).

An alternative approach for proving the existence and uniqueness of solutions is based on the theory of accretive operators and the Crandall–Ligget generation theorem for nonlinear semigroups. This is beyond the scope of this elementary text. The semigroup approach has been developed by several authors. Interested readers should consult Lions [19], Crandall and Lions [112] and [113], Tamburro [140], Burch [104], Crandall and Souganidis [114] and the current literature.

Stationary control laws

For design of a stationary optimal regulator one may consider, for example, the solution of the Dirichlet problem (6.3.18)′. Suppose there exists a continuous function $v(x, q, p)$ on $\Omega \times R \times R^n$ to U such that

$$H(x, q, p) = (f(x, v), p) + \ell(x, v) - c(x, v)q$$

for all $(x, q, p) \in \Omega \times R \times R^n$ and that the problem (6.3.18)′ has a viscosity solution $V \in C(\Omega)$. Then the optimal regulator may be defined by setting

$$u(x) = v(x, V(x), p), \qquad p \in (D^+V)(x) \cup (D^-V)(x). \tag{6.3.63}$$

If V is differentiable at x, then the control is uniquely defined there with $p=(DV)(x)$. Clearly question (c), raised in the introduction, is relevant here.

It is not clear how one should interpret the control law given by (6.3.63). One way to define a unique control law may be to introduce a sort of relaxed control or equivalently a probability measure $\lambda_x(dp)$ with support $(D^+V)(x)\cup(D^-V)(x)$, and replace (6.3.63) by

$$u^0(x)=\int_{(D^+V)(x)\cup(D^-V)(x)} v(x, V(x), p)\lambda_x(dp) \qquad (6.3.63)'$$

where λ_x is such that, for $x\in d(V)$, $\lambda_x(dp)$ is a Dirac measure with support $\{DV(x)\}$. This is a subject for the future.

6.4 Computation of optimal controls and system identification

In sections 6.1 and 6.2 we presented the necessary conditions that must be satisfied by the optimal pair $\{x^*, u^*\}$. These results can be broadly classified into three categories: (a) bang-bang conditions in the case of time optimal control, (b) Riccati type equations in the case of optimal regulators, and (c) minimum principle for general optimal control problems. In numerical applications, the Riccati equations, (6.3.29)′–(6.3.31)′, can be solved backward in time using standard methods for solution of ordinary differential equations. For large systems, application of certain transformations to the matrix $K(t)$ has been found [105] to be useful in reducing the computer time.

In time-optimal control and general optimal control problems, the optimality conditions are given in terms of a set of differential equations for some of which the initial conditions are known and for the remaining equations the final conditions are given. In most of the cases, solution of this two-point boundary-value problem is an extremely difficult task. In practice, it is relatively easy to compute the optimal solution through iterative techniques using the necessary conditions of optimality. Some of the commonly used methods are the gradient method or the method of steepest descent, second variation method, quasilinearization method, primal–dual method [131] etc. Recently Teo [141, 142] has developed a control parameterization method for iteratively computing the optimal control.

In section 6.4.1, we shall briefly discuss the gradient method, which is relatively simple and easy to implement. In section 6.4.2, we shall see that the problem of identification of unknown parameters, appearing in mathematical models of physical systems, can be formulated as a control problem and can be solved using the integral form of the minimum principle [see equation (6.2.36)]. Also we present a direct gradient algorithm without using the minimum principle.

6.4.1 Computation of optimal controls

Gradient method

The gradient method consists of updating the controls using the gradient vector, in a way such that successive iterates produce maximum reduction of the cost. To illustrate, we consider the following control problem:

$$\text{minimize}\left\{J(u)=\phi(x(T))+\int_0^T \ell(t, x(t), u(t))\,\mathrm{d}t\right\} \tag{6.4.1}$$

subject to the dynamic constraint

$$\begin{aligned}\dot{x}&=f(t, x(t), u(t)), \qquad u\in\mathcal{U},\\ x(0)&=x_0.\end{aligned} \tag{6.4.2}$$

Define the Hamiltonian $H(t, x, \psi, u)$ by

$$H(t, x, \psi, u)=\ell(t, x(t), u(t))+(f(t, x(t), u(t)), \psi(t)). \tag{6.4.3}$$

Then it follows from Corollary 6.2.6 that the optimal control u^* and the corresponding pair $\{x^*, \psi^*\}$ satisfy the system equation (6.4.2), and the adjoint equation

$$\begin{aligned}\dot{\psi}&=-\ell_x(t, x(t), u(t))-f_x'(t, x(t), u(t))\psi\\ \psi(T)&=\frac{\partial\phi}{\partial x}(x(T)),\end{aligned} \tag{6.4.4}$$

and the inequality

$$H(t, x^*(t), \psi^*(t), u^*(t))\leq H(t, x^*(t), \psi^*(t), u) \qquad \text{for all } u\in\mathcal{U}, \tag{6.4.5}$$

where $x^*\equiv x(u^*)$ and $\psi^*\equiv\psi(u^*)$.

Using the above necessary conditions, we can devise an iterative algorithm for computing the optimal control u^*. We note, however, that although the necessary conditions (6.4.2)–(6.4.5) are based on rather general assumptions on the functions f and ℓ, numerical evaluation of u^* calls for somewhat stronger conditions. We present the iterative scheme in the following.

Algorithm A

Step 1 Guess $u_1 \in \mathcal{U}$ and set $n = 1$.
Step 2 Solve the initial-value problem (6.4.2) with $u(t) = u_n(t)$, giving $x_n(t) = x_n(u_n)(t)$.
Step 3 Using the data x_n and u_n, solve the adjoint system (6.4.4) to obtain ψ_n.
Step 4 Compute the gradient vector

$$g_n(t) = \frac{\partial H}{\partial u}(t, x_n(t), \psi_n(t), u_n(t)). \tag{6.4.6}$$

Step 5
(a) If $g_n(t) \neq 0$, then modify $u_n(t)$ to $u_{n+1}(t) = u_n(t) - \varepsilon g_n(t)$ by choosing $\varepsilon > 0$ sufficiently small so that $u_{n+1}(t) \in U$ and $J(u_{n+1}) \leq J(u_n)$. A stopping criterion may be used at this stage. If $|J(u_{n+1}) - J(u_n)| \leq \delta$ for some small $\delta > 0$, then stop; otherwise set $u_n = u_{n+1}$, $n = n + 1$, and go to Step 2.
(b) If $g_n(t) = 0$ at the nth stage, then u_n is a (local) minimizing element of $J(u)$.

Convergence of the algorithm

Let u_n and u_{n+1} be two controls in the admissible class which are not optimal, and for $i = n$, $n + 1$, denote $x_i \equiv x(u_i)$ and $\psi_i \equiv \psi(u_i)$ for the solutions of (6.4.2) and (6.4.4) respectively. Using the Hamiltonian H, as defined in (6.4.3), it is easy to see that

$$\begin{aligned} J(u_{n+1}) - J(u_n) = {} & \phi(x_{n+1}(T)) - \phi(x_n(T)) \\ & + \int_0^T \{H(t, x_{n+1}, \psi_{n+1}, u_{n+1}) - H(t, x_n, \psi_n, u_n)\}\, dt \\ & + \int_0^T \{(f(t, x_n, u_n), \psi_n) - (f(t, x_{n+1}, u_{n+1}), \psi_{n+1})\}\, dt. \end{aligned} \tag{6.4.7}$$

We assume that f and ℓ are Lipschitz and C^2 in both x and u, and the second partials of f in x and u are locally integrable in t. Then using (6.4.2), (6.4.4) and continuous dependence of solutions on parameters, like controls, one can verify that, for u_{n+1} within a small neighbourhood of u_n,

$$J(u_{n+1}) - J(u_n) = \int_0^T \left(\frac{\partial H}{\partial u}(t, x_n, \psi_n, u_n), u_{n+1} - u_n \right) dt + o(\|u_{n+1} - u_n\|).$$

We can choose

$$u_{n+1}(t) = u_n(t) - \varepsilon \frac{\partial H}{\partial u}(t, x_n, \psi_n, u_n) \tag{6.4.8}$$

with $\varepsilon > 0$ sufficiently small so that $u_{n+1} \in \mathcal{U}$. Then we have

$$J(u_{n+1}) = J(u_n) - \varepsilon \int_0^T \left| \frac{\partial H}{\partial u}(t, x_n, \psi_n, u_n) \right|^2 \mathrm{d}t + o(\varepsilon).$$

Clearly, for ε sufficiently small, we have $J(u_{n+1}) < J(u_n)$. This shows that the sequence $\{u_n\}$ constructed according to the Algorithm A is convergent.

Remark 6.4.1 The gradient method has the property that the computed solutions tend to converge to a local minimum. Thus once a minimum is attained, it is necessary to change the initial guess and search for other minima in the control space. In case the optimal solution is unique, it will be attained in a few cycles of iteration.

Remark 6.4.2 The control sequence generated by the gradient method does converge to a local minimum even if the initial guess is poor. However, convergence becomes slower as the minimum is approached.

In computer applications, convergence of the iterative process in the above algorithm can be substantially improved by using the conjugate gradient method in Step 5 for updating the controls. We present this in the following flowchart, while the details can be found in [128].

Algorithm B

Step 1 Guess $u_0 \in U$ and set $n = 0$.
Step 2 Solve the state equation (6.4.2) from $t = 0$ to T.
Step 3 Solve the adjoint equation (6.4.4) from $t = T$ to 0.
Step 4 Compute the gradient g_n using (6.4.6).
Step 5 Compute the search direction S_n by

$$S_n = -g_n \qquad \text{if } n = 0,$$

or

$$S_n = -g_n + \beta_n S_{n-1} \quad \text{with} \quad \beta_n = \|g_n\|^2 / \|g_{n-1}\|^2 \qquad \text{if } n \geqslant 1.$$

Step 6 Perform a linear search for ε_n so that

$$u_n + \varepsilon_n S_n \in U \qquad \text{and} \qquad J(u_n + \varepsilon_n S_n) < J(u_n).$$

Note that convergence of the algorithm depends on the choice of ε. In practice, computing the cost for four different ε's and interpolating using a cubic polynomial to obtain the minimum cost and the corresponding ε (which is taken as ε_n) gives faster convergence.

Step 7 Update the control u_n by

$$u_{n+1} = u_n + \varepsilon_n S_n.$$

Step 8 The iteration is repeated from Step 2 until a stopping criterion is satisfied. A suitable criterion can be taken as $|J(u_{n+1}) - J(u_n)| < \delta$ for some small δ.

Example 6.4.3 Consider the nuclear reactor dynamics (1.2.63). For simplicity, we shall assume that there are only two precursors denoted R_1 and R_2. It is desired to raise the level of neutron density N (hence the power level) from the initial state N_0 to the final state N_1 in time T by increasing the reactivity ρ. In practice, the reactivity is modified by changing the position of the control rods. Because of safety requirements and physical limitations in the operation of the control rods, it is required to maintain $\dot{\rho}$ as small as possible.

Defining ρ as a new state variable, we have the following control problem:

$$\dot{N} = \frac{N}{\ell}(\rho - \beta) + \lambda_1 R_1 + \lambda_2 R_2$$

$$\dot{R}_1 = \frac{\beta\mu_1}{\ell} N - \lambda_1 R_1$$

$$\dot{R}_2 = \frac{\beta\mu_2}{\ell} N - \lambda_2 R_2$$

$$\dot{\rho} = u$$

$$J(u) = \tfrac{1}{2}\gamma_1 (N(T) - N_1)^2 + \tfrac{1}{2}\gamma_2 \int_0^T u^2 \, dt \equiv \text{minimize},$$

where γ_1 and γ_2 are suitable positive weighting factors.

Iterations were carried out using the algorithm discussed above with the initial guess $u_0(t) = 0$ for all $t \in [0, T]$. Convergence was obtained in five iterations. The following data were used in the simulation:

$\lambda_1 = 0.154\ \text{s}^{-1}$	$\beta = 0.0076$	$\gamma_1 = 10^6$
$\lambda_2 = 0.456$	$\ell = 0.0015$	$\gamma_2 = 1.0$
$\mu_1 = 0.65$	$T = 1.0$	$N(0) = 5\ \text{kW}.$
$\mu_2 = 0.35$		

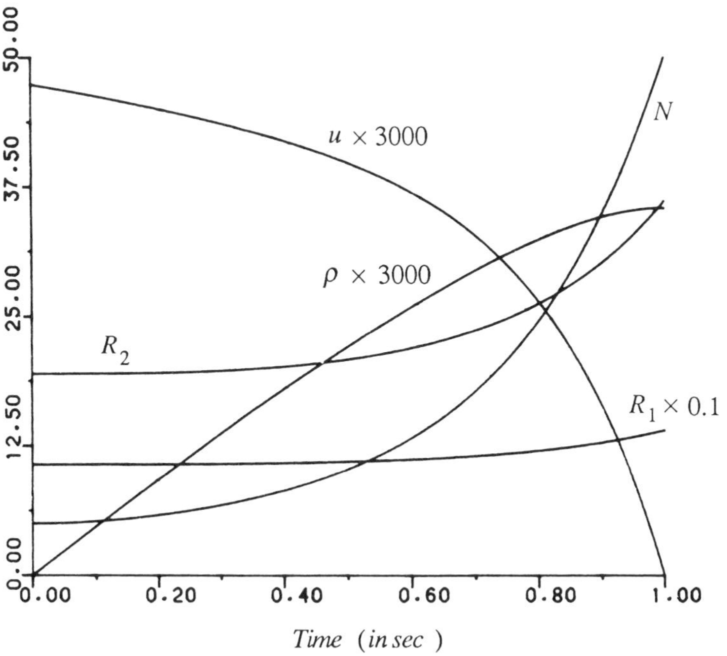

Figure 6.4 State and control trajectories for nuclear reactor

Figure 6.4 shows the optimal control and the corresponding state trajectory. For the given data, $J(u_0) = 1.0125 \times 10^9$ and $J(u_n) = 0.7757 \times 10^{-4}$ for $n = 5$.

Shooting method

Consider the problem (6.4.1)–(6.4.2). Suppose there exists a function η: $I \times R^n \times R^n \to U$ such that

$$M(t, x, \psi) \equiv \inf_{v \in U} H(t, x, \psi, v) = H(t, x, \psi, \eta(t, x, \psi))$$

for all $(t, x, \psi) \in I \times R^n \times R^n$. Using the function η and the canonical equations, $\dot{x} = H_\psi$, $\dot{\psi} = -H_x$, we can deparameterize the original control problem to obtain

$$\begin{aligned} \dot{x}(t) &= g_1(t, x(t), \psi(t)), \qquad & x(0) &= x_1 \\ \dot{\psi}(t) &= g_2(t, x(t), \psi(t)), \qquad & \psi(T) &= \varphi_x(x(T)), \end{aligned} \tag{6.4.9}$$

where

$$g_1(t, x, y) = f(t, x, \eta(t, x, y))$$

$$g_2(t, x, y) = -f_x'(t, x, \eta(t, x, y))y - \ell_x(t, x, \eta(t, x, y)).$$

This is a system of coupled $2n$ nonlinear differential equations with (mixed) n initial and n terminal conditions. Usually such problems are known as two-point boundary-value problems and are difficult to solve. In the shooting method, one chooses the missing initial conditions, $\psi(0) = \xi$, and solves the equation (6.4.9) to obain $x \equiv x(t, \xi)$, $\psi \equiv \psi(t, \xi)$. Then one introduces an error function,

$$E(\xi) \equiv |\psi(T, \xi) - \varphi_x(x(T, \xi))|^2_{R^n},$$

thereby reducing the original optimal control problem to that of minimizing the function $E(\xi)$ on R^n. Here one uses any of the gradient techniques to minimize $E(\xi)$. The problem of computing the gradient is extremely laborious and time consuming. For further details see [68], [141], [142].

6.4.2 System identification

Usually a physical system is modelled on the basis of intuition, some idealizing assumptions, and knowledge of functional relationship between the state variables and their time rate of change (see Chapter 1). In this procedure, a number of unknown parameters are introduced in the differential equations which are expected to describe the evolution of the physical system. A fundamental problem in system modelling is, then, determination of these unknown parameters so that the corresponding response of the model equation approximates as closely as possible the actual response of the physical system.

Consider the following model equation:

$$\text{M:} \quad \begin{cases} \dot{x} = f(t, x, \alpha), & t \in [0, T] \equiv I \\ x(0) = x^0 \text{ given}, \end{cases} \tag{6.4.10}$$

where $x(t) \in R^n$ is the state vector and α, taking values in a closed bounded subset $\mathcal{P}$ of R^m, is the parameter vector which is unknown. The function f: $I \times R^n \times R^m \to R^n$ is known, except for the vector α. It is assumed that the response of the physical system, which M is desired to represent, is given in the form of data $y(t)$, $t \in I$. Then the problem is to find a parameter $\alpha^* \in \mathcal{P} \subset R^m$ that minimizes the identification error

$$J(\alpha) = \frac{1}{2} \int_0^T |x(t, \alpha) - y(t)|^2 \, dt, \tag{6.4.11}$$

i.e. $J(\alpha^*) \leq J(\alpha)$ for all $\alpha \in \mathcal{P} \subset R^m$, where $x(\cdot, \alpha)$ is the response of the model equation M corresponding to the parameter α and the initial condition x^0.

Optimal control approach

In view of the above discussion, it is clear that a very natural approach for solving the identification problem is to consider it as a control problem. In fact this could be considered as a special case of the general control problem, and using the integral form of the minimum principle (see equation (6.2.36)) we can obtain a set of necessary conditions for identification of the unknown parameter.

Define the Hamiltonian $H(t, x, \psi, \alpha)$ by

$$H(t, x, \psi, \alpha) = (f(t, x, \alpha), \psi) + \tfrac{1}{2}|x(t, \alpha) - y(t)|^2. \tag{6.4.12}$$

Then by Theorem 6.2.1 it follows that the optimal parameter α^* and the corresponding pair $\{x^*, \psi^*\}$ satisfy the following set of equations:

$$\begin{cases} \dot{x} = f(t, x, \alpha) \\ x(0) = x^0, \end{cases} \tag{6.4.13}$$

$$\begin{cases} \dot{\psi} = -f'_x(t, x, \alpha)\psi + (y(t) - x(t)) \\ \psi(T) = 0, \end{cases} \tag{6.4.14}$$

and the inequality

$$\int_0^T H(t, x^*(t), \psi^*(t), \alpha^*)\, dt \leq \int_0^T H(t, x^*(t), \psi^*(t), \alpha)\, dt \tag{6.4.15}$$

for all $\alpha \in \mathscr{P}$.

For numerical computation of the optimal parameter α^*, we can use the gradient methods discussed earlier. In fact the Algorithms A or B can be modified very easily so as to apply to this case. Indeed, for the gradient method applied to identification problems, we have the following:

Algorithm C

Same as Algorithm A, except for the following modifications:

(a) In Step 4, the gradient vector is given by

$$g_n = \int_0^T \frac{\partial H}{\partial \alpha}(t, x_n, \psi_n, \alpha_n)\, dt = \int_0^T f'_\alpha(t, x_n, \alpha_n)\psi_n\, dt. \tag{6.4.16}$$

(b) The control vector $u_n \in U$ is replaced by $\alpha_n \in \mathscr{P}$, and the equation numbers are appropriately modified.

Direct variational approach

The optimal control approach for parameter identification suffers from a drawback. If the parameters appear linearly in the model equation and if there are no constraints on the values that the parameters can assume, then the minimum principle approach does not make much sense. In what follows, we shall present another method [82] which is free from such limitations.

For simplicity, we shall assume that the admissible parameter space $\mathscr{P}$ is all of R^m, i.e. $\mathscr{P} \equiv R^m$. For the function f, we assume that it satisfies conditions that assure existence of solutions of the model equation M for each parameter $\alpha \in R^m$. We also assume that f is measurable in $t \in I$ for each $x \in R^n$ and $\alpha \in R^m$, and that for almost all $t \in I$, it is a C^2 function on $R^n \times R^m$ with respect to x and α such that the second partial derivatives of f in x and α are locally integrable in t.

Under these assumptions, we can find an approximating sequence of parameters $\{\alpha_n\} \in \mathscr{P}$ through an iterative process such that $J(\alpha_{n+1}) \leq J(\alpha_n)$ and as $n \to \infty$, $\alpha_n \to \alpha^* \in \mathscr{P}$. This is discussed below.

Let $\alpha_n \in R^m$ be an arbitrary choice and suppose $x_n(t) \equiv x(t, \alpha_n)$, $t \in I$, is the response of the model equation M corresponding to the parameter α_n. By definition the corresponding error function is

$$J(\alpha_n) = \frac{1}{2} \int_0^T (x(t, \alpha_n) - y(t), x(t, \alpha_n) - y(t))\, \mathrm{d}t.$$

Let us modify $\alpha_n \to \alpha_n + \varepsilon(\delta\alpha)$, $\delta\alpha \in R^m$, $\varepsilon \in R$. Then taking the derivative of the functional $J(\alpha_n + \varepsilon(\delta\alpha))$ with respect to ε and setting $\varepsilon = 0$, we have

$$\left.\frac{\partial J}{\partial \varepsilon}\right|_{\varepsilon=0} = \int_0^T (x_\alpha(t, \alpha_n) \cdot \delta\alpha, x(t, \alpha_n) - y(t))\, \mathrm{d}t. \tag{6.4.17}$$

Denote $x_\alpha(t, \alpha_n)$ by $z(t, \alpha_n)$, an $(n \times m)$ matrix-valued function. It is easy to verify from the integral equation

$$x(t, \alpha) = x^0 + \int_0^t f(\theta, x(\theta, \alpha), \alpha)\, \mathrm{d}\theta, \qquad t \in I$$

that $z(t, \alpha_n)$, $t \in I$ satisfies the matrix differential equation

$$\begin{aligned} &\frac{\mathrm{d}}{\mathrm{d}t} z(t, \alpha_n) = f_x(t, x_n, \alpha_n) z(t, \alpha_n) + f_\alpha(t, x_n, \alpha_n), \qquad t \in I, \\ &z(0, \alpha_n) = 0. \end{aligned} \tag{6.4.18}$$

Using this change in notation in (6.4.17)′, denoting transpose by a prime and noting that α is independent of time, we have

$$\left.\frac{\partial J}{\partial \varepsilon}\right|_{\varepsilon=0} = \left(\delta\alpha, \int_0^T z'(t, \alpha_n)[x(t, \alpha_n) - y(t)]\,\mathrm{d}t\right)$$
$$= (\delta\alpha, J'(\alpha_n)).$$

The quantity

$$J'(\alpha_n) \equiv \int_0^T z'(t, \alpha_n)[x(t, \alpha_n) - y(t)]\,\mathrm{d}t \tag{6.4.19}$$

defines the gradient of $J(\alpha)$ at $\alpha = \alpha_n$. Thus it follows from the Cauchy–Schwartz inequality that $\partial J/\partial\varepsilon|_{\varepsilon=0}$ attains its maximum if $\delta\alpha$ is taken as a constant multiple of $J'(\alpha_n)$; i.e. $(\delta\alpha)_n = J'(\alpha_n)$. This gives us the direction of steepest descent through the expression

$$\alpha_{n+1} = \alpha_n - \varepsilon(\delta\alpha)_n = \alpha_n - \varepsilon J'(\alpha_n). \tag{6.4.20}$$

Then it can be verified that

$$J(\alpha_{n+1}) = J(\alpha_n) - \varepsilon(J'(\alpha_n), (\delta\alpha)_n)$$
$$+ \varepsilon^2 \int_0^1 \theta\left\{\int_0^1 J''(\alpha_n - \gamma\theta\varepsilon(\delta\alpha)_n)(\delta\alpha)_n, (\delta\alpha)_n)\,\mathrm{d}\gamma\right\}\mathrm{d}\theta. \tag{6.4.21}$$

Under the given assumptions J'' is bounded in the neighbourhood of α_n, and hence

$$J(\alpha_{n+1}) = J(\alpha_n) - \varepsilon\,|J'(\alpha_n)|^2 + o(\varepsilon) \tag{6.4.22}$$

where $\lim_{\varepsilon\to 0} o(\varepsilon)/\varepsilon = 0$. Thus it is clear that by choosing $\varepsilon > 0$ sufficiently small, one can obtain

$$J(\alpha_{n+1}) < J(\alpha_n).$$

In case $J'(\alpha_n)$ is zero (vector), α_n is a local minimum, otherwise one takes $\alpha = \alpha_{n+1}$ and repeats the procedure.

We summarize the results in the following algorithm.

Algorithm D

Step 1 Guess α_0 and set $n = 0$.
Step 2 Solve the initial value problem (6.4.10) with $\alpha = \alpha_n$, giving $x_n \equiv x(\alpha_n)$.
Step 3 Solve the matrix differential equation (6.4.18) corresponding to x_n and α_n to obtain $z_n \equiv z(\alpha_n)$.

Step 4 Compute the integral

$$(\delta\alpha)_n = \int_0^T z_n'(t)[x_n(t) - y(t)]\, dt.$$

Step 5

(a) If $(\delta\alpha)_n \neq 0$, then modify α_n to $\alpha_{n+1} = \alpha_n - \varepsilon(\delta\alpha)_n$ by choosing $\varepsilon > 0$ sufficiently small so that $\alpha_{n+1} \in \mathscr{P}$ and $J(\alpha_{n+1}) \leqslant J(\alpha_n)$. If the stopping criterion, $|J(\alpha_{n+1}) - J(\alpha_n)| \leqslant \delta$, for $\delta > 0$ small, is not satisfied, then set $n = n + 1$ and go to Step 2.

(b) If $(\delta\alpha)_n = 0$, then α_n is the position of a local minimum for the functional J.

Remark 6.4.4 Clearly, Step 4 in the above algorithm is equivalent to

$$\begin{aligned} \dot{\eta} &= z_n'(t)(x_n(t) - y(t)), \qquad t \in [0, T] \\ \eta(0) &= 0 \\ (\delta\alpha)_n &= \eta(T). \end{aligned} \tag{6.4.23}$$

Thus, in actual computer application of the algorithm, Steps 2–4 can be combined in one step, and the three sets of differential equations, (6.4.10), (6.4.18) and (6.4.23), can be solved simultaneously. The major advantage gained by doing so is the elimination of storage requirement for the trajectories $x_n(t)$ and $z_n(t)$, $t \in I$.

Example 6.4.5 Consider the predator–prey dynamics

$$\dot{x}_1 = \alpha_1 x_1 - \alpha_2 x_1 x_2$$

$$\dot{x}_2 = -\alpha_3 x_2 + \alpha_4 x_1 x_2$$

where the growth rates α_1, $\alpha_3 > 0$ and the rates of interspecies interaction α_2, $\alpha_4 > 0$ are unknown. In order to estimate the four parameters one can monitor the population level of the two species in their natural habitat over a certain period of time. In this example, the 'observation' data were generated by solving the model equations with some known parameters, which will be termed as the 'true' parameters. Table 6.1 shows the convergence of the iterative scheme shown in Algorithm C.

For other numerical results on parameter optimization see reference [101].

Remark 6.4.6 Note that the parameter optimization problem can also be viewed as a control problem where the control is the initial state.

Table 6.1

Iter	α_1	α_2	α_3	α_4	$J \times 10^{10}$
0	1.0	0.001	0.1	0.001	0.2249×10^{10}
1	2.634	0.0	0.1197	0.0	0.2480×10^{9}
3	2.9994	0.0017165	0.1260	0.0	0.1497×10^{7}
5	2.9996	0.0015998	0.4149	0.0	0.6240×10^{5}
7	2.9996	0.0016053	0.4997	0.0	0.3884
15	2.9997	0.0026717	0.50016	0.0005978	0.1506
20	3.0000	0.005989	0.5000	0.0002030	0.2721×10^{-2}
True	**3.0**	**0.006**	**0.5**	**0.0002**	—

Defining $\dot{\alpha} = 0$ and introducing the state vector

$$z = \begin{bmatrix} x \\ \alpha \end{bmatrix},$$

one has

$$\dot{z} = \tilde{f}(t, z), \qquad \tilde{f} = \begin{bmatrix} f \\ 0 \end{bmatrix}, \qquad z(0, \alpha) = \begin{pmatrix} x_0 \\ \alpha \end{pmatrix} = z_0.$$

The problem, then, is to find z_0 such that $J(\alpha) \equiv \int_I g(t, y(t), z(t, \alpha))\, dt$ is minimum.

Remark 6.4.7 The success of the control theory in application depends crucially on the efficiency of the computational algorithms. Unfortunately, there seem to be no one single method which can be considered equally satisfactory for all the problems.

For further study of the subject see [24], [131], [141], [142].

Exercises

6.1.P1 Show that for a scalar input system, $\dot{x} = Ax + bu$, normality and properness are equivalent.

6.1.P2 Verify that the system $\dot{x}_1 = x_2 + u_1$, $\dot{x}_2 = x_1 + u_1 + u_2$ is proper but not normal. Construct another example having the same property.

6.1.P3 Prove Theorem 6.1.9.

6.1.P4 Consider Example 6.1.11 and construct its switching curve and hence the optimal feedback control.

6.1.P5 Construct the phase-plane portraits for Example 6.1.12.

6.2.P6 Introducing Lagrange multipliers, give the Euler–Lagrange equations for the following two problems:

$$\min \varphi(x) = \min \int_{t_1}^{t_2} \ell(t, x(t), \dot{x}(t))\, dt, \qquad x(t_1) = x_1, \quad x(t_2) = x_2,$$

subject to

(i) differential constraints, $f_i(t, x, \dot{x}) = 0$, $i = 1, 2, \ldots, m < n$,

(ii) isoperimetric constraints, $\int_{t_1}^{t_2} g_i(\theta, x(\theta), \dot{x}(\theta))\, d\theta = c_i$, $i = 1, 2, \ldots, m < n$, where c_i are specified numbers.

6.2.P7

(a) Show that the problems of calculus of variations (i) and (ii) of (6.2.P6) can be converted into equivalent optimal control problems. (*Hint*: introduce $\dot{x} = u$, $u(t) \in R^n$).

(b) Introduce the adjoint equation and the Hamiltonian for the problem (ii). Then using transversality condition, verify that there are $2(n + m + 1)$ differential equations with $2(n + m + 1)$ boundary conditions.

6.2.P8

(a) Show that the set $K \subset R^{n+1}$ defined by the expression (6.2.45) is a convex cone.

(b) Prove that the point $\eta^* = (-1, 0, 0, \ldots, 0)'$ is not an interior point of K.

6.2.P9 Consider the system $\dot{x} = f(t, x, u)$, $t \in [t_1, t_2]$, t_2 free, $x(t_1) = x_1$, $u \in \mathcal{U}$ (as in Theorem 6.2.1). It is required to find a control $u \in \mathcal{U}$ so that the corresponding trajectory $x(t)$, $t \geq t_1$ hits the moving target $z(t)$, $t \geq t_1$ ($z \in C^1$) in minimum time ($t_2 - t_1$). Show that, in order that t_2 be the optimal hitting time it must satisfy the transversality condition:

$$\lambda_0 + M(t_2) - (\psi(t_2), \dot{z}(t_2)) = 0$$

(see Theorem 6.2.1 for notation).

6.2.P10 Using Lagrange multipliers (assuming their existence) prove the necessary conditions of optimality of Theorem 6.2.1 as suggested under the heading 'Alternative proof of Theorem 6.2.1.'

6.2.P11 Verify Remark 6.2.2.

6.2.P12 Consider the system $\ddot{\theta} = u$, $|u(t)| \leq 1$.

(a) Find the time optimal control for transfer to the zero state and sketch the trajectory

(b) Same as in (a) with desired target $\theta = 0$,

(c) Find a control that minimizes the fuel cost $J(u)=\int_0^{t_2}|u|\,dt$ assuming $t_2>t_{\min}$. Comment for $t_2\leqslant t_{\min}$.

6.2.P13 Consider the system (1.2.6) for a locomotive consisting of four cars. Suppose the locomotive is to be brought to maximum admissible speed within a prespecified distance while minimizing discomforts to passengers.

(a) Define a measure of discomfort in terms of relative velocities between adjacent cars.
(b) Define a cost functional using (a) and the cost of fuel.
(c) Write the necessary conditions of optimality. Using appropriate values for the parameters (see Problem 1.2P1) and the given algorithm, determine the optimal control f and the corresponding state trajectory $(x_1, x_2, \ldots, x_8)$.

6.2.P14 Consider the satellite dynamics (1.2.29) with flywheel accelerations and reaction jets $(\dot{\Omega}_x, \dot{\Omega}_y, \dot{\Omega}_z, T_x, T_y, T_z)$ taken as the control vector u and $(p, q, r, \Omega_x, \Omega_y, \Omega_z)$ as the state vector x. Let

$$U\equiv\{u: |u_i|\leqslant \alpha_i,\ i=1,2,\ldots,6\}$$

denote the control constraint. (Note that the equation takes the form $\dot{x}=f(x)+Bu$.) Define

$$J(u)=\frac{1}{2}\int_{t_1}^{t_2}\{[I_xp^2+I_y(q-w_0)^2 + I_zr^2+C_x\Omega_x^2+C_y\Omega_y^2+C_z\Omega_z^2]+(Ru, u)\}\,dt$$

where R is a suitable positive definite symmetric matrix and t_2 free.

(a) Write the adjoint equations and the Hamiltonian.
(b) Verify that optimal control has the form $u=-R^{-1}B'x$.
(c) Write the canonical equations as a two-point boundary-value problem.

6.2.P15 Consider the fisheries model

$$\dot{x}=f(x)-u,\quad f(x)=\alpha x\left(1-\frac{x}{k}\right),\qquad \alpha>0,\quad k>0.$$

The problem is to maximize the revenue,

$$J(u)=\int_0^T R(u(t))\exp(-\delta t)\,dt,$$

where R is a positive non-decreasing C^2 function (concave upward) and δ is the discount rate. By use of the necessary conditions of optimality, prove that the optimal control is given by the solution of the two-point boundary-value problem,

$$\dot{x} = f(x) - u,$$

$$\dot{u} = \left(\frac{R'(u)}{R''(u)}\right)(\delta - f'(x)),$$

$$x(0) = x_1, \quad x(T) = x_2.$$

6.2.P16 [96] The pacemaker battery is required to deliver during each cardiac cycle a stimulating current $I(t)$ to the cardiac cell which responds with a membrane potential $V(t)$. It is required to drive the membrane potential, $V(0) = 0$, to a threshold potential V_h at the end of each period T at which the cell fires. This must be accomplished with minimum dissipation of energy from the battery. Let R_0 denote the membrane resistance as seen by the cell. Then the energy dissipated is

$$J(I) \equiv \int_0^T R_0 I^2 \, \mathrm{d}t.$$

The membrane potential is governed by the equation

$$C\frac{\mathrm{d}V}{\mathrm{d}t} + \frac{V}{R} = I$$

which is equivalent to a parallel RC circuit driven by a current source $I(t)$. Using the minimum principle, show that the optimal current is given by

$$I(t) = \frac{V_h}{R \sinh\left(\frac{T}{\tau}\right)} \exp\left(\frac{t}{\tau}\right)$$

and

$$V(t) = V_h \frac{\sinh(t/\tau)}{\sinh(T/\tau)}$$

6.2.P17 Consider the time-optimal control problem

$$\dot{x} = \int_U f(t, x, u)\mu_t(\mathrm{d}u), \qquad x(t_1) = x_1, \quad x(t_2) = x_2$$

$\min\{t_2(\mu), \mu \in \mathcal{M}\}$ where

$$\mathcal{M} \equiv \Bigg\{\mu\colon \mu_t(\mathrm{d}u) = \sum_{i=1}^{k} \xi_i(t)\delta_{u_i}(\mathrm{d}u),$$

$$\xi = (\xi_1, \ldots, \xi_k)' \text{ measurable, } \xi(t) \in \Lambda \text{ a.e.}\Bigg\},$$

$$\Lambda = \{\alpha \in R^k\colon \quad \alpha_i \geqslant 0, \quad \sum \alpha_i = 1\}, \{u_i, \quad i = 1, 2, \ldots, k\}$$

is a set of k linearly independent vectors from $U \subset R^m$, $k < m$, $\delta_\sigma(du)$ is the Dirac measure with mass 1 at $u = \sigma$. Then in order that $\mu_t^* = \sum_{i=1}^k \xi_i^*(t)\delta_{u_i}$ be the optimal control with optimal time t_2^*, they must satisfy the following necessary conditions:

(a) $\dot{x}^*(t) = \langle H_\psi, \mu^* \rangle = \int_U f(t, x^*(t), u)\mu_t^*(du)$, a.e. on $[t_1, t_2^*]$

$$\dot{\psi}^*(t) = -\langle H_x, \mu^* \rangle = -\int_U f'_x(t, x^*(t), u) \cdot \psi^* \mu_t^*(du)$$

(b) $\int_{t_1}^{t_2^*} (\xi^*(t), H^*(t))\, dt \leqslant \int_{t_1}^{t_2^*} (\xi(t), H^*(t))\, dt$ where ξ is measurable with values $\xi(t) \in \Lambda$, $H^*(t) = (H_1^*(t), H_2^*(t), \ldots, H_k^*(t))'$, $H_i^*(t) = (f(t, x^*(t), u_i), \psi^*(t))$, $i = 1, 2, \ldots, k$.

(c) $M(t_2^*) + 1 = 0$.

Justify (a)–(c) and determine whether $\xi^*(t)$ can take values other than the vertices of the simplex Λ.

6.2.P18 Consider the system $\dot{x} = u$ (scalar), $[t_1, t_2] = [0, 1]$, $U = R$, with $x(0) = x_1 = 1$, and target $x_2 = 0$. Let the cost function be given by

$$J(u) = \int_0^1 t^2 u^2\, dt, \qquad u \in \mathcal{U} = \left\{ u\colon u(t) \in U, \int |u(t)|\, dt < \infty \right\}.$$

The problem is to find a control $u \in \mathcal{U}$ that imparts a minimum to $J(u)$. Show that there is no optimal control. (*Hint*: Use $u_\varepsilon(t) = -1/\varepsilon$ for $t \in [0, \varepsilon]$ and 0 for $t \in (\varepsilon, 1]$, and show that this is a minimizing sequence with $\lim_{\varepsilon \downarrow 0} J(u_\varepsilon) = 0$. Also $\inf\{J(u), u \in \mathcal{U}\} = 0$ but there is no control in $\mathcal{U}$ that realizes the infimum).

6.2.P19 Consider the system

$$\dot{x} = f(t, x) + G(t, x)u, \qquad t \in I,$$

with control constraints,

(a) $U_0 = \{v \in R^m\colon |v_i| = 1,\ i = 1, 2, \ldots, m\}$

or

(b) $U_r = \{v \in R^m\colon |v_i| \leqslant 1,\ i = 1, 2, \ldots, m\}$.

Let X_0 (resp. X_r) and $A_0(t)$ (resp. $A_r(t)$) denote the trajectories and attainable sets corresponding to U_0 (resp. U_r). Justify that $\bar{X}_0 = X_r$, $\bar{A}_0(t) = A_r(t)$.

6.3.P20 Justify Remark 6.3.7.

6.3.P21 Using equations (6.3.29)′–(6.3.31)′, design an optimal regulator for the system $\ddot{\theta} + \beta\dot{\theta} = u$ to minimize the quadratic cost,

$$J(u) = \frac{1}{2}\int_0^T \{\lambda_1(x_2 - w)^2 + \lambda_2 u^2\}\, dt, \qquad \lambda_1, \lambda_2 > 0.$$

For numerical results choose suitable data.

6.3.P22 Prove Theorems 6.3.8, 6.3.13.

6.3.P23 Prove that $(D^+V)(y)$ and $(D^-V)(y)$ are closed convex sets (see equation (6.3.58)).

6.3.P24 Derive the matrix Riccati differential equation (6.3.34) using Pontryagin's minimum principle (Corollary 6.2.6).

7

Elements of stochastic systems

7.0 Introduction

In this chapter, we present the fundamentals of stochastic systems. In section 7.1, we present a selection of basic concepts and results from the probability theory essential for reading this chapter. In section 7.2, we introduce stochastic integrals based on the Wiener process and the Poisson process. Linear and nonlinear stochastic systems are studied in sections 7.3 and 7.4 respectively. Here we discuss the questions of existence and uniqueness of solutions including their regularity properties, and continuous dependence on parameters. We consider both continuous diffusion and jump Markov processes and present Kolmogorov equations for various physical problems of interest. In section 7.5, we study both linear and nonlinear filter theory and identification covering linear filtering (Kalman–Bucy) with complete proofs. Some elements of stability, optimal control and reliability are studied in section 7.6. Here we discuss the questions of stabilizability, necessary conditions of optimality, and Bellman's equations. The chapter is concluded with a brief discussion of a specific class of reliability problems.

7.1 Basic concepts from probability theory

A probability space is defined as the triple $(\Omega, \mathscr{F}, P)$, where Ω is the sample space of elementary events $\omega \in \Omega$, and $\mathscr{F}$ is a class of subsets of the set Ω satisfying:

(a) the empty set $\varnothing \in \mathscr{F}$, $\Omega \in \mathscr{F}$,
(b) A' ($\equiv$complement of A) $\in \mathscr{F}$ whenever $A \in \mathscr{F}$,
(c) for any countable family of sets $\{A_i\} \in \mathscr{F}$, $\bigcup_{i=1}^{\infty} A_i \in \mathscr{F}$.

The class $\mathscr{F} \subset 2^{\Omega}$ ($\equiv$the set of all subsets of Ω) is called the σ-algebra if $\mathscr{F}$ satisfies the properties (a), (b) and (c). If (c) is valid only for finite union, then $\mathscr{F}$ is called an algebra. Throughout we are interested only in the σ-algebra.

The third member P is a set function mapping $\mathscr{F}$ to $[0, 1]$ such that

(d) $P(\varnothing) = 0$,
(e) $P(\Omega) = 1$,
(f) $P(\bigcup_{i=1}^{\infty} A_i) = \sum_{i=1}^{\infty} P(A_i)$ whenever $A_i \cap A_j = \varnothing$, $i \neq j$.

A set function P satisfying the above properties is called a *probability measure.*

Random variables (measurable functions)

Throughout we shall be dealing with random variables or elements taking values in a metric space (S, ρ). We use $\mathscr{B}(S)$ to denote the σ-algebra of Borel subsets of S, that is, the σ-algebra generated by open or closed subsets of S. For $S = R^n$, we can use any of the equivalent metrics.

Definition 7.1.1 (Random variables or measurable functions). A function $X: (\Omega, \mathscr{F}) \to S$ is called an S-valued random variable (random element or measurable function) if $\{\omega \in \Omega: X(\omega) \in \Gamma\} \equiv X^{-1}(\Gamma) \in \mathscr{F}$ for every open set $\Gamma \in S$. X is said to be Borel measurable if for every $\Gamma \in \mathscr{B}(S)$, $X^{-1}(\Gamma) \in \mathscr{F}$.

For any $\mathscr{F}$-measurable element X we shall use the symbol $\sigma(X)$ to denote the smallest subsigma algebra of $\mathscr{F}$ with respect to which X is measurable. We are mostly concerned with R^n-valued random variables and occasionally random variables taking values in general metric spaces.

It is easy to verify that the class of real or complex-valued measurable functions forms an algebra, that is, it is closed with respect to addition and multiplication, including division. We shall denote by $L_0(\Omega, \mathscr{F}, P; R^n)$ the class of R^n-valued $\mathscr{F}$-measurable random variables. It is clear that for each $X \in L_0(\Omega, \mathscr{F}, P; R^n)$, there corresponds a probability measure μ^* on R^n given by

$$\mu^*(K) = P\{\omega \in \Omega: X(\omega) \in K\} = P\{X \in K\} = PX^{-1}(K)$$

for each $K \in \mathscr{B}(R^n)$.

A random variable $X \in L_0(\Omega, \mathscr{F}, P; R^n)$ is said to belong to $L_p(\Omega, \mathscr{F}, P; R^n)$, $1 \leqslant p < \infty$, if

$$\int_{\Omega} |X(\omega)|^p \, P(d\omega) = \int_{R^n} |x|^p \, \mu^*(dx) < \infty. \tag{7.1.1}$$

For $p = \infty$, $X \in L_\infty$ if ess sup$\{|X(\omega)|, \omega \in \Omega\} < \infty$. Note that $L_\infty \subset L_p \subset L_1 \subset L_0$ for $1 < p < \infty$.

For $X \in L_1 \equiv L_1(\Omega, \mathscr{F}, \mathrm{P}; R^n)$, the mean is defined by

$$m(X) = \mathrm{E}(X) = \int_\Omega X(\omega)\mathrm{P}(\mathrm{d}\omega) = \int_{R^n} x\mu^*(\mathrm{d}x), \qquad (7.1.2)$$

and for $X \in L_2$, the covariance is defined by the operator $C(X)$ given by

$$(C(X)\xi, \xi) \equiv \mathrm{E}(X - m(X), \xi)^2 = \int_\Omega (X(\omega) - m(X), \xi)^2 \mathrm{P}(\mathrm{d}\omega)$$

$$= \int_{R^n} (x - m(X), \xi)^2 \mu^*(\mathrm{d}x) \qquad (7.1.3)$$

for $\xi \in R^n$.

Modes of convergence

Definition 7.1.2 (Convergence in law). A sequence $\{X_\ell\} \in L_0(\Omega, \mathscr{F}, \mathrm{P}; R^n)$ is said to converge to $X \in L_0$ in law if the corresponding sequence of measures $\mu^\ell \equiv \mathrm{P}X_\ell^{-1}$ converges weakly to the measure $\mu \equiv \mathrm{P}X^{-1}$ in the sense that, for every continuous bounded function f on R^n,

$$\int_{R^n} f(x)\mu^\ell(\mathrm{d}x) \to \int_{R^n} f(x)\mu(\mathrm{d}x) \qquad \text{as } \ell \to \infty. \qquad (7.1.4)$$

Definition 7.1.3 (Convergence a.s. or a.e.). A sequence $\{X_\ell\}$ is said to converge to X almost surely (a.s.), or almost everywhere (a.e.), or with probability one ($\mathrm{P} = 1$) if

$$\mathrm{P}\left\{\omega \in \Omega: \lim_{\ell \to \infty} \rho(X_\ell, X) = 0\right\} = 1, \qquad (7.1.5)$$

where ρ is any metric on R^n.

Definition 7.1.4 (Convergence in probability or measure). The sequence $X_\ell \xrightarrow{\mathrm{P}} X$ (in probability or in measure) if for every $\varepsilon > 0$,

$$\lim_{\ell \to \infty} \mathrm{P}\{\omega \in \Omega: \rho(X_\ell, X) > \varepsilon\} = 0. \qquad (7.1.6)$$

Definition 7.1.5 (Convergence in the mean of order p). Suppose X_ℓ, $X \in L_p(\Omega, \mathscr{F}, \mathrm{P}; R^n)$. The sequence $\{X_\ell\}$ is said to converge to X in the mean of order p (or L_p sense) if

$$\lim_{\ell \to \infty} \int_\Omega |X_\ell - X|^p \, \mathrm{P}(\mathrm{d}\omega) = 0. \tag{7.1.7}$$

We now present some basic results which will be frequently used throughout this chapter. For the proofs the reader may consult any standard text on measure theory and stochastic processes [64], [65], [41].

Proposition 7.1.6 *In a probability space* $(\Omega, \mathscr{F}, \mathrm{P})$, *conv.* a.e. (a.s.)$\Rightarrow$*conv. meas. (conv. in prob.)*$\Rightarrow$*conv. weak (conv. in law).*

Two random variables $X, Y \in L_1(\Omega, \mathscr{F}, \mathrm{P}; R^n)$ are said to be *independent* if for all $\alpha, \beta \in R^n$

$$\mathrm{E}\{(X, \alpha)(Y, \beta)\} = \mathrm{E}\{(X, \alpha)\}\mathrm{E}\{(Y, \beta)\}. \tag{7.1.8}$$

Two sets A_1, $A_2 \in \mathscr{F}$ are said to be independent if

$$\mathrm{E}(\chi_{A_1}\chi_{A_2}) = \mathrm{E}(\chi_{A_1})\mathrm{E}(\chi_{A_2}),$$

where

$$\chi_A \equiv \begin{cases} 1 & \text{if } \omega \in A, \\ 0 & \text{otherwise,} \end{cases}$$

is called the indicator function of the set A. Similarly, two subsigma algebras $\mathscr{F}_1$, $\mathscr{F}_2 \subset \mathscr{F}$ are said to be independent if

$$\mathrm{E}(\chi_{A_1}\chi_{A_2}) = \mathrm{E}(\chi_{A_1})\mathrm{E}(\chi_{A_2}),$$

for all $A_1 \in \mathscr{F}_1$ and $A_2 \in \mathscr{F}_2$.

Proposition 7.1.7 (Borel–Cantelli lemma). *Let* $\{A_i\}$ *be a sequence of* $\mathscr{F}$*-measurable sets and let* A *denote the set of points in infinitely many of the* A_i *(that is,* $A = \overline{\lim}\, A_i = \bigcap_{n=1}^{\infty} \bigcup_{i=n}^{\infty} A_i$*) then*

$$\mathrm{P}(A) = 0 \qquad \text{if } \sum_{i=1}^{\infty} \mathrm{P}(A_i) < \infty,$$

and

$$\mathrm{P}(A) = 1 \qquad \text{if } \sum_{i=1}^{\infty} \mathrm{P}(A_i) = \infty \quad \textit{and the } A_i \textit{ are mutually independent.}$$

Proposition 7.1.8 (Fatou's lemma). *Let* $\{X_r\} \in L_0(\Omega, \mathscr{F}, \mathrm{P}; R^n)$*; then*

$$\mathrm{E}\left(\underline{\lim}_{r} |X_r|\right) \leq \underline{\lim}_{r} \mathrm{E}(|X_r|).$$

Conditional expectation

Consider the measurable space $(\Omega, \mathscr{F})$ and suppose μ_1 and μ_2 are two finite positive measures (not necessarily probability measures) on $\mathscr{F}$. The measure μ_2 is said to be *absolutely continuous* with respect to the measure μ_1 (written $\mu_2 \ll \mu_1$) if, for every $\varepsilon > 0$, there exists a $\delta = \delta(\varepsilon) > 0$ such that $\mu_2(A) < \varepsilon$ whenever $\mu_1(A) < \delta$, for all $A \in \mathscr{F}$. Given $g \in L_1^+(\Omega, \mathscr{F}, \mu_1)$, it is clear that μ_2, defined by

$$\mu_2(A) = \int_A g(\omega)\mu_1(d\omega), \tag{7.1.9}$$

is absolutely continuous with respect to μ_1. The converse is called the Radon–Nikodym theorem, which states that if μ_1 and μ_2 are two finite positive measures on $\mathscr{F}$ and if $\mu_2 \ll \mu_1$, then there exists a function $g \in L_1^+(\Omega, \mathscr{F}, \mu_1)$ such that $\mu_2(A) = \int_A g(\omega)\mu_1(d\omega)$ for all $A \in \mathscr{F}$.

Let $X \in L_1(\Omega, \mathscr{F}, \mathrm{P}; R^n)$ and G, a (completed) subsigma algebra of $\mathscr{F}$, be given. Any G-measurable R^n-valued random variable Y satisfying

$$\int_A Y\mathrm{P}(d\omega) = \int_A X\mathrm{P}(d\omega) \quad \text{for all } A \in G, \tag{7.1.10}$$

is called the conditional expectation of X given the subsigma algebra G, and is written as

$$Y = \mathrm{E}\{X \mid G\}. \tag{7.1.11}$$

This random variable is uniquely defined by virtue of the Radon–Nikodym theorem. Any other G-measurable random variable z satisfying (7.1.10) must equal Y, P_G-a.e., that is $\mathrm{P}_G\{\omega : z(\omega) \neq Y(\omega)\} = 0$, where P_G is the restriction of the measure P on G.

Using this general result one can define the conditional expectation of any random variable X with reference to any family of random variables $\{X_1, X_2, \ldots, X_m\}$ as

$$\mathrm{E}\{X \mid \{X_1, X_2, \ldots, X_m\}\} = \mathrm{E}\{X \mid \sigma(X_1, X_2, \ldots, X_m)\}, \tag{7.1.12}$$

where $\sigma(X_1, X_2, \ldots, X_m)$ denotes the smallest σ-algebra $\subset \mathscr{F}$ with respect to which $\{X_1, X_2, \ldots, X_m\}$ are all measurable.

Note that, if X is taken as the indicator function of a set $A \in \mathscr{F}$, then

$$Y = \mathrm{E}(X \mid G) = \mathrm{E}(\chi_A \mid G) = \mathrm{P}(A \mid G), \tag{7.1.13}$$

gives the conditional probability of the event A given the σ-algebra G.

The following proposition gives us the basic properties of conditional expectation which are of considerable importance in the sequel.

Proposition 7.1.9 (Properties of conditional expectation). *The conditional expectation* $\mathrm{E}\{\cdot \mid G\}$ *operator satisfies the following properties:*

(a) $\mathrm{E}\{\alpha_1 X_1 + \alpha_2 X_2 \mid G\} = \alpha_1 \mathrm{E}\{X_1 \mid G\} + \alpha_2 \mathrm{E}\{X_2 \mid G\}$ *for all scalars* α_1, α_2 *and* $X_1, X_2 \in L_1(\Omega, \mathscr{F}, \mathrm{P}; R^n)$ *and* $G \subset \mathscr{F}$,

(b) $\mathrm{E}\{\mathrm{E}(X \mid G_2) \mid G_1\} = \mathrm{E}\{X \mid G_1\}$ *for all* $X \in L_1$ *and* $G_1 \subset G_2 \subset \mathscr{F}$,

(c) $\mathrm{E}\{\mathrm{E}(X \mid G)\} = \mathrm{E}(X)$ *for all* $X \in L_1$ *and* $G \subset \mathscr{F}$,

(d) $\mathrm{E}(X \mid G) = \mathrm{E}(X)$ *if* $\sigma(X)$ *is independent of* G *written* $\sigma(X) \perp G$,

(e) $\mathrm{E}\{(z, X) \mid G\} = (z, \mathrm{E}(X \mid G))$ *for all* G*-measurable* $z \in L_\infty(\Omega, \mathscr{F}, \mathrm{P}; R^n)$ *and* $X \in L_1(\Omega, \mathscr{F}, \mathrm{P}; R^n)$.

Note that if $z \in L_p(\Omega, \mathscr{F}, \mathrm{P}; R^n)$ and it is measurable with respect to a subsigma algebra G, then $z \in L_p(\Omega, G, \mathrm{P}_G; R^n)$ where P_G is the restriction of P to the σ-algebra G. Hence $L_p(\Omega, G, \mathrm{P}_G; R^n) \subset L_p(\Omega, \mathscr{F}, \mathrm{P}; R^n)$.

Proposition 7.1.10 *Let* $\{X_\ell\} \in L_1(\Omega, \mathscr{F}, \mathrm{P}; R^n)$ *and suppose there exists an* $\eta \in L_1^+(\Omega, \mathscr{F}, \mathrm{P})$ *such that* $|X_\ell(\omega)| \leqslant \eta(\omega)$, P-*a.s. for all* ℓ *and that* $X_\ell \to X$, P-*a.s.. Then* $X \in L_1(\Omega, \mathscr{F}, \mathrm{P}; R^n)$ *and*

(a) $\lim_{\ell \to \infty} \mathrm{E}(X_\ell) = \mathrm{E}(X)$,

(b) $\lim_{\ell \to \infty} \mathrm{E}\{X_\ell \mid G\} = \mathrm{E}\{X \mid G\}$,

for any subsigma algebra $G \subset \mathscr{F}$.

Stochastic processes

An R^n-valued stochastic process $\xi = \{\xi(t), t \in I\}$ is a measurable mapping from $I \times \Omega$ to R^n, that is, for each $\Gamma \in \mathscr{B}(R^n)$, $\{(t, \omega) \in I \times \Omega: \xi(t, \omega) \in \Gamma\} \in \mathscr{B}(I) \times \mathscr{F}$ where $\mathscr{B}(I)$ is the σ-algebra of Borel sets in I. Also one may consider $\xi = \{\xi_t(\omega), t \in I\}$ as an indexed family of R^n-valued random variables such that for each $t \in I$, $\{\omega \in \Omega: \xi_t(\omega) \in \Gamma\} \in \mathscr{F}$.

The process ξ is said to be continuous with probability 1 if

$$P\left\{\lim_{s \to t} \rho(\xi(t), \xi(s)) = 0\right\} = 1, \tag{7.1.14}$$

for all $s,t \in I$, where ρ is any equivalent metric on R^n; and it is said to be continuous in probability if, for any $\varepsilon > 0$,

$$\lim_{s \to t} \mathrm{P}\{\rho(\xi(t), \xi(s)) > \varepsilon\} = 0, \tag{7.1.15}$$

for all s, $t \in I$. Similarly, one can define continuity in the mean of order p

if

$$\lim_{s \to t} \mathrm{E}\{(\rho(\xi(t), \xi(s)))^p\} = 0, \tag{7.1.16}$$

for all $s,t \in I$. A process ξ is said to be right or left continuous if the above equalities hold when the limits are taken from the right or the left.

A stochastic process $\xi = \{\xi(t), t \in I\}$, defined on the probability space $(\Omega, \mathcal{F}, \mathrm{P})$, is said to be separable if, for every $\Gamma \in \mathcal{B}(R^n)$ and every countable dense set $\tilde{I} \subset I$, and $J \subset I$ with $\tilde{J} = J \cap \tilde{I}$, the set $\{\omega \in \Omega: \xi(t, \omega) \in \Gamma \text{ for } t \in \tilde{J}\}$ differs from the set $\{\omega \in \Omega: \xi(t, \omega) \in \Gamma \text{ for } t \in J\}$ by a set of P-measure zero.

The following criterion for continuity of stochastic processes is given by Kolmogorov.

Proposition 7.1.11 (Continuity with probability 1). *A process ξ with values in R^n and defined on a probability space $(\Omega, \mathcal{F}, \mathrm{P})$ is said to be continuous with probability* 1 *if there exist numbers β, $\delta > 0$ and a finite positive number k such that*

$$\mathrm{E}\,|\xi(t) - \xi(s)|^\beta \leqslant k\,|t-s|^{1+\delta}, \tag{7.1.17}$$

for all s, $t \in I$.

Adapted process

Let $\{\mathcal{F}_t, t \geqslant 0\}$ be an increasing family of subsigma algebras of the σ-algebra $\mathcal{F}$, that is, $\mathcal{F}_{t_1} \subset \mathcal{F}_{t_2}$ for $t_1 < t_2$. A stochastic process ξ is said to be adapted to $\mathcal{F}_t$ if for each $t \in I$, $\xi(t)$ is $\mathcal{F}_t$-measurable, that is, $\{\omega: \xi(t, \omega) \in \Gamma\} \in \mathcal{F}_t$, for every $\Gamma \in \mathcal{B}(R^n)$.

Markov process

A stochastic process $\xi = \{\xi(t), t \in I\}$ defined on a probability space $(\Omega, \mathcal{F}, P)$ is said to be an $\mathcal{F}_t$-*Markov process* if, for every $A \in \mathcal{B}(R^n)$, and $s < t$,

$$\mathrm{P}\{\xi(t) \in A \mid \mathcal{F}_s\} = \mathrm{P}\{\xi(t) \in A \mid \sigma(\xi(s))\}, \tag{7.1.18}$$

or equivalently, for any bounded measurable function f on R^n,

$$\mathrm{E}\{f(\xi(t)) \mid \mathcal{F}_s\} = \mathrm{E}\{f(\xi(t)) \mid \sigma(\xi(s))\}. \tag{7.1.19}$$

A Markov process is completely characterized by its transition probability kernel $\mathrm{P}(s, x; t, A) \equiv \mathrm{P}\{\xi(t) \in A \mid \xi(s) = x\}$, and the initial distribution $\Pi_0(A) \equiv P\{\xi(0) \in A\}$, by the relation

$$P\{\xi(t_1) \in A_1, \xi(t_2) \in A_2, \ldots, \xi(t_k) \in A_k\}$$

$$\equiv \int_{R^n} \Pi_0(dx) \int_{A_1} P(0, x; t_1, dx_1) \int_{A_2} P(t_1, x_1; t_2, dx_2) \cdots$$

$$\int_{A_{k-1}} P(t_{k-2}, x_{k-2}; t_{k-1}, dx_{k-1}) P(t_{k-1}, x_{k-1}; t_k, A_k) \quad (7.1.20)$$

for $0 < t_1 < t_2 \ldots < t_k$.

Wiener process

A process $W = \{W(t), t \in I\}$ with values in R^n is said to be a *standard Wiener process* or *Brownian motion* on the probability space $(\Omega, \mathcal{F}, \mathcal{F}_{t\geq 0}, P)$ if it is an $\mathcal{F}_t$-Markov process satisfying

(a) $P\{W(0) = 0\} = 1$, and

(b) $P\{W(t) \in A \mid W(s) = x\}$

$$= \int_A \frac{1}{(2\pi(t-s))^{n/2}} \exp\left(-\frac{1}{2(t-s)} |y-x|^2\right) dy, \qquad \text{for } s < t, \ s, \ t \in I. \quad (7.1.21)$$

Clearly, the Wiener process is both temporally and spatially homogeneous since it follows from (7.1.21) that

$$P(s, x; t, dy) = p(t-s, y-x)\, dy,$$

where

$$p(t-s, y-x) = \left(\frac{1}{(2\pi(t-s))^{n/2}}\right) \exp\left(-\frac{|y-x|^2}{2(t-s)}\right). \quad (7.1.22)$$

The Wiener process has many interesting properties:

(W)(a) the transition probability is absolutely continuous with respect to Lebesgue measure in R^n.

(W)(b) it is continuous with probability 1; in fact, with probability 1, it is Hölder continuous with exponent $\alpha < \frac{1}{2}$,

(W)(c) its increments over disjoint intervals of time are independent Gaussian random elements,

(W)(d) its sample paths are nowhere differentiable with probability 1,

(W)(e) its sample paths have unbounded variation, that is,

$$P\left\{\sup_n \sum_{i=1}^{n} |W(t_{i+1}) - W(t_i)| < r\right\} = 0 \qquad \text{for any } 0 < r < \infty,$$

(W)(f) $E\{(W(t), \eta_1)(W(s), \eta_2)\} = (t \wedge s)(\eta_1, \eta_2)$. (7.1.23)

Martingale

A process $\xi = \{\xi(t), t \in I\}$ on a probability space $(\Omega, \mathscr{F}, \mathscr{F}_{t\geq 0}, \mathrm{P}\}$ is said to be an $\mathscr{F}_t$-martingale if

$$\begin{aligned} &\text{(a)}\quad \mathrm{E}\,|\xi(t)| < \infty, && t \in I, \\ &\text{(b)}\quad \mathrm{E}\{\xi(t) \mid \mathscr{F}_s\} = \xi(s) && \text{for } s \leq t, \end{aligned} \tag{7.1.24}$$

and it is said to be a pth-power integrable martingale if, in addition $\mathrm{E}\,|\xi(t)|^p < \infty$, $t \in I$.

A scalar-valued process ξ is said to be a submartingale (resp. supermartingale) if

$$\begin{aligned} &\text{(a)}\quad \mathrm{E}\,|\xi(t)| < \infty, && t \in I, \\ &\text{(b)}\quad \mathrm{E}\{\xi(t) \mid \mathscr{F}_s\} \geq \xi(s) && (\text{resp.} \leq \xi(s)). \end{aligned} \tag{7.1.25}$$

It is easy to verify from (7.1.21) that the Wiener process is a square-integrable martingale and that the process $M(t) = |W(t)|^2$ is a submartingale.

The following proposition gives the celebrated martingale inequality due to Doob.

Proposition 7.1.12 (Martingale inequality). *If ξ is an $\mathscr{F}_t$-martingale on the probability space $(\Omega, \mathscr{F}, \mathrm{P})$ then*

(a) $$\mathrm{P}\left\{\sup_{t\in I} |\, \xi(t)| > r\right\} \leq \frac{\mathrm{E}\,|\xi(T)|}{r} \textit{ for every } r > 0, \tag{7.1.26}$$

(b) *if for some $p > 1$, $\mathrm{E}\,|\xi(t)|^p < \infty$ for all $t \in I = [0, T]$, then*

$$\mathrm{E}\left\{\sup_{t\in I} |\xi(t)|^p\right\} \leq (q)^p \mathrm{E}\,|\xi(T)|^p, \tag{7.1.27}$$

for $\left(\frac{1}{p}\right) + \left(\frac{1}{q}\right) = 1$, and, in particular, for square-integrable martingales

$$\mathrm{E}\left\{\sup_{t\in I} |\xi(t)|^2\right\} \leq 4\mathrm{E}\,|\xi(T)|^2, \tag{7.1.28}$$

(c) *the space of L_p-martingales with the norm topology $\|\xi\| \equiv (\mathrm{E}\,|\xi(T)|^p)^{1/p}$ is a Banach space denoted by M^p.*

Proposition 7.1.13 (Jensen's inequality). *Given any convex function f and $X \in L_p$,*

$$f(\mathrm{E}\,|X|^p) \leq \mathrm{E}f(|X|^p). \tag{7.1.29}$$

7.2 Stochastic integrals and differentials

In the study of systems governed by stochastic differential equations stochastic calculus plays a central role. In this section, we shall briefly introduce stochastic integrals and differentials as originally developed by Itô. We shall consider stochastic integrals based on the Wiener process as well as on the Poisson process. These fundamental processes are considered as the building blocks of stochastic differential equations.

7.2.1 Stochastic integrals based on the Wiener process

Let $(\Omega, \mathscr{F}, \mathrm{P})$ be a complete probability space and $\mathscr{F}_t$, $t \geqslant 0$, an increasing family of subsigma algebras of the σ-algebra $\mathscr{F}$. Let $W = \{W(t), t \geqslant 0\}$ be a standard $\mathscr{F}_t$ Brownian motion (or Wiener process) and let $f = \{f(t), t \geqslant 0\}$ be an $\mathscr{F}_t$-adapted (or $\mathscr{F}_t$-measurable) stochastic process and suppose $\mathscr{F}_t$ is independent of the σ-algebra $\sigma\{W(s) - W(\theta); s \geqslant \theta \geqslant t\}$. We consider integrals of the form

$$L(f) \equiv \int_I f(t)\,\mathrm{d}W(t), \qquad I = [0, T]. \tag{7.2.1}$$

This integral has the appearance of the classical Lebesgue–Stieltjes integral

$$J(f) = \int_I f(t)\,\mathrm{d}\alpha(t), \tag{7.2.2}$$

in which α is required to be a function of bounded variation. That is, for every partition $\pi \equiv \{0 = t_0 < t_1 \cdots < t_n = T\}$,

$$\sup_{\pi \in F} \left\{ \sum_{i=0}^{n-1} |\alpha(t_{i+1}) - \alpha(t_i)| \right\} < \infty, \tag{7.2.3}$$

where F denotes the class of all finite partitions of the interval $I = [0, T]$. In the case of Brownian motion

$$\mathrm{P}\left\{ \lim_n \sum_{i=0}^{n-1} |W(t_{i+1}) - W(t_i)| < \infty \right\} = 0. \tag{7.2.4}$$

Hence the integral (7.2.1) is not defined in the Lebesgue–Stieltjes sense. There is a unique way for defining the integral (7.2.1).

First we define the integral (7.2.1) for simple functions. A function f is said to be simple (or step function) if there exists a positive integer ℓ, a partition

$$\pi^{\ell} \equiv \{0 = t_0 < t_1 \cdots < t_{\ell} = T\}$$

and a sequence of random variables $\{a_i, i=0, 1, \ldots, \ell-1\}$ such that

$$f(t)=a_i \qquad \text{for } t \in (t_i, t_{i+1}], \quad i=0, 1, \ldots, \ell-1, \tag{7.2.5}$$

and a_i is $\mathscr{F}_{t_i}$-measurable. Let $S(I, \mathscr{F}_t)$ denote the class of all simple functions as defined above and $L_2^e(I, \mathscr{F}_t)$ the class of functions $\{f\}$ such that $f(t)$ is $\mathscr{F}_t$-measurable for each $t \in I$ and

$$\mathrm{E}\int_I |f(t)|^2 \,\mathrm{d}t < \infty.$$

For $f \in S(I, \mathscr{F}_t) \cap L_2^e(I, \mathscr{F}_t) \equiv S \cap L_2^e$, we define

$$L(f) = \int_I f(t)\,\mathrm{d}W(t) = \sum_{i=0}^{n-1} a_i(W(t_{i+1}) - W(t_i)). \tag{7.2.6}$$

Clearly, this is a well-defined random variable and it is easy to verify that it satisfies several properties as described below:

(a) For $f_1, f_2 \in S \cap L_2^e$, and α_1, α_2 constants

$$L(\alpha_1 f_1 + \alpha_2 f_2) = \alpha_1 L(f_1) + \alpha_2 L(f_2). \tag{7.2.7}$$

(b) For $f \in S \cap L_2^e$,

$$\begin{aligned} \mathrm{E}L(f) &= \mathrm{E}\Big\{\sum_i a_i(W(t_{i+1}) - W(t_i))\Big\} \\ &= \mathrm{E}\Big\{\sum_i \mathrm{E}\{a_i(W(t_{i+1}) - W(t_i)) \mid \mathscr{F}_{t_i}\}\Big\}. \end{aligned}$$

Since a_i is $\mathscr{F}_{t_i}$-measurable

$$\mathrm{E}\{a_i(W(t_{i+1}) - W(t_i)) \mid \mathscr{F}_{t_i}\} = a_i \mathrm{E}\{(W(t_{i+1}) - W(t_i)) \mid \mathscr{F}_{t_i}\}.$$

Hence

$$\mathrm{E}L(f) = \mathrm{E}\Big\{\sum_i a_i \mathrm{E}\{(W(t_{i+1}) - W(t_i)) \mid \mathscr{F}_{t_i}\}\Big\}.$$

Since W is an $\mathscr{F}_t$-martingale, $\mathrm{E}\{W(t_{i+1}) - W(t_i) \mid \mathscr{F}_{t_i}\} = 0$, and hence

$$\mathrm{E}L(f) = 0. \tag{7.2.8}$$

(c) For $f \in S \cap L_2^e$,

$$\mathrm{E}(L(f))^2 = \mathrm{E}\Big\{\sum_{i,j} a_i a_j (W(t_{i+1}) - W(t_i))(W(t_{j+1}) - W(t_j))\Big\}.$$

Since the increments of the Wiener process over disjoint intervals are independent, by conditioning as in (b), it is easy to verify that for $i \neq j$, the expression on the right-hand side is zero. Hence

$$\mathrm{E}(L(f))^2 = \mathrm{E}\Big\{\sum_i |a_i|^2 (W(t_{i+1}) - W(t_i))^2\Big\}$$

$$= \mathrm{E}\Big\{\sum_i |a_i|^2 \mathrm{E}\{(W(t_{i+1}) - W(t_i))^2 \mid \mathscr{F}_{t_i}\}\Big\}.$$

Since $\mathrm{E}\{(W(t_{i+1}) - W(t_i))^2 \mid \mathscr{F}_{t_i}\} = (t_{i+1} - t_i)$, we have

$$\mathrm{E}(L(f))^2 = \mathrm{E}\Big\{\sum_i |a_i|^2 (t_{i+1} - t_i)\Big\}$$

$$= \mathrm{E}\int_I |f(t)|^2 \,\mathrm{d}t. \tag{7.2.9}$$

In general, for $f,\ g \in S \cap L_2^e$,

$$\mathrm{E}(L(f)L(g)) = \mathrm{E}\int_I f(t)g(t)\,\mathrm{d}t. \tag{7.2.10}$$

(d) Define the stochastic process ξ given by

$$\xi(t) \equiv L_t(f) \equiv \int_0^t f(\theta)\,\mathrm{d}W(\theta), \qquad t \in I \tag{7.2.11}$$

for $f \in S \cap L_2^e$. Since the Wiener process W is continuous with probability 1 (P-a.s.) so also is ξ. Indeed, let $\pi^{\ell} \equiv \{0 = t_0 < t_1 < t_2 < \cdots < t_\ell = T\}$ denote the partition of I and $\{a_i,\ i = 0, 1, 2, \ldots, \ell - 1\}$ the values of f on $I_i \equiv (t_i, t_{i+1}]$. Then for $s,t \in I_k$, $0 \leq k \leq \ell - 1$, $\xi(t) - \xi(s) = \alpha_k(W(t) - W(s))$ and hence continuity follows.

(e) The process ξ is an $\mathscr{F}_t$-martingale. Indeed for $t > s$ we verify that

$$\mathrm{E}\{\xi(t) - \xi(s) \mid \mathscr{F}_s\} = 0. \tag{7.2.12}$$

Let $s \in I_r = (t_r, t_{r+1}]$, $t \in I_k = (t_k, t_{k+1}]$, $r \leq k < \ell - 1$, then

$$\mathrm{E}\{\xi(t) - \xi(s) \mid \mathscr{F}_s\} = \mathrm{E}\Big\{\sum_{r \leq i \leq k-1} a_i(W(t_{i+1}) - W(t_i)) + a_k(W(t) - W(t_k))$$

$$- a_r(W(s) - W(t_r)) \mid \mathscr{F}_s\Big\}.$$

Applying conditional expectation with respect to $\mathscr{F}_{t_k} \supset \mathscr{F}_{t_{k-1}} \supset \cdots \supset \mathscr{F}_{t_{r+1}}$ in that order, it is easy to verify that

$$\mathrm{E}\{\xi(t) - \xi(s) \mid \mathscr{F}_s\} = \mathrm{E}\{a_r(W(t_{r+1}) - W(s)) \mid \mathscr{F}_s\}$$

$$= a_r \mathrm{E}\{(W(t_{r+1}) - W(s)) \mid \mathscr{F}_s\}, \tag{7.2.13}$$

the last equality follows from the facts that a_r is $\mathscr{F}_{t_r}$-measurable and

$\mathscr{F}_{t_r} \subset \mathscr{F}_s$, since $t_r \leqslant s$. Due to the martingale property of W, (7.2.12) follows from (7.2.13).

(f) The quadratic variation process corresponding to the process $\xi(t) = L_t(f)$ is defined by

$$\langle \xi(t) \rangle = \langle L_t(f) \rangle \equiv \langle L_t(f), L_t(f) \rangle \equiv \int_0^t |f(\theta)|^2 \, d\theta. \tag{7.2.14}$$

The process $\langle L_t(f) \rangle$ is also known as the square variation or the characteristics or the compensator of the martingale $\xi(t) = L_t(f)$. The process η given by

$$\eta(t) \equiv (L_t(f))^2 - \langle L_t(f) \rangle, \qquad t \in I,$$

is an $\mathscr{F}_t$-martingale. Indeed, for $s < t$,

$$\begin{aligned} &\mathrm{E}\{\eta(t) - \eta(s) \mid \mathscr{F}_s\} \\ &\quad = \mathrm{E}\left\{(L_t(f) + L_s(f))(L_t(f) - L_s(f)) - \int_s^t |f(\theta)|^2 \, d\theta \mid \mathscr{F}_s\right\} \\ &\quad = \mathrm{E}\left\{(L_t(f) - L_s(f))^2 - \int_s^t |f(\theta)|^2 \, d\theta \mid \mathscr{F}_s\right\} = 0, \end{aligned}$$

and $(L_t(f))^2$ is a submartingale.

Summarizing the above results we have the following theorem.

Theorem 7.2.1 (Stochastic integrals for $f \in S \cap L_2^e$). *For each $f \in S \cap L_2^e$, the stochastic process $L_t(f)$, $t \in I$, is a uniquely defined random process satisfying the following properties*:

(a) $L(f) \equiv L_T(f) \in L_2(\Omega, \mathscr{F}, \mathrm{P})$.
(b) $L(\alpha_1 f_1 + \alpha_2 f_2) = \alpha_1 L(f_1) + \alpha_2 L(f_2)$ *for all* $f_1, f_2 \in S \cap L_2^e$ *and* α_1, α_2 *scalars or* $\mathscr{F}_0$*-measurable random variables.*
(c) $\mathrm{E}L(f) = 0$.
(d) $\mathrm{E}(L(f))^2 = \mathrm{E} \int_I |f(t)|^2 \, dt$.
(e) *The process* ξ *given by*

$$\xi(t) = L_t(f) = \int_0^t f(\theta) \, dW(\theta),$$

is a continuous square-integrable $\mathscr{F}_t$*-martingale.*

(f) *The process*

$$\langle \xi(t) \rangle = \langle \xi(t), \xi(t) \rangle = \int_0^t |f(\theta)|^2 \, d\theta,$$

is called the quadratic variation process, and

$$\eta(t) \equiv (\xi(t))^2 - \langle \xi(t) \rangle,$$

is an $\mathscr{F}_t$*-martingale and* $(\xi(t))^2$ *is an* $\mathscr{F}_t$*-submartingale.*

Note that the stochastic integral $L(f)=\int_I f(t)\,\mathrm{d}W(t)$, can be considered as a linear operator from $S\cap L_2^e$ to $L_2(\Omega, \mathscr{F}, \mathrm{P})$. The integral $L(f)$ is also defined for more general integrands. In particular, we can define $L(f)$ for $f\in L_2^e$ as well as for $f\in L_2^p$, where

$$L_2^e\equiv\left\{f: f(t) \text{ is } \mathscr{F}_t\text{-measurable and } \mathrm{E}\int_I |f(t)|^2\,\mathrm{d}t<\infty\right\},$$

and

$$L_2^p\equiv\left\{f: f(t) \text{ is } \mathscr{F}_t\text{-measurable and } \mathrm{P}\left\{\int_I |f(t)|^2\,\mathrm{d}t<\infty\right\}=1\right\}.$$

We introduce in L_2^e the norm $\|\cdot\|$, given by

$$\|f\|^2=\mathrm{E}\int_I |f(t)|^2\,\mathrm{d}t, \tag{7.2.15}$$

with respect to which it is a Banach space. We note that any element $f\in L_2^e$ can be approximated as closely as desired by simple functions. In fact the following result holds.

Lemma 7.2.2 *The set $S\cap L_2^e$ is dense in L_2^e, that is, for each $f\in L_2^e$ there exists a sequence $\{f_n\}\in S\cap L_2^e$ such that*

$$\lim_{n\to\infty}\|f-f_n\|=\lim_{n\to\infty}\left(\mathrm{E}\int_I |f(t)-f_n(t)|^2\,\mathrm{d}t\right)^{1/2}=0.$$

Proof We shall give only an outline of the proof. Define

$$f_m(t)=f(t)\psi_m(f(t)), \tag{7.2.16}$$

where

$$\psi_m(x)=\begin{cases}1 & \text{if } |x|\leq m,\\ 0 & \text{otherwise.}\end{cases} \tag{7.2.17}$$

Clearly, $f_m(t)$ is $\mathscr{F}_t$-measurable, $|f_m(t)|\leq|f(t)|$, $f_m(t)\to f(t)$ a.e. on I, P-a.s., and by Lebesgue dominated convergence theorem $f_m\to f$ in L_2^e. Hence, given $\varepsilon>0$, there exists an $m_0=m_0(\varepsilon)$ such that

$$\|f_m-f\|_{L_2^e}<\frac{\varepsilon}{3} \qquad \text{for } m\geq m_0. \tag{7.2.18}$$

Take f_{m_0} and smooth it out by use of the molifier φ giving

$$f_{m_0,n}(t)\equiv\int_{-\infty}^{+\infty} f_{m_0}\left(t-\frac{\tau}{n}\right)\varphi(\tau)\,\mathrm{d}\tau,$$

for positive integers n, where $\varphi(t) \geqslant 0$ for $t \in R$, $\varphi(t) = 0$ for $t \leqslant 0$, $\varphi \in C^\infty$ and

$$\int_{-\infty}^{+\infty} \varphi(t)\, dt = 1,$$

and define $f_{m_0}(t) \equiv f_{m_0}(0)$ for $t \leqslant 0$. Clearly, $\lim_{n\to\infty} f_{m_0,n}(t) = f_{m_0}(t)$ a.e. on I, P-a.s. and $\{f_{m_0,n}\}$ is $\mathscr{F}_t$-adapted for all n. Hence by choosing $n = n_0$ sufficiently large, we obtain

$$\|f_{m_0,n_0} - f_{m_0}\|_{L_2^e} < \frac{\varepsilon}{3}. \tag{7.2.19}$$

Then define

$$f_{m_0,n_0,k}(t) \equiv f_{m_0,n_0}\left(\frac{[kt]}{k}\right),$$

for positive integers k, where $[kt] \equiv$ the largest integer $\leqslant kt$. This is a sequence of simple functions adapted to $\mathscr{F}_t$, and by choosing k sufficiently large we can have

$$\|f_{m_0,n_0,k} - f\|_{L_2^e} < \varepsilon. \tag{7.2.20}$$

Thus we have constructed a sequence of simple functions $g_k = f_{m_0,n_0,k} \in S \cap L_2^e$ approximating f as closely as desired. This completes the proof. ■

Now we can extend the definition of stochastic integrals over L_2^e.

Theorem 7.2.3 (Stochastic integrals over L_2^e). *For each $f \in L_2^e$, the stochastic integral $L_t(f) = \int_0^t f(\theta)\, dW(\theta)$ is a uniquely defined random process satisfying all the properties* (a)–(f) *of Theorem* 7.2.1.

Proof We shall give a brief outline of the proof. For each $f \in L_2^e$, by Lemma 7.2.2, there exists a sequence $f_n \in S \cap L_2^e$ such that $f_n \to f$ in L_2^e. Then $z_n \equiv L_T(f_n)$ is a Cauchy sequence in $L_2(\Omega, \mathscr{F}, \mathrm{P})$ and hence there exists a unique z, independent of the choice of the approximating sequence $\{f_n\}$, such that $z_n \to z$ in $L_2(\Omega, \mathscr{F}, \mathrm{P})$ and hence in probability. The limit z is called the stochastic integral of f and is denoted by $z = L(f) = L_T(f)$.

The proof of (a)–(d), and (f) are easy. We prove (e). Note that for each $t \in I$,

$$\xi_n(t) \equiv L_t(f_n) \to L_t(f) \equiv \xi(t),$$

in $L_2(\Omega, \mathscr{F}, \mathrm{P})$ and hence in probability. Hence it follows from Proposition 7.1.10 that

$$E\{\xi_n(t) - \xi(t) \mid \mathscr{F}_s\} \to 0,$$

in probability. Since $\{\xi_n\}$ is a sequence of square-integrable martingales we have, for $0 \leq s < t$,

$$E\{\xi(t) - \xi(s) \mid \mathscr{F}_s\} = E\{\xi(t) - \xi_n(t) \mid \mathscr{F}_s\} + \xi_n(s) - \xi(s).$$

Hence letting $n \to \infty$ we have $E\{\xi(t) - \xi(s) \mid \mathscr{F}_s\} = 0$, proving that ξ is a martingale.

For continuity we use the martingale inequality (Proposition 7.1.12) and the Borel–Cantelli lemma (Proposition 7.1.7). Since both $\{\xi_n\}$ and ξ are square-integrable martingales, $\xi - \xi_n$ is also a square-integrable martingale. Hence by the Chebyshev inequality and the martingale inequality, we have

$$P\left\{\sup_{t \in I} |\xi(t) - \xi_n(t)| > c\right\} \leq \frac{4}{c^2} \|f - f_n\|^2_{L_2^e}. \tag{7.2.21}$$

Since $f_n \to f$ in L_2^e, we can choose a subsequence $\{n_k\} \subset \{n\}$ such that

$$\sum_{k=1}^{\infty} 4^k \|f - f_{n_k}\|^2_{L_2^e} < \infty. \tag{7.2.22}$$

Thus

$$P\left\{\sup_{t \in I} |\xi(t) - \xi_{n_k}(t)| > \frac{1}{2^k}\right\} \leq 4^{k+1} \|f - f_{n_k}\|^2_{L_2^e}. \tag{7.2.23}$$

Defining

$$A_k \equiv \left\{\sup_{t \in I} |\xi(t) - \xi_{n_k}(t)| > \frac{1}{2^k}\right\},$$

it follows from (7.2.22) and (7.2.23) that

$$P\left(\bigcup_{k=1}^{\infty} A_k\right) \leq \sum_{k=1}^{\infty} P(A_k) \leq 4 \sum_{k=1}^{\infty} 4^k \|f - f_{n_k}\|^2_{L_2^e} < \infty. \tag{7.2.24}$$

Hence by the Borel–Cantelli lemma, only finitely many events of the form A_k can occur. Hence with probability one, $\xi_{n_k} \to \xi$ uniformly on I. Since $\{\xi_n\}$ are continuous and the convergence is uniform on I with probability 1, the limit ξ is also continuous with probability 1. This completes the proof. ■

We can define stochastic integrals for the more general integrands L_2^p which may be given the topology of convergence as follows. The sequence $f_n \to f$ in $L_2^{\mathcal{P}}$ if and only if

$$\lim_{n\to\infty} \mathrm{P}\left\{\int_I |f_n(t) - f(t)|^2 \,\mathrm{d}t > 0\right\} = 0.$$

Theorem 7.2.4 (Stochastic integrals over $L_2^{\mathcal{P}}$). *For each $f \in L_2^{\mathcal{P}}$, the stochastic integral $L(f) = \int_I f(t)\,\mathrm{d}W(t)$, is a uniquely defined random variable satisfying (a) and (b) of Theorem 7.2.1. Further, the process ξ given by $\xi(t) = \int_0^t f(\theta)\,\mathrm{d}W(\theta)$, is a continuous, square-integrable, local $\mathcal{F}_t$-martingale.*

Proof (Outline). For each positive integer n, define

$$f_n(t) \equiv f(t)\psi_n\left(\int_0^t |f(\theta)|^2 \,\mathrm{d}\theta\right),$$

where

$$\psi_n(x) = \begin{cases} 1 & \text{for } 0 \leqslant x \leqslant n, \\ 0 & \text{otherwise.} \end{cases} \tag{7.2.25}$$

Clearly,

$$\int_I |f_n(t)|^2 \,\mathrm{d}t \leqslant n \qquad \text{with probability 1,}$$

and

$$\lim_{n\to\infty} \mathrm{P}\left\{\int_I |f_n(t) - f(t)|^2 \,\mathrm{d}t > 0\right\} = 0.$$

Then the sequence of random variables $z_n \equiv L(f_n)$ is a Cauchy sequence in probability. Indeed, for any $p \geqslant 1$,

$$\mathrm{P}\{|z_{n+p} - z_n| > 0\} \leqslant \mathrm{P}\left\{\int_I |f(t)|^2 \,\mathrm{d}t > n\right\},$$

and hence

$$\lim_{n\to\infty} \mathrm{P}\{|z_{n+p} - z_n| > 0\} = 0$$

due to the fact that $f \in L_2^{\mathcal{P}}$. Thus z_n has a unique limit z in probability. The limit z is called the stochastic integral of f and is denoted by $z = L(f)$. The process

$$\xi(t) = \int_0^t f(\theta)\,\mathrm{d}W(\theta), \qquad t \in I,$$

is a continuous square-integrable local martingale. Indeed, let

$$\tau_n \equiv \inf\left\{t \in I: \int_0^t |f(\theta)|^2 \, \mathrm{d}\theta \geq n\right\},$$

and note that $\{\tau_n \geq t\} \in \mathscr{F}_t$. Define $\xi_n(t) \equiv \xi(t \wedge \tau_n)$. Then $\{\xi_n\}$ is a sequence of continuous square-integrable martingales. This outlines the proof. ■

Corollary 7.2.5 *Let $f_n \in L_2^p$ and $f_n \to f$ a.e. on I in probability and there exists a $g \in L_2^p$ such that $|f_n(t)| \leq g(t)$, a.e. on I,* P-a.s. *Then $L(f_n) \to L(f)$ in probability.*

The results presented above also hold for the multidimensional case. Let W be an m-dimensional $\mathscr{F}_t$-Brownian motion with independent components on the probability space $(\Omega, \mathscr{F}, \mathrm{P})$. Denote again by L_2^e, the equivalence classes of $R(n \times m)$-valued processes adapted to $\mathscr{F}_t$ satisfying

$$\mathrm{E} \int_I \|\sigma(t)\|^2 \, \mathrm{d}t < \infty.$$

Then the stochastic integral

$$L(\sigma) = \int_I \sigma(t) \, \mathrm{d}W(t),$$

and the stochastic process

$$\xi(t) \equiv \int_0^t \sigma(\theta) \, \mathrm{d}W(\theta), \qquad t \in I,$$

are well defined. The operator L maps L_2^e into $L_2(\Omega, \mathscr{F}, \mathrm{P}; R^n)$ and satisfies the following properties:

(L)(a) $L(\alpha_1\sigma_1 + \alpha_2\sigma_2) = \alpha_1 L(\sigma_1) + \alpha_2 L(\sigma_2)$ for σ_1, $\sigma_2 \in L_2^e$ and α_1, α_2 scalars,

(L)(b) $\mathrm{E}L(\sigma) = 0$ for $\sigma \in L_2^e$,

(L)(c) for $x,y \in R^n$, and $\sigma_1, \sigma_2 \in L_2^e$

$$\mathrm{E}\{(L(\sigma_1), x)(L(\sigma_2), y)\} = \mathrm{E} \int_I (\sigma_2(t)\sigma_1'(t)x, y) \, \mathrm{d}t, \qquad (7.2.26)$$

(L)(d) $\xi(t)$, $t \in I$, is an R^n-valued continuous square integrable $\mathscr{F}_t$-martingale,

(L)(e) for each $x \in R^n$,

$$M_t(x) = (\xi(t), x)^2 - \int_0^t (\sigma(\theta)\sigma'(\theta)x, x) \, \mathrm{d}\theta, \qquad (7.2.27)$$

is an $\mathcal{F}_t$-martingale and $\langle \xi(t) \rangle = \int_0^t \sigma(\theta)\sigma'(\theta)\,d\theta$ is the quadratic (matrix) variation process for the martingale $\xi(t) = L_t(\sigma)$,

(L)(f) for $\sigma \in L_2^p$, the process $\xi(t) \equiv L_t(\sigma)$ is an R^n-valued local martingale.

7.2.2 Stochastic integrals based on the Poisson process

Let $\{p(t), t \in I\}$ be a Poisson process on $(\Omega, \mathcal{F}, \mathrm{P})$ and let $\mathcal{B}_0(R^n)$ denote the class of Borel sets in $R_0^n = R^n \setminus \{0\}$ and $\nu(t, \Gamma)$, $t \in I$, $\Gamma \in \mathcal{B}_0(R^n)$, denote the random counting measure giving the number of jumps of $p(\theta)$ over the interval $[0, t]$ taking values $p(\theta) - p(\theta - 0) \equiv \Delta p(\theta) \in \Gamma$.

Define

$$\lambda(t, \Gamma) \equiv \nu(t, \Gamma) - \mathrm{E}\nu(t, \Gamma), \tag{7.2.28}$$

and suppose $\mathrm{E}\nu(t, \Gamma) = t\Pi(\Gamma)$, where Π is a positive measure on $\mathcal{B}_0(R^n)$. We assume that p is a homogeneous Poisson process with the increments of the process $\nu(t, \Gamma)$ over disjoint intervals of time being independent. For each non-negative integer m, and $\Gamma \in \mathcal{B}_0(R^n)$,

$$\mathrm{P}\{\nu(t, \Gamma) = m\} = \frac{\exp\{-t\Pi(\Gamma)\}(t\Pi(\Gamma))^m}{m!}. \tag{7.2.29}$$

We can define stochastic integrals with respect to the centred Poisson random measure $\lambda(t, \Gamma)$. Note that

$$\mathrm{E}\lambda(t, \Gamma) = 0,$$

$$\mathrm{E}(\lambda(t, \Gamma))^2 = t\Pi(\Gamma),$$

and

$$\mathrm{E}\{\lambda(t, \Gamma) - \lambda(s, \Gamma) \mid \mathcal{F}_s\} = 0, \qquad s < t,$$

and hence $\lambda(t, \Gamma)$ is an $\mathcal{F}_t$-martingale. Let $S(I \times R^n, \mathcal{F}_t)$ denote the class of simple random functions g: $I \times R^n \to R^n$ such that there exist $\{\Gamma_\ell\} \in \mathcal{B}_0(R^n)$ with $\Gamma_\ell \cap \Gamma_k = \varnothing$ and $\bigcup_\ell \Gamma_\ell = R_0^n$ and a partition

$$\pi^m \equiv \{0 = t_0 < t_1 < \cdots < t_m = T\}$$

of the interval $I = \bigcup_i I_i$, $I_i \equiv (t_i, t_{i+1}]$ and a sequence of random variables $\{g_{i\ell}\}$ which are $\mathcal{F}_{t_i}$-measurable giving

$$g(t, y) = \sum_{i,\ell} g_{i\ell}\chi_{I_i \times \Gamma_\ell}(t, y). \tag{7.2.30}$$

Let $L_2^\varepsilon(I \times R^n, \mathcal{F}_t, \Pi) \equiv L_2^\varepsilon(\Pi)$ denote the class of $\mathcal{F}_t$-measurable functions $\{g\}$ defined on $I \times R^n$ such that

$$\mathrm{E}\int_{I \times R^n} |g(t, y)|^2\,dt\,\Pi(dy) < \infty. \tag{7.2.31}$$

For $g \in S(\Pi) \cap L_2^e(\Pi)$ define the stochastic integral

$$K(g) \equiv \int_{I \times R_0^n} g(t, y)\lambda(dt, dy) \equiv \sum_{i,\ell} g_{i\ell}\lambda(I_i, \Gamma_\ell)$$

$$= \sum_{i,\ell} g_{i\ell}\{\lambda(t_{i+1}, \Gamma_\ell) - \lambda(t_i, \Gamma_\ell)\}. \tag{7.2.32}$$

It is easy to verify that

$$EK(g) = 0,$$

$$E(K(g))^2 = E\sum_{i,\ell} |g_{i\ell}|^2\, |I_i|\, \Pi(\Gamma_\ell)$$

$$= E\int_{I \times R_0^n} |g(t, y)|^2\, dt\, \Pi(dy). \tag{7.2.33}$$

Hence K is a continuous linear operator from $S \cap L_2^e(\Pi)$ to $L_2(\Omega, \mathcal{F}, P)$ and $S \cap L_2^e(\Pi)$ is dense in $L_2^e(\Pi)$. Hence by the principle of continuous extension we can define K on all of $L_2^e(\Pi)$. Indeed, for every $g \in L_2^e(\Pi)$ there exists a sequence $g_m \in S \cap L_2^e(\Pi)$ such that

$$E\int_{I \times R_0^n} |g(t, y) - g_m(t, y)|^2\, dt\, \Pi(dy) \to 0,$$

as $m \to \infty$, and we have

$$K(g) = \text{P-limit } K(g_m). \tag{7.2.34}$$

Define the process ξ by

$$\xi(t) = K_t(g) \equiv \int_0^t \int_{R_0^n} g(\theta, y)\lambda(d\theta, dy), \qquad t \in I. \tag{7.2.35}$$

The process ξ is a square integrable $\mathcal{F}_t$-martingale with sample paths having discontinuities of no more than that of the first kind. Let D_1 denote the space of functions on I which are right continuous having left limits. The process $\xi \in D_1$ with probability 1 and it is a square-integrable martingale satisfying

$$E\{\xi(t) - \xi(s) \mid \mathcal{F}_s\} = 0,$$

$$E\{|\xi(t) - \xi(s)|^2 \mid \mathcal{F}_s\} = E\left\{\int_s^t \int_{R_0^n} |g(\theta, y)|^2\, d\theta\, \Pi(dy) \mid \mathcal{F}_s\right\},$$

$$E\left\{\sup_{t \in I} |\xi(t)|^2\right\} \leq 4E\int_{I \times R_0^n} |g(t, y)|^2\, dt\, \Pi\, (dy).$$

In fact, we can prove, as in the case of stochastic integrals based on Wiener process, that

(K)(a) $K(g) \equiv \int_{I\times R_0^n} g(t, y)\lambda(\mathrm{d}t, \mathrm{d}y)$ is well defined for all $g \in L_2^e(\Pi)$,
(K)(b) $K(\alpha_1 g_1 + \alpha_2 g_2) = \alpha_1 K(g_1) + \alpha_2 K(g_2)$ for α_i scalars and $g_i \in L_2^e(\Pi)$, $i = 1, 2$,
(K)(c) $K(g) \in L_2(\Omega, \mathscr{F}, \mathrm{P})$ for each $g \in L_2^e(\Pi)$,
(K)(d) $\mathrm{E}K(g) = 0$ for all $g \in L_2^e(\Pi)$,
(K)(e) $\mathrm{E}\,|K(g)|^2 = \mathrm{E}\int_{I\times R_0^n} |g|^2\,\mathrm{d}t\,\Pi(\mathrm{d}y)$, $g \in L_2^e(\Pi)$,
(K)(f) $\xi(t) = K_t(g)$ is a square-integrable $\mathscr{F}_t$-martingale with sample paths from D_1 and

$$\mathrm{E}\left\{\sup_{t\in I} |\xi(t)|^2\right\} \leqslant 4\mathrm{E}\,|\xi(T)|^2 = 4\mathrm{E}\int_{I\times R_0^n} |g(t, y)|^2\,\mathrm{d}t\,\Pi(\mathrm{d}y).$$

7.2.3 Itô formula and stochastic differentials

According to classical calculus if $f \equiv f(t, x)$ is C^1 in both t and x on $I \times R^n$, then, for any $\xi \in C^1(I, R^n)$,

$$\frac{\mathrm{d}}{\mathrm{d}t} f(t, \xi(t)) = f_t + (f_x, \dot{\xi}),$$

where $f_t \equiv \partial f/\partial t$, $f_x \equiv \nabla f$, and $\dot{\xi} = \mathrm{d}\xi/\mathrm{d}t$. Equivalently, one has

$$\mathrm{d}f = f_t\,\mathrm{d}t + (f_x, \mathrm{d}\xi). \tag{7.2.36}$$

In the case of the Itô-process this formula is no longer true and this is due to the presence of the martingale term in the expression

$$\xi(t) = \xi(0) + V(t) + M(t), \tag{7.2.37}$$

where $V(t)$ is an $\mathscr{F}_t$-measurable process with bounded variation and $M(t)$ is an $\mathscr{F}_t$-martingale. Any stochastic process which admits a decomposition of the form (7.2.37) is called a *semimartingale*. The correct rule of stochastic differentials is given by Itô's formula.

Theorem 7.2.6 (Itô's formula, continuous case). *Consider the semimartingale ξ given by* (7.2.37) *and suppose both V and M are continuous. Let $f = f(t, x)$ be C^1 in t and C^2 in x on $I \times R^n$. Then the stochastic differential of $\eta(t) \equiv f(t, \xi(t))$ is given by*

$$\begin{aligned}\mathrm{d}\eta = \mathrm{d}f &= f_t\,\mathrm{d}t + (f_x, \mathrm{d}\xi) + \tfrac{1}{2}\langle f_{xx}\,\mathrm{d}\xi, \mathrm{d}\xi\rangle \\ &= f_t\,\mathrm{d}t + (f_x, \mathrm{d}V) + (f_x, \mathrm{d}M) + \tfrac{1}{2}\langle f_{xx}\,\mathrm{d}M, \mathrm{d}M\rangle.\end{aligned} \tag{7.2.38}$$

The proof is lengthy but classical. For an elegant proof the reader is referred to the book of Ikeda and Watanabe [46].

In particular, let $a(t) \equiv a(t, \omega)$ and $\sigma(t) \equiv \sigma(t, \omega)$ be $\mathscr{F}_t$-adapted processes with values in R^n and $R(n \times m)$ respectively, and suppose ξ is given by

$$\mathrm{d}\xi(t) = a(t)\,\mathrm{d}t + \sigma(t)\,\mathrm{d}W(t),$$

or equivalently

$$\xi(t) = \xi(0) + \int_0^t a(\theta)\,\mathrm{d}\theta + \int_0^t \sigma(\theta)\,\mathrm{d}W(\theta), \tag{7.2.39}$$

where W is an m-dimensional $\mathscr{F}_t$-Brownian motion. Clearly, here we have

$$\begin{aligned} V(t) &= \int_0^t a(\theta)\,\mathrm{d}\theta, \\ M(t) &= \int_0^t \sigma(\theta)\,\mathrm{d}W(\theta). \end{aligned} \tag{7.2.40}$$

Then for any $f = f(t, x)$ which is C^1 in t and C^2 in x, the stochastic differential is given by

$$\begin{aligned} \mathrm{d}f &= f_t\,\mathrm{d}t + (f_x, a)\,\mathrm{d}t + (\sigma' f_x, \mathrm{d}W) + \frac{1}{2}\sum_{i,j} f_{x_i x_j}(\sigma\sigma')_{ij}\,\mathrm{d}t \\ &= (\mathbf{L}f)\,\mathrm{d}t + (\sigma' f_x, \mathrm{d}W), \end{aligned} \tag{7.2.41}$$

where the operator $\mathbf{L}$ is a second-order differential operator given by

$$\begin{aligned} \mathbf{L}\varphi &= \frac{\partial}{\partial t}\varphi + \sum_i a_i \varphi_{x_i} + \frac{1}{2}\sum_{i,j} (\sigma\sigma')_{ij}\varphi_{x_i x_j} \\ &\equiv \frac{\partial}{\partial t}\varphi + \mathbf{A}(t)\varphi. \end{aligned} \tag{7.2.42}$$

Note that the stochastic differential of f includes all the terms of order $(\mathrm{d}t)$ and a martingale term. Hence if ξ is an Itô process given by

$$\mathrm{d}\xi(t) = a(t)\,\mathrm{d}t + \sigma(t)\,\mathrm{d}W(t),$$

so also is the process $\eta(t) \equiv f(t, \xi(t))$, and is given by

$$\mathrm{d}\eta = (\mathbf{L}f)\,\mathrm{d}t + (\sigma' f_x, \mathrm{d}W).$$

As we shall see later, the formula (7.2.38) and, in particular, (7.2.41) have extensive applications in the study of stability, control, and filtering of stochastic systems.

As an example, let ξ_1 and ξ_2 be two Itô processes given by

$$\begin{aligned} \mathrm{d}\xi_1(t) &= a_1(t)\,\mathrm{d}t + \sigma_1(t)\,\mathrm{d}W(t), \qquad \xi_1(0) = \xi_{10}, \\ \mathrm{d}\xi_2(t) &= a_2(t)\,\mathrm{d}t + \sigma_2(t)\,\mathrm{d}W(t), \qquad \xi_2(0) = \xi_{20}, \end{aligned} \tag{7.2.43}$$

and define

$$\eta(t) = f(\xi_1(t), \xi_2(t)) = (\xi_1(t), \xi_2(t)).$$

Then

$$\begin{aligned} d\eta &= d(\xi_1, \xi_2) = (\xi_1, d\xi_2) + (d\xi_1, \xi_2) + \langle d\xi_1, d\xi_2 \rangle \\ &= \{(a_1, \xi_2) + (a_2, \xi_1) + \operatorname{tr}(\sigma_2'\sigma_1)\}\, dt + (\sigma_1'\xi_2 + \sigma_2'\xi_1, dW), \end{aligned} \tag{7.2.44}$$

where $\operatorname{tr}(A)$ denotes the trace of the matrix A.

The parameters a and σ appearing in the expression

$$d\xi(t) = a(t)\, dt + \sigma(t)\, dW(t),$$

are usually known as the drift and diffusion or dispersion parameters. It is interesting to observe that to every continuous Itô process ξ there corresponds a unique pair $\{a, \sigma\}$ of drift and dispersion parameters. Indeed, let ξ_1 and ξ_2 be two Itô processes given by (7.2.43) and suppose $P\{\xi_1(t) = \xi_2(t),\ t \in [0, T]\} = 1$, then $a_1 = a_2$, $\sigma_1 = \sigma_2$ for almost all $t \in [0, T]$ with probability 1. For the proof we note that

$$0 = d(\xi_1 - \xi_2) = (a_1 - a_2)\, dt + (\sigma_1 - \sigma_2)\, dW. \tag{7.2.45}$$

Hence

$$(a_1 - a_2, a_1 - a_2)\, dt + ((\sigma_1 - \sigma_2)'(a_1 - a_2), dW) = 0.$$

Integrating the above expression, we have

$$\int_0^t |a_1 - a_2|^2\, d\theta + \int_0^t ((\sigma_1 - \sigma_2)'(a_1 - a_2), dW) = 0 \tag{7.2.46}$$

for all $t \in I = [0, T]$. Taking the expectation and assuming that $|(\sigma_1 - \sigma_2)'(a_1 - a_2)|$ is square integrable, we have

$$E\int_0^t |a_1 - a_2|^2\, d\theta = 0, \qquad t \in I.$$

Hence $a_1 = a_2$ a.e. on I with probability one. Using this fact in (7.2.45), we have

$$\int_0^t ((\sigma_1 - \sigma_2)'y, dW) = 0,$$

for all $t \in I$ and $y \in R^n$, and hence

$$E\left(\int_0^t ((\sigma_1 - \sigma_2)'y, dW)\right)^2 = 0$$

for all $t \in I$. Clearly, it follows from the property (7.2.26) that

$$0 = \mathrm{E}\Big(\int_0^t ((\sigma_1 - \sigma_2)'y, \mathrm{d}W)\Big)^2 = \mathrm{E}\int_0^t |(\sigma_1 - \sigma_2)'y|^2 \,\mathrm{d}\theta$$

for all $t \in I$, $y \in R^n$. Hence $\sigma_1 = \sigma_2$ a.e. on I with probability one.

An elementary example illustrating the difference between the stochastic integral and the classical integral is given by

$$\eta(t) = \int_0^t W(\theta)\,\mathrm{d}W(\theta), \tag{7.2.47}$$

where W is the real Brownian motion. By use of Itô's formula, the reader may easily verify that

$$\eta(t) = \frac{W^2(t)}{2} - \frac{t}{2}. \tag{7.2.48}$$

Following the rules of classical calculus one would expect only the first term in (7.2.48).

Similarly, using Itô's formula applied to $\ln \xi$, one can easily verify that the solution of the scalar Itô equation

$$\begin{aligned} \mathrm{d}\xi &= \alpha\xi\,\mathrm{d}t + \beta\xi\,\mathrm{d}W, \\ \xi(0) &= \xi_0, \end{aligned} \tag{7.2.49}$$

is given by

$$\xi(t) = \xi_0 \exp\Big\{\int_0^t \Big(\alpha - \frac{\beta^2}{2}\Big)\,\mathrm{d}\theta + \int_0^t \beta\,\mathrm{d}W(\theta)\Big\}. \tag{7.2.50}$$

Following classical calculus, one would have the same expression with $\beta^2/2$ missing.

By application of Itô's formula, we derive a result of Paul Lévy characterizing Brownian motion.

Corollary 7.2.7 *An $\mathcal{F}_t$-measurable process $W(t)$, $t \geq 0$, with values in R^m, is a standard Brownian motion on the probability space $(\Omega, \mathcal{F}, \mathcal{F}_{t\geq 0}, \mathrm{P})$ if and only if, for all $\eta \in R^m$,*

$$\mathrm{E}\{\exp \mathrm{i}(\eta, W(t) - W(s)) \mid \mathcal{F}_s\} = \exp(-\tfrac{1}{2}|\eta|^2(t-s)). \tag{7.2.51}$$

Proof We prove the necessary condition. By Itô's formula applied to $\exp \mathrm{i}(\eta, W(t))$, for $\eta \in R^m$, and integrating over $[s, t]$, we obtain

$$\exp \mathrm{i}(\eta, W(t)) - \exp \mathrm{i}(\eta, W(s)) = \mathrm{i}\int_s^t \exp \mathrm{i}(\eta, W(\theta)) \cdot (\eta, \mathrm{d}W(\theta)) - \tfrac{1}{2}|\eta|^2 \int_s^t \exp \mathrm{i}(\eta, W(\theta))\,\mathrm{d}\theta.$$

Multiplying both sides by $\exp - \mathrm{i}(\eta, W(s))$ and taking conditional expectation, we have

$$\begin{aligned} &\mathrm{E}\{\exp \mathrm{i}(\eta, W(t) - W(s)) \mid \mathcal{F}_s\} \\ &\quad = 1 + \mathrm{i}\mathrm{E}\left\{\int_s^t \{\exp \mathrm{i}(\eta, W(\theta) - W(s))\}(\eta, \mathrm{d}W(\theta)) \mid \mathcal{F}_s\right\} \\ &\qquad - \frac{1}{2}|\eta|^2 \int_s^t \mathrm{E}\{\exp \mathrm{i}(\eta, W(\theta) - W(s)) \mid \mathcal{F}_s\}\, \mathrm{d}\theta. \quad (7.2.52) \end{aligned}$$

Since W is an $\mathcal{F}_t$-Brownian motion, the second term vanishes. Hence, defining

$$\varphi(t) \equiv \mathrm{E}\{\exp \mathrm{i}(\eta, W(t) - W(s)) \mid \mathcal{F}_s\}, \qquad t \geqslant s,$$

we arrive at the expression

$$\varphi(t) = 1 - \frac{1}{2}|\eta|^2 \int_s^t \varphi(\theta)\, \mathrm{d}\theta, \qquad t \geqslant s. \qquad (7.2.53)$$

Thus

$$\varphi(t) = \exp(-\tfrac{1}{2}|\eta|^2 (t - s)),$$

proving equation (7.2.51). The sufficiency follows from the facts that if the characteristic functionals of two measures coincide, the two measures must be identical, and further the right-hand member of (7.2.51) is the characteristic functional of a Gaussian process with independent increments. Hence it is a standard Brownian motion. ■

Theorem 7.2.8 (Itô's formula for jump processes). *Suppose the process ξ has the stochastic differential given by*

$$\mathrm{d}\xi(t) = a(t)\, \mathrm{d}t + \sigma(t)\, \mathrm{d}W(t) + \int_{R_0^n} g(t, y)\lambda(\mathrm{d}t, \mathrm{d}y) \qquad (7.2.54)$$

where

(a) *the Wiener process W and the Poisson process λ are assumed to be independent, and*
(b) *the processes a, σ, g are $\mathcal{F}_t$-adapted taking values in R^n, $R(n \times m)$ and R^n, respectively, with $a, \sigma \in L_2^e$ and $g \in L_2^e(\Pi)$.*

Then for every f, which is C^1 in t and C^2 in x, the process $\eta(t) \equiv f(t, \xi(t))$ has the stochastic differential

$$\mathrm{d}\eta = \mathrm{d}f = (\mathbf{L}_c f + \mathrm{L}_d f)\, \mathrm{d}t + \mathrm{d}M_c + \mathrm{d}M_d, \qquad (7.2.55)$$

where $\mathbf{L}_c = \mathbf{L}$ as in the continuous case (see (7.2.42)), and

$$\mathbf{L}_d f = \int_{R_0^n} \{f(t, \xi + g) - f(t, \xi) - (f_x, g)\}\Pi(dy),$$

$$M_c(t) = \int_0^t (\sigma' f_x, dW),$$

$$M_d(t) = \int_0^t \int_{R_0^n} \{f(\theta, \xi(\theta) + g(\theta, y)) - f(\theta, \xi(\theta))\}\lambda(d\theta, dy).$$

Note that M_c and M_d are the continuous and the discontinuous parts of the martingale $M = M_c + M_d$.

As an example, let

$$d\xi_i = a_i\, dt + \sigma_i\, dW + \int_{R_0^n} g_i \lambda(dt, dy), \qquad i = 1, 2, \tag{7.2.56}$$

then

$$d(\xi_1(t), \xi_2(t)) = (d\xi_1, \xi_2) + (\xi_1, d\xi_2) + \operatorname{tr}(\sigma_1\sigma_2')\, dt + \int_{R_0^n} (g_1, g_2)\Pi(dy)\, dt. \tag{7.2.57}$$

We shall have occasion to use Itô's formula for the jump processes once again in section 7.4.5, where we consider functionals of Markov processes with jumps.

7.3 Systems governed by linear stochastic differential equations

In this section we study the questions of existence and uniqueness of solutions of linear stochastic systems and continuous dependence of solutions on system parameters. The importance of linear systems can be easily appreciated from their extensive application in engineering, physical sciences and many areas of economics and social sciences.

7.3.1 Existence and uniqueness of solutions

A general linear system in R^n may be described by the following stochastic differential equation

$$\begin{aligned} d\xi &= A(t)\xi\, dt + f(t)\, dt + \sigma(t)\, dW(t), \qquad t \geqslant 0, \\ \xi(0) &= \xi_0, \end{aligned} \tag{7.3.1}$$

where ξ is an R^n-valued process, W is an R^m-valued standard Brownian motion and A, f and σ are $R(n \times n)$, $R(n \times 1) = R^n$, and $R(n \times m)$-valued processes, respectively, with ξ_0 being an R^n-valued random element.

We assume throughout that all random processes, vector or matrix valued, are defined on a complete probability space $(\Omega, \mathscr{F}, \mathscr{F}_{t \geq 0}, \mathrm{P})$ furnished with the current of σ-algebras $\mathscr{F}_{t \geq 0} \subset \mathscr{F}$ such that $\mathscr{F}_s \subset \mathscr{F}_t$ whenever $s \leq t$. In other words, $\mathscr{F}_t$ is an increasing family of subsigma algebras of the σ-algebra $\mathscr{F}$ indicated by $\mathscr{F}_t \uparrow$. Note that in case $\sigma \equiv 0$, and A, f, ξ_0 are deterministic, we have the ordinary differential equation

$$\frac{\mathrm{d}}{\mathrm{d}t} \xi = A(t)\xi + f(t),$$

$$\xi(0) = \xi_0,$$

which we studied in chapter 2.

Before considering the general case let us examine the solution of the problem

$$\begin{aligned} \mathrm{d}\xi &= A(t)\xi \, \mathrm{d}t + \sigma(t) \, \mathrm{d}W(t), \\ \xi(0) &= \xi_0, \end{aligned} \tag{7.3.2}$$

where both A and σ are deterministic. In this case, the solution can be directly written as

$$\xi(t) = \Phi(t, 0)\xi_0 + \int_0^t \Phi(t, \theta)\sigma(\theta) \, \mathrm{d}W(\theta), \tag{7.3.3}$$

for $t \geq 0$, where $\Phi(t, \tau)$, $0 \leq \tau \leq t$, is the transition operator corresponding to the plant or system matrix A. In the time invariant case ξ is given by

$$\xi(t) = \mathrm{e}^{tA}\xi_0 + \int_0^t \mathrm{e}^{(t-\theta)A}\sigma \, \mathrm{d}W(\theta). \tag{7.3.4}$$

By direct substitution, one can easily verify that the process ξ, as given by the expression (7.3.3), solves the stochastic integral equation

$$\xi(t) = \xi_0 + \int_0^t A(\theta)\xi(\theta) \, \mathrm{d}\theta + \int_0^t \sigma(\theta) \, \mathrm{d}W(\theta). \tag{7.3.5}$$

This is left as an exercise for the reader.

Since a linear transformation of a Gaussian process is Gaussian, the process $\{\xi_2(t); t \geq 0\}$ given by

$$\xi_2(t) \equiv \int_0^t \Phi(t, \theta)\sigma(\theta) \, \mathrm{d}W(\theta), \qquad t \geq 0,$$

is Gaussian, and $\xi_1(t) \equiv \Phi(t, 0)\xi_0$, $t \geq 0$, is also Gaussian if ξ_0 is. Hence if ξ_0 is Gaussian, the process $\{\xi(t); t \geq 0\}$, given by (7.3.3), is a Gaussian Markov process with mean

$$m(t) = \Phi(t, 0)m_0, \qquad m_0 = \mathrm{E}\xi_0, \qquad t \geq 0, \tag{7.3.6}$$

and covariance operator

$$C(t) = \Phi(t, 0)C_0\Phi'(t, 0) + \int_0^t \Phi(t, \theta)\sigma(\theta)\sigma'(\theta)\Phi'(t, \theta)\,\mathrm{d}\theta, \tag{7.3.7}$$

provided the initial state ξ_0 is independent of the Wiener process W, written $\sigma(\xi_0) \perp \sigma\{W(t), t \geq 0\}$. Clearly the mean m and the covariance operator C satisfy the following pair of differential equations:

$$\begin{aligned} &\frac{\mathrm{d}}{\mathrm{d}t} m = A(t)m, \\ &m(0) = m_0, \end{aligned} \tag{7.3.8}$$

and

$$\begin{aligned} &\frac{\mathrm{d}}{\mathrm{d}t} C = A(t)C + CA'(t) + \sigma(t)\sigma'(t), \\ &C(0) = C_0 = \mathrm{E}\{(\xi_0 - m_0)(\xi_0 - m_0)'\}. \end{aligned} \tag{7.3.9}$$

Since a Gaussian process is completely determined by its mean and covariance, the mean $m(t)$ and the covariance $C(t)$, determined by the solutions of (7.3.8) and (7.3.9), completely characterize the process $\{\xi(t), t \geq 0\}$ given by the solution of equation (7.3.2). Note that in (7.3.7) the last expression is precisely the controllability matrix (see chapter 5). Thus if $C_0 \geq 0$, and the system

$$\frac{\mathrm{d}}{\mathrm{d}t} y = A(t)y + \sigma(t)v, \quad v(t) \in R^m, \quad t \geq 0,$$

with v considered as the control variable, is controllable, then $\{\xi(t), t \geq 0\}$ is a non-singular Gaussian process, and for $\Gamma \in \mathscr{B}(R^n)$,

$$\mathrm{P}\{\xi(t) \in \Gamma\} = \int_\Gamma g(x, m(t), C(t))\,\mathrm{d}x, \tag{7.3.10}$$

where

$$g(x, m, C) = \left(\frac{1}{2\pi}\right)^{n/2}\left(\frac{1}{\sqrt{(\det C)}}\right)\exp -\frac{1}{2}(C^{-1}(x - m), x - m), \tag{7.3.11}$$

denotes the density of a Gaussian measure in R^n with the mean m and the covariance C.

For linear time-invariant systems one can say a little more. Consider the system

$$\begin{aligned} d\xi &= A\xi \, dt + \sigma \, dW, \\ \xi(0) &= \xi_0, \end{aligned} \tag{7.3.12}$$

with A, σ independent of time and suppose ξ_0 induces a Gaussian measure μ_0 on $\mathscr{B}(R^n)$, given by

$$\mu_0(\Gamma) \equiv \int_\Gamma g(x, m_0, C_0) \, dx, \qquad \Gamma \in \mathscr{B}(R^n).$$

Then for each $t > 0$, the measure induced by $\xi(t)$ on $\mathscr{B}(R^n)$, is given by

$$\mu_t(\Gamma) = \int_\Gamma g(\xi, m(t), C(t)) \, d\xi = \mathrm{P}\{\xi(t) \in \Gamma\},$$

where m and C are the solutions of the equations

$$\frac{dm}{dt} = Am, \quad m(0) = m_0, \qquad t \geqslant 0, \tag{7.3.13(a)}$$

and

$$\frac{d}{dt} C = AC + CA' + \sigma\sigma', \qquad C(0) = C_0, \quad t \geqslant 0, \tag{7.3.13(b)}$$

respectively.

The following result is interesting. It gives us the conditions under which the measure-valued function, $t \to \mu_t(\cdot)$, has a limit as $t \to \infty$. For ordinary dynamical systems one is often interested in the question of existence of asymptotic limit of the state. For stochastic systems, one may consider the associated measure as the state and naturally one is interested in the question of existence of its limit.

Theorem 7.3.1 *If $\sigma\sigma' > 0$ and A is a stability matrix, then for all Gaussian measures μ_0 on $\mathscr{B}(R^n)$ corresponding to the initial state ξ_0 of* (7.3.12), *there exists a unique Gaussian measure μ^* on $\mathscr{B}(R^n)$ such that, for each $\Gamma \in \mathscr{B}(R^n)$,*

$$\lim_{t\to\infty} \mathrm{P}\{\xi(t) \in \Gamma\} = \lim_{t\to\infty} \mu_t(\Gamma) = \mu^*(\Gamma).$$

Proof If A is a stability matrix and $\sigma\sigma' > 0$, then the Lyapunov equation

$$AC + CA' + \sigma\sigma' = 0,$$

has a unique symmetric positive definite solution $C = C^*$. In fact the solution is given by

$$C^* = \int_0^\infty e^{tA}\sigma\sigma' e^{tA'}\, dt,$$

which is independent of C_0 and that it is the unique limit of the solution $C(t)$ of the equation (7.3.13)(b), that is, $C^* = \lim_{t\to\infty} C(t)$. Since A is a stability matrix, $\lim_{t\to\infty} m(t) = \lim_{t\to\infty} e^{tA} m_0 = 0$ for any $m_0 \in R^n$. It is clear that $t \to m(t)$ and $t \to C(t)$ are absolutely continuous solutions of (7.3.13) and that, under the given assumptions, they are uniformly bounded and $C(t)$ is non-singular for $t > 0$. Hence the function

$$t \to g(\xi, m(t), C(t)), \qquad t > 0,$$

is continuous and bounded and, for each $\Gamma \in \mathscr{B}(R^n)$,

$$\begin{aligned}\lim_{t\to\infty} \mu_t(\Gamma) &= \lim_{t\to\infty} \int_\Gamma g(\xi, m(t), C(t))\, d\xi \\ &= \int_\Gamma g(\xi, 0, C^*)\, d\xi \\ &\equiv \mu^*(\Gamma),\end{aligned}$$

or equivalently, for any continuous bounded function f on R^n,

$$\lim_{t\to\infty} \int_{R^n} f(x)\mu_t(dx) = \int_{R^n} f(x)\mu^*(dx).$$

This proves the existence question. The uniqueness of μ^* follows from the uniqueness of C^*. ■

Note: The conclusion of the theorem remains valid for arbitrary initial probability measure μ_0 (see Problem 7.3.P7).

Now we return to the question of existence of solution of the general linear problem (7.3.1). In the following result we shall assume that the parameters $\{A, f, \sigma, \xi_0\}$ are all stochastic.

Theorem 7.3.2 *Consider the system (7.3.1) and suppose that $W \equiv \{W(t), t \in I = [0, T]\}$ is an $\mathscr{F}_t$-adapted m-dimensional standard Wiener process satisfying*

$$\sigma\{W(s) - W(t), s \geq t\} \perp \mathscr{F}_t,$$

and the processes $\{A(t), f(t)$, and $\sigma(t)$, $t \in I\}$ are all $\mathscr{F}_t$-adapted with values in $R(n \times n)$, R^n and $R(n \times m)$ respectively, and ξ_0 is $\mathscr{F}_0$-

measurable satisfying the following properties:

(a) *there exists a* $g \in L_1^+(I)$ *such that* $\mathrm{P}\{\|A(t)\|^2 \leq g(t),\ t \in I\} = 1$,
(b) $\mathrm{E}\int_I |f(t)|^2\,\mathrm{d}t < \infty$, $\mathrm{E}\int_I \operatorname{tr}(\sigma\sigma')(t)\,\mathrm{d}t < \infty$, $\mathrm{E}\,|\xi_0|^2 < \infty$.

Then the problem (7.3.1) *has a unique solution* $\xi = \{\xi(t),\ t \in I\}$ *which is* $\mathscr{F}_t$*-adapted, continuous with probability* 1, *and* $\mathrm{E}\{\sup_{t\in I} |\xi(t)|^2\} < \infty$.

Proof We prove this result by the usual Picard approximation technique applied to the stochastic integral equation

$$\xi(t) = \xi_0 + \int_0^t A(\theta)\xi(\theta)\,\mathrm{d}\theta + \int_0^t f(\theta)\,\mathrm{d}\theta + \int_0^t \sigma(\theta)\,\mathrm{d}W(\theta). \tag{7.3.14}$$

For convenience, define

$$\eta(t) \equiv \int_0^t f(\theta)\,\mathrm{d}\theta + \int_0^t \sigma(\theta)\,\mathrm{d}W(\theta),$$

and write (7.3.14) as

$$\xi(t) = \xi_0 + \int_0^t A(\theta)\xi(\theta)\,\mathrm{d}\theta + \eta(t), \qquad t \in I.$$

Consider the sequence of approximations,

$$\xi_{n+1}(t) = \xi_0 + \int_0^t A(\theta)\xi_n(\theta)\,\mathrm{d}\theta + \eta(t), \qquad n = 0, 1, 2, \ldots, \tag{7.3.15}$$

with

$$\xi_0(\theta) \equiv \xi_0, \qquad \theta \in I.$$

From the first approximation ($n = 0$) in (7.3.15), we have

$$\sup_{0\leq s\leq t\leq T} |\xi_1(s) - \xi_0|^2 \leq 2\left\{\sup_{0\leq s\leq t} |\eta(s)|^2 + T\left(\int_0^t \|A(\theta)\|^2\,\mathrm{d}\theta\right)|\xi_0|^2\right\}, \tag{7.3.16}$$

and using the martingale inequality (Proposition 7.1.12), we obtain

$$\mathrm{E}\left\{\sup_{0\leq s\leq t} |\eta(s)|^2\right\} \leq 2\mathrm{E}\left\{T\int_0^T |f(\theta)|^2\,\mathrm{d}\theta\right\} + 8\mathrm{E}\left\{\int_0^T \operatorname{tr}(\sigma\sigma')(\theta)\,\mathrm{d}\theta\right\} \equiv c_1, \tag{7.3.17}$$

for some constant $c_1 < \infty$. Hence it follows from the assumptions (a) and

(b) that there exists a constant $c > 0$ such that

$$\mathrm{E}\left\{\sup_{0\leqslant s\leqslant t} |\xi_1(s) - \xi_0|^2\right\} \leqslant 2\left(c_1 + T\mathrm{E}\,|\xi_0|^2 \int_0^t g(\theta)\,\mathrm{d}\theta\right) \equiv c < \infty,$$

for all $t \in I = [0, T]$. Defining

$$\rho_t(x, y) \equiv \mathrm{E}\left\{\sup_{0\leqslant s\leqslant t} |x(s) - y(s)|^2\right\},$$

we have

$$\rho_t(\xi_1, \xi_0) \leqslant c \qquad \text{for all } t \in I.$$

Similarly, from

$$\xi_2(s) - \xi_1(s) = \int_0^s A(\theta)[\xi_1(\theta) - \xi_0]\,\mathrm{d}\theta, \qquad s \in I,$$

we have

$$\sup_{0\leqslant s\leqslant t} |\xi_2(s) - \xi_1(s)|^2 \leqslant T\int_0^t \|A(\theta)\|^2 \left\{\sup_{0\leqslant \tau\leqslant \theta} |\xi_1(\tau) - \xi_0|^2\right\} \mathrm{d}\theta.$$

Thus

$$\rho_t(\xi_2, \xi_1) \leqslant T\int_0^t g(\theta)\rho_\theta(\xi_1, \xi_0)\,\mathrm{d}\theta \leqslant cT\int_0^t g(\theta)\,\mathrm{d}\theta.$$

Defining

$$h(t) \equiv \int_0^t g(\theta)\,\mathrm{d}\theta,$$

we have

$$\rho_t(\xi_2, \xi_1) \leqslant cTh(t), \qquad t \in I,$$

and hence

$$\rho_t(\xi_3, \xi_2) \leqslant cT^2\int_0^t g(\theta)h(\theta)\,\mathrm{d}\theta = cT^2\int_0^t h(\theta)\,\mathrm{d}h(\theta)$$
$$\leqslant \frac{cT^2}{2!}h^2(t),$$

and in general

$$\rho_t(\xi_{n+1}, \xi_n) \leqslant \frac{cT^n(h(t))^n}{n!} \leqslant \frac{cT^n(h(T))^n}{n!},$$

for all $t \in I$. Hence, for any $\varepsilon > 0$,

$$P\left\{\sup_{t \in I} |\xi_{n+1}(t) - \xi_n(t)| > \varepsilon\right\} < \left\{\frac{\rho_T(\xi_{n+1}, \xi_n)}{\varepsilon^2}\right\}.$$

Choosing $\varepsilon = 2^{-n}$, we obtain

$$P\left\{\sup_{t \in I} |\xi_{n+1}(t) - \xi_n(t)| > 2^{-n}\right\} \leqslant \left\{\frac{c(4T)^n (h(T))^n}{n!}\right\}. \tag{7.3.18}$$

Hence by the Borel–Cantelli lemma (Proposition 7.1.7), we conclude that there exists a unique process $\xi \equiv \{\xi(t), t \in I\}$ to which ξ_n converges uniformly on I with probability 1. Since the sequence $\{\xi_n\}$, as constructed, is continuous with probability 1 and the convergence is uniform on I, the limiting process ξ is also continuous with probability 1. Almost-sure convergence of ξ_n to ξ and the fact that each member of the sequence $\{\xi_n\}$ is $\mathcal{F}_t$-adapted imply that ξ is also $\mathcal{F}_t$-adapted. Letting $n \to \infty$ in the equation

$$\xi_{n+1}(t) = \xi_0 + \int_0^t A(\theta)\xi_n(\theta)\, d\theta + \eta(t),$$

we conclude that ξ is a solution of (7.3.14). The uniqueness follows from Gronwall's lemma applied to the inequality

$$\rho_t(\xi_1, \xi_2) \leqslant T \int_0^t g(\theta)\rho_\theta(\xi_1, \xi_2)\, d\theta, \qquad t \in I,$$

where ξ_1 and ξ_2 are assumed to be any two solutions of the problem (7.3.14). The fact that $E\{\sup_{\theta \in I} |\xi(\theta)|^2\} < \infty$ is obvious. This completes the proof. ■

7.3.2 Continuous dependence of solutions on parameters

Consider the system (7.3.1) with the parameters $\{\xi_0, A, f, \sigma\}$ and let $\xi(\cdot) \equiv \xi(\cdot, \xi_0, A, f, \sigma)$ denote the solution process as a function of the parameters $\{\xi_0, A, f, \sigma\}$. In the following theorem, we show that the mapping $(\xi_0, A, f, \sigma) \to \xi(\cdot, \xi_0, A, f, \sigma)$ is continuous in an appropriate sense.

Let X denote the Banach space of continuous $\mathcal{F}_t$-adapted stochastic processes on I with values in R^n furnished with the norm

$$\|\xi\| = \|\xi\|_X \equiv (E\{\sup |\xi(t)|^2, t \in I\})^{1/2}.$$

Theorem 7.3.3 (Continuous dependence of solutions). *Consider the system* (7.3.1) *corresponding to the sequence of parameters* $\{\xi_{0,k}, A_k, f_k, \sigma_k\}$ *such that, as* $k \to \infty$,

$$\begin{aligned} \xi_{0,k} &\to \xi_0 && \textit{in } L_2(\Omega, \mathscr{F}_0, \mathrm{P}; R^n) \equiv L_2(\mathscr{F}_0; R^n) \\ f_k &\to f && \textit{in } L_2^e(I, \mathscr{F}_t; R^n) \\ \sigma_k &\to \sigma && \textit{in } L_2^e(I, \mathscr{F}_t; R(n \times m)) \\ A_k &\to A && \textit{a.e. on } I, \text{ P-a.s.}, \end{aligned}$$

and there exists a $g \in L_1^+(I)$ *such that*

$$\mathrm{P}\{\|A(t)\|^2, \|A_k(t)\|^2 \leqslant g(t), t \in I\} = 1 \qquad \textit{for all integers } k \geqslant 1.$$

Then as $k \to \infty$, $\xi(\cdot, \xi_{0,k}, A_k, f_k, \sigma_k) \to \xi(\cdot, \xi_0, A, f, \sigma)$ *in* X.

Proof First we prove that the sequence of solutions $\xi_k(\cdot) = \xi(\cdot, \xi_{0,k}, A_k, f_k, \sigma_k)$ is contained in a bounded subset of X, or equivalently, $\sup_k \|\xi_k\|_X < \infty$.

Clearly, by the previous theorem, $\{\xi_k\}$ is a well-defined sequence of solutions of (7.3.1) corresponding to $\{\xi_{0,k}, A_k, f_k, \sigma_k\}$, and for each integer $k \geqslant 1$, $\xi_k \in X$. Hence, for all $t \in I$, $T < \infty$,

$$\xi_k(t) = \xi_{0,k} + \int_0^t A_k(\theta)\xi_k(\theta)\,\mathrm{d}\theta + \int_0^t f_k(\theta)\,\mathrm{d}\theta + \int_0^t \sigma_k(\theta)\,\mathrm{d}W(\theta),$$

with probability 1 (P-a.s.). Using the martingale inequality one can easily deduce from this that

$$\begin{aligned} \mathrm{E}\Big\{\sup_{0\leqslant\theta\leqslant t} |\xi_k(\theta)|^2\Big\} \\ \leqslant 4\Big\{\mathrm{E}\,|\xi_{0,k}|^2 + T\mathrm{E}\int_0^t \|A_k(\theta)\|^2\Big(\sup_{0\leqslant\tau\leqslant\theta} |\xi_k(\tau)|^2\Big)\,\mathrm{d}\theta \\ + T\mathrm{E}\int_0^t |f_k(\theta)|^2\,\mathrm{d}\theta + 4\mathrm{E}\int_0^t \mathrm{tr}(\sigma_k\sigma_k')\,\mathrm{d}\theta\Big\}. \end{aligned} \tag{7.3.19}$$

Since $\{\xi_{0,k}, A_k, f_k, \sigma_k\}$ is a convergent sequence, there exists a constant $c_1 < \infty$ such that

$$\mathrm{E}\,|\xi_{0,k}|^2 + T\mathrm{E}\int_0^T |f_k(\theta)|^2\,\mathrm{d}\theta + 4\mathrm{E}\int_0^T \mathrm{tr}(\sigma_k\sigma_k')\,\mathrm{d}\theta \leqslant c_1 < \infty,$$

for all integers $k \geqslant 1$. Using this fact and defining

$$\varphi_k(t) \equiv \mathrm{E}\Big\{\sup_{0\leqslant\theta\leqslant t} |\xi_k(\theta)|^2\Big\},$$

we can rewrite the inequality (7.3.19) as

$$\varphi_k(t) \leqslant 4\left(c_1 + T\int_0^t g(\theta)\varphi_k(\theta)\,\mathrm{d}\theta\right),$$

where we have used the assumption on $\{A_k\}$. Hence by Gronwall's lemma,

$$\varphi_k(t) \leqslant 4c_1 \exp\left(4T\int_0^t g(\theta)\,\mathrm{d}\theta\right) \leqslant c_2 < \infty,$$

for all $t \in I$ and all integers $k \geqslant 1$. This shows that

$$\sup_k \|\xi_k\|_X < \infty,$$

and hence the sequence $\{\xi_k\}$ is contained in a bounded subset of X. By the preceding theorem, (7.3.1) has a unique solution $\xi(\cdot) = \xi(\cdot, \xi_0, A, f, \sigma) \in X$ corresponding to the limit $\{\xi_0, A, f, \sigma\}$ of the sequence $\{\xi_{0,k}, A_k, f_k, \sigma_k\}$. We show that $\xi_k \to \xi$ in X.

Clearly,

$$\begin{aligned}
&|\xi(t) - \xi_k(t)|^2 \\
&\quad \leqslant 5\left\{|\xi_0 - \xi_{0,k}|^2 + T\int_0^t \|A_k(\theta)\|^2\,|\xi(\theta) - \xi_k(\theta)|^2\,\mathrm{d}\theta\right. \\
&\qquad + T\int_0^T \|A(\theta) - A_k(\theta)\|^2\,|\xi(\theta)|^2\,\mathrm{d}\theta + T\int_0^t |f(\theta) - f_k(\theta)|^2\,\mathrm{d}\theta \\
&\qquad \left. + \left|\int_0^t (\sigma(\theta) - \sigma_k(\theta))\,\mathrm{d}W(\theta)\right|^2\right\}.
\end{aligned}$$

By using the martingale inequality and the given assumptions, we obtain

$$\begin{aligned}
&\mathrm{E}\left\{\sup_{0\leqslant\theta\leqslant t} |\xi(\theta) - \xi_k(\theta)|^2\right\} \\
&\quad \leqslant 5\left\{\mathrm{E}\,|\xi_0 - \xi_{0,k}|^2 + T\int_0^t g(\theta)\mathrm{E}\left\{\sup_{0\leqslant\tau\leqslant\theta} |\xi(\tau) - \xi_k(\tau)|^2\right\}\mathrm{d}\theta\right. \\
&\qquad + T\mathrm{E}\int_0^t \|A(\theta) - A_k(\theta)\|^2\,|\xi(\theta)|^2\,\mathrm{d}\theta + T\mathrm{E}\int_0^t |f(\theta) - f_k(\theta)|^2\,\mathrm{d}\theta \\
&\qquad \left. + 4\mathrm{E}\int_0^t \mathrm{tr}\{(\sigma - \sigma_k)(\sigma - \sigma_k)'\}\,\mathrm{d}\theta\right\}.
\end{aligned}$$

Defining

$$\rho_t(\xi, \xi_k) \equiv \mathrm{E}\left\{\sup_{0\leqslant\theta\leqslant t} |\xi(\theta) - \xi_k(\theta)|^2\right\},$$

we rewrite the above inequality in the following compact form:

$$\rho_t(\xi, \xi_k) \leq \alpha_k + \beta \int_0^t g(\theta)\rho_\theta(\xi, \xi_k)\, d\theta,$$

where

$$\alpha_k = 5\Big\{E\,|\xi_0 - \xi_{0,k}|^2 + TE \int_0^T \|A(\theta) - A_k(\theta)\|^2\, |\xi(\theta)|^2\, d\theta$$
$$+ E\int_0^T |f(\theta) - f_k(\theta)|^2\, d\theta + 4E\int_0^T \operatorname{tr}(\sigma - \sigma_k)(\sigma - \sigma_k)'\, d\theta\Big\},$$
$$\beta = 5T.$$

Once again using Gronwall's lemma, we obtain

$$\rho_t(\xi, \xi_k) \leq \rho_T(\xi, \xi_k) \leq \alpha_k \exp\Big(\beta \int_0^T g(\theta)\, d\theta\Big),$$

for all $t \in I = [0, T]$. Since for almost all $t \in I$,

$$\{\|A_k(t)\|^2, \|A(t)\|^2\} \leq g(t), \qquad \text{P-a.s.},$$

where $g \in L_1^+$, it is clear that

$$\|A(t) - A_k(t)\|^2\, |\xi(t)|^2 \leq 4g(t)\, |\xi(t)|^2 \qquad \text{a.e. on } I, \quad \text{P-a.s.},$$

and

$$E\int_0^T \|A_k(\theta) - A(\theta)\|^2\, |\xi(\theta)|^2\, d\theta \leq 4\int_0^T g(\theta) E\,|\xi(\theta)|^2\, d\theta < \infty.$$

Therefore, since $A_k(t) \to A(t)$ a.e. on I P-a.s., it follows from the dominated convergence theorem that

$$\lim_{k\to\infty} E\int_0^T \|A(\theta) - A_k(\theta)\|^2\, |\xi(\theta)|^2\, d\theta = 0.$$

Thus under the given assumptions, $\lim_{k\to\infty} \alpha_k = 0$, and consequently,

$$\lim_{k\to\infty} \rho_T(\xi, \xi_k) = 0,$$

which implies that $\xi_k \to \xi$ in X. This proves the desired continuity. ■

7.4 Systems governed by nonlinear stochastic differential equations

In this section we study the questions of existence and uniqueness of solutions of nonlinear stochastic differential equations, and the continuous dependence of solutions on system parameters. We shall mainly consider strong solutions and indicate only briefly the concept of solutions in the weak sense.

7.4.1 Existence and uniqueness of solutions

The general form of a nonlinear stochastic system in R^n yielding continuous sample paths $\{\xi(t), t \geq 0\}$ is given by

$$\begin{aligned} d\xi &= a(t, \xi)\, dt + \sigma(t, \xi)\, dW, \qquad t \in I \equiv [0, T], \\ \xi(0) &= \xi_0, \end{aligned} \tag{7.4.1}$$

where $a\colon I \times R^n \to R^n$, $\sigma\colon I \times R^n \to R(n \times m)$, and $W = \{W(t), t \in I\}$ is an $\mathscr{F}_t$-Brownian motion with values in R^m.

The meaning of the system (7.4.1) is to be understood in its integral form

$$\xi(t) = \xi_0 + \int_0^t a(\theta, \xi(\theta))\, d\theta + \int_0^t \sigma(\theta, \xi(\theta))\, dW(\theta), \tag{7.4.2}$$

where the first integral is taken in the classical Lebesgue sense, and the second integral is understood in the Itô sense as discussed in section 7.2. Note that, given $\xi(t) = x$ and (Δt) small, we can approximate $\xi(t + \Delta t)$ by

$$\xi(t + \Delta t) \cong x + a(t, x)\Delta t + \sigma(t, x)(W(t + \Delta t) - W(t)).$$

This expression states that the mean instantaneous direction or velocity of motion at $\xi(t) = x$ is determined by the value of a at (t, x); and the fluctuation around this mean is determined by $\sigma(t, x)$ or more precisely by $(\sigma\sigma')(t, x)$. This is why a and σ are known as the infinitesimal mean and diffusion (dispersion) parameters. Also note that, given $\xi(t) = x$, $\xi(t + \Delta t)$ is Gaussian for small Δt, with mean $x + a(t, x)\Delta t$ and covariance $(\sigma\sigma')(t, x)$. In other words, we can consider ξ to be a locally conditionally Gaussian process, though in global terms it is generally non-Gaussian.

For the proof of existence and uniqueness of solutions of the system (7.4.1) we consider the integral equation (7.4.2). Throughout the remaining section we shall use the probability space $(\Omega, \mathscr{F}, \mathscr{F}_{t \geq 0}, P)$ without further reference.

Definition 7.4.1 (Strong solution). The system (7.4.1) is said to have a strong solution ξ if $\xi(t)$ is $\mathscr{F}_t$-adapted for each $t \in I$, $t \to \xi(t)$ is continuous with probability 1, and ξ satisfies the equality (7.4.2) for all $t \in I$ with probability 1.

According to this definition a strong solution has the representation

$$\xi(t) = F(t, \xi_0, W(\cdot)), \qquad t \in I,$$

where F is continuous in t and measurable in the rest of the variables with respect to $\mathscr{F}_0 \times \mathscr{F}_t$, and $F(t, x, \eta_1) = F(t, x, \eta_2)$ if $\eta_1(\theta) = \eta_2(\theta)$ for $0 \leqslant \theta \leqslant t$, and $\eta_i \in C(I, R^n)$, that is, F is non-anticipative in the third variable.

Theorem 7.4.2 (Existence and uniqueness). *Consider the system* (7.4.1) *and suppose there exists a $K \in L_1^+(I)$ such that for all $t \in I$ and x, $y \in R^n$,*

$$\begin{aligned} &\text{(a)} \quad T|a(t, x)|^2 + 4\,\|\sigma(t, x)\|^2 \leqslant K(t)[1 + |x|^2] \\ &\text{(b)} \quad T|a(t, x) - a(t, y)|^2 + 4\,\|\sigma(t, x) - \sigma(t, y)\|^2 \leqslant K(t)\,|x - y|^2. \end{aligned} \tag{7.4.3}$$

Then for every $\xi_0 \in L_2(\mathscr{F}_0, R^n)$, the system (7.4.1) *has a unique strong solution $\xi = \{\xi(t), t \in I\}$ which is $\mathscr{F}_t$-adapted, continuous with probability* 1 *and*

$$\mathrm{E}\left\{\sup_{t \in I} |\xi(t)|^2\right\} < \infty.$$

Proof We shall use the Banach fixed-point theorem. Define the vector space

$$X \equiv \{x = \{x(t), t \in I\} : x(t) \text{ is } \mathscr{F}_t\text{-adapted}$$
$$\text{for } t \in I, \text{ and } x \in C(I, R^n) \text{ P-a.s.}\}.$$

We consider X with the norm topology $\|\ \|_X$ given by

$$\|x\|_X^2 = \mathrm{E}\left\{\sup_{t \in I} |x(t)|^2\right\}.$$

By virtue of the Borel–Cantelli lemma (Proposition 7.1.7), one can verify that X is a Banach space. Define the operator S on X by

$$(S\xi)(t) \equiv \xi_0 + \int_0^t a(\theta, \xi(\theta))\,\mathrm{d}\theta + \int_0^t \sigma(\theta, \xi(\theta))\,\mathrm{d}W(\theta) \tag{7.4.4}$$

for $t \in I$. We prove the existence and uniqueness of a solution of the problem (7.4.1) in its integral form (7.4.2) by showing that S has a unique fixed point in X. This will follow from Theorem 3.1.3 if we prove that S: $X \to X$ and that some power of S is a contraction on X. First we prove that S: $X \to X$.

For any $\xi \in X$, $(S\xi)(t)$ is given by (7.4.4), $t \in I$, and

$$\sup_{0 \leq \tau \leq t} |(S\xi)(\tau)|^2$$

$$\leq 3\left\{|\xi_0|^2 + T\int_0^t |a(\theta, \xi(\theta))|^2 \, \mathrm{d}\theta + \sup_{0 \leq \tau \leq t}\left|\int_0^\tau \sigma(\theta, \xi(\theta)) \, \mathrm{d}W(\theta)\right|^2\right\}.$$

Taking the (mathematical) expectation and using the martingale inequality, we have

$$\mathrm{E}\left\{\sup_{0 \leq \tau \leq t} |(S\xi)(\tau)|^2\right\}$$

$$\leq 3\left\{\mathrm{E}\,|\xi_0|^2 + T\mathrm{E}\int_0^t |a(\theta, \xi(\theta))|^2 \, \mathrm{d}\theta + 4\mathrm{E}\int_0^t \|\sigma(\theta, \xi(\theta))\|^2 \, \mathrm{d}\theta\right\}. \quad (7.4.5)$$

Using assumption (a) of the theorem, it follows from the above inequality that

$$\mathrm{E}\left\{\sup_{0 \leq \tau \leq t} |(S\xi)(\tau)|^2\right\} \leq 3\left\{\mathrm{E}\,|\xi_0|^2 + \int_0^t K(\theta)\left\{1 + \mathrm{E}\sup_{0 \leq \tau \leq \theta} |\xi(\tau)|^2\right\} \mathrm{d}\theta\right\},$$

for all $t \in I$. Hence

$$\|S\xi\|_X^2 \leq c(1 + \|\xi\|_X^2), \quad (7.4.6)$$

where

$$c \equiv 3 \max\left\{\mathrm{E}\,|\xi_0|^2, \int_0^T K(\theta) \, \mathrm{d}\theta\right\}.$$

Thus S maps X into itself, that is, $S: X \to X$.

Now we show that S^n is a contraction on X for sufficiently large n. This will then imply, by virtue of Theorem 3.1.3, that S has a unique fixed point in X. Let ξ, $\eta \in X$, then

$$\sup_{0 \leq \theta \leq t} |(S\xi)(\theta) - (S\eta)(\theta)|^2$$

$$\leq 2\left\{T\int_0^t |a(\tau, \xi(\tau)) - a(\tau, \eta(\tau))|^2 \, \mathrm{d}\tau\right.$$

$$\left. + \sup_{0 \leq \theta \leq t}\left|\int_0^\theta [\sigma(\tau, \xi(\tau)) - \sigma(\tau, \eta(\tau))] \, \mathrm{d}W(\tau)\right|^2\right\},$$

and, by the martingale inequality and hypothesis (b) of the theorem, we obtain

$$\mathrm{E}\left\{\sup_{0\leqslant\theta\leqslant t}|(S\xi)(\theta)-(S\eta)(\theta)|^2\right\}$$
$$\leqslant 2\int_0^t K(\theta)\mathrm{E}\left\{\sup_{0\leqslant\tau\leqslant\theta}|\xi(\tau)-\eta(\tau)|^2\right\}\mathrm{d}\theta. \quad (7.4.7)$$

Define, for each pair $x,\ y\in X$,

$$\rho_t(x,y)\equiv\mathrm{E}\left\{\sup_{0\leqslant\theta\leqslant t}|x(\theta)-y(\theta)|^2\right\}. \quad (7.4.8)$$

Substituting (7.4.8) into (7.4.7), we obtain

$$\rho_t(S\xi,S\eta)\leqslant 2\int_0^t K(\theta)\rho_\theta(\xi,\eta)\,\mathrm{d}\theta. \quad (7.4.9)$$

Repeating this procedure n times, we find that

$$\rho_T(S^n\xi,S^n\eta)\leqslant\left(\frac{(2h(T))^n}{n!}\right)\rho_T(\xi,\eta),$$

for all $\xi,\ \eta\in X$, where

$$h(t)\equiv\int_0^t K(\theta)\,\mathrm{d}\theta, \qquad t\in I.$$

Hence

$$\|S^n\xi-S^n\eta\|_X\leqslant\alpha_n\,\|\xi-\eta\|_X, \quad (7.4.10)$$

where

$$\alpha_n\equiv\left(\frac{2^n(h(T))^n}{n!}\right)^{1/2}.$$

Therefore, for n sufficiently large, say n_0, $\alpha_{n_0}<1$ and hence S^{n_0} is a contraction on X. Thus by the Banach fixed-point theorem (Theorem 3.1.3), S^{n_0} has a unique fixed point in X and consequently S has the same fixed point. This shows that for each $\xi_0\in L_2(\mathscr{F}_0,R^n)$, there is a unique $\xi\in X$ that solves the problem (7.4.2) and hence (7.4.1). ■

The conclusion of Theorem 7.4.2 holds under more general conditions.

Theorem 7.4.3 *Consider the system* (7.4.1) *and suppose the hypothesis* (a) *of Theorem* 7.4.2 *holds and* (b) *is replaced by the condition*:
(b)′ *for every $m>0$, there exists a constant $K_m>0$ such that*

$$T\,|a(t,x)-a(t,y)|^2+4\,\|\sigma(t,x)-\sigma(t,y)\|^2\leqslant K_m\,|x-y|^2,$$

for all $x,\ y\in B_m\equiv\{x\in R^n\colon |x|\leqslant m\}$.

Then for every ξ_0 *satisfying* $P\{|\xi_0| < \infty\} = 1$, *equation* (7.4.1) *or equivalently* (7.4.2) *has a unique strong solution* $\xi \in X$.

Proof We give an outline of the proof. For each integer $m \geq 0$, define

$$\psi_m(x) = \begin{cases} 1 & \text{for } x \in B_m, \\ m + 1 - |x| & \text{for } x \in B_{m+1} \backslash B_m, \\ 0 & \text{for } x \in B'_{m+1}, \end{cases}$$

and

$$\begin{aligned} &\xi_0^m \equiv \psi_m(\xi_0)\xi_0, \\ &a^m(t, x) \equiv \psi_m(x)a(t, x), \qquad (t, x) \in I \times R^n, \\ &\sigma^m(t, x) \equiv \psi_m(x)\sigma(t, x) \qquad (t, x) \in I \times R^n. \end{aligned} \tag{7.4.11}$$

Clearly, ξ_0^m, a^m and σ^m satisfy the assumptions of Theorem 7.4.2, hence the equation

$$\xi^m(t) = \xi_0^m + \int_0^t a^m(\theta, \xi^m(\theta))\, d\theta + \int_0^t \sigma^m(\theta, \xi^m(\theta))\, dW(\theta), \tag{7.4.12}$$

has a unique solution $\xi^m \in X$. Then by use of the growth condition (a) and martingale inequality one can prove that there exists a subsequence $\{\xi^{m_k}\} \subset \{\xi^m\}$ such that

$$P\left\{\sup_{t \in I} |\xi^{m_{k+1}}(t) - \xi^{m_k}(t)| > 0\right\} \leq \frac{2}{k^2}.$$

Then by virtue of the Borel–Cantelli lemma, we conclude that there exists a process $\xi \in X$ such that $\xi^{m_k}(t) \to \xi(t)$ uniformly on I with probability 1 (P-a.s.) and

$$P\left\{\sup_{t \in I} |\xi(t)| < \infty\right\} = 1.$$

It is clear from the definition of the cut-off function ψ_m and equation (7.4.12) that

$$\sup_{t \in I} |\xi^{m_k}(t)| \leq 1 + m_k \qquad \text{P-a.s.}$$

Further, if $\sup_{t \in I} |\xi^{m_k}(t)| < m_k$, then $\xi(t) = \xi^{m_k}(t)$, and if $\sup_{t \in I} |\xi^{m_k}(t)| \geq m_k$, then $\sup_{t \in I} |\xi(t)| \geq m_k$ also. Hence

$$\sup_{t \in I} |\xi^{m_k}(t)| \leq 1 + \sup_{t \in I} |\xi(t)|, \qquad \text{P-a.s.}$$

Therefore by virtue of the growth condition (a), it follows from Corollary 7.2.5 that

$$\lim_{k\to\infty}\int_0^t \sigma^{m_k}(\theta, \xi^{m_k}(\theta))\, \mathrm{d}W(\theta) = \int_0^t \sigma(\theta, \xi(\theta))\, \mathrm{d}W(\theta),$$

with probability 1 for all $t \in I$.

Similarly, with probability 1

$$\lim_{k\to\infty}\int_0^t a^{m_k}(\theta, \xi^{m_k}(\theta))\, \mathrm{d}\theta = \int_0^t a(\theta, \xi(\theta))\, \mathrm{d}\theta,$$

for all $t \in I$, and $\lim_{k\to\infty} \xi_0^{m_k} = \xi_0$. Hence replacing m by m_k in (7.4.12) and letting $k \to \infty$ we conclude that ξ is a solution of the problem (7.4.2).

We prove the uniqueness by contradiction. Suppose η is another solution of the problem (7.4.2) with $\eta(0) = \xi_0$. Define

$$\tau_m \equiv \sup\{t \in I\colon \xi(\theta), \eta(\theta) \in B_m \text{ for all } \theta \in [0, t]\},$$

where $B_m = \{x \in R^n\colon |x| \leq m\}$. Clearly, τ_m is a stopping time and for any $t \in I$, $\{\tau_m > t\}$ is $\mathscr{F}_t$-measurable. Define

$$\chi_m(t) = \begin{cases} 1 & \text{if } t \leq \tau_m, \\ 0 & \text{otherwise.} \end{cases}$$

Then for any $t \in I$,

$$\begin{aligned}\chi_m(t)(\xi(t) - \eta(t)) = \chi_m(t) \int_0^t [a(\theta, \xi(\theta)) - a(\theta, \eta(\theta))]\, \mathrm{d}\theta \\ + \chi_m(t) \int_0^t [\sigma(\theta, \xi(\theta)) - \sigma(\theta, \eta(\theta))]\, \mathrm{d}W(\theta).\end{aligned} \tag{7.4.13}$$

Note that for $t \in I = [0, T]$,

$$\begin{aligned}&E\, |\chi_m(t) \int_0^t [a(\theta, \xi(\theta)) - a(\theta, \eta(\theta))]\, \mathrm{d}\theta|^2 \\ &\leq T\mathrm{E} \int_0^t \chi_m(\theta) |a(\theta, \xi(\theta)) - a(\theta, \eta(\theta))|^2\, \mathrm{d}\theta,\end{aligned} \tag{7.4.14}$$

and

$$\begin{aligned}&\mathrm{E}\, |\chi_m(t) \int_0^t [\sigma(\theta, \xi(\theta)) - \sigma(\theta, \eta(\theta))]\, \mathrm{d}W(\theta)|^2 \\ &\leq \mathrm{E} \left| \int_0^t \chi_m(\theta) [\sigma(\theta, \xi(\theta)) - \sigma(\theta, \eta(\theta))]\, \mathrm{d}W(\theta) \right|^2 \\ &\leq \mathrm{E} \int_0^t \chi_m(\theta)\, \|\sigma(\theta, \xi(\theta)) - \sigma(\theta, \eta(\theta))\|^2\, \mathrm{d}\theta.\end{aligned} \tag{7.4.15}$$

Hence by assumption (b)′, we have

$$\begin{aligned} &\mathrm{E}\{\chi_m(t)\,|\xi(t)-\eta(t)|^2\} \\ &\leqslant 2\mathrm{E}\int_0^t \chi_m(\theta)\{T\,|a(\theta,\,\xi(\theta))-a(\theta,\,\eta(\theta))|^2 \\ &\quad + \|\sigma(\theta,\,\xi(\theta))-\sigma(\theta,\,\eta(\theta))\|^2\}\,\mathrm{d}\theta \\ &\leqslant 2K_m\int_0^t \mathrm{E}\{\chi_m(\theta)\,|\xi(\theta)-\eta(\theta)|^2\}\,\mathrm{d}\theta. \end{aligned} \tag{7.4.16}$$

Defining

$$v_m(t) \equiv \mathrm{E}\{\chi_m(t)\,|\xi(t)-\eta(t)|^2\},$$

it follows from (7.4.16) that

$$v_m(t) \leqslant 2K_m\int_0^t v_m(\theta)\,\mathrm{d}\theta, \qquad t\in I,$$

and hence $v_m(t)\equiv 0$ for all $t\in I$ and any finite integer $m\geqslant 0$. This implies that ξ coincides with η for all $t\leqslant \tau_m$. Since

$$\mathrm{P}\Big\{\sup_{t\in I}|\xi(t)| \text{ and } \sup_{t\in I}|\eta(t)|<\infty\Big\}=1,$$

and $\chi_m(t)\to 1$, P-a.s. for all $t\in I$ as $m\to\infty$, we have

$$\mathrm{P}\{\xi(t)=\eta(t),\ t\in I\}=1. \quad \blacksquare$$

Remark 7.4.4 As in the deterministic case, the Lipschitz condition (b) and its local version (b)′ can be replaced by mere continuity of the functions a and σ. However, under this relaxed condition uniqueness is not guaranteed.

Under the assumption of atmost linear growth, as given in Theorem 7.4.2, we saw that if the initial state has second moment then the state $\xi(t)$ also has second moment. In fact, one can prove the following more general result.

Corollary 7.4.5 (Higher-order moments). *Suppose a and σ satisfy the assumptions of Theorem* 7.4.2 *with constant K and let* $\mathrm{E}\,|\xi_0|^{2p}<\infty$ *for a positive integer* $p\geqslant 1$. *Then there exists a constant* $c_T=c(T,K,p)>0$, *such that*

(a) $\mathrm{E}\,|\xi(t)|^{2p}\leqslant c_T(1+\mathrm{E}\,|\xi_0|^{2p})<\infty$,
(b) $\mathrm{E}\{\sup_{0\leqslant s\leqslant t}|\xi(s)-\xi_0|^{2p}\}\leqslant c_T(1+\mathrm{E}\,|\xi_0|^{2p}\}t^p$.

Proof Consider the system

$$d\xi = a(t, \xi)\, dt + \sigma(t, \xi)\, dW,$$

with $\xi(0) = \xi_0$, and define $f(x) = |x|^{2p}$ and $\eta(t) = f(\xi(t))$. Then applying Itô's formula to η and using Gronwall's lemma, one can verify after some elementary computations that there exists a constant $c_T = c(T, K, p)$ such that

$$(1 + \mathrm{E}\,|\xi(t)|^{2p}) \leq c_T(1 + \mathrm{E}\,|\xi_0|^{2p})$$

for all $t \in I$. Hence (a) follows and (b) is verified in the same way using (a). ■

Remark 7.4.6 In case $\xi_0 = x$ is a constant vector, it follows from the above result that $\mathrm{E}\,|\xi(t)|^{2p} < \infty$ for all finite $p > 0$ and not just the integers.

7.4.2 Continuous dependence of solutions on parameters

Consider the system (7.4.1) with parameters $\{\xi_0^r, a^r, \sigma^r, r \geq 1\}$ such that $\xi_0^r \in L_2(\mathcal{F}_0, R^n)$ and a^r, σ^r are measurable functions on $I \times R^n$ with values in R^n and $R(n \times m)$ respectively. Let ξ^r, $r \geq 1$ denote the solution of (7.4.1) corresponding to the set $\{\xi_0^r, a^r, \sigma^r\}$. Suppose $\{\xi_0^r, a^r, \sigma^r\}$ converges to $\{\xi_0, a, \sigma\}$ in some sense and let ξ denote the solution of (7.4.1) corresponding to the set $\{\xi_0, a, \sigma\}$. We shall prove in the following theorem that under certain conditions $\xi^r \to \xi$ uniformly on I with probability 1 as $r \to \infty$. This will prove the desired continuity of solutions with respect to the system parameters. We must emphasize that continuous dependence of solutions on system parameters is not automatically guaranteed unless the parameters themselves satisfy certain regularity properties.

Theorem 7.4.7 (Continuous dependence). *Consider the system (7.4.1) and suppose for each integer $r \geq 1$, ξ^r and ξ denote the solutions corresponding to the parameters $\{\xi_0^r, a^r, \sigma^r\}$ and $\{\xi_0, a, \sigma\}$ respectively. Suppose*

(a) *there exists a $K \in L_1^+(I)$ such that*

$$T\,|a^r(t, x) - a^r(t, y)|^2 + 4\,\|\sigma^r(t, x) - \sigma^r(t, y)\|^2 \leq K(t)\,|x - y|^2$$

and

$$T\,|a^r(t, x)|^2 + 4\,\|\sigma^r(t, x)\|^2 \leq K(t)(1 + |x|^2)$$

for all $t \in I = [0, T]$ *and* $x, y \in R^n$ *independently of* $r \geqslant 0$, *where* $(\xi_0^0, a^0, \sigma^0) = (\xi_0, a, \sigma)$.

(b) *the sequence* $\{\xi_0^r, a^r, \sigma^r\}$ *converges to* $\{\xi_0, a, \sigma\}$ *in the following sense*:

$$\xi_0^r \to \xi_0 \qquad \text{in } L_2(\mathscr{F}_0, R^n),$$

$$a^r \to a \qquad \text{a.e. on } I \times R^n,$$

$$\sigma^r \to \sigma \qquad \text{a.e. on } I \times R^n.$$

Then $\xi^r \to \xi$ *uniformly on I with probability* 1.

Proof By virtue of the previous theorems the solution of (7.4.1) corresponding to $\{\xi_0^r, a^r, \sigma^r\}$ and $\{\xi_0, a, \sigma\}$ are uniquely defined by the stochastic integral equation:

$$\xi^r(t) = \xi_0^r + \int_0^t a^r(\theta, \xi^r(\theta))\, d\theta + \int_0^t \sigma^r(\theta, \xi^r(\theta))\, dW(\theta),$$

and

$$\xi(t) = \xi_0 + \int_0^t a(\theta, \xi(\theta))\, d\theta + \int_0^t \sigma(\theta, \xi(\theta))\, dW(\theta),$$

respectively. Using these equations and the martingale inequality (Proposition 7.1.12), it is easy to verify that

$$E\left\{\sup_{0 \leqslant \tau \leqslant t} |\xi^r(\tau) - \xi(\tau)|^2\right\} \leqslant \alpha_r + \beta_r(t) \tag{7.4.17}$$

where

$$\alpha_r \equiv 5\left\{E\,|\xi_0^r - \xi_0|^2 + E \int_0^T [T\,|a^r(\theta, \xi(\theta)) - a(\theta, \xi(\theta))|^2 + 4\,\|\sigma^r(\theta, \xi(\theta)) - \sigma(\theta, \xi(\theta))\|^2]\, d\theta\right\}$$

and

$$\beta_r(t) \equiv 5E \int_0^t [T\,|a^r(\theta, \xi^r(\theta)) - a^r(\theta, \xi(\theta))|^2 + 4\,\|\sigma^r(\theta, \xi^r(\theta)) - \sigma^r(\theta, \xi(\theta))\|^2]\, d\theta.$$

Then by hypothesis (a),

$$\beta_r(t) \leqslant 5E \int_0^t K(\theta)\left\{\sup_{0 \leqslant \tau \leqslant \theta} |\xi^r(\tau) - \xi(\tau)|^2\right\} d\theta.$$

Defining

$$\rho_t(\xi^r, \xi) \equiv \mathrm{E} \sup_{0 \leq \tau \leq t} |\xi^r(\tau) - \xi(\tau))|^2$$

it follows from (7.4.17) that

$$\rho_t(\xi^r, \xi) \leq \alpha_r + 5 \int_0^t K(\theta) \rho_\theta(\xi^r, \xi) \, \mathrm{d}\theta, \tag{7.4.18}$$

and hence, by Gronwall's inequality, we have

$$\|\xi^r - \xi\|_X^2 = \rho_T(\xi^r, \xi) \leq c_T \alpha_r, \tag{7.4.19}$$

where $c_T \equiv \exp(5 \int_0^T K(\theta) \, \mathrm{d}\theta)$. In order to complete the proof, we show that $\lim_{r \to \infty} \alpha_r = 0$.

By the uniform growth condition (a),

$$\begin{aligned} T\, |a^r(\theta, \xi(\theta)) - a(\theta, \xi(\theta))|^2 + 4\, \|\sigma^r(\theta, \xi(\theta)) - \sigma(\theta, \xi(\theta))\|^2 \\ \leq 2\{T(|a^r(\theta, \xi(\theta))|^2 + |a(\theta, \xi(\theta))|^2) \\ + 4(\|\sigma^r(\theta, \xi(\theta))\|^2 + \|\sigma(\theta, \xi(\theta))\|^2)\} \\ \leq 4\{K(\theta)(1 + |\xi(\theta)|^2)\}. \end{aligned}$$

Thus the integrand in the expression for α_r is dominated by this integrable function. Hence by virtue of (b), it follows from the dominated convergence theorem that $\lim_{r \to \infty} \alpha_r = 0$. Therefore,

$$\lim_{r \to \infty} \|\xi^r - \xi\|_X^2 = \lim_{r \to \infty} \rho_T(\xi^r, \xi) = 0. \tag{7.4.20}$$

This completes the proof. ∎

Remark 7.4.8 The conclusion of Theorem 7.4.7 also holds under the local Lipschitz condition.

In applications the matrices a and σ more often depend on certain parameters $\alpha \in \mathcal{P}$, and the system appears in the form

$$\mathrm{d}\xi = a(t, \xi(t), \alpha) \, \mathrm{d}t + \sigma(t, \xi(t), \alpha) \, \mathrm{d}W(t). \tag{7.4.21}$$

Here one is interested in the properties of ξ considered as a function of α on $\mathcal{P}$, for example continuity, sensitivity and differentiability, etc. As a corollary of the preceding theorem, we have the following result.

Corollary 7.4.9 Consider the system (7.4.21) *and suppose $\mathcal{P}$ is a metric space and the functions a: $I \times R^n \times \mathcal{P} \to R^n$* and *$\sigma$: $I \times R^n \times \mathcal{P} \to$*

$R(n \times m)$ satisfy the following conditions:

(a)′ *there exists a $K \in L_1^+(I)$, possibly dependent on $\mathcal{P}$, such that*

$$T\,|a(t, x, \alpha) - a(t, y, \alpha)|^2 + 4\,\|\sigma(t, x, \alpha) - \sigma(t, y, \alpha)\|^2 \leq K(t)\,|x - y|^2,$$

and

$$T\,|a(t, x, \alpha)|^2 + 4\,\|\sigma(t, x, \alpha)\|^2 \leq K(t)[1 + |x|^2]$$

for all $t \in I$, and $x, y \in R^n$,

(b)′ *for each $(t, x) \in I \times R^n$ the functions $\alpha \to a(t, x, \alpha)$ and $\alpha \to \sigma(t, x, \alpha)$ are continuous from $\mathcal{P}$ to R^n and $R(n \times m)$ respectively.*

Then $\alpha \to \xi(\cdot, \alpha)$ is a continuous map from $\mathcal{P}$ to X. Further if a, σ are C^1 in x and the partials of a and σ with respect to α have at most polynomial growth in x, then $\alpha \to \xi(\cdot, \alpha)$ is C^1 in the mean square sense.

Another result of considerable interest in the study of Markov processes is the continuous dependence of solution ξ on the initial data. This is given in the following theorem.

Theorem 7.4.10 (Continuous dependence on initial data). *Consider the system $d\xi = a(t, \xi)\,dt + \sigma(t, \xi)\,dW$ and suppose the assumptions of Theorem 7.4.2 hold. Let $t_1, t_2 \in [0, T]$, $x_1, x_2 \in R^n$ and $\xi_{t_1,x_1}(t)$, $t \in (t_1, T]$, and $\xi_{t_2,x_2}(t)$, $t \in (t_2, T]$ denote the solutions corresponding to the initial data (t_1, x_1) and (t_2, x_2) respectively. Then there exists a constant $c > 0$ such that*

$$E\left\{\sup_{(t_1, T] \cap (t_2, T]} |\xi_{t_2,x_2}(t) - \xi_{t_1,x_1}(t)|^2\right\} \leq c\{|x_2 - x_1|^2 + \nu(|t_2 - t_1|)\}$$

where

$$\nu(|t_2 - t_1|) = \int_{t_1 \wedge t_2}^{t_1 \vee t_2} K(\theta)\,d\theta.$$

In case K is constant $\nu(|t_2 - t_1|)$ may be replaced by $|t_2 - t_1|$ (with c redefined).

Proof The proof essentially follows from the Gronwall lemma, and is left as an exercise for the reader. ■

As a consequence of the existence and uniqueness theorems, one can easily deduce that there exists a family of nonlinear operators $\{T_{t,s},$

$0 \leqslant s \leqslant t \leqslant T\}$ such that for each $\eta \in L_2(\mathscr{F}_s, R^n)$, the equation

$$\xi(t) = \eta + \int_s^t a(\theta, \xi(\theta))\, \mathrm{d}\theta + \int_s^t \sigma(\theta, \xi(\theta))\, \mathrm{d}W(\theta), \qquad t \in (s, T]$$

has a unique (strong) solution ξ and $T_{t,s}(\eta) = \xi(t)$ for all $t \in (s, T]$. The operator $T_{t,s}$ maps $L_2(\mathscr{F}_s, R^n)$ into $L_2(\mathscr{F}_t, R^n)$, and there exists a constant $c > 0$ such that

$$\mathrm{E}\{|T_{t,s}(\eta)|^2\} \leqslant c(1 + \mathrm{E}\,|\eta|^2). \tag{7.4.22}$$

From continuity of the solution $t \to \xi(t) = T_{t,s}(\eta)$, it follows that

$$\lim_{t \downarrow s} T_{t,s}(\eta) = \eta \quad \text{P-a.s.} \tag{7.4.23}$$

and from uniqueness, we have, for $0 \leqslant s \leqslant \theta \leqslant t \leqslant T$,

$$T_{t,s}(\eta) = T_{t,\theta}(T_{\theta,s}(\eta)) \tag{7.4.24}$$

for all $\eta \in L_2(\mathscr{F}_s, R^n)$ and, in general, for all $\mathscr{F}_s$-measurable η with $\mathrm{P}\{|\eta| < \infty\} = 1$.

Hence, in summary, $\{T_{t,s},\ 0 \leqslant s \leqslant t \leqslant T\}$ is a family of nonlinear (strongly) measurable two parameter semigroups on $L_2(\Omega, \mathscr{F}, \mathrm{P}; R^n)$ satisfying

$$\begin{aligned} &\text{(a)} \quad T_{t,s}\colon L_2(\mathscr{F}_s, R^n) \to L_2(\mathscr{F}_t, R^n) \\ &\text{(b)} \quad T_{t,t} = I \\ &\text{(c)} \quad \lim_{t \downarrow s} T_{t,s} = I, \ \text{P-a.s.} \\ &\text{(d)} \quad T_{t,\theta} T_{\theta,s} = T_{t,s}, \ \text{P-a.s.}, \ s \leqslant \theta \leqslant t. \end{aligned} \tag{7.4.25}$$

7.4.3 Solution of stochastic differential equations as Markov processes

We have already discussed that a Markov process ξ is completely characterized by its initial distribution μ_0 and the transition probability function $\mathrm{P}(s, x; t, \Gamma)$, $s < t$.

Definition 7.4.11 (Transition probability function). A transition probability function $\mathrm{P}(s, x; t, \Gamma)$, $0 \leqslant s \leqslant t \leqslant T < \infty$, $x \in R^n$, $\Gamma \in \mathscr{B}(R^n)$ is called a Markov transition function (or kernel) if it is non-negative and satisfies the following properties:

(a) for fixed s, t, Γ, $x \to \mathrm{P}(s, x; t, \Gamma)$ is Borel measurable,
(b) for fixed s, t, x, $\Gamma \to \mathrm{P}(s, x; t, \Gamma)$ is a probability measure on $\mathscr{B}(R^n)$,
(c) for $0 \leqslant s < \tau < t \leqslant T$, $\Gamma \in \mathscr{B}(R^n)$, and $x \in R^n$

$$\mathrm{P}(s, x; t, \Gamma) = \int_{R^n} \mathrm{P}(s, x; \tau, \mathrm{d}y) \mathrm{P}(\tau, y; t, \Gamma) \tag{7.4.26}$$

is called the Chapman–Kolmogorov equation.

Now we can define a Markov process. Let $(\Omega, \mathscr{F}, \mathrm{P})$ be a probability space and $\mathscr{F}_t$ an increasing family of subsigma algebras of the σ-algebra $\mathscr{F}$. A process $\xi \equiv \{\xi(t), t \geqslant 0\}$ defined on this probability space is called an *$\mathscr{F}_t$-Markov process* with values in R^n, with the transition kernel $\mathrm{P}(s, x; t, \Gamma)$, if for every bounded measurable function f on R^n, and $0 \leqslant s < t$,

$$\begin{aligned} \mathrm{E}\{f(\xi(t)) \mid \mathscr{F}_s\} &= \mathrm{E}\{f(\xi(t)) \mid \sigma(\xi(s))\} \\ &= \int_{R^n} \mathrm{P}(s, \xi(s); t, \mathrm{d}y) f(y). \end{aligned}$$

From the properties (7.4.25) and the fact that $T_{t,s}(\xi(s))$ is independent of $\mathscr{F}_s$, given $\sigma(\xi(s))$, we have, for every bounded measurable function f on R^n,

$$\begin{aligned} \mathrm{E}\{f(\xi(t)) \mid \mathscr{F}_s\} &= \mathrm{E}\{f(T_{t,s}(\xi(s))) \mid \mathscr{F}_s\} \\ &= \mathrm{E}\{f(T_{t,s}(\xi(s))) \mid \sigma(\xi(s))\} \\ &= \mathrm{E}\{f(\xi(t)) \mid \sigma(\xi(s))\} \end{aligned}$$

for $0 \leqslant s < t$. In particular, for $f = \chi_\Gamma$, the indicator function of $\Gamma \in \mathscr{B}(R^n)$, we have

$$\mathrm{P}\{\xi(t) \in \Gamma \mid \mathscr{F}_s\} = \mathrm{P}\{\xi(t) \in \Gamma \mid \sigma(\xi(s))\}.$$

Therefore ξ is a continuous Markov process with the transition kernel given by $\mathrm{P}(s, x; t, \Gamma) \equiv \mathrm{P}\{\xi(t) \in \Gamma \mid \xi(s) = x\}$. From (7.4.25)(d), one can easily verify that the transition kernel $\mathrm{P}(s, x; t, \Gamma)$ satisfies the Chapman–Kolmogorov equation (7.4.26). Moreover, if the mapping

$$(s, x) \rightarrow (\Pi_{s,t} f)(x) \equiv \int_{R^n} \mathrm{P}(s, x; t, \mathrm{d}y) f(y),$$

is continuous on $[0, t) \times R^n$ for each $f \in C_b(R^n)$, the space of bounded continuous functions on R^n, then $\{\Pi_{s,t}, 0 \leqslant s \leqslant t \leqslant T\}$ is a family of linear semigroup operators in $C_b(R^n)$. In this case the corresponding Markov process is called a *Feller process*. Therefore, under the assumptions of Theorem 7.4.2 or Theorem 7.4.3, the solution ξ of any stochastic differential equation of the form

$$\mathrm{d}\xi(t) = a(t, \xi(t))\,\mathrm{d}t + \sigma(t, \xi(t))\,\mathrm{d}W(t), \qquad t \geqslant 0,$$

is a Feller (Markov) process.

An R^n-valued Markov process with the transition probability $\mathrm{P}(s, x; t, \Gamma)$ is called a *diffusion process* if P satisfies the following properties:

(D1) $\lim_{h \downarrow 0} 1/h \int_{N'_\varepsilon(x)} \mathrm{P}(t, x; t+h, \mathrm{d}y) = 0$ for every $\varepsilon > 0$, where N'_ε is the complement of $N_\varepsilon(x) \equiv \{y \in R^n\colon |y - x| \leqslant \varepsilon\}$.

(D2) there exists a vector $a(t, x)$: $I \times R^n \to R^n$ and a matrix $b(t, x)$: $I \times R^n \to R(n \times n)$ such that

(a) $\lim_{h\downarrow 0} 1/h \int_{N_\varepsilon(x)}(y - x, z)\mathrm{P}(t, x; t+h, \mathrm{d}y) = (a(t, x), z)$,

(b) $\lim_{h\downarrow 0} 1/h \int_{N_\varepsilon(x)} (y - x, z)^2\mathrm{P}(t, x; t+h, \mathrm{d}y) = (b(t, x)z, z)$ for all $z \in R^n$.

The vector a is called the drift (infinitesimal mean) and b the diffusion matrix (infinitesimal covariance).

Using Corollary 7.4.5, we can verify the following result.

Theorem 7.4.12 (Diffusion process). *Under the assumptions of Theorem 7.4.2 or Theorem 7.4.3, the process ξ determined by the solution of any stochastic differential equation of the form $\mathrm{d}\xi(t) = a(t, \xi(t))\,\mathrm{d}t + \sigma(t, \xi(t))\,\mathrm{d}W(t)$, $t \geq 0$, is a diffusion process in R^n with a being the drift and $b = (\sigma\sigma')$ the diffusion matrix.*

We mention that diffusion processes have been widely studied in the literature under more general situations using semigroup theory and the Strook–Varadhan approach based on martingale theory [46].

The differential generator of the Markov semigroup $\Pi_{s,t}$ is given by the limit

$$\lim_{h\downarrow 0} \left\{\frac{(\Pi_{t,t+h}\varphi) - \varphi}{h}\right\} = G_t\varphi$$

for all $\varphi \in C(R^n)$ for which the limit exists. Indeed, for $\varphi \in C^2(R^n)$, we can write

$$\varphi(y) = \varphi(x) + (\varphi_x, y - x) + \tfrac{1}{2}(\varphi_{xx}(y - x), y - x) + o(|y - x|^2);$$

then

$$\begin{aligned} G_t\varphi(x) &= \lim_{h\downarrow 0} \left\{\frac{(\Pi_{t,t+h}\varphi)(x) - \varphi(x)}{h}\right\} \\ &= \lim_{h\downarrow 0} \frac{1}{h}\int_{R^n} (\varphi(y) - \varphi(x))\mathrm{P}(t, x; t+h, \mathrm{d}y) \\ &= \lim_{h\downarrow 0} \frac{1}{h}\int_{R^n} \mathrm{P}(t, x; t+h, dy)\{(\varphi_x, y - x) + \tfrac{1}{2}(\varphi_{xx}(y - x), (y - x)) \\ &\quad + o(|y - x|^2)\}. \end{aligned}$$

Hence it follows from (D1) and (D2) that

$$(G_t\varphi)(x) = (\varphi_x, a(t, x)) + \tfrac{1}{2}\mathrm{tr}(\sigma\sigma')(t, x) \cdot \varphi_{xx} = (A(t)\varphi)(x)$$

where $A(t)$ is the differential operator that arises in the Itô formula (7.2.41)–(7.2.42).

So far we have considered strong solutions, that is, solutions which are $\mathscr{F}_t$-adapted under rather strong assumptions on a and σ. In fact, existence of solutions can be proved under much more relaxed conditions. It is sufficient if a, σ are non-anticipative, a measurable, and σ is continuous and non-degenerate. In this case, the notion of a solution must be weakened sacrificing the need of $\mathscr{F}_t$-measurability with reference to a fixed probability space $(\Omega, \mathscr{F}, \mathscr{F}_{t\geqslant 0}, \mathrm{P})$. A process ξ with $\xi(0)=x$ is said to be a weak solution if there exists a probability space $(\tilde{\Omega}, \tilde{\mathscr{F}}, \tilde{\mathscr{F}}_{t\geqslant 0}, \tilde{\mathrm{P}})$ on which a Brownian motion $\tilde{W}$ is defined such that the pair $(\xi, \tilde{W})$ satisfies the equation

$$\begin{aligned} \mathrm{d}\xi &= a(t, \xi)\,\mathrm{d}t + \sigma(t, \xi)\,\mathrm{d}\tilde{W}, \\ \xi(0) &= x. \end{aligned} \tag{7.4.27}$$

Such a solution is constructed through the celebrated Cameroon–Martin–Girsanov formula giving the Radon–Nikodym derivative of one measure with respect to another. Since we know that solutions of equations of the form (7.4.27) are continuous with probability one, we can choose $C \equiv C(I, R^n)$ as the sample space for Ω, and for $\mathscr{F}$ we choose $\mathscr{B}(C)$, the Borel σ-algebra of subsets of the set C containing all cylinder sets of the form

$$G \equiv \{\xi \in C\colon \xi(t_i) \in \Gamma_i,\ t_i \in I,\ \Gamma_i \in \mathscr{B}(R^n),\\ i = 1, 2, \ldots, k,\ k \text{ a positive integer}\}.$$

Then $(C, \mathscr{B}(C))$ is a measurable space called the canonical sample space. Clearly, we can define different measures $\{\mu\}$ on this space turning $(C, \mathscr{B}(C))$ into different measure spaces $\{(C, \mathscr{B}(C), \mu)\}$. Let $\mathscr{B}_t(C)$ denote the smallest σ-algebra contained in $\mathscr{B}(C)$ and containing all cylinder sets based on $[0, t]$. Consider the system

$$\begin{aligned} \mathrm{d}\xi_0(t) &= \sigma(t, \xi_0(t))\,\mathrm{d}W(t), \qquad t \geqslant 0, \\ \xi_0(0) &= x, \end{aligned} \tag{7.4.28}$$

with a removed and σ assumed to be continuous and non-degenerate. Let μ_0 denote the measure induced by ξ_0 on $(C(\mathscr{B}(C))$. Define

$$\rho_0^T = \exp\left\{\int_0^T ((\sigma^{-1}a)(\theta, \xi_0(\theta)), \mathrm{d}W(\theta)) - \frac{1}{2}\int_0^T |\sigma^{-1}a|^2\,\mathrm{d}\theta\right\}. \tag{7.4.29}$$

If $\|\sigma^{-1}a\|$ is bounded, then

$$\mathrm{E}_{\mu_0}\{\rho_0^T\} = \int_C \rho_0^T(\eta)\,\mathrm{d}\mu_0(\eta) = 1 \tag{7.4.30}$$

and μ, given by $\mathrm{d}\mu = \rho_0^T\,\mathrm{d}\mu_0$, is a probability measure on $(C, \mathscr{B}(C))$. Then μ is associated with the solution of

$$\begin{aligned} \mathrm{d}\xi(t) &= a(t, \xi(t))\,\mathrm{d}t + \sigma(t, \xi(t))\,\mathrm{d}W(t), \\ \xi(0) &= x, \end{aligned} \tag{7.4.31}$$

and is called the weak solution of the problem (7.4.31). Further, $\tilde{W}$ defined by

$$\tilde{W}(t) = W(t) - \int_0^t (\sigma^{-1}a)(\theta, \xi_0(\theta))\,\mathrm{d}\theta, \tag{7.4.32}$$

is a Brownian motion on $(C, \mathscr{B}(C), \mu)$. Hence

$$\mathrm{d}W = \sigma^{-1}a\,\mathrm{d}t + \mathrm{d}\tilde{W}, \qquad t \geq 0,$$

or equivalently,

$$\begin{aligned} \mathrm{d}\eta &= a(t, \eta)\,\mathrm{d}t + \sigma(t, \eta)\,\mathrm{d}\tilde{W}, \qquad t \geq 0, \\ \eta(0) &= x. \end{aligned} \tag{7.4.33}$$

Comparing (7.4.33) with (7.4.31), we see that η and ξ are solutions of one and the same equation but driven by two different Brownian motions $\tilde{W}$ and W respectively. This shows that weak solutions exist under the conditions that a is measurable as a function of $(t, x) \in I \times R^n$ and that $\|\sigma^{-1}a\|$ is bounded. This last condition can be further relaxed [46]. This is beyond the scope of this book. Before closing we mention that the Radon–Nikodym derivative ρ_0^t, considered as a function of t, is a μ_0-martingale on the probability space $(C, \mathscr{B}(C), \mu_0)$ with reference to $\mathscr{B}_t(C)$. That is,

$$\begin{aligned} &\text{(a)}\quad \mathrm{E}_{\mu_0}\{\rho_0^t\} = 1,\ t \in I, \\ &\text{(b)}\quad \mathrm{E}_{\mu_0}\{\rho_0^t \mid \mathscr{B}_s(C)\} = \rho_0^s,\ s < t, \\ &\text{(c)}\quad \mathrm{E}_{\mu_0}\{\rho_t^T \mid \mathscr{B}_t(C)\} = 1,\ \mu_0\text{-a.s.} \end{aligned} \tag{7.4.34)(a}$$

For any $\mathscr{B}(C)$-measurable and μ-integrable function h, we shall write $\mathrm{E}_\mu(h) = \int_C h(\xi)\,\mathrm{d}\mu(\xi)$. This gives us the expected value of a functional h on C under the measure μ, that is, the expected value of any functional of the process ξ generated by the system (7.4.31). Such functionals are denoted by $L_1(C, \mathscr{B}(C), \mu)$. Since $\mathrm{d}\mu = \rho_0^T\mu_0$, we have

$$\text{(d)}\quad \mathrm{E}_\mu\{h\} = \int_C h(\xi)\,\mathrm{d}\mu(\xi) = \int_C h(\xi)\rho_0^T(\xi)\,\mathrm{d}\mu_0 = \mathrm{E}_{\mu_0}\{h\rho_0^T\}. \tag{7.4.34)(b}$$

7.4.4 Kolmogorov's equations

In the study of stability, optimal control, and filtering, quantities of practical interest are often given by (mathematical) expectation of certain functionals of the solution process ξ. These quantities can be computed by solving a class of initial- or initial–boundary-value problems involving second-order partial differential equations of parabolic or elliptic type. In other words, stochastic problems are converted into equivalent deterministic problems which can be numerically solved to determine the desired quantities. Before we can derive these equations, we need some preparation.

Definition 7.4.13 (Mean square differentiability). A random function $f = f(x, \omega)$ mapping R^n into $L_2(\Omega, \mathcal{F}, \mathrm{P}; R^n)$ is said to be differentiable in the mean square sense if for each coordinate variable x_i, there exists a random function g_i: $R^n \to L_2(\Omega, \mathcal{F}, \mathrm{P}; R^n)$ such that for each $x \in R^n$

$$\int_\Omega \left| \frac{f(x + he_i; \omega) - f(x, \omega)}{h} - g_i(x, \omega) \right|^2 \mathrm{P}(\mathrm{d}\omega) \to 0$$

as $h \to 0$. Then $g_i \equiv \partial f / \partial x_i$ is the partial of f in the mean square sense. Similarly one can define higher-order partials. Partial derivatives in the mean of order $p \geq 1$ are also defined in the same way with p taking the place of 2.

A function f: $R^n \to R^n$ is said to have at most a polynomial growth if there exist constants β, $\delta > 0$ such that

$$|f(x)| \leq \beta(1 + |x|^\delta) \qquad \text{for all } x \in R^n.$$

Let $\alpha = (\alpha_1, \alpha_2, \ldots, \alpha_n)$ be a multi-index with α_i being non-negative integers. Define

$$|\alpha| \equiv \sum \alpha_i, \qquad D^{\alpha_i}\varphi \equiv \frac{\partial^{\alpha_i}\varphi}{\partial x_i^{\alpha_i}}, \qquad D^\alpha \varphi \equiv D^{\alpha_1} D^{\alpha_2} \cdots D^{\alpha_n}\varphi,$$

and for each integer $\ell \geq 0$,

$$C^\ell \equiv \{\varphi: D^\alpha \varphi \in C^0 \text{ for all } |\alpha| \leq \ell\}.$$

For further development, we need the following result that gives sufficient conditions for differentiability of solutions of stochastic differential equations with respect to the initial state.

Theorem 7.4.14 (Differentiability of $x \to \xi_{t,x}(\cdot)$). *Consider the system* $\mathrm{d}\xi(\theta) = a(\theta, \xi)\,\mathrm{d}\theta + \sigma(\theta, \xi)\,\mathrm{d}W$, $\theta \geq t$, *with* $\xi(t) = x$. *Suppose the coefficients* a, σ *satisfy the assumptions of Theorem* 7.4.2 *with constant* K. *Then*

(a) *if* a, σ *are* C^1 *in* x, *then* $x \to \xi_{t,x}(\cdot)$ *is also* C^1 *in* x *in the mean square sense; and in general,*

(b) *if* a, $\sigma \in C^{\ell}$ *and* $D^{\alpha}a$, $D^{\alpha}\sigma$ *for* $|\alpha| \leq \ell$, *are continuous on* $I \times R^n$, *and have at most polynomial growth, then* $D_x^{\alpha}\xi_{t,x}(\cdot)$ *exists in the mean square sense for* $|\alpha| \leq \ell$,

(c) *if* $f \in C^2(R^n)$, *and* $D^{\alpha}f$, $|\alpha| \leq 2$, *has at most a polynomial growth and the parameters* a, σ *satisfy the conditions of* (b) *for* $\ell = 2$, *then* $x \to \varphi(t, x) \equiv \mathrm{E}\{f(\xi_{t,x}(\tau))\}$, $0 \leq t \leq \tau$, *is also* C^2 *in* x.

Proof The proof is based on direct but somewhat lengthy computation and uses Gronwall's lemma, the Lebesgue dominated convergence theorem, and the following facts:

(i) $\xi_{t,x}(\cdot)$ has moments of all finite order (Corollary 7.4.5),

(ii) $\mathrm{P}\{\|a_x(\theta, \xi_{t,x}(\theta))\|, \|\sigma_{x_i}(\theta, \xi_{t,x}(\theta))\| \leq K\} = 1$,

(iii) $D_x^{\alpha}a$, $D_x^{\alpha}\sigma$ have at most polynomial growth for $|\alpha| \leq \ell$. ■

In the following theorems, we shall prove that many practically useful functionals of the process ξ are given by solutions of certain associated parabolic and elliptic partial differential equations.

Theorem 7.4.15 (Functionals of terminal state). *Suppose the assumptions of Theorem* 7.4.14 *hold. Then for every* $f \in C^2$ *with* $D^{\alpha}f$, $|\alpha| \leq 2$, *having at most polynomial growth, the function* φ *given by*

$$\varphi(t, x) \equiv \mathrm{E}\{f(\xi(T)) \mid \xi(t) = x\} = \mathrm{E}\{f(\xi_{t,x}(T))\} \tag{7.4.35}$$

satisfies the Kolmogorov equation

$$\begin{aligned} &\frac{\partial \varphi}{\partial t} + \mathbf{A}(t)\varphi = 0 \qquad \text{in } [0, T) \times R^n, \\ &\varphi(T, x) = f(x) \qquad \text{in } R^n, \end{aligned} \tag{7.4.36}$$

where

$$\mathbf{A}(t)\varphi = \sum_{i=1}^{n} a_i(t, x)\varphi_{x_i} + \frac{1}{2}\sum_{i,j}^{n} (\sigma\sigma')_{ij}(t, x)\varphi_{x_i x_j}$$

is the infinitesimal generator of the Markov process ξ.

Proof For $\Delta t > 0$, $t > 0$, $t - \Delta t \geq 0$, and $x \in R^n$,

$$\begin{aligned} \varphi(t - \Delta t, x) &= \mathrm{E}\{f(\xi_{t-\Delta t,x}(T))\} = \mathrm{E}\{f(\xi(T)) \mid \xi(t - \Delta t) = x\} \\ &= \mathrm{E}\{\mathrm{E}\{f(\xi(T)) \mid \mathscr{F}_t^{t-\Delta t}\} \mid \xi(t - \Delta t) = x\}, \end{aligned}$$

where $\mathscr{F}_t^s = \sigma\{\xi(\theta), s \leq \theta \leq t\}$. Hence by the Markov property,

$$\begin{aligned} \varphi(t - \Delta t, x) &= \mathrm{E}\{\mathrm{E}\{f(\xi(T)) \mid \xi(t)\} \mid \xi(t - \Delta t) = x\} \\ &= \mathrm{E}\{\varphi(t, \xi(t)) \mid \xi(t - \Delta t) = x\}. \end{aligned} \tag{7.4.37}$$

Under the given assumptions φ is C^2 in x; hence by Itô's formula applied to $\varphi(t, \xi(t))$ for fixed t, we have

$$\varphi(t, \xi(t)) = \varphi(t, x) + \int_{t-\Delta t}^{t} \mathbf{A}(\theta)\varphi(t, \xi(\theta))\, d\theta + M_{t,t-\Delta t}, \tag{7.4.38}$$

where

$$M_{t,t-\Delta t} = \int_{t-\Delta t}^{t} (\sigma' \varphi_x, dW(\theta)). \tag{7.4.39}$$

Since σ and φ_x have at most polynomial growth and $\xi_{t-\Delta t,x}(\theta)$ has all moments of finite order, $\sigma'\varphi_x$ is square integrable along the path $\xi_{t-\Delta t,x}(\theta)$. Hence

$$\mathrm{E}\{M_{t,t-\Delta t} \mid \xi(t-\Delta t) = x\} = 0. \tag{7.4.40}$$

Using (7.4.38) and (7.4.40) into (7.4.37), we obtain

$$\varphi(t-\Delta t, x) = \varphi(t, x) + \mathrm{E} \int_{t-\Delta t}^{t} \mathbf{A}(\theta)\varphi(t, \xi_{t-\Delta t,x}(\theta))\, d\theta. \tag{7.4.41}$$

Since the coefficients of $\mathbf{A}$ are continuous having at most polynomial growth, and φ, φ_x, φ_{xx} also have at most polynomial growth, and $\xi_{t-\Delta t,x}(\theta)$ has all moments of finite order, there exists a constant c such that

$$\left| \mathrm{E} \int_{t-\Delta t}^{t} \mathbf{A}(\theta)\varphi(t, \xi_{t-\Delta t,x}(\theta))\, d\theta \right| \leq c\, \Delta t. \tag{7.4.42}$$

Hence

$$|\varphi(t-\Delta t, x) - \varphi(t, x)| \leq c\, \Delta t,$$

and therefore $t \to \varphi(t, x)$ is absolutely continuous and differentiable almost everywhere for any $x \in R^n$. Hence

$$\begin{aligned} \lim_{\Delta t \to 0} \left(\frac{\varphi(t-\Delta t, x) - \varphi(t, x)}{\Delta t} \right) &= -\frac{\partial \varphi}{\partial t} = \lim_{\Delta t \to 0} \frac{1}{\Delta t} \mathrm{E} \int_{t-\Delta t}^{t} \mathbf{A}(\theta)\varphi(t, \xi_{t-\Delta t,x}(\theta))\, d\theta. \\ &= \mathbf{A}(t)\varphi(t, x) \end{aligned}$$

Since the coefficients of $\mathbf{A}$ are continuous in all the variables, the equality holds for $(t, x) \in [0, T) \times R^n$. Thus we have shown that φ satisfies the equation

$$\frac{\partial \varphi}{\partial t} + \mathbf{A}(t)\varphi = 0 \qquad \text{for } (t, x) \in [0, T) \times R^n.$$

Further, it follows from the continuity of $t \to \xi_{t,x}(\cdot)$ (see Theorem 7.4.10) and that of f on R^n that

$$\lim_{t\to T} \varphi(t, x) = \lim_{t\to T} \mathrm{E}\{f(\xi_{t,x}(T))\}$$
$$= f(x).$$

Thus (7.4.35) is satisfied. This completes the proof. ■

Similarly we can prove the following result.

Theorem 7.4.16 (Functionals of trajectories ξ). *Under the assumptions of Theorem 7.4.15, for any function g continuous in t and C^2 in x with $D_x^\alpha g$, $|\alpha| \leq 2$, having at most polynomial growth, the function φ, given by*

$$\varphi(t, x) = \mathrm{E}\left\{f(\xi_{t,x}(T)) + \int_t^T g(\theta, \xi_{t,x}(\theta))\, \mathrm{d}\theta\right\}, \tag{7.4.43}$$

is a solution of the Cauchy problem

$$\begin{aligned} &\frac{\partial \varphi}{\partial t} + \mathbf{A}(t)\varphi + g = 0 \qquad \text{in } (t, x) \in [0, T) \times R^n \\ &\qquad \varphi(T, x) = f(x) \qquad \text{in } x \in R^n. \end{aligned} \tag{7.4.44}$$

Further for any function $\gamma = \gamma(t, x)$, continuous and bounded on $I \times R^n$ and non-positive, the function φ given by

$$\begin{aligned} \varphi(t, x) = \mathrm{E}\Bigg\{ & f(\xi_{t,x}(T)) \exp \int_t^T \gamma(\theta, \xi_{t,x}(\theta))\, \mathrm{d}\theta \\ & + \int_t^T g(\theta, \xi_{t,x}(\theta))\left\{\exp \int_t^\theta \gamma(s, \xi_{t,x}(s))\, \mathrm{d}s\right\} \mathrm{d}\theta \Bigg\}, \end{aligned} \tag{7.4.45}$$

satisfies the Cauchy problem,

$$\begin{aligned} &\frac{\partial \varphi}{\partial t} + \mathbf{A}(t)\varphi + \gamma\varphi + g = 0 \qquad \text{in } (t, x) \in [0, T) \times R^n, \\ &\qquad \varphi(T, x) = f(x) \qquad \text{in } x \in R^n. \end{aligned} \tag{7.4.46}$$

The functionals (7.4.35), (7.4.43) and (7.4.45) represent many physical quantities and are useful in control theory. Consider the system

$$\begin{aligned} &\mathrm{d}\xi = a(t, \xi)\, \mathrm{d}t + \sigma(t, \xi)\, \mathrm{d}W, \\ &\xi(0) = \xi_0, \end{aligned} \tag{7.4.47}$$

and suppose the initial state is random, having a distribution μ_0 with compact support in R^n. Suppose f satisfy the assumptions of Theorem 7.4.15. Then

$$\int_{R^n} \varphi(0, x)\mu_0(\mathrm{d}x) = \mathrm{E}\{f(\xi(T))\} \tag{7.4.48}$$

provided φ is a solution of the Cauchy problem (7.4.36). The expression on the left of the equation (7.4.48) is, in fact, common for all the problems (7.4.35), (7.4.43) and (7.4.45) with φ being the solutions of the corresponding Cauchy problems (7.4.36), (7.4.44) and (7.4.46) respectively. As an example, if f is taken as the indicator function of any set $\Gamma \in \mathscr{B}(R^n)$, then

$$\int_{R^n} \varphi(0, x)\mu_0(\mathrm{d}x) = \mathrm{P}\{\xi(T) \in \Gamma\} \equiv \mu_T(\Gamma).$$

Unfortunately $f = \chi_\Gamma(\cdot)$ does not satisfy the C^2 assumption of Theorem 7.4.15. Hence one may choose a sequence of functions f_n which are C^2 in x and converge to the indicator function of Γ as $n \to \infty$, and solve the Cauchy problem (7.4.36) with $f = f_n$ giving φ_n. Then $\varphi = \lim \varphi_n$ may be considered as a weak solution of the problem (7.4.36), giving $\mu_T(\Gamma) = \lim_n \int \varphi_n(0, x)\mu_0(\mathrm{d}x)$.

Another class of problems that can be solved by use of Kolmogorov's equations are the Markov time problems. Let D be an open bounded set in R^n with smooth boundary and $x \in D$. Consider the system

$$\mathrm{d}\xi = a(t, \xi)\,\mathrm{d}t + \sigma(t, \xi)\,\mathrm{d}W,$$

$$\xi(0) = x.$$

Let

$$\tau_x \equiv \inf\{t \in [0, T]\colon \xi_{0,x}(t) \in \partial D\} \tag{7.4.49}$$

be the first time ξ hits the boundary ∂D of D. If the indicated set is empty let $\tau_x = T$. We are interested in the average life time of the process ξ in the set D, that is, $\mathrm{E}(\tau_x)$. We shall see that this is given by the solution of an initial boundary value problem.

Theorem 7.4.17 (Hitting time problem). *Let D be as stated above and φ a solution of the initial boundary-value problem,*

$$\begin{aligned} \frac{\partial \varphi}{\partial t} + \mathrm{A}(t)\varphi + 1 &= 0, && (t, x) \in [0, T) \times D, \\ \varphi(t, x) &= 0, && (t, x) \in [0, T] \times \partial D, \\ \varphi(T, x) &= 0, && x \in D, \end{aligned} \tag{7.4.50}$$

which is C^1 in t and C^2 in x. Then $\mathrm{E}(\tau_x) = \varphi(0, x)$.

Proof Applying Itô's formula to $\varphi(t, \xi(t))$, we have

$$\varphi(\tau_x, \xi(\tau_x)) = \varphi(0, x) + \int_0^{\tau_x} \left\{\left(\frac{\partial\varphi}{\partial\theta} + \mathbf{A}(\theta)\varphi\right)\right\} d\theta + \int_0^{\tau_x} (\sigma'\varphi_x, dW) \quad (7.4.51)$$

all evaluated along the path $\xi(\theta)$, $0 \leqslant \theta \leqslant T$, with $\xi(0) = x$. Since

$$\mathrm{E}\int_0^{\tau_x} (\sigma'\varphi_x, dW(\theta)) = \mathrm{E}\int_0^T \chi_{\theta\leqslant\tau_x}(\sigma'\varphi_x, dW(\theta)) = 0,$$

we have

$$\mathrm{E}\varphi(\tau_x, \xi(\tau_x)) = \varphi(0, x) + \mathrm{E}\int_0^T \chi_{\theta<\tau_x}\left(\frac{\partial\varphi}{\partial\theta} + \mathbf{A}(\theta)\varphi\right) d\theta. \quad (7.4.52)$$

Since $\xi(\tau_x) \in \partial D$ for $T > \tau_x$ and $\xi(\tau_x) \in D \cup \partial D$ for $\tau_x = T$, it follows from the lateral boundary condition

$$\varphi(t, x) = 0 \qquad \text{for } (t, x) \in ([0, T] \times \partial D) \cup (\{T\} \times D)$$

that $\mathrm{E}\varphi(\tau_x, \xi(\tau_x)) = 0$. Again for $0 \leqslant \theta < \tau_x$, $\xi(\theta) \in D$, and hence it follows from the first equation of (7.4.50) that

$$\mathrm{E}\int_0^T \chi_{\theta<\tau_x}\left(\frac{\partial\varphi}{\partial\theta} + \mathbf{A}(\theta)\varphi\right) d\theta = -\mathrm{E}(\tau_x).$$

Using these facts, it follows from (7.4.52) that $\mathrm{E}(\tau_x) = \varphi(0, x)$. This completes the proof. ■

Corollary 7.4.18 Suppose **A** *is time invariant. Then* $\mathrm{E}(\tau_x) = \varphi(x)$, *where* φ *is the solution of the Dirichlet problem*

$$\begin{aligned} \mathbf{A}\varphi + 1 &= 0, \qquad x \in D, \\ \varphi(x) &= 0, \qquad x \in \partial D. \end{aligned} \quad (7.4.53)$$

Remark 7.4.19 Consider the system (7.4.47) with the initial state having a distribution μ_0 with support contained in D. Define

$$\tau_{\mu_0} = \int_D \tau_x \mu_0(dx).$$

Then the average life time of the process ξ in D is given by

$$\mathrm{E}(\tau_{\mu_0}) = \int_D \varphi(0, x)\mu_0(dx).$$

We conclude this section with the following result. The proof is essentially based on Itô's formula.

Theorem 7.4.20 (Dirichlet problem). *Suppose a, σ are Lipschitz continuous on $[0, T] \times \bar{D}$, where D is an open bounded set with C^2 boundary. Suppose*

(a) $(\sigma\sigma')(t, x) > 0$ *for* $(t, x) \in [0, T] \times D$,
(b) f *is continuous on* $\bar{D}$, ψ *is continuous on* $[0, T] \times \bar{D}$ *with* $f(x) = \psi(T, x)$ *for* $x \in \partial D$,
(c) g *is Hölder continuous on* $[0, T] \times \bar{D}$. *Then* φ *given by*

$$\varphi(t, x) = \mathrm{E}\Big\{f(\xi(T))\chi_{\tau_x = T} + \psi(\tau_x, \xi(\tau_x))\chi_{\tau_x < T} + \int_t^{\tau_x} g(\theta, \xi(\theta))\, \mathrm{d}\theta \mid \xi(t) = x\Big\}, \quad (7.4.54)$$

is the unique solution of the initial boundary-value problem

$$\begin{aligned} \frac{\partial\varphi}{\partial t} + \mathbf{A}(t)\varphi + g = 0 \qquad & \text{for } (t, x) \in [0, T) \times D, \\ \varphi(t, x) = \psi(t, x) \qquad & \text{for } (t, x) \in [0, T] \times \partial D, \\ \varphi(T, x) = f(x) \qquad & \text{for } x \in D. \end{aligned} \quad (7.4.55)$$

7.4.5 *Jump Markov processes*

Systems admitting state transition by jumps are governed by differential equations of the form

$$\mathrm{d}\xi = a(t, \xi)\, \mathrm{d}t + \sigma(t, \xi)\, \mathrm{d}W + \int_{R_0^n = R^n \setminus \{0\}} b(t, \xi, y)\lambda(\mathrm{d}t, \mathrm{d}y), \quad (7.4.56)$$

where λ is the Poisson random measure as introduced in section 7.2.2. For this system we can also prove existence and uniqueness of solutions in the same way as in the continuous case (see Theorem 7.4.2 etc). We define the operator S by

$$(S\xi)(t) = \xi_0 + \int_0^t a(\theta, \xi)\, \mathrm{d}\theta + \int_0^t \sigma(\theta, \xi)\, \mathrm{d}W + \int_0^t \int_{R_0^n} b(\theta, \xi, y)\lambda(\mathrm{d}\theta, \mathrm{d}y), \quad (7.4.57)$$

and introduce the Banach space

$$X_1 \equiv \{\xi \colon t \to \xi(t, \omega) \in D_1(I, R^n) \text{ P-a.s. and } \xi(t) \text{ is } \mathscr{F}_t\text{-adapted}\}$$

with the norm

$$\|\xi\|_1 = \left(\mathrm{E}\left(\sup_{t\in I} |\xi(t)|^2\right)\right)^{1/2}. \tag{7.4.58}$$

Theorem 7.4.21 (Existence of solutions). *Suppose a, b, σ satisfy*

(a) $|a(t, x)|^2 + \|\sigma(t, x\|^2 + \int_{R_0^n} |b(t, x, y)|^2 \Pi(\mathrm{d}y) \leq K^2(1 + |x|^2)$,

(b) $|a(t, x) - a(t, z)|^2 + \|\sigma(t, x) - \sigma(t, z)\|^2$
$$+ \int_{R_0^n} |b(t, x, y) - b(t, z, y)|^2 \Pi(\mathrm{d}y) \leq K^2 |x - z|^2,$$

(c) $\mathrm{E}\, |\xi_0|^2 < \infty$.

Then the operator S has a unique fixed point in X_1 and hence the system (7.4.56) *has a unique solution $\xi \in X_1$ for any given initial state ξ_0 satisfying* (c).

This result is also valid under the local Lipschitz condition.

As in the continuous case we can also develop Kolmogorov's equations for various functionals of the solution process. Here we present only one result concerning the terminal state. Let C_b^2 denote the class of functions on R^n whose partials up to order 2 are uniformly bounded.

Theorem 7.4.22 (Functionals of terminal state). *Suppose a, b, σ, f satisfy the following assumptions*:

(a) *a, b, σ are C^2 in x and their first partials in x are uniformly bounded on $I \times R^n$,*

(b) *for r* = 3, 4 *and* $|\alpha| \leq 2$, $\int_{R_0^n} |D_x^\alpha b_i|^r \Pi(\mathrm{d}y) < \infty$ *on* $I \times R^n$,

(c) $f \in C_b^2$.

Then the function φ, given by

$$\varphi(t, x) \equiv \mathrm{E}\{f(\xi_{t,x}(T))\}, \tag{7.4.59}$$

satisfies the following integro-partial differential equation,

$$\begin{aligned} &\frac{\partial \varphi}{\partial t} + \mathbf{A}_c(t)\varphi + \mathbf{A}_d(t)\varphi = 0, \qquad && (t, x) \in [0, T) \times R^n \\ &\qquad \varphi(T, x) = f(x) && x \in R^n, \end{aligned} \tag{7.4.60}$$

where

$$\begin{aligned} \mathbf{A}_c(t)\varphi &= (a, \varphi_x) + \tfrac{1}{2}\,\mathrm{tr}(\sigma\sigma' \varphi_{xx}), \\ \mathbf{A}_d(t)\varphi &= \int_{R_0^n} \{\varphi(t, x + b(t, x, y)) - \varphi(t, x) - (\varphi_x, b(t, x, y))\}\Pi(\mathrm{d}y). \end{aligned} \tag{7.4.61}$$

If the initial state ξ_0 *has the distribution* μ_0, *then*

$$Ef(\xi(T)) = \int_{R^n} \varphi(0, x)\mu_0(dx) \tag{7.4.62}$$

with φ *being the solution of* (7.4.60).

Remark 7.4.23 Results similar to those of Theorems 7.4.16, 7.4.17, 7.4.20 are also valid in this case.

7.5 Filtering and identification

Over the past decade there has been substantial development of filtering theory for both continuous and discontinuous Markov processes governed by linear and nonlinear systems. We shall present here the classical Kalman–Bucy filter theory [40] and refer the reader to the current literature for general theory [130], [146].

Consider the system

$$\begin{aligned} d\xi &= a(t, \xi)\, dt + \sigma(t, \xi)\, dW, \qquad \xi(0) = \xi_0 \\ dy &= h(t, \xi)\, dt + \sigma_0(t, \xi)\, dW_0, \qquad y(0) = 0, \end{aligned} \tag{7.5.1}$$

where ξ is the state (process) and y the output process. Let $\mathscr{F}_t^y$ denote the σ-algebra generated by the output process y up to time t, $t \in I$. Let f be any Borel-measurable function on R^n taking values in R^k. We wish to estimate $f_\xi(t) \equiv f(\xi(t))$ given the history of y up to time t, that is, $\mathscr{F}_t^y$. Let $L_2(\mathscr{F}_t^y, R^k)$ and $L_2(\mathscr{F}_t, R^k)$ denote the Hilbert spaces of R^k-valued random process which are measurable with respect to the σ-algebras of $\mathscr{F}_t^y$ and $\mathscr{F}_t$ respectively. The filtering problem can be stated as follows: Given $f_\xi(t) \in L_2(\mathscr{F}_t, R^k)$, find $\tilde{f}(t) \in L_2(\mathscr{F}_t^y, R^k)$ such that

$$E\{|f_\xi(t) - \tilde{f}(t)|^2 \,|\, \mathscr{F}_t^y\} \tag{7.5.2}$$

is minimum. Define

$$\hat{f}_\xi(t) \equiv E\{f_\xi(t) \mid \mathscr{F}_t^y\}. \tag{7.5.3}$$

Using the properties of conditional expectations (Proposition 7.1.9(e)) it is easy to verify that

$$E\{|f_\xi(t) - \tilde{f}(t)|^2 \mid \mathscr{F}_t^y\} = E\{|f_\xi(t) - \hat{f}_\xi(t)|^2 \mid \mathscr{F}_t^y\} + |\hat{f}_\xi(t) - \tilde{f}(t)|^2. \tag{7.5.4}$$

Clearly, it follows from this expression that the quantity given by (7.5.2) attains its minimum for

$$\tilde{f}(t) = \hat{f}_\xi(t) = E\{f(\xi(t)) \mid \mathscr{F}_t^y\}. \tag{7.5.5}$$

This shows that the conditional expectation gives the best estimate in the mean square sense. Since

$$\bar{f}(t) = \mathrm{E}\{f(\xi(t)) \mid \mathscr{F}_t^y\} = \int_{R^n} f(x)\mathrm{P}\{\xi(t) \in \mathrm{d}x \mid \mathscr{F}_t^y\} \tag{7.5.6}$$

$$\equiv \int_{R^n} f(x) Q_t^y(\mathrm{d}x), \tag{7.5.7}$$

where Q_t^y is a measure-valued $\mathscr{F}_t^y$-measurable random process. If the measure-valued process Q is determined, theoretically the whole filtering problem is resolved, since all quantities of interest can be computed once Q is known. In fact, general filter theory aims at determining Q. In the following section, we are mainly concerned with the filtering problem for linear systems.

7.5.1 Linear filtering

Consider the system given by:

$$\begin{aligned} \mathrm{d}\xi &= A(t)\xi\,\mathrm{d}t + \sigma(t)\,\mathrm{d}W, \qquad & \xi(0) &= \xi_0, \\ \mathrm{d}y &= H(t)\xi\,\mathrm{d}t + \sigma_0(t)\,\mathrm{d}W_0, \qquad & y(0) &= 0, \end{aligned} \tag{7.5.8}$$

where W and W_0 are independent standard Wiener processes independent of ξ_0. We are interested in the least squares estimate of the state $\xi(t)$ given $\mathscr{F}_t^y$, that is,

$$\hat{\xi}(t) = \mathrm{E}\{\xi(t) \mid \mathscr{F}_t^y\}. \tag{7.5.9}$$

For the estimate $\hat{\xi}$, the Kalman–Bucy theory gives a stochastic differential equation which is driven by the observed process $\{y(\theta),\ \theta \leq t\}$. For practical applications this is what is most desirable.

Theorem 7.5.1 (Kalman–Bucy filter)

(a) *The process* $\hat{\xi}$ *satisfies the stochastic differential equation*

$$\begin{aligned} \mathrm{d}\hat{\xi} &= A(t)\hat{\xi}\,\mathrm{d}t + G(t)[\mathrm{d}y(t) - H(t)\hat{\xi}\,\mathrm{d}t], \qquad t \in [0, T], \\ \hat{\xi}(0) &= E\xi_0, \end{aligned} \tag{7.5.10}$$

where

$$G(t) = P(t)H'(t)\Gamma_0^{-1}(t), \qquad \Gamma_0(t) \equiv (\sigma_0\sigma_0')(t). \tag{7.5.11}$$

(b) *The error covariance*

$$P(t) \equiv \mathrm{E}\{(\xi(t) - \hat{\xi}(t))(\xi(t) - \hat{\xi}(t))'\} \tag{7.5.12}$$

satisfies the matrix Riccati differential equation

$$\frac{dP}{dt} = A(t)P + PA'(t) - PH'(t)\Gamma_0^{-1}(t)H(t)P + \Gamma(t) \tag{7.5.13}$$
$$P(0) = E(\xi(0) - \hat{\xi}(0))(\xi(0) - \hat{\xi}(0))' \equiv P_0, \qquad \Gamma(t) = \sigma(t)\sigma'(t)$$

(c) *The process* $\hat{W}$ *defined by*

$$\hat{W}(t) \equiv \int_0^t \sigma_0^{-1}(\theta)[dy(\theta) - H(\theta)\hat{\xi}(\theta)\, d\theta] \tag{7.5.14}$$

is a standard Brownian motion taking values in the same space as y *and is known as the innovation process.*

Proof Let $\Phi(t, \theta)$ denote the transition operator corresponding to the matrix A; then

$$\xi(t) = \Phi(t, 0)\xi_0 + \int_0^t \Phi(t, \theta)\sigma(\theta)\, dW(\theta). \tag{7.5.15}$$

Assuming ξ_0 to be Gaussian with mean zero, it is clear that ξ is Gaussian and also conditionally Gaussian relative to $\mathscr{F}_t^y$. Hence we can choose $\tilde{f}(t)$ from the class of $\mathscr{F}_t^y$-measurable functions given by the linear transformation

$$\tilde{f}(t) = \int_0^t K(t, \theta)\, dy(\theta), \tag{7.5.16}$$

where K must be chosen so as to minimize the error $E\,|\xi(t) - \tilde{f}(t)|^2$, or equivalently K must be such that

$$E(\xi(t) - \tilde{f}(t), \eta)^2 \tag{7.5.17}$$

is minimum for all $\eta \in R^n$. From (7.5.8), (7.5.15) and (7.5.16), we obtain

$$\xi(t) - \tilde{f}(t) \equiv \xi(t) - \tilde{f}(K, t)$$
$$= R(t, 0)\xi_0 + \int_0^t R(t, \theta)\sigma(\theta)\, dW(\theta) - \int_0^t K(t, \theta)\sigma_0(\theta)\, dW_0, \tag{7.5.18}$$

where

$$R(t, \theta) \equiv \Phi(t, \theta) - \int_\theta^t K(t, \tau)H(\tau)\Phi(\tau, \theta)\, d\tau. \tag{7.5.19}$$

Using (7.5.18) and recalling that ξ_0, W and W_0 are independent, we have

$$\begin{aligned}\mathrm{E}(\xi(t)-\tilde{f}(K,t),\eta)^2 &= (P_0R'(t,0)\eta, R'(t,0)\eta)\\ &\quad+\int_0^t(\sigma(\theta)\sigma'(\theta)R'(t,\theta)\eta, R'(t,\theta)\eta)\,\mathrm{d}\theta\\ &\quad+\int_0^t(\sigma_0(\theta)\sigma_0'(\theta)K'(t,\theta)\eta, K'(t,\theta)\eta)\,\mathrm{d}\theta. \end{aligned}\tag{7.5.20}$$

By direct differentiation, it is easy to verify that R satisfies the differential equation

$$\begin{aligned}&\frac{\partial}{\partial\theta}R(t,\theta) = -R(t,\theta)A(\theta)+K(t,\theta)H(\theta), \qquad 0\leqslant\theta<t,\\ &R(t,t)=I \qquad \text{(identity operator)},\end{aligned}\tag{7.5.21}$$

or equivalently, for any $\eta\in R^n$,

$$\begin{aligned}&\frac{\partial}{\partial\theta}R'(t,\theta)\eta = -A'(\theta)R'(t,\theta)\eta+H'(\theta)K'(t,\theta)\eta, \qquad 0\leqslant\theta<t,\\ &R'(t,t)\eta=\eta.\end{aligned}\tag{7.5.22}$$

Defining

$$\begin{aligned}&x(\tau)=R'(t,t-\tau)\eta, \qquad x(t)=R'(t,0)\eta,\\ &u(\tau)=K'(t,t-\tau)\eta, \qquad 0\leqslant\tau\leqslant t,\end{aligned}\tag{7.5.23}$$

and

$$\Gamma(t)=\sigma(t)\sigma'(t), \qquad \Gamma_0(t)=\sigma_0(t)\sigma_0'(t), \tag{7.5.24}$$

it follows from (7.5.22) and (7.5.20) that

$$\begin{aligned}&\frac{\mathrm{d}x(\tau)}{\mathrm{d}\tau}=A'(t-\tau)x(\tau)-H'(t-\tau)u(\tau), \qquad 0\leqslant\tau\leqslant t,\\ &x(0)=\eta,\end{aligned}\tag{7.5.25}$$

and

$$\begin{aligned}J_\eta(u)&\equiv \mathrm{E}(\xi(t)-\tilde{f}(K,t),\eta)^2\\ &=(P_0x(t),x(t))+\int_0^t(\Gamma(t-\tau)x(\tau),x(\tau))\,\mathrm{d}\tau\\ &\quad+\int_0^t(\Gamma_0(t-\tau)u(\tau),u(\tau))\,\mathrm{d}\tau.\end{aligned}\tag{7.5.26}$$

Thus the problem of finding the optimal K, and hence the optimal filter, reduces to the problem of finding a control u for the problem (7.5.25) that minimizes the quadratic cost functional $J_\eta(u)$. This is precisely the quadratic regulator problem discussed in chapter 6. Hence the optimal feedback control is given by

$$u^*(\tau) = \Gamma_0^{-1}(t-\tau)H(t-\tau)M(\tau)x, \qquad x = x(\tau), \tag{7.5.27}$$

where M satisfies the matrix Riccati differential equation

$$\begin{aligned} -\frac{\mathrm{d}M}{\mathrm{d}\tau} &= A(t-\tau)M(\tau) + M(\tau)A'(t-\tau) \\ &\quad - M(\tau)H'(t-\tau)\Gamma_0^{-1}(t-\tau)H(t-\tau)M(\tau) \\ &\quad + \Gamma(t-\tau) \qquad 0 \leqslant \tau < t, \\ M(t) &= P_0. \end{aligned} \tag{7.5.28}$$

Defining $t - \tau = \theta$, $M(t-\theta) \equiv P(\theta)$, the equation (7.5.28) takes the standard form

$$\begin{aligned} \frac{\mathrm{d}P}{\mathrm{d}\theta} &= A(\theta)P(\theta) + P(\theta)A'(\theta) \\ &\quad - P(\theta)H'(\theta)\Gamma_0^{-1}(\theta)H(\theta)P(\theta) + \Gamma(\theta), \\ P(0) &= P_0, \end{aligned} \tag{7.5.29}$$

and the optimal control is given by

$$\begin{aligned} u^*(\theta) &= \Gamma_0^{-1}(\theta)H(\theta)P(\theta)x(t-\theta) \\ &= \Gamma_0^{-1}(\theta)H(\theta)P(\theta)R'(t,\theta)\eta. \end{aligned} \tag{7.5.30}$$

Equating the expressions for u from (7.5.23) and (7.5.30), we obtain

$$K'(t,\theta)\eta = \Gamma_0^{-1}(\theta)H(\theta)P(\theta)R'(t,\theta)\eta \tag{7.5.31}$$

for all $\eta \in R^n$ and $0 \leqslant \theta \leqslant t \leqslant T$. Hence the optimal K is given by

$$K(t,\theta) = R(t,\theta)P(\theta)H'(\theta)\Gamma_0^{-1}(\theta) \equiv R(t,\theta)G(\theta)$$

where

$$G(\theta) \equiv P(\theta)H'(\theta)\Gamma_0^{-1}. \tag{7.5.32}$$

Therefore, the optimal estimate is given by

$$\hat{\xi}(t) = \int_0^t R(t,\theta)G(\theta)\,\mathrm{d}y(\theta). \tag{7.5.33}$$

Taking the Itô differential of $\hat{\xi}$, we obtain

$$\mathrm{d}\hat{\xi}(t) = \left(\int_0^t \left(\frac{\partial}{\partial t}R(t,\theta)\right)G(\theta)\,\mathrm{d}y(\theta)\right)\mathrm{d}t + G(t)\,\mathrm{d}y(t). \tag{7.5.34}$$

Using (7.5.19) and (7.5.32), it is easy to verify that

$$\frac{\partial}{\partial t}R(t, \theta) = (A(t) - G(t)H(t))R(t, \theta), \qquad 0 \leq \theta \leq t \leq T. \qquad (7.5.35)$$

Substituting (7.5.35) into (7.5.34) we obtain (7.5.10). Thus (a) and (b) have been verified. For (c), note that we can write

$$\mathrm{d}\hat{W} = \sigma_0^{-1}H(\xi - \hat{\xi})\,\mathrm{d}t + \mathrm{d}W_0,$$

and hence one can verify that $\hat{W}$ is an $\mathscr{F}_t^y$-martingale and

$$\mathrm{E}\{\exp \mathrm{i}(\eta, \hat{W}(t) - \hat{W}(s)) \mid \mathscr{F}_s^y\} = \exp(-\tfrac{1}{2}|\eta|^2(t-s)).$$

This proves (c). To complete the proof, we must eliminate the assumption that $\mathrm{E}\xi_0 = 0$. This is done by adding $m(t) \equiv \mathrm{E}\xi(t) = \Phi(t, 0)\hat{\xi}_0$ to the estimate obtained under the assumption of zero mean. ■

Remark 7.5.2 Note that the estimator equation (7.5.10) can also be written as

$$\mathrm{d}\hat{\xi} = A(t)\hat{\xi}\,\mathrm{d}t + G(t)\sigma_0(t)\,\mathrm{d}\hat{W} \qquad (7.5.36)$$

which is again a stochastic differential equation of the original form.

Writing (7.5.10) in the form

$$\mathrm{d}\hat{\xi} = (A(t) - G(t)H(t))\hat{\xi}\,\mathrm{d}t + G(t)\,\mathrm{d}y, \qquad (7.5.37)$$

we observe that $\hat{\xi}(t)$ should be computed in real time as the data y become available. This recursive property makes the Kalman–Bucy filter so attractive in applications.

The true estimate in the case of nonzero mean is given by

$$\hat{\xi}(t) = m(t) + \int_0^t K_0(t, \theta)\,\mathrm{d}y(\theta) \qquad (7.5.38)$$

where $K_0(t, \theta) = \Phi_{A-GH}(t, \theta)G(\theta)$ and Φ_{A-GH} is the transition operator corresponding to $A - GH$. Since G is zero whenever H is zero, it follows from (7.5.38) that the filter rejects (purely) noisy signal whenever the state information is blocked.

For time-invariant systems (A, σ, H, σ_0 constant), it is natural to look for a steady-state filter. Such a filter exists provided the matrix Riccati differential equation (7.5.13) has a steady-state solution.

7.5.2 Identification

Consider the system in R^n:

$$\begin{aligned} \mathrm{d}\xi &= a(t, \xi, \alpha)\,\mathrm{d}t + \sigma(t, \xi)\,\mathrm{d}W, \\ \xi(0) &= \xi_0, \end{aligned} \qquad (7.5.39)$$

where α is an unknown element from a finite-dimensional space, say R^m. The problem is to identify the true parameter α^* from the observation of the process ξ. For simplicity we consider $a(t, \xi, \alpha) = \sum_{i=1}^m \alpha_i a_i(t, \xi)$ where, for each i, a_i is an n-vector valued function. Clearly, we can write $a(t, \xi, \alpha) = A(t, \xi)\alpha$, where A is an $n \times m$-matrix-valued function. We present here a maximum likelihood estimate of the true parameter α^*.

Let μ^1 be the measure induced by the process ξ on $(C, \mathscr{B}(C))$ and μ^0 the measure induced by the process ξ^0 given by

$$\begin{aligned} d\xi^0 &= \sigma(t, \xi^0)\, dW, \qquad t \geq 0, \\ \xi^0(0) &= \xi_0. \end{aligned} \tag{7.5.40}$$

Define, for each $t > 0$, the Radon–Nikodym derivative

$$\rho_0^t(\alpha) = \exp\left\{\int_0^t (\sigma^{-1}(\theta, \xi^0(\theta))A(\theta, \xi^0(\theta))\alpha, \sigma^{-1}(\theta, \xi^0(\theta))\, d\xi^0) - \frac{1}{2}\int_0^t |\sigma^{-1}(\theta, \xi^0(\theta))A(\theta, \xi^0(\theta))\alpha|^2\, d\theta\right\} \tag{7.5.41}$$

The function $\rho_0^T(\alpha)(\cdot)$, considered as a function on $(C, \mathscr{B}(C))$, is the Radon–Nikodym derivative of μ^1 with respect to the measure μ^0, written

$$\left(\frac{d\mu^1}{d\mu^0}\right)(\cdot) = \rho_0^T(\alpha)(\cdot). \tag{7.5.42}$$

Under the assumption that $\|\sigma^{-1}A\alpha\|$ is bounded, we know that $\mu^1(C) = 1$ (see equation (7.4.30)) where, for each $\Gamma \in \mathscr{B}(C)$,

$$\mu^1(\Gamma) = \int_\Gamma \rho_0^T(\alpha)(\xi)\mu^0(d\xi) \equiv E_0\{\rho_0^T(\alpha)(\xi)\chi_\Gamma(\xi)\}.$$

It is clear that, for each $\Gamma \in \mathscr{B}(C)$, $\mu^1(\Gamma)$ is maximized if the function $\alpha \to \rho_0^T(\alpha)(\cdot)$ is maximized by appropriate choice of $\alpha \in R^m$. In other words, the element $\alpha = \alpha_T$ that maximizes $\rho_0^T(\alpha)$ is closest to the true parameter, given $\mathscr{F}_T^\xi$. Maximizing $\rho_0^T(\alpha)$ is equivalent to maximizing $\ln \rho_0^T(\alpha)$. Thus, we find α by setting $d(\ln \rho_0^T)/d\alpha = 0$. For each $t > 0$, it is easy to verify that

$$\ln \rho_0^t(\alpha) = (\beta_t, \alpha) - \tfrac{1}{2}(\Gamma_t\alpha, \alpha), \tag{7.5.43}$$

where

$$\beta_t \equiv \int_0^t A'(\sigma\sigma')^{-1}\, d\xi^0, \qquad \Gamma_t \equiv \int_0^t A'(\sigma\sigma')^{-1}A\, d\theta. \tag{7.5.44}$$

For $t > 0$,

$$(\Gamma_t\eta, \eta) = \int_0^t |\sigma^{-1}A\eta|^2\, d\theta \geq 0 \qquad \text{for all } \eta \in R^m.$$

Hence Γ_t is symmetric and positive semidefinite for all $t \geq 0$. If there exists a finite $T > 0$ for which $\Gamma_T > 0$, then $\Gamma_t > 0$ for all $t \geq T$, and in this case we can write

$$\alpha_t = \Gamma_t^{-1}\beta_t, \qquad t \geq T, \tag{7.5.45}$$

which maximizes $\rho_0^t(\alpha)$ and hence gives the maximum likelihood estimate of the true element α^* given the history $\mathscr{F}_t^\xi$. Note that in this section C represents the space of continuous functions on $[0, \infty)$ with values in R^n furnished with the topology of uniform convergence on compact intervals.

In order for the estimate (7.5.45) to be practically useful it is necessary that $\alpha_t \to \alpha^*$ in some sense as $t \to \infty$. We prove that this is indeed true. We know from (7.4.32) that $\hat{W}$ given by

$$\hat{W}(t) = W(t) - \int_0^t \sigma^{-1} A\alpha^* \, d\theta, \qquad t \geq 0, \tag{7.5.46}$$

is a Brownian motion on the probability space $(C, \mathscr{B}(C), \mu^1)$ and hence

$$d\xi^0 = (A\alpha^*)\, dt + \sigma\, d\hat{W}. \tag{7.5.47}$$

Substituting this in the expression for β_t, we have

$$\beta_t = \Gamma_t \alpha^* + \int_0^t (\sigma^{-1}A)' \, d\hat{W} \tag{7.5.48}$$

which, combined with the equation (7.5.45), gives

$$\alpha_t - \alpha^* = \Gamma_t^{-1}\left(\int_0^t (\sigma^{-1}A)' \, d\hat{W}\right). \tag{7.5.49}$$

The quadratic (matrix) variation process of the martingale $\int_0^t (\sigma^{-1}A)' \, d\hat{W}$ is given by Γ_t. Under certain assumptions it follows from this expression that $\alpha_t \to \alpha^*$ in probability as $t \to \infty$. The situation here is analogous to the scalar case where

$$\lim_{0<t\to\infty} P\left\{\frac{|W(t)|}{t} > \varepsilon\right\} = 0 \qquad \text{for any } \varepsilon > 0.$$

Remark 7.5.3 If $\Gamma_t > 0$ for $t \geq t_0$, then we can define α_t as the solution of the stochastic differential equation

$$d\alpha_t = -\Gamma_t^{-1}(\sigma^{-1}A)'(\sigma^{-1}A)\alpha_t \, dt + \Gamma_t^{-1}(\sigma^{-1}A)' \, dW, \qquad t \geq t_0. \tag{7.5.50}$$

7.5.3 Nonlinear filtering

Since the appearance of the classical Kalman–Bucy (linear) filter theory, substantial progress has been made in the area of nonlinear filtering. It would take us too far even to briefly indicate the many notable contributions of Kushner, Zakai, Kallianpur, Varaya, Bénes, Davis, Boel and many others [130], [146], [48], [103], [116], [97]. Here we briefly mention the Kushner and Zakai equations for nonlinear filters of diffusion processes.

Consider the system

$$\begin{aligned} d\xi &= a(t, \xi)\,dt + \sigma(t, \xi)\,dW \qquad \text{in } R^n \\ dy &= h(t, \xi)\,dt + \sigma_0(t, y)\,dW_0 \qquad \text{in } R^k, \end{aligned} \tag{7.5.51}$$

with ξ being the state and y the output process, and assume that $\Gamma_0^{-1} \equiv (\sigma_0\sigma_0')^{-1}$ is uniformly bounded on $\{t \geq 0\} \times R^k$. Let $Q_t^y(dx)$ denote the conditional probability measure $P\{\xi(t) \in dx \mid \mathscr{F}_t^y\}$.

Kushner's equation is a nonlinear stochastic partial differential equation for the density q_t^y corresponding to the measure Q_t^y and is given by

$$\begin{aligned} dq_t^y &= \mathbf{A}^* q_t^y\,dt + (h - \hat{h}, \Gamma_0^{-1}(dy - \hat{h}\,dt))q_t^y, \qquad t \geq 0, \\ q_0^y &= q_0, \end{aligned} \tag{7.5.52}$$

where q_0 is the (initial) density of ξ_0, and

$$\hat{h} \equiv \hat{h}(t) = \int_{R^n} h(t, x) q_t^y(x)\,dx = \int_{R^n} h(t, x) Q_t^y(dx) \equiv Q_t^y(h),$$

and $\mathbf{A}^*$ is the adjoint of the operator $\mathbf{A}$, given by $\mathbf{A}\varphi = (a, \varphi_x) + \frac{1}{2}\operatorname{tr}(\sigma\sigma'\varphi_{xx})$. The equation (7.5.52) is better understood in the weak sense. This is given by

$$\begin{aligned} dQ_t^y(f) &= Q_t^y(\mathbf{A}f)\,dt + (Q_t^y(f \cdot (h - \hat{h})), \Gamma_0^{-1}(dy - \hat{h}dt)), \\ Q_0^y(f) &= Q_0(f) = \int_{R^n} f(x) Q_0(dx), \end{aligned} \tag{7.5.53}$$

which holds for all C^2 functions f with compact support.

If this equation is satisfied for all C^2 functions, then by choosing f appropriately one gets the moment equations. However, the first moment depends on the second, and the second moment depends on the third, and so on, resulting in an infinite system of coupled moment equations. Since, in the linear case with Gaussian initial distribution, the conditional distribution is also Gaussian, the first- and second-order moments contain all the information about the process ξ. Using this fact one can easily derive the filter equations (7.5.10) and (7.5.13) from the Kushner equation (7.5.53) [47].

Another equation which is theoretically more convenient was obtained by Zakai [146], which is a linear stochastic partial differential equation describing the flow of unnormalized conditional distribution v_t^y of $\xi(t)$

given $\mathscr{F}_t^y$. This is written in the weak form as

$$\begin{aligned} \mathrm{d}\nu_t^y(f) &= \nu_t^y(\mathbf{A}f)\,\mathrm{d}t + (\nu_t^y(hf), \Gamma_0^{-1}\,\mathrm{d}y) \\ \nu_0(f) &= Q_0(f), \qquad f \in C_0^2, \end{aligned} \tag{7.5.54}$$

from which the conditional distribution is determined by the relation $Q_t^y(f) = \nu_t^y(f)/\nu_t^y(1)$.

Till now application of these equations have been rather limited because of the difficulties encountered in the real-time solution of stochastic partial differential equations in multidimensional spaces. It is expected that this problem will be resolved with the development of supercomputers.

In recent years filter theory for jump processes has been extensively studied by Varaiya, Wong, Davis, Boel, Vaca, Snyder, Ahmed, Dabbous and many others [103], [115], [116], [143], [94], [85]. This subject is beyond the scope of this book.

Remark 7.5.4 By use of Itô's formula one can easily verify that the Kushner and Zakai equations are equivalent.

We can apply Kushner's equation to the following problem. Consider the system

$$\mathrm{d}y = h(t, y, \xi(t))\,\mathrm{d}t + \sigma_0(t, y)\,\mathrm{d}W_0, \tag{7.5.55}$$

where ξ is a finite-state temporally homogeneous Markov chain with state space $\Sigma \equiv \{e_1, e_2, \ldots, e_N\}$. Let $\Lambda \equiv \{\lambda_{ij}, i, j = 1, 2, \ldots, N\}$ denote the infinitesimal generator (matrix) of the process ξ. That is, for $\Delta t > 0$,

$$p_{ij}(t, t + \Delta t) = \delta_{ij} + \lambda_{ij}\Delta t + o(\Delta t), \tag{7.5.56}$$

where δ_{ij} is the Kronecker delta, $o(\Delta t)$ means of order smaller than (Δt), and

$$p_{ij}(t, t + \Delta t) = \mathrm{P}\{\xi(t + \Delta t) = e_j \mid \xi(t) = e_i\}. \tag{7.5.57}$$

Define

$$p_i(t) \equiv \mathrm{P}\{\xi(t) = e_i \mid \mathscr{F}_t^y\}, \qquad i = 1, 2, \ldots, N, \tag{7.5.58}$$

and p = column vector $(p_i, i = 1, 2, \ldots, N)$. Then the best mean square estimate of $\xi(t)$ given the process $\{y(\theta), \theta \leqslant t\}$ is given by

$$\hat{\xi}(t) = \mathrm{E}\{\xi(t) \mid \mathscr{F}_t^y\} = \sum_{i=1}^{N} e_i p_i(t). \tag{7.5.59}$$

Using Kushner's equation (7.5.52) one can easily verify that p_i satisfies the stochastic differential equation

$$dp_i(t) = (\Lambda^* p)_i \, dt + p_i(t) \cdot (\sigma_0^{-1}(a(t, y, e_i) - \hat{a}(t, y)), d\hat{W}),$$
$$i = 1, 2, \ldots, N, \quad (7.5.60)$$

where

$$d\hat{W} \equiv \sigma_0^{-1}(dy - \hat{a} \, dt)$$

$$\hat{a}(t, y) \equiv \sum_{i=1}^{N} a(t, y, e_i) p_i(t).$$

This is a system of stochastic ordinary differential equations, whereas (7.5.52) is a stochastic partial differential equation. Note that this equation is also nonlinear.

Remark 7.5.5 Equation (7.5.60) has interesting application in N-ary feedback communication (see Problem 7.5.P18).

Linear filter revisited

The Kalman–Bucy filter theory can be derived in a very elementary but instructive way by looking at the filtering problem from a different point of view. One demands that the estimator be linear and be driven by the observed process y (equation (7.5.8)) in the form

$$\begin{aligned} d\eta &= \Gamma(t)\eta \, dt + B(t) \, dy \\ &= \Gamma(t)\eta \, dt + B(t)H\xi \, dt + B\sigma_0 \, dW_0 \end{aligned}$$

where Γ and B are to be chosen so as to obtain an unbiased ($E\eta = E\xi$) and minimum variance filter. Then comparing the above equation with the state equation $d\xi = A(t)\xi \, dt + \sigma(t) \, dW$ one has

$$\begin{aligned} d(\xi - \eta) = {} & (A - BH)(\xi - \eta) \, dt \\ & + (A - BH - \Gamma)\eta \, dt + \sigma \, dW - B\sigma_0 \, dW_0. \end{aligned}$$

Hence for unbiased estimate Γ must equal $A - BH$ leading to the estimator equation:

$$\begin{aligned} d\eta &= (A - BH)\eta \, dt + B \, dy = A\eta \, dt + B(dy - H\eta \, dt) \\ &= (A - BH)\eta \, dt + BH\xi \, dt + B\sigma_0 \, dW_0. \end{aligned}$$

Subtracting this from the state equation one obtains

$$d(\xi - \eta) = (A - BH)(\xi - \eta) \, dt - B\sigma_0 \, dW_0 + \sigma \, dW,$$

and the problem reduces to finding B that minimizes $E(\xi(t) - \eta(t), z)^2$

for every $z \in R^n$. This is equivalent to the control problem on the space of matrix-valued functions $\{B\}$,

$$\frac{\mathrm{d}}{\mathrm{d}t}P = (A - BH)P + P(A - BH)' + (B\sigma_0\sigma_0'B') + \sigma\sigma',$$

$$P(0) = P_0,$$

$$J(B) \equiv \int_0^T \mathrm{tr}(P\Omega)\,\mathrm{d}t = \min.$$

for any positive definite symmetric matrix Ω. Using the calculus of variations one arrives at the optimal $B_0 = P_0H'\ (\sigma_0\sigma_0')^{-1}$, which when substituted in the P-equation gives the matrix Riccati differential equation (7.5.13) and the estimator equation (7.5.10) with $B_0 = G$.

The same approach applies to jump processes.

7.6 Stability and optimal control

There is an extensive literature on the subject of stability and optimal control of stochastic systems. There are also several books written on the subject. Here we can only briefly touch upon some of the well-known results that have proved to be useful in applications. For stability the reader may consult the books of Kushner [50] and Has'minskii [45], and for optimal control the books of Krylov [49], Fleming and Rishel [42], Gihman and Skorokhod [44] and the review of the author [37].

7.6.1 Stability

Consider the system

$$\mathrm{d}\xi = a(t, \xi)\,\mathrm{d}t + \sigma(t, \xi)\,\mathrm{d}W, \qquad t \geq 0 \tag{7.6.1}$$

and suppose we wish to study the stability of the zero state $\xi \equiv 0$. Without loss of generality, we can assume that $a(t, 0) = 0$, and $\sigma(t, 0) = 0$.

Definition 7.6.1 (Stability in probability). The system (7.6.1) is said to be stable in probability with respect to the zero state for $t \geq 0$, if for any $s \geq 0$ and $\varepsilon > 0$

$$\lim_{x \to 0} \mathrm{P}\left\{\sup_{t > s} |\xi_{s,x}(t)| \geq \varepsilon\right\} = 0. \tag{7.6.2}$$

This says that the sample paths of the process $\xi_{s,x}$ starting from a point x at time s will always remain within the ε-neighbourhood of the origin with probability tending to 1 as $|x| \to 0$.

Recall that a function $V(t, x)$ is said to be positive definite in the sense of Lyapunov if $V(t, 0)=0$ and $V(t, x)>W(x)$ in the neighbourhood of the origin with $W(x)>0$ for $x \neq 0$.

Theorem 7.6.2 (Stability in probability). *Let G be an open subset of R^n with $0 \in \operatorname{Int} G$ and $D \equiv \{(t, x)\colon t>0,\ x \in G\}$. Suppose there exists a $V \in C_0^{1,2}(D)$ ($\equiv$ the space of all real-valued functions which are C^1 in t and C^2 in x on D and vanish at $x=0$) which is positive definite in the sense of Lyapunov and satisfies*

$$LV=\frac{\partial}{\partial t} V+(a, V_x)+\tfrac{1}{2} \operatorname{tr}(\sigma\sigma' \cdot V_{xx}) \leqslant 0 \tag{7.6.3}$$

for $x \in G$. Then the trivial solution $\xi(t) \equiv 0$ is stable in probability.

Proof For $r>0$, define $G(r)=\{x \in R^n\colon |x|<r\}$ and set $\tau_r=\inf\{t \geqslant s\colon \xi_{s,x}(t) \in \partial G(r)\}$ for $x \in G(r)$. Take r sufficiently small so that $\bar{G}(r)=G(r) \cup \partial G(r) \subset G$. Define

$$V_r=\inf\{V(t, x),\ x \in G \backslash G(r),\ t \geqslant s\}.$$

Since V is positive definite, $V_r>0$. Define $\tau_r(t)=\tau_r \wedge t$, then by Itô's formula, one can easily verify that

$$\mathrm{E} V(\tau_r(t), \xi_{s,x}(\tau_r(t))) \leqslant V(s, x) \qquad \text{for all } t \geqslant s.$$

On the other hand, we can write

$$\mathrm{E} V(\tau_r(t), \xi_{s,x}(\tau_r(t)))=\int_{\{\tau_r(t)<t\}} V \mathrm{P}(d\omega)+\int_{\{\tau_r(t) \geqslant t\}} V \mathrm{P}(d\omega) \tag{7.6.4}$$

and hence

$$\begin{aligned} \mathrm{P}\left\{\sup_{s<\theta \leqslant t} |\xi_{s,x}(\theta)| \geqslant r\right\} &\leqslant \frac{\mathrm{E} V(\tau_r(t), \xi_{s,x}(\tau_r(t)))}{V_r} \\ &\leqslant \frac{V(s, x)}{V_r} \qquad \text{for all } t>s. \end{aligned}$$

Hence letting $t \rightarrow \infty$, we obtain

$$\mathrm{P}\left\{\sup_{\theta>s} |\xi_{s,x}(\theta)| \geqslant r\right\} \leqslant \frac{V(s, x)}{V_r}.$$

Since $V(s, 0)=0$ and $x \rightarrow V(s, x)$ is continuous, it follows from the preceding inequality that

$$\lim_{x \rightarrow 0} \mathrm{P}\left\{\sup_{\theta>s} |\xi_{s,x}(\theta)| \geqslant r\right\}=0.$$

Hence the system is stable in probability with respect to the zero state. This completes the proof. ■

Using the above result one can verify that the scalar equation (one-dimensional)

$$d\eta = a\eta \, dt + \sigma\eta \, dW, \qquad \eta(0) = x,$$

is stable in probability with respect to the origin if merely $a < \sigma^2/2$. In fact this problem has a closed-form solution

$$\eta_x(t) = x \exp(a - \sigma^2/2)t \cdot \exp(\sigma W(t)), \quad t \geqslant 0,$$

obtained from Itô's formula applied to $\ln \eta$. Hence one can also directly verify that

$$\lim_{x \to 0} \mathrm{P}\left\{ \sup_{t>0} |\eta_x(t)| > \varepsilon \right\} = 0 \qquad \text{for any } \varepsilon > 0.$$

Definition 7.6.3 (Stability in the mean p). The solution $\xi(t) \equiv 0$ of the system (7.6.1) is said to be stable in the mean of order $p > 0$ (or p-stable) if, for $t \geqslant 0$, and $\delta > 0$,

$$\sup\{\mathrm{E}\, |\xi_{s,x}(t)|^p, \, |x| \leqslant \delta, \, t \geqslant s\} \to 0 \qquad \text{as } \delta \to 0, \tag{7.6.5}$$

and it is said to be asymptotically p-stable if

(a) it is p-stable,

and (7.6.6)

(b) $\lim_{t \to \infty} \mathrm{E}\, |\xi_{s,x}(t)|^p = 0$

The trivial solution is said to be exponentially p-stable if there exist constants c_1 and c_2 such that

$$\mathrm{E}\, |\xi_{s,x}(t)|^p \leqslant c_1 \, |x|^p \exp(-c_2(t-s)) \qquad t \geqslant s. \tag{7.6.7}$$

Theorem 7.6.4 (Exponential p-stability). *The trivial solution is exponentially p-stable for $t \geqslant 0$ if there exist a function $V \in C_0^{1,2}$ and positive constants c_1, c_2, c_3 such that*

(a) $c_1 \, |x|^p \leqslant V(t, x) \leqslant c_2 \, |x|^p$,

(b) $(LV)(t, x) \leqslant -c_3 \, |x|^p$. (7.6.8)

Proof The proof is straightforward and follows from Itô's formula. ■

Remark 7.6.5 In fact under the given assumptions the system is actually almost surely exponentially p-stable.

As an application of the above result, consider the stochastic differential equation

$$d\begin{bmatrix} y \\ z \end{bmatrix} = \begin{bmatrix} 0 & 1 \\ -\frac{k}{m} & -\frac{c}{m} \end{bmatrix}\begin{bmatrix} y \\ z \end{bmatrix} dt + \begin{bmatrix} 0 \\ -\frac{\sigma}{m} z \end{bmatrix} dW \tag{7.6.9}$$

which describes the dynamics of a galloping transmission line. Here W is a one-dimensional standard Brownian motion, m, k, c, σ are line parameters with c being the damping factor. For a Lyapunov function, one can choose

$$V(t, x) \equiv V(t, y, z) = \frac{c^2}{2m} y^2 + ky^2 + mz^2 + cyz. \tag{7.6.10}$$

Noting that $k, m > 0$, we define $|x| = \sqrt{(ky^2 + mz^2)}$, and take $p = 2$. Then one can verify that, for

$$c_1 \equiv 1/[1 + (1 + 4km/c^2)^{-1/2}],$$
$$c_2 \equiv 1/[1 - (1 + 4km/c^2)^{-1/2}],$$

we have

$$c_1 |x|^2 \leqslant V(t, x) \leqslant c_2 |x|^2, \tag{7.6.11}$$

and

$$(LV)(t, x) = -\left\{\left(\frac{ck}{m}\right)y^2 + \left(\frac{c}{m} - \left(\frac{\sigma}{m}\right)^2\right)mz^2\right\}. \tag{7.6.12}$$

If the damping factor c is chosen sufficiently large such that $\left(\left(\frac{c}{m}\right) - \left(\frac{\sigma}{m}\right)^2\right) > 0$, then for $c_3 = \left(\left(\frac{c}{m}\right) - \left(\frac{\sigma}{m}\right)^2\right)$, we have

$$(LV)(t, x) \leqslant -c_3 |x|^2. \tag{7.6.13}$$

Hence the system is exponentially mean square stable ($p = 2$). For details see [136].

Feedback stabilization

Consider the controlled system

$$d\xi = a(t, \xi, u(t, \xi))\, dt + \sigma(t, \xi)\, dW \equiv a^u\, dt + \sigma\, dW, \tag{7.6.14}$$

and suppose $a(t, 0, 0) = 0$, $\sigma(t, 0) = 0$, and the system is generally unstable without control (i.e. $u \equiv 0$). The problem is to find a feedback control law, $u = u(t, \xi)$, that makes the system stable. Let $\mathcal{U}$ denote the class of all Borel-measurable control laws $\{u\}$ from $[0, \infty) \times R^n$ to R^m. Suppose there exist a $u_0 \in \mathcal{U}$ and a constant $\gamma \leqslant 0$ such that

$$L^{u_0} V \leqslant \gamma V \tag{7.6.15}$$

has at least one solution $V \in C_0^{1,2} \equiv C_0^{1,2}(R^n)$ which is positive definite in the Lyapunov sense. Then by Itô's formula, we have

$$\mathrm{E}V(t, \xi^0_{s,x}(t)) \leqslant V(s, x) \exp \gamma(t - s) \qquad \text{for all } t \geqslant s, \tag{7.6.16}$$

where $\xi^0_{s,x}(t)$, $t \geqslant s$ is a solution of the problem (7.6.14) corresponding to the control u_0 initiating at (s, x). Hence the system is stable in probability. Thus the problem of stabilizability is equivalent to the question of existence of a positive definite solution $V \in C_0^{1,2}$ of the differential inequality (7.6.15) for some (admissible) control law $u \in \mathcal{U}$ and any constant $\gamma \leqslant 0$.

We shall present here a result due to Gao and the author. For details and other results in this direction see [118, 119].

Consider the system

$$\mathrm{d}\xi = (F(t)\xi + B(t)u(t, \xi))\,\mathrm{d}t + \sum_{i=1}^{\ell} \sigma_i(t)\xi(t)\,\mathrm{d}W_i, \tag{7.6.17}$$

and suppose the system is generally unstable in the absence of controls ($u \equiv 0$).

Theorem 7.6.6 (Stabilizability). *Suppose F, B, σ_i are continuous and there exist positive numbers $\nu < \lambda$ such that*

$$\nu M \leqslant \sum_{i=1}^{\ell} \sigma_i' M \sigma_i \leqslant \lambda M$$

for all symmetric positive matrices M. Further suppose the matrix Riccati differential equation

$$\frac{\mathrm{d}P}{\mathrm{d}t} + P\left(F + \frac{\lambda}{2} I\right) + \left(F + \frac{\lambda}{2}\right)' P - PBB'P + Q = 0, \qquad t \geqslant 0, \tag{7.6.18}$$

has at least one symmetric positive solution P for some $Q(t) > 0$. Then for the control law

$$u_0(t, x) = -\tfrac{1}{2} B'(t) P(t) x, \tag{7.6.19}$$

the system is asymptotically stable in probability.

Proof Define $V(t, x) = (P(t)x, x)$ and $L^{u_0} = L^0$. Then

$$(L^0V)(t, x) = \left(\left(\dot{P} + PF + F'P - PBB'P + \sum_{i=1}^{\ell} \sigma_i' P \sigma_i\right)x, x\right)$$

$$\leqslant \left(\left(\dot{P} + P\left(F + \frac{\lambda}{2}I\right) + \left(F + \frac{\lambda}{2}I\right)' P - PBB'P\right)x, x\right)$$

$$= -(Qx, x). \tag{7.6.20}$$

Let $\xi^0_{s,x}(t)$ denote the solution of (7.6.17) corresponding to the control law u_0 given by (7.6.19). Then it follows from Itô's formula that

$$\mathrm{E}V(t, \xi^0_{s,x}(t)) = V(s, x) + \mathrm{E}\int_s^t (L^0V)(\theta, \xi^0_{s,x}(\theta))\, \mathrm{d}\theta. \tag{7.6.21}$$

Hence it follows from (7.6.20) and (7.6.21) that

$$\mathrm{E}V(t, \xi^0_{s,x}(t)) \leqslant V(s, x) - \mathrm{E}\int_s^t (Q(\theta)\xi^0_{s,x}(\theta), \xi^0_{s,x}(\theta))\, \mathrm{d}\theta$$

for all $t \geqslant s$, and $x \in R^n$ except for $x = 0$. Hence V is a Lyapunov function, and it follows from the above inequality that the system is asymptotically stable in probability. ■

Remark 7.6.7 In fact for the given control law u_0, $V(t, \xi^0_{s,x}(t))$ is a positive supermartingale and one can show that it converges to zero as $t \to \infty$ with probability 1. Hence the system (7.6.17) with control u_0 is also almost surely asymptotically stable.

7.6.2 Optimal control

As in the deterministic case, in stochastic control also one is concerned with the questions of existence of optimal controls and necessary conditions of optimality.

Consider the system

$$\begin{aligned} \mathrm{d}\xi &= a^u(t, \xi)\, \mathrm{d}t + \sigma(t, \xi)\, \mathrm{d}W, \qquad t \in I = [0, T], \\ \xi(0) &= \xi_0, \qquad a^u(t, x) \equiv a(t, x, u(t, x)) \end{aligned} \tag{7.6.22}$$

and let $\mathscr{U}$ denote the class of Borel-measurable functions on $I \times R^n$ to U, where U is a compact subset of R^m. Suppose it is required to find a control $u_0 \in \mathscr{U}$ such that

$$J(u_0) = \mathrm{E}\{\ell(\xi^{u_0}, u_0)\} \leqslant \mathrm{E}\{\ell(\xi^u, u)\} = J(u) \tag{7.6.23}$$

for all $u \in \mathcal{U}$, where l may be given by

$$\ell(\xi^u, u) \equiv \int_0^T g(t, \xi^u(t), u(t, \xi^u(t)))\, dt + f(\xi^u(T)). \tag{7.6.24}$$

By use of the Girsanov transformation, one can write

$$J(u) = \int_C \ell(\xi, u)\rho_0^T(u)(\xi)\, d\mu_0(\xi) \tag{7.6.23$'$}$$

where $\rho_0^T(u)$ is given by the expression (7.4.29) with a replaced by a^u and μ_0 is the measure induced by the process η given by $d\eta = \sigma(t, \eta)\, dW$, $\eta(0) = \xi_0$. Recall that $d\mu^u = \rho_0^T(u)\, d\mu_0$, and $(C, \mathcal{B}(C), \mu_0)$ and $(C, \mathcal{B}(C), \mu^u)$ are probability spaces.

In fact in the above formulation, one can take for the admissible controls $\mathcal{U}$, the class of all $\mathcal{B}_t(C)$ measurable functions $\{u(t, x(\cdot))\}$ assuming values from U. But we do not intend this generality.

Under fairly general assumptions, such as

(a) $\mathrm{E}\,|\xi_0|^{2\beta} < \infty$ for some $\beta(\beta > 1)$ sufficiently large,

(b) there exists a positive $K < \infty$ such that, for $x \in R^n$ and $u \in \mathcal{U}$.

$$|a^u(t, x)|^2 \leqslant K(1 + |x|^2),$$
$$\|\sigma(t, x)\|^2 \leqslant K(1 + |x|^2),$$
$$|\sigma^{-1}a^u|^2 \leqslant K(1 + |x|^2),$$

(c) there exist positive numbers $K < \infty$ and $0 < \alpha < 2\beta$ such that

$$|\ell(\xi, u)| \leqslant K(1 + \|\xi\|^\alpha) \qquad \text{for } \xi \in C,$$

and

(d) $a(t, x, U) \equiv \{y \in R^n\colon\ y = a(t, x, z) \text{ for some } z \in U\}$ is a convex set-valued function on $I \times R^n$.

one can prove the existence of a feedback control law $u_0 \in \mathcal{U}$ such that $J(u_0) \leqslant J(u)$ for all $u \in \mathcal{U}$. If the process starts from a fixed point $\xi_0 = x$, then ξ has moments of all finite order (see Corollary 7.4.5), and (a) is satisfied for any $\beta \in [0, \infty)$ and hence for any given α one can freely choose a β satisfying (c). This is a probabilistic approach and is based on arguments involving selection theorems and conditions leading to weak compactness of the set

$$Y \equiv \{\zeta \in L_1(C, \mathcal{B}(C), \mu_0)\colon\ \zeta = \ell(\cdot, u)\rho_0^T(u)(\cdot),\ u \in \mathcal{U}\}.$$

Further discussion of this topic is beyond the scope of this elementary book.

The same problem can be attacked by use of PDE. In this approach the optimal control problem of a stochastic system reduces to an equivalent problem of a deterministic distributed-parameter system.

Define

$$\varphi^u(t,x) = \mathrm{E}\left\{\int_t^T g(\theta, \xi_{t,x}^u(\theta), u(\theta, \xi_{t,x}^u(\theta)))\,\mathrm{d}\theta + f(\xi_{t,x}(T))\right\}. \quad (7.6.25)$$

Then by Theorem 7.4.16, φ^u satisfies the Kolmogorov equation

$$\begin{aligned} &\frac{\partial \varphi^u}{\partial t} + \mathbf{A}^u \varphi^u + g^u = 0, \qquad (t,x) \in [0,T) \times R^n, \\ &\varphi^u(T,x) = f(x), \qquad x \in R^n, \end{aligned} \quad (7.6.26)$$

where

$$\begin{aligned} \mathbf{A}^u \varphi &\equiv (a^u, \varphi_x) + \tfrac{1}{2}\operatorname{tr}(\sigma\sigma'\varphi_{xx}), \\ a^u(t,x) &\equiv a(t,x,u(t,x)), \\ g^u(t,x) &\equiv g(t,x,u(t,x)). \end{aligned}$$

In this formulation the cost functional is given by

$$J(u) = \int_{R^n} \varphi^u(0,x) Q_0(\mathrm{d}x), \quad (7.6.27)$$

where Q_0 is the distribution of the initial state ξ_0. Under similar assumptions one can prove the existence of optimal controls. The stochastic problem (7.6.22) and (7.6.23) has been transformed into an equivalent deterministic problem (7.6.26) and (7.6.27). For a discussion of the two approaches the reader may refer to the original papers of Varaiya, Davis, Benes, Fleming, Rishel, Ahmed, Teo as listed in Ahmed [37].

Here we are mainly interested in the necessary conditions of optimality which are used to determine optimal (more precisely extremal) policies. Throughout the rest of this section, we shall assume that an optimal control $u^* \in \mathcal{U}$ exists.

Define the Hamiltonian

$$H(t,x,v,p) = (a(t,x,v), p) + g(t,x,v) \quad (7.6.28)$$

for $(t,x,v,p) \in I \times R^n \times U \times R^n$. We assume throughout the rest of this section that $((\sigma\sigma')(t,x)\lambda, \lambda) \geqslant \gamma\,|\lambda|^2$ for some $\gamma > 0$ and all $\lambda \in R^n$, and $(t,x) \in G = I \times R^n$.

Theorem 7.6.8 *Consider the system* (7.6.22) *with the cost functional* (7.6.23)–(7.6.24) *and suppose there exists an optimal control* $u^* \in \mathcal{U}$ *with* φ^* *denoting the solution of* (7.6.26) *corresponding to the control* u^*.

Suppose the distribution Q_0 has a density q_0. Then the following minimum principle holds:

$$H(t, x, u^*(t, x), \varphi_x^*(t, x)) \leqslant H(t, x, v, \varphi_x^*(t, x)) \quad a.e. \text{ on } G \qquad (7.6.29)$$

for all $v \in U$. Further, for any bounded measurable subset $G_0 \subset G$, the pointwise minimum principle is equivalent to the minimum principle in its integral form

$$\int_{G_0} H(t, x, u^*(t, x), \varphi_x^*(t, x))\, dx\, dt \leqslant \int_{G_0} H(t, x, u(t, x), \varphi_x^*(t, x))\, dx\, dt. \qquad (7.6.30)$$

Proof Consider the system

$$\begin{aligned} \frac{\partial q}{\partial t} - \Lambda^*(q) &= 0 \qquad \text{on } G, \\ q(0, x) &= q_0(x) \qquad \text{on } R^n, \end{aligned} \qquad (7.6.31)$$

where Λ^* is the formal adjoint of the operator Λ given by $\Lambda\varphi \equiv \frac{1}{2}\operatorname{tr}((\sigma\sigma')\varphi_{xx})$. Under the given assumptions this problem has a positive solution $q(t, x)$.

Let φ^u be any solution of (7.6.26) corresponding to the control $u \in \mathcal{U}$. Then multiplying (7.6.31) by φ^u and denoting the scalar product $\int_{R^n} f_1(x) f_2(x)\, dx$ by $\langle f_1, f_2 \rangle$, we have

$$0 = \int_0^T \left\langle \frac{\partial q}{\partial t} - \Lambda^*(q), \varphi^u \right\rangle dt$$

$$- \langle q, \varphi^u \rangle \big|_0^T - \int_0^T \left\langle q, \frac{\partial \varphi^u}{\partial t} + \Lambda(\varphi^u) \right\rangle dt.$$

Hence

$$\begin{aligned} \langle q_0, \varphi^u(0, \cdot) \rangle &= \langle q_T, f(\cdot) \rangle - \int_0^T \left\langle q, \frac{\partial \varphi^u}{\partial t} + \Lambda(\varphi^u) \right\rangle dt \\ &= \langle q_T, f \rangle + \int_0^T \langle q, B^u \varphi^u + g^u \rangle\, dt \end{aligned}$$

where $B^u \varphi = (a^u, \varphi_x)$. Hence

$$J(u) = \langle q_T, f \rangle + \int_0^T \langle q, B^u \varphi^u + g^u \rangle\, dt. \qquad (7.6.32)$$

Since f is independent of u, we can disregard the first term and find a control that minimizes the functional

$$\tilde{J}(u)=\int_0^T \langle q, B^u\varphi^u+g^u\rangle\, dt=\int_0^T \langle q, H(t, \cdot, u, \varphi_x^u(t, \cdot))\rangle\, dt. \tag{7.6.33}$$

Clearly, if u^* is the optimal control then $\tilde{J}(u^*)\leq \tilde{J}(u)$.

Let $(t_0, x_0)\in G$ be a Lebesgue density point and $\{E_i\}$ a sequence of Lebesgue-measurable subsets of the set G, each containing (t_0, x_0) in such a way that $\lim_{i\to\infty} E_i=\{(t_0, x_0)\}$ (one point set).

Define

$$u_i(t, x)=\begin{cases} u^* & \text{if } (t, x)\in G\backslash E_i,\\ v & \text{if } (t, x)\in E_i,\end{cases}$$

where v is an arbitrary element of U.

Using u_i for u, we have

$$\tilde{J}(u^*)\leq \tilde{J}(u_i) \qquad \text{for all } i. \tag{7.6.34}$$

Hence

$$\begin{aligned}\int_{G\backslash E_i} & q(t, x)[H(t, x, u^*(t, x), \varphi_x^*(t, x))\\ &\quad - H(t, x, u^*(t, x), \varphi_x^i(t, x))]\, dx\, dt\\ &\quad + \int_{E_i} qH(t, x, u^*(t, x), \varphi_x^*(t, x))\, dx\, dt\\ &\leq \int_{E_i} qH(t, x, v, \varphi_x^i(t, x))\, dx\, dt.\end{aligned} \tag{7.6.35}$$

Under the given assumptions, it can be shown (Ahmed and Teo [53], Theorem 3.1.4), that

$$\left.\begin{aligned}\varphi^i&\to\varphi^*\\ \varphi_x^i&\to\varphi_x^*\end{aligned}\right\} \quad \text{uniformly on } G, \tag{7.6.36}$$

and that

$$\lim_{i\to\infty}\frac{1}{\ell(E_i)}\int_{G\backslash E_i} q(t, x)[H(t, x, u^*, \varphi_x^*) - H(t, x, u^*, \varphi_x^i)]\, dx\, dt=0 \tag{7.6.37}$$

where $\ell(E_i)$ denotes the Lebesgue measure of E_i. These facts are related to continuous dependence of solutions on parameters. Dividing the expression (7.6.35) by $\ell(E_i)$ and letting $i\to\infty$ it follows from (7.6.36) and

(7.6.37) that

$$q(t_0, x_0)H(t_0, x_0, u^*(t_0, x_0), \varphi_x^*(t_0, x_0)) \leqslant q(t_0, x_0)H(t_0, x_0, v, \varphi_x^*(t_0, x_0)) \quad (7.6.38)$$

for all $v \in U$. Since all the functions in the product qH are Lebesgue measurable and almost all $(t, x) \in G$ are Lebesgue density points, and since (t_0, x_0) is arbitrary and $q \geqslant 0$, the inequality (7.6.29) follows from (7.6.38). The proof of equivalence of (7.6.29) and (7.6.30) is obvious. This completes the proof. ■

Remark 7.6.9 If the measure Q_0 does not have a density, then the necessary condition (7.6.29) holds ν-a.e., where ν is the measure on G determined by the relation

$$\nu(E) = \int_E \mathrm{d}t\, Q(t, \mathrm{d}x), \qquad E \in \mathscr{B}(G). \quad (7.6.39)$$

Remark 7.6.10 Use of the measure $Q(t, \mathrm{d}x)$ or its density $q(t, x)$ corresponding to the system $\mathrm{d}\xi = \sigma(t, \xi)\,\mathrm{d}W$, $\xi(0) = \xi_0$, instead of $Q^u(t, \mathrm{d}x)$ or its density $q^u(t, x)$ corresponding to the system (7.6.22) does not affect the conclusion of the theorem. This follows from the fact that μ^u is absolutely continuous with respect to the measure μ^0 and hence the corresponding marginal distribution $Q^u(t, \cdot)$ is absolutely continuous with respect to $Q(t, \cdot)$ on $\mathscr{B}(R^n)$. Hence it is irrelevant which of the equations (7.6.31) or $-\frac{\partial}{\partial t} Q^u + (\mathbf{A}^u)^* Q^u = 0$ is used in the derivation.

According to the above theorem, one can compute the optimal control following the steps given below.

Step 1 Define

$$H_m(t, x, p) \equiv \inf_{v \in U} H(t, x, v, p) \quad (7.6.40)$$

Since U is compact, and a, g are continuous on U, the infimum exists in U and there exists a Borel measurable function $\eta = \eta(t, x, p)$ on $I \times R^n \times R^n$ to U such that

$$H_m(t, x, p) = H(t, x, \eta(t, x, p), p). \quad (7.6.41)$$

Step 2 Solve the semilinear parabolic equation

$$\begin{aligned} &\frac{\partial \varphi}{\partial t} + \mathbf{A}(\varphi) + H_m(t, x, \varphi_x) = 0, \qquad (t, x) \in G \\ &\varphi(T, x) = f(x), \qquad x \in R^n. \end{aligned} \quad (7.6.42)$$

Step 3 Define $u^*(t, x) = \eta(t, x, \varphi_x(t, x))$.

Uniqueness

The assumptions that U is compact and the functions a, g are continuous do not guarantee the uniqueness of the law u^*. However, if the function $v \to H_{t,x,p}(v) \equiv H(t, x, v, p)$ for fixed (t, x, p) is strictly convex in the sense that

$$H_{t,x,p}((1-\alpha)v_1 + \alpha v_2) < (1-\alpha)H_{t,x,p}(v_1) + \alpha H_{t,x,p}(v_2)$$

for $0 < \alpha < 1$ for all $(t, x, p) \in G \times R^n$, and the set U is also convex, then there exists a unique Borel measurable η such that $H(t, x, \eta(t, x, p), p) = H_m(t, x, p)$.

Corollary 7.6.11 *Under the assumptions of convexity of U and strict convexity of $v \to H_{t,x,p}(v)$, the optimal control u^* is unique whenever it exists. If the set U is not compact, say $U = R^m$, then the infimum in* (7.6.40) *exists provided $H(t, x, v, p) > -\infty$ for all $v \in U$.*

Quadratic regulator problem

We can apply the above result to the linear quadratic regulator problem:

$$\begin{aligned} d\xi &= A(t)\xi\, dt + B(t)u\, dt + \sigma(t)\, dW, \\ J(u) &= \frac{1}{2} \mathrm{E} \int_0^T \{(R(t)\xi, \xi) + (N(t)u, u)\}\, dt, \end{aligned} \tag{7.6.43}$$

where one seeks a control law $u^* \in \mathcal{U}$ such that $J(u^*) \leq J(u)$ for all $u \in \mathcal{U}$. In this case the Hamiltonian is given by

$$H(t, x, v, p) = (A(t)x + B(t)v, p) + \tfrac{1}{2}\{(R(t)x, x) + (N(t)v, v)\}. \tag{7.6.44}$$

Assuming that the matrices R and N are symmetric and $R(t) \geq 0$ and $N(t) > 0$, we have

$$\eta(t, x, p) = -N^{-1}(t)B'(t)p, \tag{7.6.45}$$

and

$$u^*(t, x) = \eta(t, x, \varphi_x) = -N^{-1}B'(t)\varphi_x, \tag{7.6.46}$$

where

$$\varphi(t, x) = \mathrm{E}\Big\{\frac{1}{2} \int_t^T [(R(\theta)\xi^*, \xi^*) + (N(\theta)u^*(\theta, \xi^*), u^*(\theta, \xi^*))]\, d\theta \mid \xi^*(t) = x\Big\},$$

and ξ^* is the solution of (7.6.43) corresponding to the control u^*. Since the cost is quadratic, we stipulate that φ has the form

$$\varphi(t, x) = \tfrac{1}{2}(P(t)x, x) + \alpha(t), \tag{7.6.47}$$

where P is a symmetric positive $R(n \times n)$-valued function and α is a nonnegative real-valued function. Substituting this in the Kolmogorov equation (7.6.42) with $f \equiv 0$, we obtain

$$\begin{aligned} \frac{\partial \varphi}{\partial t} + \tfrac{1}{2}\operatorname{tr}(\sigma\sigma'\varphi_{xx}) + (A(t)x - BN^{-1}B'\varphi_x, \varphi_x) \\ + \tfrac{1}{2}\{(Rx, x) + (Nu^*, u^*)\} = 0 \\ \varphi(T, x) = 0. \end{aligned} \tag{7.6.48}$$

Hence substituting (7.6.47) into (7.6.46) and (7.6.48), one can easily deduce that

$$\begin{aligned} \dot{P} + A'P + PA - PBN^{-1}B'P + R = 0, \qquad & P(T) = 0, \\ \dot{\alpha} + \tfrac{1}{2}\operatorname{tr}(\sigma\sigma'P) = 0, \qquad & \alpha(T) = 0. \end{aligned} \tag{7.6.49}$$

This shows that, for the stochastic regulator problem, the optimal feedback control law is identical to that of the deterministic regulator problem (see section 6.3.2). The only difference is that in the stochastic case the cost functional is different and is given by

$$\varphi(t, x) = \frac{1}{2}(P(t)x, x) + \frac{1}{2}\int_t^T \operatorname{tr}(\sigma\sigma'P)\, d\theta, \tag{7.6.50}$$

with $J(u^*) = \varphi(0, x)$, given that $\xi^*(0) = x$.

Summarizing this result, we have

Theorem 7.6.12 *The stochastic linear quadratic regulator problem has a solution given by the following set of equations:*

(a) $\dot{P} + A'(t)P + PA(t) - PB(t)N^{-1}(t)B'(t)P + R(t) = 0$ *with* $P(T) = 0$.

(b) $u^*(t, x) = -N^{-1}(t)B'(t)P(t)x,$ (7.6.51)

(c) $J(u^*) = \dfrac{1}{2}(P(0)x, x) + \dfrac{1}{2}\displaystyle\int_0^T \operatorname{tr}(\sigma\sigma'P)\, d\theta$ *given* $\xi^*(0) = x$,

(d) *for random initial state* ξ_0 *with* $E(\xi_0\xi_0') \equiv \Gamma_0$,

$$J(u^*) = \frac{1}{2}\operatorname{tr}(P(0)\Gamma(0)) + \frac{1}{2}\int_0^T \operatorname{tr}(\sigma\sigma'P)\, d\theta.$$

Remark 7.6.13 It follows from the general existence theorems for solutions of ordinary differential equations and the fact that

$$\varphi(t, x) = \frac{1}{2}(P(t)x, x) + \frac{1}{2}\int_t^T \operatorname{tr}(\sigma\sigma' P)\, d\theta$$

$$\leq \frac{1}{2} E \int_t^T (R(\theta)\xi^0_{t,x}(\theta), \xi^0_{t,x}(\theta))\, d\theta < \infty, \tag{7.6.52}$$

where ξ^0 is the solution corresponding to $u \equiv 0$, that the matrix Riccati differential equation (a) has a solution on any finite interval.

In case the system parameters A, B, R, N are constant, one can expect a stationary control law

$$u^*(t, x) \equiv u^*(x) = -N^{-1}B'Px, \tag{7.6.53}$$

where P is the solution of the algebraic Riccati equation

$$A'P + PA - PBN^{-1}B'P + R = 0. \tag{7.6.54}$$

The question of existence of solution of this equation has been discussed in section 6.3.2.

Bellman's equation

We have seen in the previous chapter that Bellman's equation plays a central role in the design of optimal feedback control policies which are highly desirable in many engineering applications. In this chapter we consider similar problems for stochastic systems.

For simplicity we consider the system

$$d\xi = a^u(t, \xi)\, dt + \sigma^u(t, \xi)\, dW, \qquad u \in \mathcal{U}, \tag{7.6.55}$$

along with the performance (cost) functional

$$J(u) = E\left\{\int_0^T g^u(\theta, \xi^u(\theta))\, d\theta + f(\xi^u(T))\right\},$$

where

$$a^u(t, x) = a(t, x, u(t, x)), \qquad \sigma^u(t, x) = \sigma(t, x, u(t, x)),$$

$$g^u(t, x) = g(t, x, u(t, x)), \qquad u \in \mathcal{U}.$$

For $0 \leq t < T$, $x \in R^n$ and $u \in \mathcal{U}$, define

$$\varphi^u(t, x) = E\left\{\int_t^T g^u(\theta, \xi^u_{t,x}(\theta))\, d\theta + f(\xi^u_{t,x}(T))\right\}. \tag{7.6.56}$$

The function φ^u determines the cost of future control policy u, given the current state $\xi(t)=x$. The function φ_B, defined by

$$\varphi_B(t, x)=\inf\{\varphi^u(t, x),\ u \in \mathcal{U}\}, \tag{7.6.57}$$

is called the value function and, loosely speaking, it gives the minimum cost of future operation of the system starting from $(t, x) \in [0, T) \times R^n$.

We give here a formal derivation of Bellman's equation. Its justification is nontrivial and depends on intricate estimates of various generalized derivatives of φ_B. Interested readers may refer to Krylov [49].

Let $s \in (t, T)$, and u be an admissible control restricted to the interval $[t, s]$. Then the principle of optimality states that

$$\varphi_B(t, x)=\inf_{u \in \mathcal{U}} \mathrm{E}\left\{\int_t^s g^u(\theta, \xi_{t,x}^u(\theta))\, \mathrm{d}\theta+\varphi_B(s, \xi_{t,x}^u(s))\right\}. \tag{7.6.58}$$

In other words, whatever may have been the control policy over the initial period of time $[t, s]$, the subsequent policy must be such that it minimizes the cost of operation for the remaining period $[s, T]$.

Assume that φ_B has first partial in t and second partial in x in the generalized sense of

$$W_{\mathrm{loc}}^{1,2} \equiv\{\varphi: \varphi_t, \varphi_{x_i}, \varphi_{x_i x_j} \text{ are locally square integrable}\}.$$

Then using Itô's formula, we can write

$$\mathrm{E} \varphi_B(s, \xi_{t,x}^u(s))=\varphi_B(t, x)+\mathrm{E} \int_t^s(L^u \varphi_B)(\theta, \xi_{t,x}^u(\theta))\, \mathrm{d}\theta, \tag{7.6.59}$$

where

$$L^u \varphi=\frac{\partial}{\partial t} \varphi+\mathbf{A}^u \varphi \equiv \frac{\partial}{\partial t} \varphi+\frac{1}{2} \operatorname{tr}((\sigma^u)(\sigma^u)' \varphi_{xx})+(a^u, \varphi_x).$$

Hence it follows from (7.6.58) and (7.6.59) that

$$0=\inf_{u \in \mathcal{U}} \mathrm{E}\left\{\int_t^s g^u(\theta, \xi_{t,x}^u(\theta))\, \mathrm{d}\theta+\int_t^s(L^u \varphi_B)(\theta, \xi_{t,x}^u(\theta))\, \mathrm{d}\theta\right\} \tag{7.6.60}$$

for all $t<s \leqslant T$. Dividing this expression by $(s-t)$ and letting $s \downarrow t$, we obtain Bellman's equation,

$$\begin{aligned} \inf\{L^u \varphi_B+g^u\}&=0 && \text{on } G=[0, T) \times R^n, \\ \varphi_B(T, x)&=f(x) && \text{on } R^n, \end{aligned} \tag{7.6.61}$$

or

$$\begin{aligned} \frac{\partial \varphi_B}{\partial t}+\inf_{u \in \mathcal{U}}\{\mathbf{A}^u \varphi_B+g^u\}&=0 && \text{on } G, \\ \varphi_B(T, x)&=f(x) && \text{on } R^n. \end{aligned} \tag{7.6.62}$$

The expression

$$J^* = \inf_{u\in\mathcal{U}} J(u) = \varphi_B(0, x) \tag{7.6.63}$$

gives the lowest possible cost of operation of the system over the period $[0, T]$ starting from the state x provided the infimum is attained in $\mathcal{U}$.

For jump processes, given by Itô's equations of the form (7.4.56) with both a and b dependent on $\mathcal{U}$, the Bellman equation is given by

$$\begin{aligned} &\frac{\partial\varphi}{\partial t} + \inf\{\mathbf{A}_c^u\varphi + \mathbf{A}_d^u\varphi + g^u\} = 0 \qquad (t, x) \in G, \\ &\varphi(T, x) = f(x), \qquad x \in R^n, \end{aligned} \tag{7.6.64}$$

where the operators $\mathbf{A}_c^u$ and $\mathbf{A}_d^u$ are given in (7.4.61) with $a = a^u$ and $b = b^u$.

In fact Bellman's equation also holds for Markov time problems in which the process is stopped as soon as it hits the boundary ∂D of an open bounded set $D \subset R^n$,

$$\begin{aligned} &\frac{\partial\varphi}{\partial t} + \inf\{\mathbf{A}_c^u\varphi + \mathbf{A}_d^u\varphi + g^u\} = 0, \qquad (t, x) \in [0, T) \times D, \\ &\varphi(T, x) = f(x), \qquad x \in D, \\ &\varphi(t, x) = \psi(t, x), \qquad (t, x) \in [0, T] \times \partial D, \end{aligned} \tag{7.6.65}$$

where f and ψ are prescribed functions satisfying $f(x) = \psi(T, x)$ on ∂D.

Because of mathematical difficulties, in most of the control literature, σ is assumed to be independent of control. In this situation the Bellman equations are semilinear (that is linear in the highest-order derivatives). In case σ is dependent on u, they are nonlinear. In general, the Bellman equations do not have classical solutions ($\varphi \notin C^{1,2}$). However, one may look for generalized solutions in Sobolev spaces like $W^{1,2}_{\text{loc}}$.

The usefulness of Bellman's equations can be appreciated from the following result:

Theorem 7.6.14 *If φ^* is a solution of the Bellman equation:*

$$\begin{aligned} &\frac{\partial\varphi}{\partial t} + \inf_{u\in\mathcal{U}}\{\mathbf{A}^u\varphi + g^u\} = 0 \qquad (t, x) \; G = [0, T) \times R^n \\ &\varphi(T, x) = f(x), \qquad x \in R^n \end{aligned} \tag{7.6.66}$$

and if the infimum is attained at some control $u^0 \in \mathcal{U}$, then

$$\varphi^* = \varphi_B \equiv \inf\{\varphi^u : u \in \mathcal{U}\},$$

that is, the solution of the Bellman equation coincides with the value function.

Proof If φ^* is a solution of the equation (7.6.66), then for any $u \in \mathcal{U}$

$$L^u\varphi^* + g^u \equiv \frac{\partial}{\partial t}\varphi^* + \mathbf{A}^u\varphi^* + g^u \geqslant 0.$$

Let $\xi^u_{t,x}(\theta)$, $\theta \geqslant t$, denote the solution of Itô's equation corresponding to the control u, then by Itô's formula

$$\mathrm{E}\varphi^*(s, \xi^u_{t,x}(s)) = \varphi^*(t, x) + \mathrm{E}\int_t^s (L^u\varphi^*)(\theta, \xi^u_{t,x}(\theta))\,\mathrm{d}\theta$$

$$\geqslant \varphi^*(t, x) - \mathrm{E}\int_t^s g^u(\theta, \xi^u_{t,x}(\theta))\,\mathrm{d}\theta.$$

Letting $s \to T$, and using the terminal condition of (7.6.66), we obtain

$$\mathrm{E}f(\xi^u_{t,x}(T)) + \mathrm{E}\int_t^T g^u(\theta, \xi^u_{t,x}(\theta))\,\mathrm{d}\theta \geqslant \varphi^*(t, x). \tag{7.6.67}$$

Since, by definition, the left member is $\varphi^u(t, x)$, we have

$$\varphi^u(t, x) \geqslant \varphi^*(t, x), \qquad (t, x) \in [0, T) \times R^n = G \tag{7.6.68}$$

for all $u \in \mathcal{U}$. Hence

$$\varphi_B(t, x) \equiv \inf \varphi^u(t, x) \geqslant \varphi^*(t, x), \qquad (t, x) \in G. \tag{7.6.69}$$

In order to prove the equality $\varphi^* = \varphi_B$, we must show that $\varphi_B \leqslant \varphi^*$. Suppose the infimum in (7.6.66) is attained at $u^0 \in \mathcal{U}$, let $\xi^0_{t,x}(\theta)$, $t \leqslant \theta \leqslant T$ denote the corresponding solution and define

$$\varphi^0(t, x) = \mathrm{E}\left\{f(\xi^0_{t,x}(T)) + \int_t^T g^{u^0}(\theta, \xi^0_{t,x}(\theta))\,\mathrm{d}\theta\right\}$$

Since φ^* is the solution of the Bellman equation and the infimum is attained at u^0 we must have

$$0 = \frac{\partial}{\partial t}\varphi^* + \inf\{\mathbf{A}^u\varphi^* + g^u\}$$

$$= \frac{\partial}{\partial t}\varphi^* + \{\mathbf{A}^{u^0}\varphi^* + g^{u^0}\}.$$

Hence

$$\varphi^{u^0} \equiv \varphi^0 = \varphi^*,$$

and

$$\varphi_B \equiv \inf \varphi^u \leqslant \varphi^{u^0} = \varphi^*. \tag{7.6.70}$$

From (7.6.69)–(7.6.70), we have $\varphi_B = \varphi^*$. Hence under the given assumptions the solution of the Bellman equation is the value function. ■

The major questions associated with Bellman's equations are existence, uniqueness, and smoothness of solutions. These questions have been extensively studied in Krylov [49].

7.6.3 Reliability

We discuss here a simple machine-lifetime problem. Let D be an open, bounded and connected subset of R^n representing the set of (good) states in which a physical system (a machine) is known to perform efficiently and reliably. Let $\xi(t)$, $t \in [0, T]$ denote the actual state of the system at time t and suppose it is governed by a controlled diffusion process,

$$d\xi = a^u(t, \xi)\,dt + \sigma(t, \xi)\,dW, \qquad u \in \mathcal{U}.$$

Initially the system starts from the state $x_0 \in D$. One wishes to maximize the lifetime of the system in D by an appropriate choice of a control policy (for example a maintenance policy).

Suppose $\tau^u_{x_0}$ denotes the first time $t \leqslant T$ such that $\xi^u_{0,x_0}(t)$ hits the boundary ∂D at which the process is stopped (machine retired). If the process ξ^u never reaches the boundary, set $\tau^u_{x_0} = T$. For maximum operating life, one must maximize $E(\tau^u_{x_0})$ over $\mathcal{U}$.

Define

$$H_M(t, x, p) = \sup_{v \in U} \{(a(t, x, v), p) + 1\} \tag{7.6.71}$$

and suppose there exists a Borel-measurable function $\eta = \eta(t, x, p)$ such that the expression within the parentheses attains its supremum at $v = \eta$.

By using Theorem 7.4.17 and following a similar procedure, or alternatively using Bellman's equation, one can verify that the optimal policy, if it exists, is given by the solution of the semilinear boundary-value problem

$$\begin{aligned} \frac{\partial \varphi}{\partial t} + \Lambda(\varphi) + H_M(t, x, \varphi_x) &= 0 \qquad \text{for } (t, x) \in [0, T) \times D, \\ \varphi(t, x) &= 0 \qquad \text{for } (t, x) \in [0, T] \times \partial D, \\ \varphi(T, x) &= 0 \qquad \text{for } x \in D. \end{aligned} \tag{7.6.72}$$

Hence

$$\varphi(0, x_0) = \max \mathrm{E}\{\tau_{x_0}^u, u \in \mathcal{U}\}, \tag{7.6.73}$$

and the control policy is given by

$$u^*(t, x) = \eta(t, x, \varphi_x). \tag{7.6.74}$$

In fact a similar result holds also for jump processes.

Remark 7.6.15 In the time-invariant case, one is required to solve the nonlinear boundary-value problem:

$$\Lambda(\varphi) + H_M(x, \varphi_x) = 0 \quad \text{in } D,$$

$$\varphi = 0 \quad \text{in } \partial D,$$

giving $\varphi(x_0)$ as the maximum lifetime.

Remark 7.6.16 If the system is linear, $a^u(t, x) = A(t)x + B(t)u$ and $\sigma(t, x) = \sigma(t)$, and U is the unit ball in R^m, then equation (7.6.72) is given by

$$\frac{\partial \varphi}{\partial t} + \Lambda(\varphi) + (A(t)x, \varphi_x) + |B'(t)\varphi_x| + 1 = 0 \tag{7.6.75}$$

with the same initial and boundary conditions.

Exercises

7.1.P1 Verify the properties of the Wiener process (W)(b)–(W)(f).

7.2.P2 Let W be a standard Brownian motion and $t \to \sigma(t)$ an orthogonal matrix valued function. Prove that $\eta(t) \equiv \int_0^t \sigma(\theta)\, \mathrm{d}W(\theta)$ is a standard Brownian motion.

7.2.P3 Let $M \equiv \{M(t), t \geq 0\}$ be a square integrable $\mathcal{F}_t$-martingale with square variation $\langle \mathrm{d}M, \mathrm{d}M \rangle = \mathrm{d}\beta$, where β is a non-decreasing function of bounded variation over $I = [0, T]$, $T < \infty$, with $\beta(0) = 0$. Let $L_2^{\mathcal{F}}(I, \mathrm{d}\beta)$ denote the class of all $\mathcal{F}_t$-adapted stochastic processes $\{f\}$ such that $\mathrm{E} \int_I |f(t)|^2\, \mathrm{d}\beta(t) < \infty$. Prove that

(a) the stochastic integral $L(f) \equiv \int_I f(t)\, \mathrm{d}M(\theta)$ is a well-defined random variable for each $f \in L_2^{\mathcal{F}}(I, \mathrm{d}\beta)$,
(b) $\xi(t) \equiv L_t(f) = \int_0^t f(\theta)\, \mathrm{d}M(\theta)$ is a square integrable $\mathcal{F}_t$-martingale,
(c) if β is also continuous, then $\xi(t)$ is a square-integrable continuous martingale,
(d) if $\mathrm{P}\{\int_I |f(\theta)|^2\, \mathrm{d}\beta(\theta) < \infty\} = 1$, then ξ is a local martingale.

7.2.P4 Prove properties similar to (L)(a)–(L)(f) for any $R(n \times m)$-valued $\mathscr{F}_t$-adapted random process $\sigma \in L_2^e(I, \mathrm{d}\beta)$ defining the stochastic integral $L(\sigma) = \int_I \sigma(\theta)\, \mathrm{d}M(\theta)$, where M is an R^m-valued $\mathscr{F}_t$-martingale having β as its quadratic matrix variation process.

7.2.P5 Let $1 \leq p < \infty$, $T < \infty$, $\mathrm{E} \int_0^T |f(t)|^{2p}\, \mathrm{d}t < \infty$ and ξ be given by $\xi(t) = \int_0^t f(\theta)\, \mathrm{d}W(\theta)$. Show that

$$\mathrm{E} \int_0^T |\xi(t)|^{2p}\, \mathrm{d}t \leq K(T, p) \mathrm{E} \int_0^T |f(t)|^{2p}\, \mathrm{d}t,$$

where K is a constant depending only on T and p.

7.2.P6 Verify the properties (K)(a)–K(f) for the stochastic integral with respect to the Poisson random measure $\lambda(\mathrm{d}t, \mathrm{d}y)$.

7.3.P7 Show that the conclusion of Theorem 7.3.1 remains valid for arbitrary initial probability measure μ_0 and not for Gaussian measure only, provided μ_0 has compact support.

Let U_t denote the mapping $\mu_0 \to \mu_t$ from $PM(R^n)$ to $PM(R^n)$ ($\equiv$ the space of probability measures on R^n). The above result shows that $\lim_{t\to\infty} U_t(\mu_0) = \mu^*$ for all $\mu_0 \in PM(R^n)$, implying the asymptotic stability of the Gaussian measure μ^* (see Theorem 7.3.1).

7.4.P8 Suppose the matrix A of the system (7.3.12) is unstable and the output is given by $y(t) = H\xi(t)$ for $H \in R(k \times n)$. Find conditions under which the measure γ_t^y induced by $y(t)$ on $\mathscr{B}(R^k)$ has a limit as $t \to \infty$.

7.4.P9 Prove that for the system $\mathrm{d}\xi = a(t, \xi)\, \mathrm{d}t + \sigma(t)\, \mathrm{d}W$, with σ deterministic and $(a(t, x), x) \leq K(1 + |x|^2)$,

$$\sup_{t\in[0,T]} (1 + \mathrm{E}\, |\xi(t)|^2) \leq \left\{(1 + \mathrm{E}\, |\xi_0|^2) + \int_0^T \mathrm{tr}(\sigma\sigma')\, \mathrm{d}\theta\right\} \exp 2KT.$$

7.4.P10 Prove that if a, σ of the system

$$\mathrm{d}\xi = a(t, \xi)\, \mathrm{d}t + \sigma(t, \xi)\, \mathrm{d}W, \qquad t \in [0, T],$$

$$\xi(0) = x$$

are uniformly Lipschitz and has at most linear growth, then $\xi(t)$ has moments of all finite order. (*Hint*: Show that $\mathrm{E}\, |\xi(t)|^{2p} \leq c(T, p)(1 + |x|^{2p})$).

7.4.P11 Verify Corollary 7.4.5.

7.4.P12 Prove Remark 7.4.8.

7.4.P13 Prove Theorem 7.4.10.

7.4.P14 Prove Corollary 7.4.18 and Theorem 7.4.21.

7.5.P15 Derive the Kalman–Bucy filter (equations (7.5.10)–(7.5.14)) from Kushner's equation (7.5.52).

7.5.P16 Prove that the maximum likelihood estimate of the scalar parameter α in the scalar equation $\mathrm{d}\xi = \alpha a(\xi(t))\,\mathrm{d}t + \sigma\,\mathrm{d}W$, σ constant, is given by

$$\hat{\alpha}_T = \frac{\displaystyle\int_0^T a(\xi(t))\,\mathrm{d}\xi(t)}{\displaystyle\int_0^T a^2(\xi(t))\,\mathrm{d}t}.$$

7.5.P17 Prove Remark 7.5.4.

7.5.P18 Let $\mathrm{d}y = \xi(t)\,\mathrm{d}t + \mathrm{d}W$, where ξ is a binary process taking values from $\Sigma \equiv \{e_1, e_2\}$, $e_1 = 0$, $e_2 = 1$ with transition rates given by $\lambda_{11} = -\lambda$, $\lambda_{12} = \lambda$, $\lambda_{21} = \lambda$, $\lambda_{22} = -\lambda$ for some $\lambda > 0$. Show that

$$p_2(t) \equiv \mathrm{P}\{\xi(t) = 1 \mid \mathcal{F}_t^y\}$$

satisfies the stochastic differential equation [152],

$$\mathrm{d}p_2 = \lambda(1 - 2p_2)\,\mathrm{d}t + p_2(1 - p_2)(\mathrm{d}y - p_2\,\mathrm{d}t).$$

This problem arises in communication engineering where a binary signal ξ is to be detected in the presence of white Gaussian noise.

7.6.P19 Derive the regulator equations (7.6.51) from the Bellman equation (7.6.66).

7.6.P20 Verify Remark 7.6.15.

Bibliography

The bibliography is not exhaustive. Some of the references have not been directly used in the text, but added for further studies.

Books: Finite-dimensional systems

[1] Berkovitz L.D., *Optimal Control Theory,* Springer-Verlag, New York, Heidelberg, Berlin, 1974.

[2] Brogan W.L., *Modern Control Theory,* Prentice-Hall, Englewood Cliffs, New Jersey, 1982.

[3] Bondarel R., Delmas J. and Guichet P., *Dynamic Programming and its Application to Optimal Control,* Academic Press, New York, London, 1971.

[4] Cesari L., *Optimization – Theory and Applications: Problems with Ordinary Differential Equations,* Springer-Verlag, New York, Heidelberg, Berlin, 1984.

[5] Chen C-T., *Linear System Theory and Design,* Holt, Rinehart and Winston, New York, Montreal, London, Sydney, Tokyo, 1984.

[6] Clarke F.H., *Optimization and Nonsmooth Analysis,* John Wiley, New York, Chichester, Brisbane, Toronto, Singapore, 1983.

[7] Gamkrelidze R.V., *Principles of Optimal Control Theory,* Plenum Press, New York and London, 1975.

[8] Hermes H. and LaSalle J.P., *Functional Analysis and Time Optimal Control,* Academic Press, New York and London, 1969.

[9] Jamshidi M., *Large Scale Systems: Modelling and Control,* North Holland, New York, Oxford, 1983.

[10] Jury E.I., *Inners and Stability of Dynamic Systems,* John Wiley, New York, London, Sydney, Toronto, 1974.

[11] Kailath T., *Linear Systems,* Prentice Hall, Englewood Cliffs, New Jersey, 1980.

[12] Kuo B.C., *Automatic Control Systems,* Prentice-Hall, Englewood Cliffs, New Jersey, 1962.

[13] Larson R.E. and Casti J.L., *Principles of Dynamic Programming,* Marcel Dekker, New York, Basel, 1978.
[14] LaSalle J.P., *The Stability of Dynamical Systems,* SIAM Publications, Philadelphia, 1976.
[15] LaSalle J.P. and Lefschetz S., *Stability by Lyapunov's Direct Method with Applications,* Academic Press, New York and London, 1961.
[16] Lee E. and Markus L., *Foundations of Optimal Control Theory,* John Wiley, New York, 1967.
[17] Leitmann G., *Topics in Optimization,* Academic Press, New York and London, 1967.
[18] Leitmann G., *An Introduction to Optimal Control,* McGraw-Hill, New York, 1969.
[19] Lions P.L., *Generalized Solutions of Hamilton–Jacobi Equations,* Pitman Advanced Publishing Program, Boston, London, Melbourne, 1982.
[20] Luenberger D.G., *Optimization by Vector Space Method,* John Wiley, New York, London, Sydney, Toronto, 1969.
[21] Ogata K., *State Space Analysis of Control Systems,* Prentice Hall, Englewood Cliffs, New Jersey, 1967.
[22] Oğuztöreli M.N., *Time-Lag Control System,* Academic Press, New York and London, 1966.
[23] Petrov I.P., *Variational Methods in Optimum Control Theory,* Academic Press, New York and London, 1968.
[24] Polak E., *Computational Methods in Optimization,* Academic Press, New York and London, 1971.
[25] Pontryagin L.S., Boltyanskii V.G., Gamkrelidze R.V. and Mishchenko E.F., *The Mathematical Theory of Optimal Processes,* John Wiley, New York, London, Sydney, 1962.
[26] Porter B and Crossley R., *Modal Control: Theory and Applications,* Taylor and Francis, London, 1972.
[27] Rubio J.E., *The Theory of Linear Systems,* Academic Press, New York and London, 1971.
[28] Russell D.L., *Mathematics of Finite Dimensional Control Systems: Theory and Design,* Marcel Dekker, New York, Basel, 1979.
[29] Sethi S.P. and Thompson, G.L., *Optimal Control Theory,* Martinus Nijhoff, Boston, The Hague, London, 1981.
[30] Shinners S.M., *Modern Control System Theory and Applications,* Addison Wesley, Reading, Massachusetts, London, Amsterdam, Sydney, 1978 (2nd edition).
[31] Skowroński J.M., *Multiple Nonlinear Lumped Systems,* Polish Scientific Publishers, Warsaw, 1969.
[32] Skowroński J.M., *Applied Liapunov Dynamics,* Systems and Control Engineering Consultants, Brisbane, 1984.
[33] Swan G.W., *Applications of Optimal Control Theory in Biomedicine,* Marcel Dekker, New York and Basel, 1984.
[34] Warga, J., *Optimal Control of Differential and Functional Equations,* Academic Press, New York and London, 1972.
[35] Young, L.C., *Calculus of Variations and Optimal Control,* Chelsea Publishing Co., New York, 1980.
[36] Zadeh, L.A. and Desoer C.A., *Linear System Theory,* McGraw-Hill, New York, San Francisco, Toronto, London, 1963.

Books: Stochastic systems

[37] Ahmed N.U., Optimal control of stochastic systems, in *Probabilistic Analysis and Related Topics,* Ed. A.T. Bharucha-Reid, Academic Press, New York, San Francisco, London, 1979.

[38] Bertsekas D.P., *Dynamic Programming and Stochastic Control,* Academic Press, New York, San Francisco, London, 1976.

[39] Bharucha-Reid A.T., *Random Integral Equations,* Academic Press, New York and London, 1972.

[40] Bucy R.S. and Joseph P.D., *Filtering of Stochastic Processes with Application to Guidance,* Wiley Interscience, New York, 1968.

[41] Doob J.L., *Stochastic Processes,* John Wiley, New York, London, 1964 (5th edition).

[42] Fleming W.H. and Rishel R.W., *Deterministic and Stochastic Optimal Control,* Springer-Verlag, Berlin, Heidelberg, New York, 1975.

[43] Friedman A., *Stochastic Differential Equations and Applications,* Vol. I and II, Academic Press, New York, London, 1975.

[44] Gihman I.I. and Skorokhod A.V., *Controlled Stochastic Processes,* Springer-Verlag, New York, Heidelberg, Berlin, 1979.

[45] Has'minskii R.Z., *Stochastic Stability of Differential Equations,* Sijthoff & Noordhoff, The Netherlands, 1980.

[46] Ikeda N. and Watanabe S., *Stochastic Differential Equations and Diffusion Processes,* North Holland, Amsterdam, Oxford, New York, 1981.

[47] Jazwinski A.H., *Stochastic Processes and Filtering Theory,* Academic Press, New York, London, 1970.

[48] Kallianpur G., *Stochastic Filtering Theory,* Springer-Verlag, New York, Heidelberg, Berlin, 1980.

[49] Krylov N.V., *Controlled Diffusion Processes,* Springer-Verlag, New York, Heidelberg, Berlin, 1980.

[50] Kushner H.J., *Introduction to Stochastic Control Theory,* Holt, Rinehart and Winston, New York, 1971.

[51] Skorokhod A.V., *Studies in the Theory of Random Processes,* Addison-Wesley, London, New York, 1965.

[52] Wonham W.M., Random differential equations in control theory, in *Probabilistic Methods in Applied Mathematics,* Vol. II, ed. A.T. Bharucha-Reid, Academic Press, New York and London, 1970.

Books: Distributed systems

[53] Ahmed N.U. and Teo K.L., *Optimal Control of Distributed Parameter Systems,* North Holland, New York, Oxford, 1981.

[54] Balakrishnan, A.V., *Applied Functional Analysis,* Springer-Verlag, Berlin, Heidelberg, New York, 1976.

[55] Butkovskiy A.G., *Distributed Control Systems,* American Elsevier, New York, 1969.

[56] Curtain R.F. and Pritchard A.J., *Infinite Dimensional Linear Systems Theory,* Springer-Verlag, Berlin, Heidelberg, New York, 1978.

[57] Ladyženskaja O.A., Solonikov V.A. and Ural'ceva N.N., *Linear and*

Quasilinear Equations of Parabolic Type, Trans. American Mathematical Society, Providence, Rhode Island, 1968.
[58] Lions J.L., *Optimal Control of Systems Governed by Partial Differential Equations,* Springer-Verlag, Berlin, Heidelberg, New York, 1971.
[59] Teo K.L. and Wu Z.S., *Computational Methods for Optimizing Distributed Systems,* Academic Press, New York and London, 1984.

Books: General

[60] Anderson P.M. and Fouad A.A., *Power System Control and Stability,* Iowa University Press, Iowa, 1977.
[61] Bowen J.H. and Masters E.F.O., *Nuclear Reactor Control and Instrumentation,* Temple Press, London, 1959.
[62] Clark C.W., *Mathematical Bioeconomics,* John Wiley, New York, London, Sydney, Toronto, 1976.
[63] Greensite A.L., *Analysis and Design of Space Vehicle Flight Control Systems,* Spartan Books, New York, Washington, 1970.
[64] Halmos P.R., *Measure Theory,* Van Nostrand, Princeton, New Jersey, Toronto, London, New York, 1950.
[65] Hewitt E. and Stromberg K., *Real and Abstract Analysis,* Springer-Verlag, New York, Berlin, Heidelberg, 1965.
[66] Krasnosel'skii M.A., *Topological Methods in the Theory of Nonlinear Integral Equations,* Macmillan, New York, 1964.
[67] Pascali D. and Sburlan S., *Nonlinear Mapping of Monotone Type,* Sijthoff & Noordhoff, The Netherlands, 1978.
[68] Roberts S.M. and Shipman J.S., *Two-Point Boundary Value Problems: Shooting Methods,* Elsevier, New York, London, Amsterdam, 1972.
[69] Taylor, A.E., *Introduction to Functional Analysis,* John Wiley, New York, 1958.
[70] Timoshenko S., Young D.H. and Weaver W., *Vibration Problems in Engineering,* John Wiley, New York, 1974.
[71] Tsypkin Y.Z., *Foundations of the Theory of Learning Systems,* Academic Press, New York, London, 1973.

Papers

[72] Ahmed N.U., Existence of optimal controls for a class of hereditary systems with lagging control, *Information and Control,* **26,** 178–185, 1974.
[73] Ahmed N.U., Existence of optimal bandlimited controls without convexity condition, *Information and Control,* **31,** 201–215, 1976.
[74] Ahmed N.U., L_p and Orlicz stability of a class of nonlinear time varying feedback control systems, *Information and Control,* **19,** 114–123, 1971.
[75] Ahmed N.U., Stochastic control on Hilbert space for linear evolution equation with random operator-valued coefficients, *SIAM Journal on Control and Optimization,* **19,** 401–430, 1981.
[76] Ahmed N.U., Optimal control of stochastic dynamical systems, *Information and Control,* **22,** 13–30, 1973.

[77] Ahmed N.U., Fourier analysis on Wiener measure space, *Journal of Franklin Institute,* **286,** 143–151, 1968.
[78] Ahmed N.U., Some novel properties of Wiener's canonical expansion, *IEEE Trans. of Systems Science and Cybernetics,* **SSC-5,** 140–144, 1969.
[79] Ahmed N.U., Strong and weak synthesis of nonlinear systems with constraints on the system space G_λ, *Information and Control,* **23,** 71–85, 1973.
[80] Ahmed N.U., A class of stochastic nonlinear integral equations on L_p spaces and its applications to optimal control, *Information and Control,* **14,** 512–523, 1969.
[81] Ahmed N.U., A new class of generalized nonlinear functionals of white noise with application to random integral equations, *Stochastic Analysis and Applications,* **1,** 139–158, 1983.
[82] Ahmed N.U., A simple gradient algorithm for least square estimation of system parameters, *International Journal of Systems Science,* **7,** 673–677, 1976.
[83] Ahmed N.U., Control theory and its prospects for developing nations, *The Muslim Scientists,* **10,** 361–372, 1981.
[84] Ahmed N.U. and Ahsan Q., A dynamic model for generation expansion planning, *Electric Power Systems Research,* **9,** 79–86, 1985.
[85] Ahmed N.U. and Dabbous T.E., Nonlinear filtering of systems governed by Itô differential equations with jump parameters, *Journal of Mathematical Analysis and Applications,* **115,** 76–92, 1986.
[86] Ahmed N.U. and Dabbous T.E., Modelling and on-line control of reliability dynamics of large interconnected power systems, *International Journal of Systems Science,* **14,** 1321–1353, 1983.
[87] Ahmed N.U., Dabbous T.E. and Biswas S.K., Modelling of reliability dynamics of large interconnected power systems, *Proc. 4th International Symposium on Large Engineering Systems,* University of Calgary, Calgary, Canada, 173–178, June 1982.
[88] Ahmed N.U. and Garcia F., Mathematical model of the dynamics of a living lake and optimal control to prevent its eutrophication, *Proc. Canadian Conference on Auto. Control,* University of New Brunswick, Fredricton, September, 1973.
[89] Ahmed N.U. and Georganas N.D., On optimal parameter selection, *IEEE Trans. on Auto Control,* **AC-18,** 313–314, 1973.
[90] Ahmed N.U. and Georganas N.D., Comments on 'Invariance in linear estimators', *Journal of Mathematical Analysis and Applications,* **44,** 98–99, 1973.
[91] Ahmed N.U., Jutras G.J. and Yeo S.H., Optimal control of an economic system modelled on the basis of cyclical growth theory, *Int. Journal of Control,* **25,** 827–835, 1977.
[92] Ahmed N.U. and Teo K.L., An existence theorem on optimal control of partially observable diffusion, *SIAM Journal on Control,* **12,** 351–355, 1974.
[93] Ahmed N.U. and Teo K.L., Optimal control of stochastic Itô differential systems with fixed terminal time, *Advances in Applied Probability,* **7,** 154–178, 1975.
[94] Ahmed N.U. and Wong H.W., A minimum principle for systems governed by Itô differential equations with Markov jump parameters, in *Differential*

Games and Control Theory II, Eds. E.O. Roxin, P-T. Liu and R.L. Sternberg, Marcel Dekker, New York, Basel, 1977.

[95] Bass R. W., *Lecture notes in control synthesis and optimization,* presented at the NASA Langley Research Center, August 1961.

[96] Bell J., Optimizing pacemaker battery life, *SIAM Review,* **28,** 73–77, 1986.

[97] Beneš V.E. and Karatzas I., On the relation of Zakai's and Mortensen's equations, *SIAM Journal on Control and Optimization,* **21,** 472–489, 1983.

[98] Bensoussan A. and Tapiero C.S., Impulsive control in management: prospects and applications, *Journal of Optimization Theory and Applications,* **37**(4), 419–442, 1982.

[99] Biswas S.K. and Ahmed N.U., Modelling of flexible spacecraft and their stabilization, *International Journal of Systems Science,* **16,** 535–551, 1985.

[100] Biswas S.K. and Ahmed N.U., Stabilization of a class of hybrid systems arising in flexible spacecraft, *Journal of Optimization Theory and Applications,* **50,** 83–108, 1986.

[101] Biswas S.K. and Ahmed N.U., Optimal feedback control of power systems governed by nonlinear dynamics, *Optimal Control Applications and Methods,* **7,** 289–303, 1986.

[102] Biswas S.K. and Ahmed N.U., Optimal voltage regulation of power systems under transient conditions, *Electric Power Systems Research,* **6,** 71–77, 1983.

[103] Boel R., Varaiya P. and Wong P., Martingales on jump processes – II, Applications, *SIAM Journal of Control,* **13,** 1022–1061, 1975.

[104] Burch C., A semi-group treatment of the Hamilton–Jacobi equation in several space variables, *Journal of Differential Equations,* **23,** 107–124, 1977.

[105] Casti J., Reduction of dimensionality for systems of linear two point boundary value problems with constant coefficients, *Journal of Mathematical Analysis and Applications,* **45,** 522–531, 1974.

[106] Chen Y.C. and Ahmed N.U., An application of optimal control theory in fisheries management, *International Journal of Systems Science,* **14,** 453–462, 1983.

[107] Chukwu E., Finite time controllability of nonlinear control processes, *SIAM Journal on Control,* **13,** 807–816, 1975.

[108] Colmenares W.R., *On the controllability of Linear and Nonlinear Systems,* M.A. Sc. Thesis, Department of Electrical Engineering, University of Ottawa, 1983.

[109] Colmenares W.R. and Ahmed N.U., A controllability counter example, *Int. Journal of Systems Science,* **15,** 837–840, 1984.

[110] Conti, R., *Teoria del Controllo e del Controllo Ottimo,* UTET, Torino, Italy, 1974.

[111] Crandall M.G., Evans L.C. and Lions P-L., Some properties of viscosity solutions of Hamilton–Jacobi equations, *Trans. of American Mathematical Society,* **282,** 487–502, 1984.

[112] Crandall M.G. and Lions P-L., Viscosity solutions of Hamilton–Jacobi equations, *Trans. of American Mathematical Society,* **277,** 1–42, 1983.

[113] Crandall M.G. and Lions P-L., Two approximations of solutions of Hamilton–Jacobi equations, *Mathematics of Computation,* **43,** 1–19, 1984.

[114] Crandall M.G. and Souganidis P.E., Developments in the theory of nonlinear first-order partial differential equations, in *Differential Equations,* Eds. I.W. Knowles and R.T. Lewis, pp. 131–142, North Holland, Amsterdam, New York, Oxford, 1984.

[115] Dabbous T.E. and Ahmed N.U., Nonlinear filtering of diffusion processes with discontinuous observations, *Stochastic Analysis and Applications,* **2,** 87–106, 1981.

[116] Davis, M.H.A., Pathwise nonlinear filtering, in *Stochastic Systems: The Mathematics of Filtering, Identification and Applications,* Eds. M. Hazewinkel and J.C. Willems, 505–528, D. Reidel, 1981.

[117] Davison E.J. and Wang S.H., On pole assignment in linear multivariable systems using output feedback, *IEEE Trans. on Auto. Control,* **AC-20,** 516–518, 1965.

[118] Gao Z.Y. and Ahmed N.U., Stabilizability of certain stochastic systems, *Int. Journal of System Science,* **17**(8), 1175–1185, 1986.

[119] Gao Z.Y. and Ahmed N.U., Feedback stabilizability of nonlinear stochastic systems with state-dependent noise, *Int. Journal of Control,* **45**(2), 729–737, 1987.

[120] Georganas N.D. and Ahmed N.U., Optimal initial data and control for a class of linear hereditary systems, *Information and Control,* **22,** 394–402, 1973.

[121] Gopinath B., On the control of linear multiple input–output systems, *Bell System Technical Journal,* **50,** 1063–1080, 1971.

[122] Hestenes M.G., Multipliers and gradient methods, in *Computing Methods in Optimization, Problem-2,* Eds. L.A. Zadeh, L.W. Neustadt, A.V. Balakrishnan, Academic Press, New York, London, 1969.

[123] Kalman R.E., A new approach to linear filtering and prediction problems, *Trans. AME, Journal of Basic Engineering,* **82,** 34–45, 1960.

[124] Kelley H.J., Method of gradients, in *Optimization Techniques,* Ed. G. Leitmann, Academic Press, New York, London, 1962.

[125] Ladde G.S., Stability and oscillations in single species processes with past memory, *Int. Journal of Systems Science,* **10,** 621–647, 1979.

[126] Ladde G.S., Cellular systems – II, Stability of compartmental systems, *Mathematical Bioscience,* **30,** 1–21, 1976.

[127] Ladde G.S. and Sambandham M., Stochastic versus deterministic, *Mathematics and Computers in Simulation,* **24,** 507–514, 1982.

[128] Lasdon L.S., Mitter S.K. and Warren A.D., The conjugate gradient method for optimal control problems, *IEEE Trans. on Auto. Control,* **AC12,** 132–138, 1967.

[129] Luenberger D.G., An introduction to observers, *IEEE Trans. on Auto. Control,* **16,** 596–602, 1971.

[130] Kushner H.J., Dynamical equations for optimal nonlinear filtering, *J. of Differential Equations,* **3,** 179–190, 1967.

[131] Miele A. and Wang T., *Primal–dual properties of sequential gradient-restoration algorithms for optimal control problems: Part I, Basic formulation, Part II, General Problem,* Aero-Astronautics Reports, No. 183–184, Rice University, 1985.

[132] Molinari B.P., The stabilizing solution of the algebraic Riccati equation, *SIAM Journal on Control,* **11,** 262–271, 1973.

[133] Pandolfi, L., Linear control systems: controllability with constrained controls, *Journal of Optimization Theory and Applications,* **19,** 557–585, 1976.

[134] Phillips C.R., Haidar N.I. and Poon Y.C., Kinetic models for the thermal cracking of athabasca bitumen, *Fuel,* **64,** 678–691, 1985.

[135] Pindyck R.S., Adjustment costs, uncertainty, and the behavior of the firm, *American Economic Review,* **72,** 415–427, 1982.

[136] Riaz H., Biswas S.K. and Ahmed N.U., Stochastic modelling and stabilization of galloping transmission lines, *Electric Power Systems Research,* **10**(2), 137–143, 1986.

[137] Roy S. and Ahmed N.U., Power system voltage regulation based on voltage feedback using dynamic capacitor, *Electric Power Systems Research,* **6,** 81–100, 1983.

[138] Schmitendorf W. and Barmish B., Null controllability of linear systems with constrained controls, *SIAM Journal on Control and Optimization,* **18,** 327–345, 1980.

[139] Souganidis P.E., Existence of viscosity solutions of Hamilton–Jacobi equations, *Journal of Differential Equations,* **56,** 345–390, 1985.

[140] Tamburro M.B., The evolution operator approach to the Hamilton–Jacobi equation, *Israel Journal of Mathematics,* **26,** 232–264, 1977.

[141] Teo K.L. and Womersley R.S., A control parametrization algorithm for optimal control problems involving linear systems and linear terminal inequality constraints, *Numerical Functional Analysis and Optimization,* **6,** 291–313, 1983.

[142] Teo K.L., Wong K.H. and Clements D.J., A feasible directions algorithm for time-lag optimal control problems with control and terminal inequality constraints, *Journal of Optimization Theory and Applications,* **46,** 295–317, 1985.

[143] Vaca M.V. and Snyder D.L., Estimation and decision for observations derived from Martingales: Part II, *IEEE Trans. on Information Theory,* **IT24,** 32–45, 1978.

[144] Varaiya P., Optimal control of a partially observed stochastic system, in *Stochastic Differential Equations,* Eds. J.B. Keller, H.P. McKean, American Mathematical Society, Providence, Rhode Island, 1973.

[145] Yavin Y., *An Alternative Approach to Nonlinear Filtering: Jump Process Observations,* Technical Report, National Research Institute for Mathematical Sciences, Pretoria, South Africa, 1981.

[146] Zakai M., On the optimal filtering of diffusion processes. *z. Wahrsch. verw. Geb.,* **11,** 230–243, 1969.

Additional references

[147] Spong M.W., Throp J.S. and Kleinwaks J.M., The control of robot manipulators with bounded input, *IEEE Trans. on Auto. Control,* **AC-31,** 483–490, 1986.

[148] Shin K.G. and McKay N.D., A dynamic programming approach to trajectory planning of robotic manipulators, *IEEE Trans. on Auto. Control,* **AC-31,** 491–500, 1986.

[149] Geering H.P., Guzzella L., Hepner S.A.R. and Onder C.H., Time-optimal Motions of Robots in Assembly Tasks, *IEEE Trans. on Auto. Control,* **AC-31,** 512–518, 1986.

[150] Zubov V.I., *Methods of A.M. Lyapunov and Their Application,* Noordhoff, Groningen, Netherlands, 1964.

[151] Chandra Jagadish (Ed.), Chaos in nonlinear dynamical systems, *Proc. of the Workshop held at the U.S. Army Research Office,* North Carolina, 1984, published by SIAM, Philadelphia, USA, 1984.

[152] Stoyanov J., On the estimation of partially observable stochastic processes, *Mathematica Balkanica,* **2,** 235–250, 1972.

[153] Ahmed N.U., Finite-time null controllability for a class of linear evolution equations on a Banach space with control constraints, *Journal of Optimization Theory and Applications,* **47,** 129–158, 1985.

[154] Gourishankar V.G., Trybus L. and Rink R.E., An adaptive filter for dynamic positioning, *Optimal Control Applications and Methods,* **7,** 271–287, 1986.

[155] Pai M.A., *Power System Stability Analysis by the Direct Method of Lyapunov,* North Holland, Amsterdam, New York, Oxford, 1981.

Index